COMPUTER MODELING FOR ENVIRONMENTAL MANAGEMENT SERIES

COMPUTER GENERATED PHYSICAL PROPERTIES

Stan Bumble, Ph.D.

LEWIS PUBLISHERS

Boca Raton London New York Washington, D.C.

Project Editor: *Susan Fox*
Cover Design: *Dawn Boyd*

Library of Congress Cataloging-in-Publication Data

Bumble, Stan.
Computer generated physical properties / Stan Bumble.
p. cm. — (Computer modeling for environmental management)
Includes bibliographical references and index.
ISBN 1-56670-329-0 (alk. paper)
1. Chemistry, Physical and theoretical —Data processing.
2. Chemistry, Physical and theoretical — Computer simulation.
I. Title. II. Series.
QD455.3.E4B86 1999
541'.0285—dc21

99-19200
CIP

International Standard Book Number 1-56670-329-8
Library of Congress Card Number 99-19200
Printed in the United States of America 1 2 3 4 5 6 7 8 9 0
Printed on acid-free paper

Preface

This book treats the physical properties of matter as generated by computers. It is very interdisciplinary in scope and will review the literature and introduce new material. Some of the new material will be made available to you on a disk for your computer. As well, there are a great many figures for your inspection and use.

Although it is assumed that your main interest is in chemistry, there is considerable mathematics and biology also involved but none of it is very difficult. As an added bonus, many references are furnished as URLs to the World Wide Web so that you can easily obtain more information using your computer. There are three parts to the book and there are over 225 subjects discussed. However, they are meant to be introductory and you may, because of your need or interest, easily explore these subjects further with aid of the materials given in the book.

A diagram is shown of the main subjects discussed as they develop. These, of course, are not all the subjects, but they show trends connecting some of the more important subjects.

Part I begins with the emergence of computations chemistry and its connection with mathematics. Then the development of molecular modeling tools (as a later way to obtain physical properties) is introduced. Computer simulation involves molecular dynamics, Monte Carlo methods, and molecular mechanics of liquids and solids. This is extended to material science and, of great importance to biology as well as chemistry, environmental fate. General systems theory is discussed and it will lead to a neat organization of the science as it develops. Artificial intelligence also has much use in the development of physical and environmental properties.

The development of quantitative structure–activity relationships (QSAR) has played a major role in development of the ability to obtain physical properties, whether with computers or otherwise. The development of comparative molecular field analysis (CoMFA), three-dimensional structure–activity relationships, and combinatorial chemistry has followed in fairly rapid succession. Combinatorial chemistry has been used by drug companies in a combination computer–laboratory venture to explore all possible chemical structures for biological activity quickly and efficiently. The work of Kier and Hall used molecular structure to obtain the information content of molecules, and then computers were used to generate molecular structure for significant purposes. Databases of structures were incorporated into computers, and QSARs were used not only for physical but for environmental transport, degradation, and exposure, as well. Computers were used to find safe substitutes in pollution prevention. They were used to synthesize molecules, and new mathematical techniques such as genetic programming were used to generate molecules. Kinetic theory was used as well as quantum mechanics, statistical mechanics, and physical chemistry to define the terms in the equations for QSARs.

Outstanding programs are discussed: THERM, which supplies the much-needed thermodynamic data for many programs; LSENS of NASA and Chemkin of Sandia, which find the chemical kinetics of complex chemical systems; Kintecus and the CKS Package, which serve a similar purpose; Process Synthesis, which optimizes processes for cost, efficiency, and environmental health. Modeling of groundwater, troposphere, and stratosphere are discussed as well as global programs (such as global warming), and finally the use of supercomputers is discussed for specialized complex problems.

Part II defines a number of chemical groups of which molecules are composed. By using statistical regressions performed by R. D. Cramer for the dependence of physical properties upon such groups for many hundreds of compounds taken from the literature, very good statistical fits and predictions are attained. Many toxicological and biological activity fits are added and incorporated into spreadsheets to which an optimizer program was added.

This combined program was given the name SYNPROPS. The structure of the spreadsheet allows for many constraints and variations on the field of the search for optimized molecules, and the optimization runs are very fast with the program called Quattro Pro 7. It was then found that if one expresses the optimization in matrix mechanics, one could find the properties another way and also perform the inverse solution to the problem using matrix algebra on a computer. A very large matrix was also found that assigned the contribution of each of a large number of groups to the physical properties, which is a very valuable contribution to the synthetic chemist. Applications of these techniques were used for substitution of chemicals, replacing toxic ones with benign ones. Substitutions for ozone-depleting chemicals were also performed. Toxicological constraints and better biological activities were found for antineoplastic drugs. Better process chemicals at lower cost with greater profit and fewer health problems were found, etc.

A study of computer-assisted toxicity from the literature is discussed. QSARs are discussed again, and it is interesting to note that the SYNPROPS method does not use the theory discussed in Part I, as it is purely a statistical fit and achieves very good results. However, studies are shown here that the equation that ties the molecular structure parameters to the properties is very similar to a QSAR equation and the structure parameters can be tied to the characteristics of the chemical groups. Another method for obtaining physical properties from groups is discussed that depends only on mathematical combinatorics, as found by Drs. Friedler and Fan. There is a description of a program called Envirochemkin in some detail; the program finds the results of chemical kinetics for 50 species and 100 equations on a personal computer in a jiffy. Path probability, which finds the best path from A to B, is discussed. Finally, process design programs are discussed as they depend on physical properties that are derived often by computer methods.

Part III has a paper on cancer and attempts to delineate its cause and cure. This is a property that is dependent on molecular structure. A chapter on environmental risk and its control is presented. More background on quantum mechanics and the need to communicate the contents of this book to others is stressed. Kinetic paths are extended and the chemical substitution and clean programs of the EPA and others are discussed. The material is then extended from pure organics to mixtures, biogenics, natural materials, and polymers. More material on the simulation of processes by computers is presented. The game of life is described as it relates to many of the processes in this book, and the fingerprints of molecules and the similarity between molecules are used to show other methods for deriving properties. The price change as an environmental consequence is discussed. The value of the total ecology from a publication is deemed to be of importance and the cost of environmental damage is presented.

The Author

Stan Bumble Ph.D., has guided research, development, and engineering at DuPont and Dow Corning with computer programs that optimized the best products and properties. He has used computer programs for assisting the United States Government with the recovery of disaster victims and the development of their missile program. He has helped (with the assistance of computers) the United States Department of Justice and the Environmental Protection Agency at many hazardous sites such as Love Canal.

Table of Contents

Part I

Part II

Part III

List of Figures

List of Tables

Part I

1.1 The Emergence of Computational Chemistry

Computational chemistry had its roots in early attempts by theoretical physicists, beginning in 1928, to solve the Schrodinger equation using hand-cranked calculating machines. The calculations verified that solutions to the Schrodinger equation were correct for the helium atom and the hydrogen molecule. Electronic digital computers were invented during the World War II. Enormous speed and decrease in cost resulted, and there were major developments in algorithms and methodology in the 1950s. In the 1970s there was an explosion in applications to chemical problems, which predicted and explained the structure and reactivity of molecules and complemented many types of spectral measurements. Meanwhile, the ability to do calculations on molecules with more atoms (the many-body problem) kept increasing. The representation of energy also made possible the modeling of biological systems and rational drug design. Three-dimensional computer graphics and molecular mechanics became prevalent. Then, study of many-particle systems led to the study of liquids, solid materials, and biological macromolecules, which utilized statistical mechanics, physical chemistry and thermodynamics, and the ideas of intramolecular and intermolecular forces. Phase transition studies developed and are still prevalent. Phenomena away from equilibrium and disordered systems have been attractive to workers. Molecular dynamics and Monte Carlo techniques were applied to polymers and proteins. The absolute chemical reaction rate theory as solved by computer became popular. Group theory, geometry, and other mathematics were used, and then topology, number theory, algebraic geometry, and graph theory were utilized. Success with signal processing, sound and image compression, flow problems, and electromagnetic theory followed. Recently, linear algebra, geometric topology, distance geometry, and symbolic computation are playing roles in chemical studies.

1.2 Cross-Fertilization between Mathematics and Chemistry

It saves much time if research investigators can correctly forecast the properties of compounds before their synthesis. Quantitative structure–activity relationship (QSAR) methods are used on computers to do this. The Hansch–Fujita QSAR method (1964) was developed in the 1960s and is widely used by medicinal and agricultural chemists. Molecules are described in terms of their physical properties, and then statistical methods uncover the relationship between some combination of these physical properties and biological potency. The relationships can be examined with multiple linear or nonlinear regression, classical multivariate statistical techniques. However, discriminant analysis, physical components regression, factor analysis, and neural networks can be used as well. More recently, the partial least-squares (PLS) method has found wide use. These methods have led to real success. New research in QSAR has been to calculate the descriptors of the molecules from their three-dimensional arrangements of atoms and groups in space. The popular comparative molecular field analysis (CoMFA) generates thousands of descriptors for each molecule, which may be too many to handle, and so a new method is needed to analyze matrices with thousands of correlated predictors, some of which are irrelevant to the end point. A fairly new company, Arris, has produced QSAR software that examines the three-dimensional properties of molecules using artificial intelligence.

Another application is the synthesis of organic chemicals. There are two aspects: design of the synthetic pathway (what precursors and general reaction condition will be used) and laboratory execution of the actual synthesis. Experienced synthetic chemists design the synthesis of new molecules based on their knowledge of the hundreds of types of reactions that can be done, the limitations in terms of the other structural features of the molecules that will survive the reaction conditions, and the usual success of the reactions in terms of the side products and yield. The complication with using the computer is that synthetic organic chemists rely on information contained in two-dimensional chemical structure diagrams. A reaction is simply a transformation of one diagram into another by combining and transforming starting materials into products. Hence, the expertise is organized intellectually in pictures, not numbers. The key insight that these structure diagrams can be described as labeled graphs enabled the use of the computer to process structure diagrams. The computer uses structure searching to ascertain if a proposed precursor is commercially available. Substructure searching identifies labile bonds to be broken in a retrosynthetic fashion. Maximal common subgraph algorithms have been used on the two-dimensional structures of products and reactants to perceive the part of a molecule unchanged in a chemical reaction. Two approaches have been used to apply the computer to the design of chemical syntheses. The first simply catalogs literature, chemical reactions and transformations, with the associated starting materials and product(s), conditions, yield, and literature reference. Here, the chemical reaction can be input as normally written and the computer can be used to detect which parts have changed. By definition, then, the parts of the molecular graph that change represent the chemical reaction. The unchanged part represents the chemical context in which the reaction occurs. In graph theory terms, the unchanged parts are the maximum common subgraph of the reactants and products. The second type of computer program to aid the planning of synthesis involves the actual disconnection of the synthetic target into the proposed starting materials. Such computer programs treat molecules as labeled graphs but use rules encoded by experts to guide the proposed synthetic route. Because of the large number of rules to be encoded, such retrosynthetic programs are much less complete and of less general use than the reaction library programs.

1.3 Introduction to Computational Chemistry

The term *ab initio*, Latin for "from the beginning," is used for computations derived directly from theoretical principles. The most common type of this calculation is called a Hartree–Fock (HF) calculation, and the primary approximation is called the central field approximation. The coulombic electron–electron repulsion is not taken into account, but its net effect is included. It is a variational calculation and the approximate energies calculated are all equal to or greater than the exact energy. Second, the wave function is described most often as a linear combination of Slater-type orbitals ($\exp(-ax)$ or Gaussian-type orbitals ($\exp(-ax^2)$). The wave function is formed from linear combinations of basis functions and the energy tends to be greater than the HF limit. The electron–electron repulsion called a correlation is corrected by any number of methods, such as the generalized valence bond (GVB), etc. A method that avoids the HF mistakes in the first place is called quantum Monte Carlo (QMC). An alternative *ab initio* method is density functional theory (DFT), in which the total energy is expressed in terms of the electron density, rather than the wave function. These *ab initio* methods give very good qualitative answers but are very expensive because of the computer time needed.

Semiempirical calculations have the general HF structure but use curve fitting in a few parameters or numbers, and the results can be erratic if the molecule being computed is different in the parameterization set.

The energy structure of an infinite structure is defined by a band structure plot, which gives energies of electron orbitals for each point in k-space, called the Brillouin zone. It is much more difficult to do this than the molecular calculations and requires a lot of time.

Molecular mechanics avoids quantum mechanics totally. The classical energy equations consist of the harmonic oscillator equation for the energy associated with bond stretching, bending, rotation, and intermolecular forces, such as van der Waals interactions and hydrogen bonding. The parameterization of the method (the force field) is crucial to its success. Molecular mechanics allows the modeling of enormous molecules such as proteins and segments of DNA. It cannot, however, handle some chemical properties, such as electronic excited states.

Molecular dynamics examines the time-dependent behavior of a molecule, such as vibrational motion or Brownian motion. Molecular dynamics allows the computation of properties such as diffusion coefficients or radial distribution functions for use in statistical mechanical treatments of solvent/solute systems. In order to analyze the vibrations of a single molecule, many dynamics steps are done; then the data is Fourier-transformed into the frequency domain. A given peak can be chosen and transformed back to the time domain, in order to see what the motion at that frequency looks like.

Statistical mechanics treats many problems outside of this domain. However, it is the mathematical means to extrapolate thermodynamic properties of bulk materials from a molecular description of the material. It is used at the end of *ab initio* calculations for gas-phase properties. It has also been successfully used in solid-state physics.

Thermodynamic treatment is most often not performed by computer since many aspects of the chemical applications can be described with simple mathematical expressions.

Structure–property relationships can be qualitative or quantitative in defining relationships between molecular structure and observed properties. They need not be based on any rigorous theoretical principles. The simplest cases of structure–property relationships are qualitative thumb rules. For example, an experienced polymer chemist may be able to predict whether a polymer will be soft or brittle based on the geometry and bonding of the monomers. When structure–property relationships are mentioned in the current literature, a quantitative mathematical relationship is usually implied. These are most often derived by using curve-fitting software to find the linear combination of molecular properties that best reproduces the desired property. The molecular properties are usually obtained from molecular modeling computations. Other molecular descriptors, such as molecular weight or topological descriptions, are also used.

Symbolic calculations are performed when the system is just too large for an atom-by-atom description to be viable at any level of approximation. An example might be the description of a membrane by describing the individual lipids as some representative polygon with some expression for the energy of interaction. This sort of treatment is used for computational biochemistry and even microbiology.

Techniques invented by computer scientists interested in artificial intelligence have mostly been applied to drug design in recent years; these are called *de novo* or rational drug design. Usually some functional site has been identified and it is desired to come up with a structure for a molecule that will interact with that site in order to hinder its functionality. Rather than have a chemist try hundreds or thousands of possibilities with a molecular mechanics program, the molecular mechanics is built into an artificial intelligence program, which tries enormous numbers of "reasonable" possibilities in an automated fashion. The number of techniques for describing the "intelligent" part of this operation is so diverse that the operation is impossible to describe.

1.4 Computational Chemistry and Material Science

If we were to calculate properties starting from the ultimate basis or foundation, we would follow a path as discussed below. After this is described we will show a more empirical method (the QSAR method) which has led to success more frequently.

In the more fundamental method, atomic and molecular systems are covered in a wide range of computational methods, approximations, and procedures to calculate structure, reactivity, and other properties of atomic and molecular systems. They apply to atoms, small molecules, macromolecules, polymers, and solids. The *ab initio* work covers Schrodinger's equation, the HF approximation, semiempirical methods, density (electron) functional theory, optimization of molecular geometries, quantum methods, Dirac–Fock method, etc. Classical modeling, including force field function (potential energy function) describes the internal energy of the molecule, the force field parameters (interaction strengths), molecular dynamics, free energy perturbation, etc. Classical and quantum methods are combined to solve large molecular systems and solids with improved accuracy. Visualization is used to improve the computational science. Then computational chemistry is used to predict the properties and behavior of molecular systems.

The methodologies to obtain some important properties are given below:

- Optimized molecular geometries
- Electron population analysis
- Electron multipole analysis
- Spectroscopy, including infrared and vibrational spectra
- Harmonic force constants
- Energy barriers of reactions
- Mapping sections of the Born–Oppenheimer potential surface
- Chemical reactivity and reaction energetics
- Ionization potentials solvation

Increasing accuracy and complexity involves molecular mechanics, molecular dynamics, semiempirical and quantum mechanics. One knows when one is finished by (1) comparison to experimental values, (2) tolerance of experimental values, (3) tolerance of computed values, and (4) when computer time runs out.

Some points to consider are how it is useful and how it is calculated. These might include Koopman's theorem and DFT, transition state theory under ionization potentials, and infrared, visible, and ultraviolet under electronic spectra.

Reaction energetics combine computing the energy and following a particular path (the reaction path) along the Born–Oppenheimer energy surface.

Representative Computational Programs for Computational Chemistry and Material Science:

Ab initio third party codes
- GAUSSIAN
- GAMESS
- HONDO7

SPARTAN
CADPAC
DFT codes
DMOL-BIOSYM Technologies, Inc.
DGAUSS-MSI Research
Semiempirical codes
MOPAC
AMPAC 2.1
MNDO
Classical modeling codes
DISCOVER-BIOSYM Technologies, Inc.
SYBYL-Tripos Associates
CLAMPS

A chronology of molecular modeling is presented below.

1860 - Structural stereochemistry first considered (structural formulae used)
1874 - Tetrahedral carbon discovered by van't Hoff
1953 - Three-dimensional (3D) structure of myoglobin solved by X-ray crystallography
1959 - Drieding stick models developed
1965 - CPK space-filling models developed
1970s - Computer models began to be tested

There was a progression from flat structures to correct stereochemistry or 3D representations. The advantages of computer models include that they:

Are geometrically accurate;
Are capable of precise manipulation;
Permit calculation of conformational energies (quantum mechanics and molecular mechanics);
Are readily superimposed (compare conformations, properties, volumes, etc.);
Allow precise positioning of one molecule with respect to a set of spatial requirements (enzyme substrate, activated complex, etc.);
Are easily and conveniently stored, edited, and duplicated.

Computer models are basically mathematical representations/models of molecules and properties, such as:

Atomic positions (Cartesian coordinates, internal coordinates — bond lengths, angles, and torsions);
Molecular surfaces that are mathematical functions based on atomic position and atomic radii;
Energies that are sets of equations involving atomic distances, atom type, bonding arrangement, etc.

Any field where the 3D shape of molecules and their properties are important can usually benefit from modeling. Modeling can involve computation and/or graphics.

Typical modeling exercises could involve

Prediction and visualization of shape/properties;
Comparison of shape/properties;
Examination/prediction of molecular interactions and reactions;
Investigation of unstable or excited molecules;
Modeling of dynamic systems (vibrations, diffusion, conformational changes, reactions).

Basically, modeling can aid in understanding and prediction of molecular phenomena. Some applications of modeling are

1. New drugs
2. Chemical sensors and probes
3. Chromatographic agents
4. Enzymes/catalysts
5. Stereoselective synthesis
6. Pesticides
7. Microelectronics
8. Advanced materials

In order to obtain a good model, the structure must be refined. The process is called optimization or energy minimization. There are two different methods for the calculation of molecular energies:

1. Quantum mechanics (QM)
2. Molecular mechanics (MM)

They differ in their approach to the prediction of the energy associated with an arrangement of atoms (i.e., a molecule). QM calculations solve approximations to Schrodinger's equation in order to determine electron density, energy, and other molecular properties, whereas MM treats a molecule as a series of balls and springs, and energies are determined in accordance with Hooke's law.

1.5 Computational Chemistry at the ARC

Quantum chemical *ab initio* programs attempt to use principles of quantum mechanics to obtain molecular properties of interest to a chemist or physicist. Such calculations can obtain molecular geometries, vibrational frequencies and barriers to rotation, ionization potentials and electron affinities, and many other properties. Quantum chemistry calculations provide a rapid and inexpensive alternative to experimental techniques for obtaining such properties. They are widely used as a predictive tool and to confirm or explain experimental results. Quantum chemistry calculations are applicable to a wide range of chemical problems, including enzymology, thermochemistry of combustion, homogeneous catalysis, cluster science, and organic reaction mechanisms.

Packages are available with varying capabilities, ease of use, and performance. They are often large and complex. A typical source code may be several hundred megabytes, while executable files may be as large as several tens of megabytes. State-of-the-art calculations using a typical workstation can require as much as a hundred or more megabytes of RAM, a gigabyte or more disk space, and take several days to several weeks to complete. Even larger calculations can be done with more powerful computers or clusters of workstations. Parallelization of these codes is an important step in increasing the capabilities and performance of the programs.

1.6 Desktop Chemist and Molecular Modeling Tools

Amber-Simulation — Tools for energy minimization, molecular dynamics, and nuclear magnetic resonance (NMR) structure refinement, particularly for protein and DNA modeling.

AMSOL — Semiempirical quantum mechanics for calculation of structures and energies in gas phase and in solution.

Anaconda — Calculates surface properties of molecules by the technique of gnomonic projection. These are used to compare and align molecules for 3D QSAR.

Asp — Calculates a similarity index for molecules based on their alignment, shape, and properties. The index is used to improve alignments, as a predictive metric in QSAR.

AutoDock — A suite of programs that predicts the bound conformations of a flexible ligand to a macromolecular target of known structure, like an enzyme or nucleic acid.

Cache Products — A family of desktop modeling tools widely used by experimental chemists. It includes a property-driven interface and 3D visualization. Models include mechanics, dynamics, extended Huckel, ZINDO, MOPAC. Cache systems can compute spectra, perform conformational searching, and model all elements up to 103.

Cobra — Automated conformational analysis and 3D structure generation.

Corina — Rapidly provides accurate 3D structures from 2D input descriptions, including SMILES and SD formats.

TOPKAT — A computational toxicology tool for assessing various toxic metrics of chemicals solely from molecular structures using QSTRs. TOPKAT QSTRs for Tsar accesses toxicity from Tsar spreadsheets.

Tsar — Fully integrated QSAR package that can store and analyze structures, properties, and associated data in one straightforward chemical spreadsheet. The Tsar interface helps find trends hidden in data by combining complex statistical analyses with simple visual representations of the results. Also works as front-end to ASP, Corina, and Vamp.

Unichem Series — High-performance, high-accuracy molecular modeling package. The Unichem tool allows users to build molecules from scratch, predict molecular properties, see 3D displays of molecular orbitals, charge density, potentials. Includes an easy-to-use graphical user interface (GUI) and three computer servers: MNDO94, CADPAC, and Dgauss.

Vamp — Extremely fast. Accurate and reliable semiempirical molecular orbital calculations.

1.7 The Computer Simulation by Molecular Dynamics, Monte Carlo, and Molecular Mechanics of Liquid and Solid Phases

CCP5 is a collaborative computational project for computer simulation of condensed phases. The U.K. has a large and successful community engaged in this work. The flagship project is the development of a molecular dynamics code DL POLY for the simulation of ionic, polymeric, and macromolecular systems on parallel computers and workstations. The project began in 1980, when it concentrated on the computer simulation of liquids using the Monte Carlo and molecular dynamics methods. Much of the early work of CCP5 was directed toward exploiting the power of the Cray 1s computer then in Daresbury, and a very useful library was developed. In 1983, the scope of CCP5 was extended to include the calculation of lattice properties by energy minimization methods. The skills required for these two problems were complementary, and a combined approach was beneficial.

CCP5 works with understanding complex fluids and heterogeneous systems through the interpretation of macroscopic phenomena in terms of atomistic models. The aim is to understand the orientational order and the thermodynamic and dynamic properties of liquid–crystal systems in terms of simplified models which treat the molecules as hard or soft nonspherical particles. Monte Carlo and molecular dynamics are the main tools.

Ab initio electronic structure methods are playing an ever-increasing role in the construction of models. The use of distributed multipole analysis (DMA) for representing electrostatic interactions obtained from *ab initio* calculations has become very important.

In the molecular dynamics simulation of macromolecular systems, it is a major issue to include fully the aqueous solvent, which often plays a key role in stabilizing the different polymorphs of DNA. The explicit and realistic simulation of large proteins and sections of DNA is making an important impact on studies of protein stability, the role of solvent in stabilizing the A and B forms of DNA, and the effect of drug molecules intercalated in the DNA helix.

The work of CCP5 materials is, first, toward the simulation of large complex systems and, second, toward the increasing use of *ab initio* methods. Work on complex systems is illustrated by extensive dynamic simulations on ionic systems of up to 3000 ions, used to study superior conduction in beta alumina and the structure and dynamic properties of silicate glass.

1.8 Chemical Modeling — SPARC Computational Approach

SPARC does not do "first principles" computation; rather, it analyzes chemical structure relative to a specific reactivity query much as an expert chemist might. It uses an extensive knowledge base of organic chemistry. To obtain physical properties, intermolecular interactions are expressed as a summation over all the interaction forces between molecules (i.e., dispersion, induction, dipole, and H-bonding). Each of these are expressed by molecular-level descriptors (density-based volume, molecular polarizability, molecular dipole, and H-bonding parameters) that are in turn calculated from molecular structure. For chemical reactivity, molecular structure is broken into functional units. Reaction centers with known intrinsic reactivity are identified and the impact on reactivity of appended molecular structure is quantified using mechanistic perturbation models.

A toolbox was developed with resonance models from light absorption spectra, electrostatic models from ionization equilibrium constants, solvation models from physical properties (vapor pressure, solubilities, distribution coefficient, gas chromatographic retention times, etc.).

Computational approaches in SPARC are a blend of conventional LFER, SAR, and perturbed molecular orbital (PMO) methods. SPARC uses LFER to compute thermodynamic or thermal properties and PMO theory to describe quantum effects such as delocalization energies or polarizabilities of p electrons. In reality, every chemical property involves both quantum and thermal contributions and necessarily requires the use of both perturbation methods for prediction. The SPARC predictive methods were designed for engineering applications involving physical/chemical process modeling.

1.9 AHSystems Group

AHSystems Group is a commercial organization that has a number of software items, among which are those having to do with thermodynamics and kinetics modeling.

HSC Chemistry for Windows (version 3.0) — HSC Chemistry for Windows is a chemical reaction and equilibrium software package that utilizes an extensive thermochemical database of over 8000 compounds. The program includes seven different calculation options for calculating Gibbs free energies and equilibrium constants for reaction equations, heat and material balances, equilibrium compositions, formula weights, electrochemical cell equilibria, and phase stability diagrams.

Kinetics (version 1.04) — Kinetics numerically integrates the rate constants for reversible multistep chemical reactions to determine the concentrations of the chemical species in the system as a function of time. The program provides an easy-to-use interface that allows users to create complex reaction systems by entering the reactions of interest and the associated rate constants. No other input is required by the program. The program formulates the required rate equations from the list of reactions. Graphical output of results can be obtained, and an output file suitable for use in a spreadsheet program is generated.

ChemSage — ChemSage is a user-friendly and flexible computational thermochemistry package for the calculation of phase equilibria and energy changes associated with a wide variety of technical problems in research, industry, and environmental impact studies.

TAPP — TAPP is a Windows or Macintosh materials property database with thermophysical data on the structural, physical, and thermodynamic properties for over 17,000 solid, liquid, and gas compounds and a collection of 1500 phase diagrams. The solids database has 11,000 compounds and includes information on transformation temperatures, crystal structure, density, thermal expansion, elasticity modulus, and thermal conductivity. The liquid and gas databases have 6000 compounds and contain data on transition temperatures, critical properties, densities, viscosities, thermal conductivities, and surface energies. Also included are thermodynamic data and vapor pressures for over 5000 compounds and over 1000 solid and liquid solutions. Over 1500 inorganic phase diagrams are also included.

TEP (Thermal Equilibrium Program) — TEP is an interactive combustion analysis tool for assessing the products of combustion under equilibrium conditions using the minimization of the Gibbs free energy. The NASA Lewis chemical equilibrium computer program is used as the underlying solver.

MAPP — MAPP is a Win/Mac access program to the ASM International Mat.DB engineering materials property databases. It will find properties of a material when you know a specification or common name. It will find materials that match your input for a set of property values and/or compositions. It will edit ASM data or enter new materials and it can use a simple graphic interface.

FLO-SERIES and adjunct tools — This chemical engineering software streamlines the piping design process. FLO-SERIES allows design and analysis of complete piping systems.

PROCEDE for Windows — This is a process design program that can be used to create engineering drawings and specification sheets. There is a symbol library manager, allowing users to add custom symbols to flow sheets and equipment data into specification sheets.

1.10 Environmental Fate Data Base (ENVIROFATE)

The Environmental Fate Data Base (EFDB) is a data bank from the EPA that contains data on physical/chemical properties and the environmental fate of several thousand chemicals. It consists of six interrelated parts. CASLST is a listing of CAS Registry numbers, molecular formulae, and names for chemicals in the other parts except for XREF. They are linked by CAS Registry Number; DATALOG provides a pointer to various data reported in the open literature pertaining to the environmental fate of chemicals. BIOLOG provides a pointer to data on microbial degradation or toxicity in the open literature. CHEMFATE contains literature-derived data for the environmental fate of several hundred chemicals. BIODEG is an experimental file containing (1) summaries of all available mixed-culture biodegradation studies; (2) summary judgments of biodegradability in the literature; and (3) judgments of the reliability of the summary biodegradability codes. All references are linked to the XREF file to provide complete citations for the journal articles referenced.

1.11 Materials Science

Materials science is concerned with the nature, properties, and use of materials. One materials science area is solid-state physics, focusing on the physical properties of solid materials. Such properties result from the distribution of electrons in metals, semiconductors, and insulators; oscillation in crystals; energy bounds; magnetic phenomena; dielectrics; ferromagnetics and dislocations. Another materials science area is that of polymers and liquid flow. For solid, liquid, or some transitory setting, materials science seeks an understanding of the macromolecular structure and properties of a material by drawing on knowledge of its atomic and molecular constituents.

Originally, *materials science* denoted empirical study, fundamental research, synthesis, and production in metallurgy and ceramics. Today, the term is much broader and involves interdisciplinary interactions among scientists, engineers, mathematicians, physicists, chemists, and biologists concerned with the four basic elements of modern materials science: properties, structure and composition, synthesis and processing, and performance. Interdisciplinary studies encompass all classes of materials, including biomaterials and bimolecular materials, ceramics, composites, electronic materials, magnetic materials, metals, optical and photonic materials, polymers, and superconducting materials.

1.12 General Systems Theory

General systems theory is not a theory of systems of a more or less special kind, but of universal principles applying to systems in general. Its subject matter is the formulation and derivation of those principles that are valid for "systems" in general. It is a reaction against reductionism and attempt to revive the unity of science. Systems are open to and react with their environment. Rather than reducing an entity to the properties of its parts, systems theory focuses on the arrangement of and the relations between the parts that connect them into a whole. This particular organization determines a system, which is independent of the concrete substance of the elements. Thus, the same concepts and principles of organization underlie the different disciplines, providing a basis for their unification.

Systems theory can then be defined as the transdisciplinary study of the abstract organization of phenomena, independent of their substance, type, or spatial or temporal scale of existence. It investigates both the principles common to all complex entities and the (usually mathematical) models that can be used to describe them. Some of the balancing concepts used in general systems theory include:

open ^ closed Information ^ entropy causality ^ teleology

Feedback ^ homeostasis allometry ^ surface rule

A discussion of these matters can be found in Ludwig von Bertalanffy's book, *General System Theory*.

> There are various approaches that provide for the problems concerned with applied systems research. They are systems engineering, operational research, linear and nonlinear programming, classical systems theory (utilizing "classical" mathematics such as calculus), computerization and simulation, compartment theory, set theory, graph theory, net theory, cybernetics, information theory, theory of automata, game theory, decision theory and queuing theory.

This was the original list in Bertalanffy's book. To this we now add system dynamics (models changes in a network of coupled variables, sciences of complexity, self-organization and heterogeneous networks of interacting actors, and associated domains such as far-from-equilibrium thermodynamics, chaotic dynamics, artificial life, artificial intelligence, and neural networks).

The recent explosive growth in computer power and connectivity has led to technological advances with unprecedented opportunities for providing rapid and efficient access to enormous amounts of knowledge and information; for studying vastly more complex systems than was hitherto possible; and for advancing in fundamental ways understanding of learning and intelligence in engineered systems. The National Science Foundation (NSF) knowledge and distributed intelligence (KDI) theme is a foundation-wide effort to promote the realization of the opportunities.

The NSF aims to achieve, across the scientific and engineering communities, the capability to generate, model, and represent more complex and cross-disciplinary scientific data from new sources and at enormously varying scales. KDI has three foci in 1998: knowledge networking (KN); learning and intelligent systems (LIS); and new computational challenges (NCC).

KN focuses on the integration of knowledge from different sources and domains across space and time. LIS seeks to stimulate multidisciplinary research that will unify experimentally and theoretically derived concepts related to learning and intelligent systems, and that will promote the use and development of information technologies in learning and discovery across a wide variety of fields. NCC focuses on research and tools needed to discover, model, simulate, analyze, display, or understand complex phenomena, to control resources and deal with massive volumes of data in real time, and to predict the behavior of complex systems.

The Division of Mathematical Sciences (and other parts of NSF) and the Defense Sciences Office of the Defense Advanced Projects Research Agency (DAPRA) plan jointly to support research and development of new approaches to the design and creation of efficient algorithms and optimized libraries for large-scale numerical modeling and simulation of physical phenomena arising in industrial applications.

The Optimized Portable Algorithm and Application Libraries (OPAAL) initiative is an opportunity for researchers in the mathematical sciences to join with other scientists and engineers in the development of innovative mathematical techniques applicable to the simulation of complex physical processes. Fresh ideas, approaches, techniques, the formation of new teams, and the participation of researchers new to such problem areas are particularly encouraged. Areas of mathematics that are not currently considered "applied" are a likely source of such fresh ideas, approaches, and techniques.

An essential component of this initiative is the development of mathematical formulations to enable automatic compilation of scalable, high-performance software libraries of key numerical kernels. In addition, algorithmic approaches should facilitate systematic and effective machine-specific tailoring of computational constructs such as data structures and communication protocols. Applications of particular interest are those having an impact on modeling and simulation of advanced manufacturing processes and systems, especially those involving multiple scales and complex physical and chemical processes.

1.13 Physical and Environmental Property Data Banks Management System and Simulation

Chemical Properties Tool (StEPP/PPMS)

Tony Rogers with Andrew Kline at Michigan Technological University (MTU) continue the development of a CPAS (Clean Process Advisory System) Tool that takes care of the collective environmental property expertise available in the MTU Design for Physical Property Data. The project is evolving toward the creation of a general Physical Properties Management System (PPMS) that will serve as an expandable framework for adding estimation algorithms and other third-party databases. Within PPMS is a module called PEARLS (Property Estimation and Retrieval System) a stand-alone data display for process engineers. PEARLS is a second-generation CPAS software tool expanding the Software to Estimate Physical Properties (StEPP). Pure component properties available from PEARLS are vapor pressure, molar volume at the normal boiling point, liquid density, activity coefficient, Henry's law constant, aqueous solubility, octanol–water partition coefficient, soil–water partition coefficient, and partitioning onto the organic carbon portion of biomass.

Environmental Fate and Risk Assessment Tool (EFRAT)

This project has the goal of developing a process design software tool which will estimate environmental and health impacts of chemical process design options through a combination of screening-level fate and transport calculations and risk assessment indexes. Included are the environmental and health effects for Global Warming Index, Ozone Depletion Index, Smog Formation Index, Acid Rain Index, Ingestion Non-carcinogenic Index, Ingestion Carcinogenic Index, Inhalation Non-carcinogenic Index, Inhalation Carcinogenic Index, and Fish Toxicity Index. As an example, the Global Warming Index is

$$I_{GW} = (D_A A t_{RT})_{CHEMICAL}/(D_A A t_{RT})CO_2$$

Here I_{GW} is the Global Warming Index and t_{RT} is the reaction residence time of a compound in the troposphere. The fraction of the released chemical distributed to the atmospheric compartment relative to the water and the soil is (D_A), and A is the infrared adsorptivity of the compound, which is a measure of the ability of the compound to trap radiative heat.

Physical Property Predictive Driver

The objective of this research is to develop a simple additive group contribution predictive model for the pure component vapor pressure. When coupled with an activity coefficient model for dilute aqueous mixtures, the vapor pressure model will be able to estimate other physical properties critical to process design. These include bubble point, K-factors, and relative volatilities.

Reactive Chemistry Screening Tool

The goal here is to produce a CPAS tool to assist the designer in determining the hazards associated with reactive chemicals. Design decisions motivated to reduce the pollution and emissions from a chemical plant may increase the hazard and risk. This tool will provide the designer with the ability to determine the relative safety of a process, due to reactive chemistry, during the incremental design stage.

Design Option Ranking Tool (DORT)

Conceptual design and pollution control have traditionally been done at different stages in the development of a process. However, if the designer were given the tools to view the environmental impact of a process at the very beginning of the design process, emphasis could be placed on pollution prevention and the selection of the environmentally sound alternatives. This could help eliminate total pollution output as well as reduce the costs of the end-of-the-pipe treatment which is currently done. Not only is it important to help the designer create multiple alternatives in the design process, but we must also provide the necessary tools that facilitate rapid and effective evaluation of the alternatives.

Multiattribute Making Aid

An Advanced Hierarchical Procedure (AHP) decision-making aide has been programmed in Visual Basic. This model picks the optimum of up to five processes based on indexes provided in three "level-one" areas (economics, safety, and pollution areas were assumed). The user is required to supply the number (up to five in each area), name and values for the indexes (the values represent the relative importance of each variable to the others) for each "level-two" index. The user must then provide the appropriate input values from each process being evaluated. AHP returns the process number that is the most attractive based on the input indexes.

Development of the Environmental Technologies Design Options Tool (ETDOT)

The goal of the ETDOT is the compilation of stand-alone software tools that may be used to evaluate treatment strategies for gaseous, aqueous, organic, and solid waste by-product streams. ETDOTs will assist CPAS in evaluating the technical and economic feasibility of source reduction, versus end-of-pipe treatment and waste segregation, versus treatment at a central facility. The ETDOT or its component parts should be integrated with process simulators, or other manufacturing design software.

Tools developed include (1) software for predicting the fate of volatile organic (FaVOr) compounds in wastewater treatment facilities; (2) software for predicting the fate of metals (FaMe) compounds in wastewater treatment facilities; (3) gas- and liquid-phase adsorption simulation software, and (4) aeration system analysis program (ASAP) software for removal of volatile organic compounds from water by surface, bubble, and countercurrent packed tower aeration techniques. Another tool that has been in development is (5) advanced oxidation process software (AdOx).

Process Safety and Risk Evaluation Tool

The goal here is to assist the designer in determining the relative safety of the design. Design decisions motivated to reduce the pollution and emissions from a chemical plant may increase the hazard and risk. This tool will provide the designer with the ability to determine the relative safety of a process during the incremental design stage.

Data for Waste Inventory Factors (WIFs)

CenCITT (Center for Clean Industrial and Treatment Technologies) funds a design software tool that evaluates emissions from chemical processes on a unit-operations-specific basis. This database of evaluated EPA emission factors is to be incorporated into EFRAT within the CPAS effort in order to generate noneconomic environmental indexes for evaluating designs for pollution prevention. These design tools can be used in conjunction with commercially available chemical process simulators to judge the optimum pollution prevention design alternative.

In addition to the EPA database evaluation effort, emission factors for several unit operations and a select number of chemicals (benzene, hexachlorobenzene, trichloroethylene, 1,1,1-trichloroethane, methyl chloride, and formaldehyde) were incorporated into the EFRAT software. Unit operations included in the EFRAT emission factor database are reactors, distillation columns, storage tanks, transfer and handling operations, and fugitive emissions from valves, fittings, compressors, and flanges.

Recovery of Organics from Streams Using Adsorption and Distillation

The goal of this project is to develop two software modules: (1) adsorption for recovery (AdRecover) of organics and (2) multi-component distillation (MC-Dist). These modules are developed for implementation in the separation technnologies modeling tool group of the CPAS.

Environmentally Conscious Design for Construction

This goal is for the development of a CD-ROM-based multimedia system for designers to infuse environmental considerations into the design process for constructed facilities, while simultaneously evaluating other facility performance criteria. This system will also serve as a resource database for manufacturers desiring to understand better field problems associated with installing their products.

1.14 Abstract Intelligent Agents

An abstract intelligent agent (AIA) is a hypothetical model that captures the essence of all systems which we accept as "intelligent" in the common sense. The modeling formalism is a collection of independent agents interacting via discrete events. An agent is any actor in a system, any entity that can generate events that affect itself and other agents. A program that demonstrates this is called SWARM. Simulations consist of groups of many interacting agents. Simulation of discrete interactions between agents stands in contrast to continuous simulations, where simulated phenomena are quantities in a system of coupled equations. These electronic life-forms, known as *software agents*, are computer algorithms that function autonomously and behave like living creatures. They have evolved to a point where they can be used — or domesticated — for practical purposes. They can tackle complexity problems, which often represent the seemingly unpredictable behavior of whole populations. They seem to actually learn. SWARM is an adaptive learning mechanism that is flexible and that can be used for a variety of applications. SWARM is a body of a number of different interacting elements, and they can be applied to a whole number and variety of emergent phenomena that actually arise from the dynamics of these interacting agents. An agent is every goal-driven system with the ability to control its own information acquisition. An intelligent agent is an agent with the ability to change and to evaluate its own preferences and knowledge; i.e., it has the ability to change goals if the initial intervention goal is not reachable. One of the central goals of AI (artificial intelligence) is to develop artificial agents that embody all the components of intelligence.

Stuart Kauffman defines an autonomous agent as a collectively autocatalytic system that performs one or more thermodynamic cycles. In a lecture in Santa Fe, he discussed the general nature of "fitness landscapes," as well as coevolutionary processes, in which the adaptive moves of one agent deform the fitness landscapes of its partners, and tried to show that an endogenous coevolutionary process allows agents, each adapting for its own selfish fitness, to tune the couplings and ruggedness of its fitness landscapes such that the entire system achieves a specific *self-organized critical state.* The self-organized critical state of a coevolving system of agents emerges as one part of a candidate general law characterizing coevolutionarily constructible systems of agents.

In a second lecture, he discussed a theory of life as a phase transition in chemical reaction networks from subcritical to supracritical behavior. There is an understanding of phase transitions in chemical reaction graphs as the basic underpinning of the emergence of collectively self-reproducing systems. As the diversity of organic molecules in a system increases, the diversity of reactions by which they transform to one another increases very much faster. Then at a sufficient diversity, so many reactions are catalyzed that a collectively autocatalytic set of molecules is virtually certain to emerge. Thus, the emergence of collectively reproducing systems of molecules is an expected phase transition in complex chemical systems.

If we consider the species of organic molecules now on Earth, we can also consider those species of organic molecules that do not exist, but that are one reaction step away from molecules that currently do exist. The set of organic molecules that are one reaction step away from those that do exist constitutes the chemically "adjacent possible." A feature of this biosphere is that the diversity of polymers and organic molecules in the biosphere has increased as the biosphere has expanded into the adjacent possible over the past 4 billion years. It is then a matter of considerable importance to understand what processes might govern and drive the flow into the molecular adjacent possible.

In pursuing the persistent generation of novel small molecules in the biosphere, we ask whether the biosphere as a whole is supracritical, while cells must remain subcritical to avoid the lethal consequences of the uncontrolled synthesis of novel molecules at the entry of any novel molecular species. Then we broach the possibility that causally local metabolic communities, say microbial communities, are poised on the subcritical-supracritical boundary. Such poised communities would graft into themselves useful novel molecules without being destroyed by uncontrolled supracritical behavior. This suggests that the biosphere may be poised to advance, on average, into the molecular adjacent possible as rapidly as possible. Indeed, this supposition becomes part of the tentative candidate law later proposed for coevolving self-constructing communities of autonomous agents.

1.15 The Development of Quantitative Structure–Activity Relationships

In 1968 Hansch introduced a multiple regression analysis of several physical properties of the lead molecule that statistically correlated with its biological actions. This can be summarized in the equation below.

$$\log 1/C = -k(\log P)^2 + k_2(\log P) + k_3\sigma\rho + k_3Es + k_5 \quad (1.15.1)$$

In this equation, C is the lowest concentration, P is the partition coefficient, σ and ρ are symbols we may know from Hammett's work and Es is the steric descriptor, which describes the gross bulk and the regional details of shape, or Taft's steric parameter, which he has extracted from experimental work.

Now Hammett discovered that the relative reaction rate of sets of congeners could be discussed in numerical terms. Still, vast areas of biochemical and biological reactions were not described until the early 1960s when the laboratories of Corwin Hansch extended Hammett's concepts using a hydrophobic parameter and showed that a linear combination of electronic and steric parameters allowed quantitative treatment (QSAR) of all biological reactions.

Linear free energy relationships (LFERs) were developed by Hammett and helped to describe the effects of substituents and groups that leave under ester hydrolysis. They showed how a parameter describing molecular structure (σ, and the slope, ρ) can yield properties such as equilibrium constant and rate constant and provide a thermodynamic basis for the work of QSARs, as shown in the relations:

$$\Delta G^0 = -RT \ Ln \ K = \Delta H^0 - T\Delta S^0 \quad (1.15.2)$$

QSAR developed from the LFER approach. Most popular of the QSAR approach has been the LOG (Octanol/Water) partition coefficient (or LOG P). However, many regressions do not work well with LOG P, because other effects, such as steric and electronic, are important. Thus, others used are descriptors such as Taft's Es, MR, etc., and theoretical descriptors, such as those computed from estimation programs; graph theoretical, information indexes (such as Kier and Hall's) and molecular orbital–based parameters.

Kamlet and Taft developed a relationship for studying LFERs of solute/solvent interactions, called the linear solvation energy relationship (LSER):

$$\begin{aligned}\text{Property} = {} & \text{bulk/cavity} + \text{dipolarity/polarizability} \\ & + \text{hydrogen bonding acidity} \\ & + \text{hydrogen bonding basicity} \quad (1.15.3)\end{aligned}$$

The two primary approaches today to the computation of physical and chemical properties are (1) direct computation using quantum mechanical or statistical mechanical means and (2) the more empirical linear energy relationships (LFER) and the structure activity–property relationships (QSAR/QSPR). The first, although aesthetically pleasing, is limited, and the second provides more flexibility in the computation of physical, chemical, and biological properties as used with experimental data. There is also a method called (TLSER), the theoretical solvation energy relationship that tries to conform to the same characteristics as the LSER but under a theoretical basis. It has a parameter that can be computed from molecular orbital methods.

Tsar is an integrated system for identification of QSARs. The system allows medicinal chemists, statisticians, and modelers to read molecular structures and activity data into a spreadsheet. From within the spreadsheet users have access to many of the latest univariate and multivariate statistical techniques, including neural network algorithms for developing QSARs. Tsar also enables users to calculate molecular and substituent properties to create additional data for QSAR.

In 1990, Lyman et al. published a very fine book, *Handbook of Chemical Property Estimation Methods*, which has been used frequently.

Methods for estimating the following environmental fate properties are reviewed in a paper by Howard and Meylan:

Octanol–water partition coefficient
Water solubility
Boiling point, melting point, vapor pressure
Dissociation constants (pKa)
Henry's law constant
Atmospheric oxidation rates
Hydrolysis rates
Biodegradation
Soil/sediment adsorption coefficient
Bioconcentration factor

The paper concludes that it is now possible to predict most physical/chemical/transport/degradation processes that are important to the environmental fate of a chemical with a reasonable indication of the amount of the error expected for a new chemical, without having to select a model based upon structural similarity. Most of these properties can be predicted using available software with chemical structure input.

Tratnyek has used QSARs for oxidation and reduction rate constants and states that they are of enormous importance in environmental chemistry and toxicology because of their predictive power. He states that they also reveal a great deal about reaction mechanisms and the nature of substituent effects. Usually, QSARs are empirically derived, but there are a few that are theoretically based. Of these, the Marcus equation describing rates of electron-transfer mechanisms is one.

Biotransformation affects the bioavailability and toxicity of a compound and, thus, is now considered in the early stages of computer-aided drug discovery, which requires predictive methods and experimental measurements. The factors that influence drug distribution and metabolism are complex. Metabolic experimental studies are expensive and time-consuming, so predictive judgments about the sites for metabolic activities in the absence of experimental data become relevant for the compound in question. These judgments are incorporated into toxicity predictions in both pharmacological and environmental toxicology, synthesis programs in the pharmaceutical industry, and in the choice of labeling sites and reference compounds in metabolic research. Thus, work has been done toward formulating computational chemistry–based models, largely reliant upon molecular similarity arguments, for deriving such arguments. Computer-based approaches to metabolic prediction utilize databases with information on the biotransformation pathways of xenobiotics. The predictive ability of computer-based methods for suggesting xenobiotic biotransformations will increase as further improvements in determining structure–reactivity relations are made. Metabolic pathways are represented as a series of pair-wise connected molecules. An arrow between the two molecules indicates that a biotransformation from the reactant to the product has been observed. Reactivity defined in a simple sense as the observed occurrence or the nonoccurrence of a given reaction may be thought of as a molecular activity. By defining reactivity as a molecular activity, the data analysis techniques that are used in structure–activity relationship (SAR) studies become applicable. The structure maps presented in this study for a particular analysis apply to SARs in general. An atlas of structure–activity maps constructed from a metabolic pathway database would provide a useful tool for researchers interested in predicting drug metabolism. By examining these structure maps it may be possible to determine which structural modifications are necessary to influence the metabolism of a new compound.

The thrust of new research in QSAR has been to calculate the descriptors of the molecules from their 3D arrangements of atoms and electrons in space. One of the popular methods, the comparative molecular field analysis (CoMFA), generates thousands of descriptors for each molecule, whereas the data set typically contains biological activity for only 10 to 30 members. QSAR needs a new method to analyze matrices with thousands of correlated predictors.

The QSAR database has been assembled at MedChem (Pomona College) over a period of more than 20 years. The database covers both biological and physical-organic chemistry, for a total of over 10,000 data sets. It may well be the most comprehensive collection of QSAR data available

The QSAR and Modelling Society exists in the U.S. The society publishes a journal, *Quantitative Structure-Activity Relationships*, and a newsletter and has about 370 members (1995). The society is recommended to all scientists who investigate QSARs in medicinal, agricultural, or environmental chemistry.

The Hansch–Fujita QSAR method (Hansch and Fujita, 1964) was developed in the early 1960s and has become widely used by medicinal and agricultural chemists. In this method, one first describes each molecule in terms of its physical properties and then uses statistical methods to uncover the relationship between some combination of these physical properties and biological potency.

The partition coefficient is the equilibrium concentration of solute in a nonpolar solvent divided by the concentration of the same species in a polar solvent. The solvent is usually water. The logarithm of the partition function, log P, has been successfully used as a hydrophobic parameter in "extrathermodynamic" Hammett methodology. 1-Octanol has much to recommend it as the choice of the nonpolar phase (1), and log P(ow) has been used successfully in QSARs in the following special fields: drug and pesticide design, pharmacokinetics, anesthesiology, environmental transport and soil binding, toxicology, bioaccumulation, protein folding, enzyme binding, enzymic reaction in nonaqueous solvents and host-guest.

The C2.QSAR+ module provides intuitive and easy-to-use access to innovative QSAR technology. A QSAR model is a multivariant mathematical relationship between a set of physicochemical properties (descriptors) and a property of the system being studied, such as the biological activity, solubility, or mechanical behavior. The MSI sophisticated C2QSAR+ software facilitates the graphical analysis of QSAR models to assist perception of subtle chemical relationships in fields as diverse as drug discovery, polymers, and materials science. It integrates a wide range of regression and analysis technologies in a "chemically aware" spreadsheet interface.

ADAPT (Automated Data Analysis Using Pattern Recognition Toolkit) is a software system designed to allow the user to develop SARs or structure–property relationships (SPRs). ADAPT provides facilities for graphical entry and storage of molecular structures and their associated data, generation of 3D molecular models, molecular descriptor calculations, and the analysis of the descriptors using multivariate statistical, pattern recognition, or neural network methods to build predictive equations. ADAPT has topological, geometric, electronic, and physicochemical molecular descriptor generation routines and can generate hybrid descriptions that combine features. It supports many, many statistical approaches.

ACD/Labs has a program that calculates log P, log D, the substituent constant σ, B.P. and "Vaporization," pK_a, solubility, bioconcentration factor, and adsorption coefficient. It can obtain the physical properties as well. It uses QSARs to obtain some properies in terms of log P. Examples are

$$\log(1/S) = 1.214 * \log P - 0.85 + 0.0095 * (T_m - 25) \quad (1.15.4)$$

$$\log BCF = 0.76 * \log P - 0.23 \quad (1.15.5)$$

$$\log K_{oc} = 0.544 * \log P + 1.377 \quad (1.15.6)$$

for solubility, bioconcentration factor, and adsorption coefficient, respectively, where T_m is the melting point and P is the partition coefficient. The database for ACD/sigma has 850 substituents with over 3000 derived experimental electronic constants. The pK_a has more than 8800 compounds. The database for log P contributions has increments for over 500 groups. The fragmental contributions differ as a result of their chemical structure; the carbon atoms are different through their state of hybridization (i.e., sp, sp^2, sp^3), number of attached hydrogens, branching, cyclization, and aromaticity; and the database of the intramolecular interaction contributions contains well-characterized increments for over 2000 different types of pair-wise group interactions. They differ from each other by the type of interacting terminal groups.and the length and type of the fragmental system between the interacting groups.

SYBYL is a suite of molecular modeling programs for building, editing, and analyzing small-molecule organics, biopolymers, polymers, and materials. Modules include Base SYBYL, Biopolymer, QSAR, Advanced Comp — for conformation searching and interfaces to QCPE (quantum chemistry program exchange) programs. Additional functionality is also available. SYBYL is available from Tripos, Inc.

A single QSAR standing alone has little value. Different laboratories have derived QSARs having different meanings for the same data set. Confidence in a QSAR develops when it can be related in a variety of ways to many others, which is a primary motivation for constructing a data management system. Comparative QSARs are only possible via a common set of widely used parameters. At present, the thousands of published QSARs are based on Hammett–Taft-type models for SARs. Within the next few years it will be possible to collect a large enough sample of the published QSARs so that there will be examples with which to compare most newly developed equations.

1.16 Comparative Molecular Field Analysis

CoMFA combines QSARs with molecular modeling. It is particularly suited for mapping receptors with unknown structures.

Since its introduction, several years ago, CoMFA has become one of the most powerful tools for QSARs and drug design. It is particularly suited for mapping receptors with unknown structures. CoMFA has pioneered a new paradigm of 3D QSARs where the shapes, properties, etc. of molecules are related to specific molecular features (substituents, etc.) and their spatial relationship. Thus, molecular modification to improve biological performance based on QSAR can be more rooted in the actual chemistry of involved molecules. While steric and electrostatic properties of molecules are major physicochemical properties related to biological activity, they are purely enthalpic. It is desirable in many cases to characterize additional properties on a 3D basis. Efforts to include entropic properties within a CoMFA framework have been to characterize the hydrophobic nature of molecules. Recently, reactivity-based fields such as those of molecular orbitals have also been imported into CoMFA studies. The field generated and included in a CoMFA model depends on the creativity of the research and the validity of the underlying theory. In 3D QSAR studies, the steric and electronic effects are approximated using the molecular shape (Lennard–Jones) and charge distribution (coulombic) potential energy fields. Missing is the hydrophobic binding component. Approaches used for this are empirically based hydropathic interaction (HINT) field, a large variety of lipophilic potential fields, the molecular mechanically based water probe for the description of hydrogen bonds, and the Poisson–Boltzmann finite difference approach–based calculation of desolvation energy fields. It is apparent that the creation of new field for 3D QSAR is a fairly straightforward process. Treating atomistic QSAR parameters as 3D fields makes it possible not to restrict investigations to common "atoms." In conventional QSAR, most valid comparisons involving atomistic parameters are those where the atom is conserved by all molecules in the data set. This is quite restrictive, as it disallows many compounds that may have different backbones, etc., but that are clearly related chemically to other molecules in the set. However, with fields, the data parameters are now the values at grid points. This means that any molecule (accounting for superposition issues, of course) can be included in the model.

The goal of QSARs is to derive a function relating biological activity (or other properties) with parameters describing a feature of the molecule

$$\text{Biological activity} = f\,(\text{parameters}) \qquad (1.16.1)$$

One can use the "classical" QSAR with physicochemical descriptors (hydrophobic constants, Hammett electronic parameters, Taft's steric, Swain–Lupton coefficients, etc.). If the molecules are more complex, we try 3D QSAR. The procedure is

1. Select a set of molecules with known activity.
2. Align molecules in expected binding conformation.
3. Calculate spatially dependent parameters in "receptor space" around each molecule.
4. Derive QSAR.
5. Establish self-consistency and predict ability of QSAR; refine model, if necessary.

The CoMFA technique, developed by Dick Cramer, has become hugely popular and successful. The ligand/receptor interaction is represented as the electrostatic and steric interaction between the ligand and a probe atom (very similar to GRID, but with a simpler force field). Cramer assumed the relationship between the spatial parameters and biological activity was, for simplicity, linear, e.g.,

$$\begin{aligned}\text{Activity} = {} & aS1 + bS2 + \cdots + mSn + nE1 \\ & + \cdots + zEn + \text{constant} \qquad (1.16.2)\end{aligned}$$

where Si and Ei are the steric and electrostatic interaction energies at the grid point I and a to z are coefficients. Typically, there will be hundreds or thousands of descriptors but only tens of molecules from which to derive the QSAR. The QSAR equation is highly underdetermined and many possible solutions will exist. Traditional least-square regressions will find a solution that will not be the best.

1.17 Three-Dimensional Structure–Activity Relationships

In a paper on 3D SARs, Marshall and Cramer state that attempts to correlate biological activity and chemical structure have had limited success unless the study has confined itself to congeneric series. However, recent advances in computational chemistry, molecular graphics, and problem formulation appear to promise general solutions to the problem of comparison of structurally diverse molecules interacting at a receptor site. Two major problem classes may be distinguished: those in which the 3D structure of the receptor site has been determined by crystallography, or inferred by model building from known homologous structures, and those in which the receptor structure has not been determined.

If one knows the 3D structure of the therapeutic target, the techniques of molecular dynamics and Monte Carlo simulation can be combined with the thermodynamic cycle perturbation approach to yield predicted affinities with a firm physical chemical basis and reasonable expectation of quantitative agreement, when the postulated binding mode is correct. The more common situation of an operational receptor of an unknown receptor is more problematic, but rational strategies for proceeding have evolved, based on the traditional concepts of pharmacophore and complementarity between drug and receptor.

In a paper by van Drie, it is stated that experience with 3D database searching and *de novo* design indicates that these methods are complementary, with neither obviating the other. 3D database searching is more appropriate for lead finding, because our knowledge of the receptor–ligand recognition is usually poor at that and one does not want to invest much time synthesizing new molecules to test one's ideas. *De novo* design is most appropriate for lead optimization, to progress past the point where some patentable lead structure is already in hand by modifying it optimally to complement the receptor, and where one is at the point where one's understanding justifies a significant investment of synthetic efforts.

The future of 3D database searching appears to be quite promising. Representation of the molecules in databases has shifted from explicit definition of atoms to a more fuzzy atom-type description. This should have the effect of increasing the number of new classes of hits discovered via 3D search. Software tools for analyzing the results of 3D searches, however, have yet to make a dramatic impact. Since many 3D pharmacophore-based queries yield hundreds to thousands of hits, scoring functions need to be further improved to rank these in a useful fashion. Techniques capable of performing pharmacophore-based search in the context of a binding site followed by energy minimization should increase the likelihood of finding new leads. Furthermore, a new generation of techniques must be developed to analyze the results of 3D searches when no target structural information is available. Two key challenges that remain are concerned with conformational flexibility (of both the ligand and the macromolecular target) and the development of scoring functions that can rapidly and accurately assess the free energy of binding of the candidate structures.

The work at the University of Milan of Todeschini et al. involves QSARs and chemometrics in chemistry, toxicology, pharmacology, and the environment. Chemometrics is the discipline concerned with the application of statistical and mathematical methods to chemical data. They use WHIM (weighted holistic invariant molecular) descriptors, which are 3D molecular descriptors, as chemical information. These contain information about the whole 3D molecular structure in terms of size, shape, symmetry, and atom distribution, calculated from (x,y,z)-coordinates of a molecule within different weighting schemes. The algorithm consists of performing a principal components analysis (PCA) on the centered atom coordinates by using a weighted covariance matrix. WHIM theory has been extended for energy interaction fields (G-WHIM descriptors) and for local molecular properties (SWIM descriptors).

1.18 Combinatorial Chemistry

Combinatorial chemistry offers a means either to supplement existing compound databases in the search for new lead compounds or to replace them. Automation methods, together with a greater list of chemical reactions, amenable to combinatorial chemistry, make it possible to synthesize huge libraries of compounds in a very short period of time.

Processes are now designed to make molecules with chemical functionality of immediate great value in a lead discovery program. These molecules are sufficiently different from each other so that redundancy of effort is avoided. The identification of novel compounds with biological activity against significant therapeutic targets is an essential component in the drug discovery process. The lead structures enable us to cross a critical threshold of no chemical information to some chemical information. If the information is useful, then it sets us on a path to lead optimization and ultimately, it is hoped, to the development of pharmaceutical or environmental agents of commercial value.

Ferguson et al. state that the selection of compounds for use in high-throughput biological assays is one of the critical factors that dictates the likelihood of detecting exploitable biological properties. They use a process that delivers molecules that contain functionality of immediate value in a lead discovery program and that are sufficiently different from each other to assure that redundancy of effort is avoided. The system they use is one of high-throughput screening (HTS), and tens of thousands of reaction products have been added to the Optiverse library.

Combinatorial chemistry required chemists to think not only of synthesizing single, well-characterized compounds, but in terms of simultaneously synthesizing large populations of compounds. It also required that those people involved with information management and computational chemistry systems address the same issues as the chemists. The information systems, too, had until then dealt essentially with one molecule at a time. Technology such as combinatorial chemistry and high-throughput screening generates masses of relatively unrefined data. It is becoming more important than ever for companies to be able to handle this flood of data. Information management is absolutely critical in today's research environment.

Combinatorial chemistry initially focused on the synthesis of very large libraries of biological oligomers such as peptides and oligonucleotides. But drug developers generally prefer to focus on small organic molecules with molecular weights of about 500 Da or less, the class of compounds from which most successful drugs have traditionally emerged. So combinatorial chemistry researchers are concentrating on small organic compounds as well. Drug discovery is the primary goal of most combinatorial chemistry research, but combinatorial methods can also have potential applications for development of advanced materials and catalysts in organic chemistry and also beyond the realm of organic chemistry. Physicist Xiao-Dong Xiang of Lawrence Berkeley National Laboratory and chemistry professor Peter G. Schultz of U.C. Berkeley and co-workers recently devised a combinatorial strategy for finding advanced materials with novel chemical or physical properties — extending the combinatorial approach from biological and organic molecules to the remainder of the periodic table. Xiang et al. used thin-film deposition and physical masking techniques to synthesize libraries of solid-state materials. The properties of the resulting materials were then evaluated to identify promising candidates for further development. Organic libraries are now screened for catalytic as well as noncovalent binding. Hundreds or thousands of compounds are made very quickly and then tested for catalytic power. Potential catalysts are polymers that have multiple functional groups in different proportions and different sequences, plus a metal ion. Any reaction of interest can be chosen in screening. Once a polymer with activity is found, one begins to tinker with the proportions to fine-tune it until it gets faster and faster. Robotic combinatorial synthesizers have been developed. Many researchers now use some form of tagging or encoding to label compounds in large combinatorial libraries. Radiofrequency tags have been developed that use a technique in which memory devices are associated with derivatized polymer during combinatorial synthesis. The chips encode relevant information about the synthetic pathway, including not only reagents used, but also reaction conditions such as temperature and pH. The device can then report this information to a receiver via radiofrequency transmission.

Published work still centers on solid-support synthesis, although considerable momentum is building for chemistry in solution. Combinatorial chemistry has created challenges in collateral fields, such as chemical information and computational analysis. These have arisen in the fields of pharmaceutical and agrochemical research communities: how to store and manipulate combinatorial libraries (definition), how to make intelligent selections of subsets from combinatorial libraries (dissection), and how to evaluate the performance of the metrics used in making such selections (validation). Pattern recognition, partitioning, dissimilarity-based selection, cluster-based selection, hierarchical clustering, categorical validation, and neighborhood behavior of metrics are just some of the techniques corresponding to the above divisions. Combinatorial libraries are intrinsically large and cumbersome objects, but they can be managed efficiently when defined and handled appropriately. Many powerful computational and information management tools are currently available, but one still must take considerable care in deciding which tool-product or reactant-based library building in Legion — or dissimilarity — or cluster-based selection in Selector — is best suited to any particular task.

An alternative to the serial approach to structure–property refinement involves the development of methods for parallel synthesis of compounds that meet specific geometric requirements of a target receptor binding site. Custom chemical scaffolds can be designed that are directed to fit a particular binding site and that can be synthetically elaborated through the combinatorial reaction with commercially available reagents. Although such targeted combinatorial schemes can produce libraries of druglike compounds with thousands to millions of members, parallel automated synthesis methods are at present capable of synthesizing libraries containing on the order of a hundred discrete compounds. This practical limitation on synthetic throughput motivates the development of effective computer search strategies that can iteratively define and refine the selection of sublibraries that will best investigate the structure–property relationships for a given library–target combination. This has been achieved through the computer generation of large "virtual" libraries of synthetically accessible compounds which are designed to explore specific features suggested from a 3D structural model and/or other SAR features thought to be important in the ultimate development of a successful drug (e.g., features relating to bioavailability, toxicology, etc.). Computer codes are then used to select library sub-sets for rounds of automated chemical synthesis and bioassay testing. Test data resulting after each round are used both as selection criteria for additional 3D target-ligand structure determinations, and to iteratively refine molecular properties using more traditional SAR methods. In many cases, extensive 3D structural data generated from a chemically "easy" combinatorial library can be combined with elements of pure structure- based designs to rapidly develop novel compounds.

1.19 Molecular Connectivity and Information Content of Molecules

Kier and Hall have published information about their molecular connectivity method. Index values based on electron and orbital counts have led to information about topology and electronic structure. This provides a basis for the interpretation that goes beyond the development of statistical relationships. Specific molecular fragments are counted and encoded into the index values calculated. Information relating to topological, geometric, and spatial attributes, branching, cyclization, etc. translates into an encoding of molecular bulk or volume and molecular surface area. Counts of lone pair electrons and elements of electronic structure influencing physical and biological properties are used. The molecular connectivity method for QSAR is designed to provide numerical indexes of molecular structure. The indexes in the QSAR equations can be interpreted in terms of structural fragments, which can be identified and quantified in terms of their contribution to the measured biological activity. The ultimate value is a rationale for drug design or directed synthesis of molecules with desired properties. Molecular connectivity was developed to provide indexes of molecular structure for use in biological QSAR. However, examination of relations to physicochemical properties was undertaken. The ability to estimate physicochemical properties became a matter of increasing importance in environmental and engineering applications, and it was important to do so.

In a paper by Kier, it is stated that since a drug molecule is an information source, the information content is available to receptive tissue and this can have quality as judged by the receptor. Quantization of the information content using Shannon's equation gives the molecular negentropy. This index ranks molecules according to symmetry and to encode characteristics influencing physical properties and biological activity in certain cases.

Another paper by Hall and Kier states that the relation between biological activity and molecular structure may be viewed mathematically as a surface. The maximum or minimum position on the surface is associated with optimum activity and structure parameters. A useful approach to QSAR is to identify the optimum structure and relate it to drug design. This was applied to the neurotoxicity of 21 substituted phosphorus compounds using molecular connectivity indexes.

Bonchev centered attention on the information theoretic description of molecules and atoms. He presented various information indexes that characterize molecular chemical composition, topology, symmetry, etc., as well as the electronic distributions in atoms and molecules. He regarded a biologically active molecule as a message, composed of atoms and bonds (or atomic nuclei and electrons). The information contained in this message transmits to a corresponding receptor, causing a biological response. He noted, however, that the specific biological activities, determined by some molecular fragments or functional groups, need to be described by new types of indexes that are not global and he expected that the fruitfulness of the information approach to QSAR studies would be demonstrated soon.

Many papers were presented at a symposium held at the University of Georgia entitled Chemical Applications of Topology and Graph Theory. These are reviewed in the two paragraphs below, the first of which concerns individual molecules and the second of which concerns assemblies of molecules.

Rouvray expressed the topological structure of chemical species by indexes that reflect their overall shape and size. He pointed out, however, that there are many such indexes and that they correlate well with a broad range of physicochemical parameters. He stated that the applications were in the areas of physicochemical parameter evaluation and pharmaceutical drug design and that the latter warranted further major study in view of its implications for the pharmaceutical industry. Magnuson et al. derived topological indexes based on neighborhood symmetry for chemical and biological applications. These were information-theoretic invariants used as quantitative molecular descriptors in SAR studies of physicochemical and biological properties of molecules. The indexes are based on the neighborhood symmetry of the vertices in a chemical graph. Topological indexes based on neighborhood symmetry are shown to compare favorably with other indexes such as Wiener number and molecular connectivity and log P. Randic et al. consider graphs that are represented by sequences of counts of the number of paths (self-avoiding walks). The approach is particularly suitable when the family of structures having different substituents is considered. The study includes structures reported as "compounds that don't fit" the widely used QSAR approach of Hansch, which is essentially a multiparametric regression scheme using empirical parameters. A mathematical model of molecular complexity based on both concepts from graph theory and information theory was used by Bertz. This is utilized together with information theory.

There have been a number of papers on chemical applications of topology and graph theory. Dugundji et al. pointed out that Cayley used graph theory for the treatment of chemical constitution in 1857 and Balaban demonstrated the versatility of graph theory in the treatment of chemical problems. The authors wrote that the applications are not confined to the representation of molecules but can include the relations among different chemical entities, e.g., ensembles of molecules. A straightforward foundation for the representation of the mutual relations between chemical systems by graphs is shown and an attempt is made to present a foundation for the logical structure of chemistry. Not only is there a representation for molecules and an assembly of molecules, but also there are mathematical schemes for synthesis design and reaction networks as well. Peusner studied the electrical network representation of n-dimensional chemical manifolds and Hass and Plath studied the multidimensional λ-model, which is a graph theoretical/algebraic approach to describe mechanistic aspects of complex chemical reactions. They also studied the logic of chemical ideas and stress the possibility of describing mathematically all the logically different aspects of reactions using Boolean sublattices and graph theory.

Topological indexes (TIs) and geometric (3D) parameters have been used for quantifying chemical structure and in developing QSARs. The relative effectiveness of these two techniques has been analyzed using graph theoretic indexes and geometric parameters like a 3D Wiener index and volume for estimating normal boiling point for 140 hydrocarbons, lipophilicity (log P, octanol/water) of a diverse set of 254 chemicals and mutagenic potency (ln R, R being the number of revertants per nanomole in the Ames test) for a set of 95 aromatics and heteroaromatics amines. The results show that topological indexes give reasonably good QSARs for these properties. Geometric parameters like 3D Wiener index and volume make some improvements in the predictive ability of QSARs, indicating that 3D variables encode information not present in topological indexes.

1.20 Computer-Aided Molecular Design

The complication with using a computer to synthesize new molecules is that synthetic organic chemists rely primarily on information contained in a two-dimensional chemical structure diagram. A chemical reaction is simply a transformation of one diagram into another by combining and transforming starting materials into products. Hence, the expertise is organized intellectually in pictures, not numbers. The key insight that these structure diagrams can be described as labeled diagrams enabled the use of the computer to process structure diagrams.

Balbes et al. have documented computer-assisted design (CADD, CAMD), computer-aided drug design, molecular modeling, molecular graphics, computational chemistry, structure-based drug design, and computational chemistry in a comprehensive article in 1994. Comparing QSAR with other methods brings out the statement that the methods, especially QSAR, which were touched upon only lightly in their article, have helped to bring products to market. It is also true that the time and consumption of computer resources for QSAR are low compared with other methods.

1.21 The WWW Chemical Structures Database

Codessa is an advanced, fully featured QSAR program that ties information from Ampac with experimental information. Together these programs allow researchers to answer the questions that are important to their efforts. Many papers showed the efficacy of Codessa and Ampac in predicting such macroscopic chemical properties as gas chromatography retention indexes, polymer glass transition temperatures, critical micelle concentration, toxicity, boiling/melting points, and flash points.

The QSAR and QSPR (quantitative structure–property relationship) functions are in the Molecular Orbital Calculation Engine (MOE) system provided in a package called QuaSAR. The first step in the analysis is to compute molecular properties. The QuaSAR graphical interface allows the user to select which descriptors to calculate. The interface scans MOE and automatically detects QuaSAR modules, including those introduced by the user. Users can quickly and easily build descriptors of their own custom design and add them into the system; the source code of the built-in descriptors is distributed with QuaSAR and can be used as a template. After descriptors are calculated, they are sent to the suite of MOE QuaSAR model-building functions for model fitting and evaluation. In the model refinement stage, further descriptor calculations may be required. This cycle continues until a satisfactory model is found, or until the current model is rejected and the process is begun anew.

A study by Carotti et al. shows further evidence of the potential of LFER and CoMFA approaches in understanding the mechanism of chiral recognition in chromatography and other related areas.

There has been constructive cross-fertilization between the mathematical sciences and chemistry. Usually in QSAR methods, multiple linear or nonlinear regression and classical multivariate statistical techniques were used. Then, discriminant analysis, principal components regression, factor analysis, and neural networks were used. More recently, partial least-squares (PLS), originally developed by a statistician for use in econometrics, has been used, and this has prompted additional statistical research to improve its speed and its ability to forecast the properties of new compounds and to provide mechanisms to include nonlinear relations in the equations. QSAR workers need a new method to analyze matrices with thousands of correlated predictors, some of which are irrelevant to the end point. A new company was formed called Arris with a close collaboration of mathematicians and chemists that produced QSAR software that examines the 3D properties of molecules using techniques from artificial intelligence.

Published work still centers on solid-support synthesis, although considerable momentum is building for chemistry in solution. Combinatorial chemistry has created challenges in collateral fields, such as chemical information and computational analysis. These have arisen in the fields of pharmaceutical and agrochemical research communities: how to store and manipulate combinatorial libraries (definition), how to make intelligent selections of subsets from combinatorial libraries (dissection), and how to evaluate the performance of the metrics used in making such selections (validation). Pattern recognition, partitioning, dissimilarity-based selection, cluster-based selection, hierarchical clustering, categorical validation, and neighborhood behavior of metrics are just some of the techniques corresponding to the above divisions. Combinatorial libraries are intrinsically large and cumbersome objects, but they can be managed efficiently when defined and handled appropriately. Many powerful computational and information management tools are currently available, but one still must take considerable care in deciding which tool-product or reactant-based library building in Legion — or dissimilarity- or cluster-based selection in Selector — is best suited to any particular task.

SPROUT is a program that assists the chemist in several steps of the structure-based rational drug design process. It is a program for constrained structure generation that can generate molecules that fit the steric and chemical constraints of a pharmacophore hypothesis or a specified protein receptor site. In the latter case, the steric constraints consist of a volume from the active site of a protein together with some interaction sites within the volume. These interaction sites are called target sites and the program uses a sophisticated representation of the different types of target sites. Structure generation is supported by additional modules, which assist other stages of the module design process.

Certain molecular properties can be assessed before the molecule is even synthesized. However, the most desirable chemical may not be the one with maximum activity but rather the one that has the maximum activity-to-toxicity ratio, so that it is essential to be equipped with reliable computer-assisted tools. Two approaches have been followed to develop such tools: the quantitative structure-toxicity relationship (QSTR) model approach and the expert system approach. An expert system is a collection of rules derived from existing subject knowledge and stored in computer memory. When a chemical is presented to such a system for assessment, the rules associated with the presented chemical are identified from the array of stored rules and a decision is displayed. There are some inherent problems in this system. In principle, an LFER-based model is a quantitative relationship between a numerical measure of toxicity and structure descriptors, i.e., $T = f(S)$, where T is a measure of toxicity, such as acute median lethal dose (LD_{50}), lowest observed adverse effect level (LOAEL), indicator of carcinogenicity, etc.; S is a set of numerical quantities representing different structural attributes; and f is a mathematical function. These structure–toxicity relationships are generally known as QSTRs. A program called TOPKAT 3.0 is commercially available that contains robust and parsimonious QSTR models with OPS (optimum prediction space) algorithms, a tool to test whether a submitted structure is within the proper space of the model. TOPKAT 3.0 can generate, from molecular structure, a variety of mammalian, aquatic, and environmental toxicity measures, and can provide an ability to follow the toxicity profile as molecular structure is modified during the design process, provided it satisfies the main components of a reliable toxicity prediction system.

Naser and Fournier use the UNIFAC model to calculate activity coefficients and liquid–liquid equilibrium and then use their program called LEDMS, which has an algorithm that does the reverse.

Daylight is a program providing computer algorithms for chemical information processing. Their manual (*Daylight Theory Manual*, Daylight 4.51) consists of the following sections: Smiles, Smarts, Chuckles, Chortles and Charts, Thor, Merlin, and Reactions. Smiles is a language that specifies molecular structure and is a chemical nomenclature system. Smarts is a substructure searching and similarity tool. It reveals the principles of substructure searching, NP-complete problems and screening, structural keys, fingerprinting, and similarity metrics. Chuckles, Chortles, and Charts are for mixtures. They are languages representing combinatorial libraries which are regular mixtures of large numbers of compounds. Thor is a chemical-information database system consisting of fundamental chemical nomenclature, chemical identifiers, etc. Merlin is a chemical information database exploration system and a pool of memory-resident information, and Reactions has all of the features for reaction processing.

HASL (hypothetical active site lattice) modeling converts a molecule to a set of regularly spaced points (lattice) defined by Cartesian coordinates (x,y,z) and atom type. Thus, individual molecules can be compared with each other and their lattices merged to form a composite description called HASL. The activity (ED_{50}, pki, etc.) of each molecule is distributed among the points in the HASL in such a way that the sum of the partial activity values associated with a set of points in the HASL belonging to each molecule is equal to the activity for the molecule. The result is a model of the active (receptor) site consisting of points in space capable of predicting the activities of as yet untested molecules. A recent innovation, the construction of a predictive 3D pharmacophore, is based on the selective editing of a HASL model.

AutoDock is the only available program that allows a ligand to dock flexibly to its proposed receptor. The entire procedure is based on a grid-based Monte Carlo simulated annealing procedure and as such is unbiased by the user. Ludi is a program for proposing fragments to bind to a receptor. Both an actual receptor as well as a pseudoreceptor based on ligand comparison can be used. It is easy to use and highly successful in predicting the fragments that constitute the endogenous ligand of receptors. Chem3DBS can detect a pharmacophore in a set of ligands and do a flexible search in the 3BDS. It is fast and rather easy to use. LigPlot utilities help in analyzing interactions in PDB files. The "schematic representation" is especially helpful for larger ligands like peptides. X-Site is a program for describing preferment interactions surrounding a protein, based on protein side chain packing analysis.

SynTree is a set of programs for exploring chemical synthesis on a personal computer. It is designed to facilitate the discovery of possible synthetic routes to a given target structure by the repeated application of retrosynthetic transforms.

Kohonen neural networks are used to analyze combinatorial libraries. For this analysis, the library compounds are represented by a vector of spatial autocorrelation coefficients. The high-dimensional descriptor space spanned by these vectors is projected on to a Kohonen map. On the basis of the Kohonen maps, libraries can be compared regarding their similarity and diversity. This is demonstrated by analyzing the combinatorial libraries.

An approach to *de novo* molecular design, PRE_LIGAND, has in the environment of a large integrated molecular design and simulation system provided a unified framework for the generation of novel molecules, which are either similar or complementary to a specified target. The approach is based on a methodology that has proved to be effective — placing molecular fragments upon target information sites — but incorporates many novel features such as the use of a rapid graph theoretical algorithm for fragment placing, a generalized driver for structure generation which offers a large variety of fragment assembly strategies to the user and the prescreening of library fragments.

Golender et al. published a paper on APEX-3D Expert System for Drug Design. It represents, elucidates, and utilizes knowledge on structure–activity relationships. It also builds 3D SAR and 3D QSAR models that can be used for activity classification and prediction. There is an emulation of the intelligence of the researcher who establishes the relationships between the structural parameters of the compound and its activity. Automated identification of pharmacophores that are used for building qualitative activity prediction rules and for creating search queries to identify new leads in a 3D database takes place. Prediction of activity for novel compounds requires the pharmacophore be present: the activity levels calculated from the QSAR equation.

DBMAKER is a set of programs that generates 3D databases based upon user-specified criteria. It does this with the aid of another program called CONCORD. Numerous user-defined parameters monitor content, composition, size, and connectivity information, but allow the program to generate random compounds within the scope of these constraints. SMILES string representations are generated, and conversion to 3D is performed by CONCORD. This yields high-quality 3D structures and portability to many proprietary storage formats. Users are provided with mechanisms to tailor the database to their specific requirements as to chemical classes considered; the candidate structures can be both diverse and/or focused. There is no limit to the variety of structures that can be generated with DBMAKER. This provides a flexible alternative to commercially available products. DBMAKER allows investigators, who had been hindered by a lack of 3D information, to employ molecular design techniques.

Cho et al. propose a novel two-dimensional QSAR method which combines chemical graph theory, PLS, and genetic algorithms (GA). Molecules are described with topological indexes generated with the Molcom-X program. A population of 100 QSARs are generated by random selection of 100 subsets of indexes from the whole set and treating each subset of variables in PLS analysis. GAs are employed to evolve the initial population of QSAR equations. Three simple functions with variables that are cross-validated are examined for the ability to distinguish data sets with actual and randomly generated activities. Each parent represents a QSAR equation with randomly chosen variables, and the calculations evolve the initial population of the QSAR equations to the population with the average value of the fitness function and the original function. In the course of the GA-PLS process, the initial number of members of the population (100) is maintained, while the average value of the fitness function for the whole population converges to a higher number. Although CoMFA is a useful lead optimization tool, many efforts in drug design have been concentrated in the area of lead generation. New methods of structural database searching and combinatorial chemistry are becoming the most popular approaches to suggest new leads. Thus, the QSAR equations obtained with GA-PLS or other fast 2D-QSAR methods can be used as the means of the QSAR approach to drug design, which can be formulated as follows. Given the QSAR equation for known compounds, screen the existing structural databases or databases generated *de novo* with the help of computational combinatorial chemistry for compounds with high predicted biological activity.

Once a lead has been identified, hundreds to thousands of additional compounds are designed and synthesized to optimize the biological profile. The cost and environmental impact of synthesis and patentability are also important issues. If the 3D structure of the biological target is known, then molecular modeling might be used in the design. As testing data are accumulated, statistical 3D QSAR may help set priorities for synthesis. Any attractive compounds found are tested in more detail through advanced protocols that more reliably forecast therapeutic and toxicity potential. Last, the surviving compounds are prioritized and the best compound known at that time is prepared for clinical trial. New mathematical techniques could have an impact on the rate of new compound discovery if the potency of compounds could be forecast more quickly and accurately before their synthesis.

1.22 Advanced Computational Methods for Rational Drug Design

The Center for Computational Engineering (CCE) was created some years ago at Sandia National Laboratories with the principles of the HPCCI to form a multidisciplinary research team coupling academia, the national laboratories, and private industry to enable major advances in high-performance computational modeling. CCE has built a strong reputation in developing and applying advanced computational chemistry methods for biotechnology. Some of the CCE accomplishments include developing the first production-quality quantum chemistry program for massively parallel computers and performing the largest all-electron no-symmetry *ab initio* quantum chemical calculation. Their strength has been in successful collaborative partnerships with biotechnology groups at American companies, universities, and other national laboratories. This constitutes a new grand challenge proposal for the advanced software technology and algorithms program.

The proposed effort is to integrate the parallel quantum chemistry and molecular dynamics methods into an object-oriented computational chemistry tool kit for distributed and massively parallel computers. This tool kit will allow rapid implementation for efficient computational methods for specific biotechnological applications, which will significantly accelerate the application of computer-aided design to real-world biotechnology problems by U.S. industries. The purpose of the proposed work is to use the expertise of the CCE in computational chemistry, parallel algorithms, advanced computing, and visualization to build the next generation of computational chemistry tools and to apply these tools to collaborative research efforts in biomedicine and agricultural chemistry.

The proposed new software tool kit for computational chemistry is a multidisciplinary project that will involve parallelizing and encapsulating existing algorithms using state-of-the-art computer science methods, developing new chemical simulation methods, and implementing new fundamental algorithms. Much of the computationally intensive computer code will be written and optimized by the computer. It is planned to develop methods to simulate the effects of aqueous solvation efficiently. Additionally, powerful molecular dynamics methods are required to predict the complex chemical properties such as MRI relaxivities and *in vivo* transport rates, which are of central importance to medicinal chemists.

1.23 Evaluation of QSARs for Physical, Transport, Degradation, Environmental Fate, and Exposure Assessments

Estimating physical/chemical/fate properties using QSARs was reviewed by Howard and Meylan after Lyman et al.'s book *Handbook of Chemical Property Estimation Methods* was reprinted in 1990. Such methods were validated on large numbers of structurally diverse chemicals. They have also been used in SYNPROPS for certain properties as well as Cramer's regression methods. The following properties are assessed.

1. The *octanol-water partition coefficient (Log P)* is a physical property that describes the lipophilic or hydrophilic properties of a chemical. It is the ratio of the concentration of a chemical in the octanol phase to its concentration in the aqueous phase of a two-phase system at equilibrium. Log P is valuable in many fate and exposure QSARs for water solubility, soil/sediment adsorption, and bioconcentration. Of the QSARs tried, the Meylan and Howard method has been tested on a validation data set of 8855 chemicals, and it yields excellent results.

2. *Water solubility* is very important in the fate and transport of a chemical in the environment as well as the toxicity of a chemical. It provides insight into the tendency for a chemical to partition to soil and sediment and to bioconcentrate. It also affects photolysis, hydrolysis, oxidation, biodegradation, and washout from the atmosphere by rain and fog. At present, the best means of estimating water solubility involves regression-derived correlations using log P. These correlations can also use melting point and molecular weight.

3. The best estimation methods for *boiling point, melting point, and vapor pressure* are group contribution methods. The Stein and Brown method has a large database (>10,000 chemicals) with an average error of 4.3%. It is in a computer program (MPBPVP) which has added a few new groups and correction factors.

4. *The dissociation constant (pK_a)* provides an indication of the amount of ionized and unionized species of the chemical that will be available at a given pH. Most approaches attempted to estimate the negative log of the dissociation constant (pK_a) have relied on using linear free energy relationship (LFER) approaches. A computer program developed by the U.S. EPA and the University of Georgia (SPARC) relies on a combination of LFER, SARs, and perturbed molecular orbital methods. It has been validated with 4000 chemicals with a root mean square in pK_a units of 0.35.

5. *Henry's law constant* is the air/water equilibrium partition coefficient for a chemical. The author has used it often on Superfund sites for the EPA Biological Technical Assistance Group. It is a most important physical property for the transport of a chemical in the environment and determines major processes in treatment methods (e.g., volatilization during activated sludge wastewater treatment). The Hine and Mookerjee method of bond or group contributions have been updated by Meylan and Howard and are available in a computer program that provides the most comprehensive and best validated estimation method.

6. *Atmospheric oxidation rate* is the most important degradation process, because, for most chemicals in the vapor phase in the atmosphere, reaction with photochemically generated hydroxyl radicals is most important. Occasionally, reaction in the atmosphere with ozone (olefins and acetylenes), nitrate radicals at night, and direct photolysis are significant for some chemicals. The Atkinson fragment and functional approach method is validated on the largest number of chemicals (641).

7. *Hydrolysis rates.* Chemicals that have functional groups that are potentially hydrolyzable, such as alkyl halides, amides, carbamates, carboxylic acid esters, epoxides and lactones, phosphate esters, and sulfonic acid esters, exhibit hydrolysis at pHs that are normally found in the environment (i.e., pHs 5 to 9). The only method that predicts hydrolysis rate constants for esters, carbamates, epoxides, and halogenated alkanes has been developed using LFER (Taft and Hammett constant) methodology.

8. *Biodegradation.* A number of comprehensive approaches using fragment constants have been attempted to predict biodegradability qualitatively. Many use a weight-of-evidence biodegradation database (BIODEG). This was specifically developed for structure/biodegradability correlations. Boethling used both the BIODEG database as well as results of an expert survey to develop four models that all used the same structural fragments. Biodegradation of organic chemicals is the only process that can result in complete mineralization (i.e., formation of carbon dioxide and water). Prediction of biodegradability is severely limited because of the lack of reproducibility of biodegradation data as well as the numerous protocols that have been used for degradation tests.

9. *Soil/sediment adsorption coefficients* (K_d) are normalized to the organic carbon content of the soil or sediment by dividing by the organic content to give the K_{oc}, since of the numerous soil properties that affect sorption (organic carbon content, particle size, clay mineral composition, pH, cation exchange, capacity) organic carbon is the most important for undissociated organic chemicals. The K_{oc} value provides considerable insight into the mobility and partitioning of the chemical between water and suspended solids/sediments and is directly related to the ability of a chemical to leach through soil. Adsorption can have a considerable impact on volatility and rates of degradation. For years, the correlation between K_{oc} and log P or water solubility was used. Meylan used an MCI and polar fragment correction method to develop good training and validation statistics on a large database (394 chemicals) of K_{oc} values.

10. *Bioconcentration factor*, which is the ability of a chemical to bioconcentrate in aquatic organisms, is of concern because it can lead to toxic concentrations being reached when the organism is consumed by higher organisms such as wildlife and humans. This ability to bioconcentrate is generally dependent upon the hydrophobic nature of the chemical as well as the ability of the organism to metabolize and excrete the chemical. Most estimation methods have used log P correlations and were developed in the 1980s.

Summary. Much work has been done since the work of Lyman et al. Books since then can now predict most physical/chemical/transport/degradation processes that are important to the environmental fate of a chemical with a reasonable indication of the amount of error expected for a new chemical, without having to select a model based on chemical similarity. Most of these properties can be predicted using available software with chemical structure input.

1.24 Physical and Environmental Properties

A repository and disseminator of evaluated data called DIPPR (Design Institute for Physical Property Data) is part of the American Institute of Chemical Engineers and sponsors research essential to industry, chemical engineers, and chemists. Computer-generated properties are compared with their data to test its validity. The DIPPR work is divided into a number of projects: 801 is Evaluated Process Design Data, which has a large collection of physical, thermodynamic, and transport properties for use in chemical process end equipment design; 805 is Experimental Data on Mixtures, which has contributed greatly to increasing the amount of phase equilibrium data available in the literature; 821 is Pure Component Liquid Vapor Pressure, which is an acute need for accurate vapor pressure measurements used in the design of industrial processes; 851 is Critical Properties of Pure Compounds, required for the correlation prediction of essential physical and thermodynamic properties used for process design and operation as they are used to estimate liquid density, viscosity, heat capacity, heat of vaporization, thermal conductivity, diffusion coefficients, surface tension, etc.; 871 is Determination of Pure Component Ideal Gas Heat of Formation, used for the evaluation of the energy hazard potential of organic compounds, for making heat balances around process equipment, and for calculating Gibbs energies and equilibrium constants for reactions; 911 is Environmental, Safety and Health (ESH) Data Compilation, in response to a critical need for information required to make environmentally responsible engineering decisions and to meet regulatory requirements. It focuses on a database of properties of chemicals applicable to the environment, health, and safety concerns. Project 931 is called Data Prediction Methods and has involved physical and thermodynamic property-estimating techniques using nontraditional correlating parameters. These properties are required for chemical process and equipment design and the predictive methods are applicable to industrially important chemical compounds. Molecular descriptors rather than group contribution techniques have been used to predict normal boiling temperatures of structurally diverse sets of organic compounds and several other more homogeneous groups of chemicals, as well as Henry's law constants, surface tensions, and autoignition temperatures for a wide variety of compounds. Mathematical and statistical procedures are achieved with ADAPT software. Verification of these correlations is done by comparison of predictions with data from Project 801 and other reliable data sources. Comparisons are also made with existing estimation techniques.

Hatchet Crossfire plus Reactions is a quarterly updated database with over 7.2 million organic compounds (not counting biopolymers, polymers, and metal-organic compounds) with structures, chemical and physical data, and the appropriate literature starting from 1779, as well as 4.7 million reactions. The Hatchet Beilstein is considered primarily as a source of information to organic chemists and is the most extensive collection of spectroscopic data references, thermodynamic data, and information about natural substances. It is also then valuable to the physical chemist physicist, chemical engineer, and the pharmacist. Gmelin Crossfire has 1 million metal-organic and inorganic compounds (with physical data with literature from 1772). It has about 465,000 coordination connections, 55,000 alloys, 13,500 glasses and ceramic materials, 10,000 polymers, and 3000 minerals.

Rogers et al. have been working on a physical property predictive driver for clean process design guidelines. This project focuses on creating structure-based models of adsorption and vapor–liquid equilibrium as key elements of an overall physical property prediction resource. More specifically, this project delivers a data set of equilibrium capacities for representative organic chemicals adsorbed onto two commercial granular activated carbons, a linear solvation energy relationship (LSER) model for activated carbon adsorption and a training and validation data set of vapor pressure data to be used for model development.

Hand et al. have been involved with the Environmental Technologies Design Options Tools (ETDOT), a compilation of self-contained tools for use in assessing and implementing effective treatment strategies for gaseous, aqueous, organic, and solid waste by-product streams. ETDOT will assist the Clean Process Advisory System (CPAS) in evaluating the technical and economic feasibility of source reduction, versus end-of-pipe treatment and waste segregation, versus treatment at a central facility. Some tools that are near completion are (1) software for predicting the fate of volatile organic (FaVOR) compounds in wastewater treatment facilities, (2) software for predicting the fate of metals (FaMe) in wastewcter treatment facilities, (3) gas- and liquid-phase Adsorption Simulation (AdSim) software, (4) Software to Estimate Physical Properties (StEPP), and (5) Aeration System Analysis Program (ASAP) software for removal of volatile organic compounds from water by surface, bubble, and counter-current packed tower aeration techniques.

Shonnard et al. have generated an Environmental Fate and Risk Assessment Tool (EFRAT). This software estimates environmental and health impacts of chemical process design options through a combination of emission rate estimations, screening-level fate and transport calculations, risk assessment indexes, and govermental regulatory guidance. The software will be combined with a graphical user interface (GUI), an environmental risk indicator calculator (ERIC), and a multimedia fate and transport model (MMFAT).

Halada et al. in Japan have found that MLCA (Materials environmental Life-Cycle Analysis), an assessment measure of ECOMATERIALS helps Eco-Design of materials. They design materials with higher recyclability and with lower environmental load.

Work has been under way for computer-aided design for demanufacturing and remanufacturing. Here computer-based tools for the computer-aided design of products that can be recycled and reused through demanufacturing and remanufacturing are under investigation. In order to do this the issues are characterized, available CADs are benchmarked together with information management systems, a prototype CAD tool is developed, and it integrates weight, cost, manufacturability, and trade-off assessments.

1.25 Evaluating the Potential for Safe Substitutes in Pollution Prevention

In the EPA report EPA/600/R-94/170, The Product Side of Pollution Prevention: Evaluating the Potential for Safe Substitutes, from the RREL Office of Research and Development, Cincinnati, OH, it is stated that regulations have not sufficiently reduced environmental releases nor have they protected workers from effects of toxic agents in the workplace or consumers from toxic chemicals found in products. It further states that substitution of products and practices reducing use of toxic chemicals can reduce hazardous waste generation, mitigate toxic air and water pollution, and reduce worker exposure and public exposure to toxic chemicals. There must be a focus on safe substitutes for products that contain or use toxic chemicals in their manufacturing process. The generation of hazardous waste and toxic pollutants comes as a choice made in the design of products and the manufacturing processes in their production. Substitution of products that do not require the use of toxic chemicals can reduce hazardous waste generation, mitigate toxic air and water pollution, and reduce both worker and public exposure to toxic chemicals.

Major efforts initiated by the EPA and other federal and state agencies to explore the use of safe substitutes for priority toxic chemicals now far surpass the modest first studies in resources and industry participation. The chlorofluorocarbon substitutes evaluation initiated by the Clean Air Act Amendments of 1990, the EPA-sponsored Cleaner Technology Demonstrations for 33/50 Chemicals, and the EPA Design for the Environment Program are only three examples out of many. Also, the use of the life cycle concept, and the methodology of life cycle assessment to evaluate the relative environmental impact of products, has been an important asset. The latter not only includes direct impacts of toxic chemical exposure but other major impacts, such as global climate change, acid rain, ozone depletion, and resource depletion. The latter approach looks at the whole manufacturing process for products and not just at the products themselves. The life cycle approach and this report place the use of toxic chemicals in the context of whole product systems to illuminate patterns of materials use and environmental impacts that are not easily seen when focusing only on one product or one chemical use.

The generation of hazardous waste and toxic pollutants is the result of choices made, choices that were not made lightly. The chemicals chosen that are now priority pollutants have properties that made them well suited for their uses. The web of production of numerous products and materials that consumers have come to depend upon is based in large measure on toxic chemicals that have become industrial building blocks. They include plastics, synthetic fibers, electronic components and batteries, paints and coatings, and plated metal materials used in automobiles and household appliances. The necessity of massive toxic chemical releases is challenged, but it is recognized that there are no across-the-board solutions and that each chemical use is unique. The objective of the study was to identify priority products that contain or use certain priority chemicals in their production, to identify and evaluate the existing substitutes for these priority products, and to determine technical impediments to the use of safe substitutes for priority products. Also required are determination of the future research needs for safe substitutes for priority products and development of a method for identifying priority chemicals for substitute evaluation. (This is exactly what SYNPROPS and Envirochemkin does!) The identification of priority products in the above EPA report focused on those included in the EPA 33/50 Program. This is a voluntary pollution prevention initiative intended to achieve reductions in pollution in a short time frame. The program enlisted companies in important industry sectors to make voluntary commitments to reduce reported releases and transfers of 17 priority chemicals. Participating companies were asked to develop their own reduction goals contributing toward national reduction goals of 33% by 1992 and 50% by 1995. Reductions are measured against a 1989 baseline of information reported to the EPA under the Toxic Release Inventory (TRI).

The 33/50 chemicals include the following:

- Metal and Metal Compounds
  - Cadmium
  - Chromium
  - Lead
  - Mercury
  - Nickel
- Organic Chemicals
  - Benzene
  - Toluene
  - Xylenes
  - Methyl ethyl ketone
- Halogenated Organic Chemicals
  - Carbon tetrachloride
  - Chloroform
  - Tetrachloroethylene
  - Dichloromethane (methylene chloride)
  - 1,1,1-Trichloroethane
  - Trichloroethylene
- Cyanide Compounds

1.26 Interlude

It occurs to this writer that the game of three-dimensional *Life* would be useful in some drug work. By making the rules such that the geometry of the docking surface would fit the geometric surface and volume of the constellation of cells that would be "on," more drug candidates could be discovered. The use of Lagrangian functions in phase space could also lead to differential equations whose momentum and position could be minimized at the same time allowing a more fortuitous docking arrangement. Furthermore, the ergodic hypothesis leads to the average over time as equivalent to that over space allowing statistical mechanics to play a role in the docking procedural description. Programs are also necessary to elucidate the shape of the ligand molecule to assure a proper fit into the cavity during docking.

1.27 Molecular Synthesis

Molgen is a program that generates mathematically possible constitutional formulae and in addition the possible stereoisomers compatible with a given set of chemical conditions. Molgen allows all configurational isomers to a given structural isomer as well as their realizations in space. All atoms in Molgen together with their multiplicity in the molecule and valence are input. A huge number of structures will occur. For example, there exist 38,496,539 isomers to $C_8H_9NO_2$.

There are three major programs in computer-aided organic synthesis: LHASA, SECS, and SYNCHEM. SECS extends the methods of LHASA and the discussion applies to both programs. SECS is demonstrably transportable to other computers. The chemist represents the structure graphically and relates it to simpler chemicals via known chemical reactions. The chemist relates those to still simpler ones, until he or she reaches a state of commands, comparable to starting materials, that are readily available. The synthetic route may involve dozens of separate reactions. A simple steroid composed of about 20 atoms has over 10^{18} possible direct routes. Synthesis routes can be viewed in terms of an AND/OR tree. The tree descends from the goal node, representing the target molecule, to the terminal nodes, equivalent to the starting materials. The branches connecting the nodes are chemical reactions. Since a synthesis plan involves combining compounds in reactions, the AND links of a tree are present in any one synthesis route; alternative ways of making a compound anywhere within the plan are represented by OR nodes. Transforms are used. All three programs have a large knowledge base of reverse chemical reactions called transforms — production rules of the condition–action form. Both LHASA and SECS have a large procedural knowledge of transforms, are interactive, and have high performance. SECS has a special interactive language for defining new transforms and interactive graphics. SYNCHEMS is motivated by interactive search and AI problems, and the evaluation is done by the computer, not by the chemist.

Polymers are designed for specialty applications such as medical implants, piezoelectric and photonic materials, or self-assembling systems, and this is challenging since such materials must meet multifaceted requirements. As an example, the field of tissue engineering hinges on the development of degradable scaffolds that promote cell proliferation and expression of desired physiological behaviors through careful control of the polymer surface properties. The concept of permutationally designed monomer systems creates libraries of structurally related polymers. Libraries created in this way can be used to increase the number of available polymeric candidate materials for any specific application and systematize the study of correlations among polymer structure, material properties, and performance. Thus, the design of two sets of monomers having small but systematic variations and their copolymerization to yield strictly alternating A–B type copolymers can lead to the rapid preparation of extensive polymer libraries. This combinatorial design of polymers will accelerate progress for preparing and evaluating new polymers for specialty applications and has the potential for accelerating the identification of important structure–property correlations. The further study of combinatorial polymer libraries will facilitate the identification of new biomaterials that are optimally matched to the requirements of a specific medical application.

The maximum common substructure (MCSS) is a similarity measure that allows the study of a variety of chemical problems. The relationships among structure and biological activity, appropriate starting materials for the target of a synthesis, and biosynthetic pathway can be investigated from the MCSS or the 3D MCSS. The determination of the largest substructure contained twice in a single molecule allows one to derive synthesis precursors that can be used several times to make a synthesis short and convergent. The determination of an MCSS is a computationally expensive task, which we solve by a genetic algorithm, an optimization method that imitates the adaptations methods of nature.

1.28 Computer Programs

Free radicals are important intermediates in natural processes involved in cytotoxicity, control of vascular tone, and neurotransmission. The chemical kinetics of free-radical reactions control the importance of competing pathways. Equilibria involving protons often influence the reaction kinetics of free radicals important in biology. Free radicals are very important in atmospheric chemistry and mechanisms, yet little is known about their physical or biological properties.

In 1958 White et al. (at Rand) published an article entitled "Chemical Equilibrium in Complex Mixtures." It was a method that calculated chemical equilibrium by the method of the minimization of free energy. It was an optimization problem in nonlinear programming and was used in industry and in defense work on mainframe computers. Personal computers were not available at that time. Also, environmental matters were not so much of a concern as they are now.

The literature and computer sites on Geographic Information Systems (GIS) are rife with a tremendous amount of information. The number of such maps is increasing greatly every day as exploration, assessment, and remediation proceed across the world wherever environmental work is taking place.

There are many software programs for geotechnical, geo-environmental, and environmental modeling. They are in the category of contaminant modeling, and most of them are in the DOS platform and are in the public domain.

Massively parallel computing systems provide an avenue for overcoming the computational requirements in the study of atmospheric chemical dynamics. The central challenge in developing a parallel air pollution model is implementing the chemistry and transport operators used to solve the atmospheric reaction–diffusion equation. The chemistry operator is generally the most computationally intensive step in atmospheric air quality models. The transport operator (advection equation) is the most challenging to solve numerically. Both of these have been improved in the work of Dabdub and Seinfeld at CalTech and have been improved in the next generation of urban and regional-scale air quality models. HPCC (high-performance computing and communications) provides the tools essential to develop our understanding of air pollution further.

EPA has three main goals for its HPCC program activities:

- Advance the capability of environmental assessment tools by adapting them to a distributed heterogeneous computing environment that includes scalable massively parallel architectures;
- Provide more effective solutions to complex environmental problems by developing the capability of performing multipollutant and multimedia pollutant assessments;
- Provide a computational and decision support environment that is easy to use and responsive to environmental problem-solving needs of key federal, state, and industrial policy-making organizations.

Thus, the EPA participates in the NREN, ASTA, IITA, and BRHR components of the HPCC program, where NREN: increasing access to a heterogeneous computing environment; ASTA: environmental assessment grand challenges; IITA: enhancing user access to environmental data and systems; BRHR: broadening the user community.

Environmental modeling of the atmosphere is most frequently performed on supercomputers. UAMGUIDES is an interface to the Urban Airshed Model (UAM). An ozone-compliance simulator is required by the Clean Air Act of 1990, so that modeling groups across the U.S. have asked the North Carolina Supercomputing Center (NCSC) to develop a portable version. The NCSC Environmental Programs Group used the CRAY Y-MP system, a previous-generation parallel vector system from Cray Research, to develop UAMGUIDES as a labor-saving interface to UAM. Running UAM is very complex, and the Cray supercomputers have been upgraded since then. Computational requirements for modeling air quality have increased significantly as models have incorporated increased functionality, covered multiday effects, and changed from urban scale to regional scale. In addition, the complexity has grown to accommodate increases in the number of chemical species and chemical reactions, the effects of chemical particle emissions on air quality, the effect of physical phenomena and to extend the geographic region covered by the models.

The effects of coal quality on utility boiler performance are difficult to predict using conventional methods. As a result of environmental concerns, more utilities are blending and selecting coals that are not the design coals for their units. This has led to a wide range of problems, from grindability and moisture concerns to fly ash collection. To help utilities predict the impacts of changing coal quality, the Electric Power Research Institute (EPRI) and the U.S. Department of Energy (DOE) have initiated a program called Coal Quality Expert (CQE). The program is undertaken to quantify coal quality impacts using data generated in field-, pilot-, and laboratory-scale investigations. As a result, FOULER is a mechanistic model placed into a computer code that predicts the coal ash deposition in a utility boiler and SLAGGO is a computer model that predicts the effects of furnace slagging in a coal-fired boiler.

In Europe, chemists, particularly Prof. Mike Pilling and Dr. Sam Saunders at the Department of Chemistry at the University of Leeds, England, have worked on tropospheric chemistry modeling and have had a large measure of success. They have devised the MCM (Master Chemical Mechanism), a computer system for handling large systems of chemical equations and were responsible for quantifying the potential that each volatile organic compound (VOC) exhibits to the development of the photochemical ozone creation potential (POCP) concept. The goal is to improve and extend the Photochemical Trajectory Model for the description of the roles of VOC and NOx in regional-scale photo-oxidant formation over Europe. In their work, they use Burcat's "Thermochemical Data for Combustion Calculations" in the NASA format.

Statistical methods, pattern recognition methods, neural networks, genetic algorithms, and graphics programming are being used for reaction prediction, synthesis design, acquisition of knowledge on chemical reactions, interpretation and simulation of mass spectra analysis and simulation of infrared spectra, analysis and modeling of biological activity, finding new lead structures, generation of 3D molecular models, assessing molecular similarity, prediction of physical, chemical, and biological properties, and databases of algorithms and electronic publishing. Examples include the course of a chemical reaction and its products for given starting materials using EROS (elaboration of reactions for organic synthesis), where the knowledge base and the problem-solving techniques are clearly separated. Another case includes methods for defining appropriate, easily obtainable starting materials for the synthesis of desired product. This includes the individual reaction steps of the entire synthesis plan. It includes methods to derive the definition of structural similarities between the target structure and available starting materials, finding strategic bonds in a given target, and rating functions to assign merit values to starting materials. Such methods are integrated into the WODCA system (Workbench for the Organization of Data for Chemical Application).

In 1992, the NSF was already looking to support work for CBA (computational biology activities): software for empirical analysis and/or simulation of neurons or networks of neurons; for modeling macromolecular structure and dynamics using X-ray, NMR, or other data; for simulating ecological dynamics and analyzing spatial and temporal environmental data; for improvement of instrument operation; for estimation of parameters in genetic linkage maps; for phylogenetic analysis of molecular data; and for visual display of biological data. They were looking for algorithm development for string searches; multiple alignments; image reconstruction involving various forms of microscopic, X-ray, or NMR data; techniques for aggregation and simplification in large-scale ecological models; optimization methods in molecular mechanics and molecular dynamics, such as in the application to protein folding; and spatial statistical optimization. They sought new tools and approaches such as computational, mathematical, or theoretical approaches to subjects like neural systems and circuitry analysis, molecular evolution, regulatory networks of gene expression in development, ecological dynamics, physiological processes, artificial life, or ion channel mechanisms.

There has been constructive cross-fertilization between the mathematical sciences and chemistry.

Usually in QSAR methods, multiple linear, or nonlinear regression, classical multivariate statistical techniques are used. Then, discriminant analysis, principal components regression, factor analysis, and neural networks are used. More recently, partial least-squares (PLS), originally developed by a statistician for use in econometrics, has been used, and this has prompted additional statistical research to improve its speed and its ability to forecast the properties of new compounds and to provide mechanisms to include nonlinear relations in the equations. QSAR workers need a new method to analyze matrices with thousands of correlated predictors, some of which are irrelevant to the end point. A new company was formed, called Arris, with a close collaboration of mathematicians and chemists that produced QSAR software that examines the 3D properties of molecules using techniques from artificial intelligence.

Historically, mathematical scientists have worked more closely with engineers and physicists than with chemists, but recently many fields of mathematics, such as numerical linear algebra, geometric topology, distance geometry, and symbolic computation, have begun to play roles in chemical studies. Many problems in computational chemistry require a concise description of the large-scale geometry and topology of a high-dimensional potential surface. Usually, such a compact description will be statistical, and many questions arise as to the appropriate ways of characterizing such a surface. Often, such concise descriptions are not what is sought; rather, one seeks a way of fairly sampling the surface and uncovering a few representative examples of simulations on the surface that are relevant to the appropriate chemistry. An example is a snapshot or typical configuration or movie of a kinetic pathway. Several chemical problems demand the solution of mathematical problems connected with the geometry of the potential surface. Such a global understanding is needed to be able to picture long-timescale, complex events in chemical systems. This includes the understanding of the conformation transitions of biological molecules. The regulation of biological molecules is quite precise and relies on sometimes rather complicated motions of a biological molecule. The most well studied of these is the so-called allosteric transition in hemoglobin, but, indeed, the regulation of most genes also relies on these phenomena. These regulation events involve rather long timescales from the molecular viewpoint. Their understanding requires navigating through the complete navigation space. Another such long-timescale process that involves complex organization in the configuration space is bimolecular folding itself.

Similarly, specific kinetic pathways are important. Some work has been done on how the specific pathways can emerge on a statistical energy landscape. These ideas are, however, based on the quasi-equilibrium statistical mechanics of such systems, and there are many questions about the rigor of this approach. Similarly, a good deal of work has been carried out to characterize computationally pathways on complicated realistic potential energy surfaces. Techniques based on path integrals have been used to good effect in studying the recombination of ligands in biomolecules and in the folding events involved in the formation of a small helix from a coiled polypeptide. These techniques tend to focus on individual optimal pathways, but it is also clear that sets of pathways are very important in such problems. How these pathways are related to each other and how to discover them and count them are still open computational challenges.

The weak point in the whole scenario of new drug discovery has been identification of the "lead." There may not be a good lead in the collection of a company. The wrong choice can doom a project to never finding compounds that merit advanced testing. Using only literature data to derive the lead may mean that the company abandons the project because it cannot patent the compounds found. These concerns have led the industry to focus on the importance of molecular diversity as a key ingredient in the search for a lead. Compared to orders of magnitude just 10 years ago, more compounds can be designed, synthesized, and tested with newly developed strategies. These changes present an opportunity for the imaginative application of mathematics.

There are three aspects to the problem of selecting samples from large collections of molecules: First, what molecular properties will be used to describe the compounds? Second, how will the similarity of these properties between pairs of molecules be quantified? Third, how will the molecules be paired or quantified?

For naturally occurring biomolecules, one of the most important approaches is the understanding of the evolutionary relationships among macromolecules, the study of which has given rise to a variety of mathematical questions in probability theory and sequence analysis. Biological macromolecules can be related to each other by various similarity measures, and, at least in simple models of molecular evolution, these similarity measures give rise to an ultrametric organization of the proteins. A good deal of work has gone into developing algorithms that take the known sequences and infer from these a parsimonious model of their biological descent.

An emerging technology is the use of multiple rounds of mutation, recombination, and selection to obtain interesting macromolecules or combinatorial covalent structures. Very little is known as yet about the mathematical constraints on finding molecules in this way, but the mathematics of such artificial evolution approaches should be quite challenging. Understanding the navigational problems in a high-dimensional sequence space may also have great relevance to understanding natural evolution. Is it punctuated or is it gradual as many have claimed in the past? Artificial evolution may obviate the need to understand completely and design biological molecules, but there will be a large number of interesting mathematical problems connected with the design.

Drug leads binding to a receptor target can be directly visualized using X-ray crystallography. There is physical complexity because the change in free energy is complex, as it involves a multiplicity of factors including changes in ligand bonding (with both solvent water and the target protein), changes in ligand conformation or flexibility, changes in ligand polarization, as well as corresponding changes in the target protein. Now structural–property refinement uses parallel synthesis to meet geometric requirements of a target receptor binding site, although custom chemical scaffolds are directed to fit receptor binding sites synthetically elaborated through combinatorial reactions. This may lead to thousands to millions of members, while parallel automated synthesis is capable of synthesizing libraries containing on the order of a hundred discrete compounds. Structure–property relationships are then supplied to refine the selection of sublibraries. 3D structural models, SAR bioavailability, and toxicology are also used in such searches. Additional 3D target–ligand structure determinations are used to refine molecular properties iteratively using more traditional SAR methods.

In the Laboratory for Applied Thermodynamics and Phase Equilibria Research an account of a computer-aided design of technical fluids is given. The environmental, safety, and health restrictions impose limitations on the choice of fluids for separation and energy processes. Group contribution methods and computer programs can assist in the design of desired compounds. These compounds and mixtures have to fulfill requirements from an integrated point of view. The research program includes both the design of the components and the experimental verification of the results.

The Molecular Research Institute (MRI) is working in many specific areas, among which are interdisciplinary computer-aided design of bioactive agents and computer-aided risk assessment and predictive toxicology, and all kinds of models for complicated biological molecules. The first area designs diverse families of bioactive agents. It is based on a synergistic partnership between computational chemistry and experimental pharmacology, allowing a more rapid and effective design of bioactive agents. It can be adapted to apply to knowledge of the mechanisms of action and to many types of active systems. It is being used for the design of CNS-active therapeutic agents, particularly opioid narcotics, tranquilizers, novel anesthetics, and the design of peptidomimetics. In computer-aided risk assessment they have produced strategies for the evaluation of toxic product formation by chemical and biochemical transformations of the parent compound, modeling of interactions of putative toxic agents with their target biomacromolecules, determination of properties leading to toxic response, and use of these properties to screen untested compounds for toxicity.

1.29 A Genetic Algorithm for the Automated Generation of Molecules within Constraints

A genetic algorithm has been designed that generates molecular structures within constraints. The constraints may be any useful function, such as molecular structures from pattern recognition or rule-induction analyses. The starting point may be random or may utilize known molecules. These are modified to grow into families of structures which, using the evolutionary operators of selection, crossover and mutate and evolve to fit the constraints better. Wide applications in chemistry include new polymers, materials (e.g., superconducting materials), and synthetic enzymes.

The central problem of creating new chemical structures is a complex task, however, and is frequently rule based; many of the fundamental operations may be embedded in an expert system. Thus, the structures must satisfy valency rules (hexavalent carbons not allowed). One example of this approach is DENDRAL, an expert system for the simulation of mass spectra.

Besides the generation of chemical structures, an optimization method and constraints that direct the structural modifications toward some goal are necessary. The chemical genesis algorithm uses a series of rules that consistently produce realistic molecules in three dimensions. The assumption that molecular properties useful in similarity searching would be a good starting point for optimization of molecular structures within constraints seems logical. The structure-generation algorithm can produce very large and diverse sets of reasonable chemical structures for searching 3D database programs.

A genetic algorithm utilizing a knowledge base for chemical-structure generation evolves molecular structures within constraints. The molecule generation is fast, effective, and stable, resulting in novel compounds that satisfy the rules of chemistry

Further development will concentrate on implementing constraints that better mimic nature (e.g., in the case of evolution within an active site, to generate scores which approximate binding energies). Also, metals and transition states may be useful additions. Because of the complexity of the software and the enormous variation in the molecules produced, unexpected algorithmic bugs do turn up, e.g., the azido (–N=N=N) group appeared and required the assignment algorithm to be modified.

1.30 EPA Programs

The new EPA chemical program under the Toxic Substances Control Act (TSCA) plays an important role in preventing pollution. TSCA requires that the EPA review new chemicals for the risks they may pose to human health and the environment before they enter the market. Anyone who plans to manufacture or import a new chemical substance must provide the EPA with a premanufacture notice (PMN) at least 90 days prior to the activity. To determine whether a substance is "new," the company must consult the EPA Inventory of Chemical Substances (the TSCA Inventory). If the substance is not listed, it is a new chemical. EPA personnel are also evaluating computer-based software for synthesis design and new chemical substances that may be safer substitutes for chemicals currently used, or that will be created via pollution-prevention processes. During the course of PMN, the EPA has identified the following desirable criteria:

- Test data on the PMN substance itself (toxicity and fate);
- No reports of adverse effects;
- Safer substitute;
- Less toxic or fewer toxic associated chemicals;
- Safer pollution prevention, source reduction, or recycling processes that reduce exposure/releases;
- Successful implementation or a recommendation resulting from the EPA "Alternative Synthetic Pathway;"
- Use of the chemical substance should be environmentally beneficial rather than harmful;
- Conservation of energy and water during manufacturing, processing, or use;
- No TSCA enforcement actions filed against the PMN submitter within the past year.

The solvent software SAGE recommends solvent replacement in cleaning and degreasing operations. The system leads the user through a question-and-answer session. The user's responses allow the system to develop a list of alternative chemistries and processes, and the user may then review brief descriptions of the alternatives on the computer screen. If any of the alternatives are of interest, the user may request SAGE to generate a detailed report for each.

Friedler et al. have written a paper entitled "A Combinatorial Approach for Generating Candidate Molecules with Desired Properties Based on Group Contributions." Although this will be discussed later at greater length, a synopsis is given here. The problem is considered to be a combinatorial problem in which the desired properties constitute constraints on the integer variables assigned to the functional groups. The feasible region defined by these constraints is determined by an algorithm involving a branching strategy. The algorithm generates those collections of the functional groups that can constitute structurally feasible molecules or compounds satisfying the constraints on the given properties. The molecular structures can be generated for any collection of the functional groups. One advantage of this approach is that it considers only the feasible partial problems and solutions in the procedure, thus resulting in a substantial reduction in search space. Examples are shown in the paper by Friedler et al. In the first, compounds are selected with both the boiling point and the melting point in a narrow range. This is done from among six groups. The second example is like the first but extends the groups to nine. The third problem fixes the molecular weight, the normal boiling point, the normal melting point, the density of the crystalline solid, and the liquid volume. The choice of groups rises to 15. The fourth problem is for polyhalogenated biphenyls (PHBs). Constraints are imposed for GC and HPLC retention indexes. The proper PHBs are selected. The last example is to design a perfume by selecting the flavor constituents having desired odors and odor-threshold values; 26 groups were offered for selection. These five examples have been solved on a personal computer in time ranging from $^1/_2$ to 15 min. This indicates the efficacy of the approach. The examples have amply demonstrated the ability of the combinatorial approach and its computer realization to compose chemical compounds in an optimal fashion based on group contributions. The approach can be recommended as a useful tool for selecting compounds with desired properties and for identifying chemicals using chromatographic data. The approach can also facilitate the first step of the quantitative approach to the design of flavor constituents on the basis of the structure–odor relationship (QSOR).

In a program I designed, called SYNPROPS, that will be discussed in more detail later, optimal molecular structures for most desired physical and environmentally benign process chemicals are generated simultaneously. Molecules are designed with a constellation of optimum physical properties and with low-risk-based toxic properties. Such molecules could replace unsatisfactory molecules with better ones in industrial processes and serve as target molecules for organic syntheses. This study found a procedure for screening out those molecules that will contaminate the environment and a way to find those molecules that are environmentally benign, in addition to possessing the most desirable physical properties. Data based on regressions of published physical properies and risk-based toxic properties for given compounds, as well as the Quatro Pro program (version 5.0) operating on a personal computer, expedite this procedure.

In another part of this procedure the statistically derived QSAR-like formula is resolved into two matrices: one structure dependent, the other a structure-to-property transition matrix. The multiplication of these matrices yields properties. Using matrix algebra one can then assign properties and find structures that correspond to such properties or, vice versa, assign structures and find corresponding properties.

Many examples of drugs (without side reactions) and environmentally benign compounds that may serve as substitutes for compounds in the process industries are shown in Sections 2.16 and 2.25-2.32.

Progress of neural network technology offers new computational tools equipped by learning features of nonlinearity. A paper by Shuzo Ohe presents the technology for providing a potential solution to the problems in the area of predicting and processing thermodynamic properties. One such property was the boiling point of hydrocarbons and alcohols. The model of backpropagation contained 14 input units, 8 hidden units, with 1 output unit. The prediction of the critical temperature of paraffins was determined by the association ability of neural networks trained by the backpropagation method, giving network carbon numbers, molecular weight, boiling point, and acentric factor as input. The expression of the solid–liquid phase diagram was improved with a modified method with selective adoption of training, and then the cuspidal point area was remarkably close, reproducing a solid–liquid phase diagram with good accuracy.

Leete et al. have introduced a new type of algorithm for constructing DNA molecules which encode answers to mathematical problems. They imagine that an efficient computer could combine biomolecular and conventional architecture, instead of working entirely in one domain. They have shown that DNA computation of expanded symbolic determinants, given their zero pattern, is both practical and useful. The limits of DNA computation technology lie within the bounds of our ability to capture the mathematical structures of certain problems in the molecular structures of the biomolecules and in creating algorithms that manipulate these relationships.

Glen and Payne write of a genetic algorithm designed to generate molecular structures within constraints; which was described earlier. The constraints may be any useful function, such as an enzyme active site, a pharmacore, or molecular properties from pattern recognition or rule-induction analyses. The starting point may be random or may utilize known molecules. These are modified to grow into families of structures which, using the evolutionary operators of selection crossover and mutation, evolve to fit the constraints better. They describe the basis of the algorithm and applications in lead generation, 3D database construction, and drug design. Genetic algorithms of this type may have wider applications in chemistry, for example, in the design and optimization of new polymers, materials (e.g., superconducting materials), or synthetic enzymes. This genetic algorithm uses a knowledge base. The molecule generation is fast, effective, stable, and results in novel compounds that satisfy the rules of chemistry. The optimizing ability of the genetic algorithm is powerful, even with relatively small population sizes. Current constraints are rather simple, and many of the molecules generated require modification to produce synthesizable and sensible druglike molecules.

Meusinger and Moros used structural information from NMR spectra to determine the research octane number (RON) of gasolines. This was done by the application of genetic algorithms and neural networks in the analysis of multicomponent mixtures using NMR spectroscopy.

1.31 Data

CrossFire contains 15 gigabytes of chemical information, once available only in the 400-volume Beilstein *Handbook*. Chemists can use ChemOffice to search over 400 different properties of over 6 million compounds. Searching will be done by registry number, text, or substructure drawn in ChemDraw. Boolean searches are also supported. Results will be displayed in ChemFinder for convenient data management and retrieval.

Region III of the U.S. EPA has a Table of Risk-Based Concentrations, created by Roy L. Smith, Ph.D. The manner in which each item is calculated is listed in detail. Results for both carcinogenic and noncarcinogenic compounds are given, and results for the various modes and media for exposure are also given. These numbers have been used for CERCLA and RCRA and the table is available on the Web on the site for Region III of the EPA.

BRENDA is a database for enzymes from Dr. Dietmar Schomburg. The database is implemented in a relational database and covers some 40 data fields with information about nomenclature, reaction and specificity, enzyme structure, isolation/ preparation, stability, literature references, and cross-references to sequence and 3D structure data banks. During the development of the data bank the book *Enzyme Handbook* represents the printed version of this data bank. The database is becoming available via the Internet through cooperation with the European Bioinformatics Institute.

The Medicinal Chemistry Project has been collecting log P from the chemical literature for 20 years. The database called Master contains over 46,000 measured log P values with literature citations. Also included are activity type, measured log P, MR, and measured pKa. Master must remain a commercial product; however, the 37,000 names (representing 31,000 structures) are available for searching. The names on this server contain the structures (as SMILES) and names for all of the compounds in the Master database.

In 1997, the data on "The Ideal Gas Thermodynamic Data for Combustion and Air-Pollution Use," by Alexander Burcat, became available on the Internet. Bonnie McBride issued a print version with Alexander Burcat in 1995. Sources for the data include calculations by M. B. Colket, the former Chemkin database, JANAF Tables, NASA Lewis Research Center, Thermodynamic Research Center (formerly American Petroleum Institute), Russian Tables (TSIV), and Technion-Israel Institute of Technology, etc. The tables are calculated using methods of statistical mechanics and use experimental measurement results for the parameters, and actual values for these parameters are given. Unfortunately, the documentation is placed in positions such that the thermodynamic data in the table cannot be used directly in the Chemkin program. This table is extremely valuable and it and its uses will be discussed at great length later.

Data files of reactions and reaction rates are also contained on the Internet. Much of this was prepared by NIST. Rate constant data are given for 177 reactions involving C, H, and O alone, and approximately another 100 reactions are added with the inclusion of N.

CAMEO is a program that employs a mechanistic approach in making its predictions and uses data tables of specific reactions. Consequently, it is capable of predicting novel chemistry. Interaction with the program is via a graphical interface. The reactant molecules are drawn on a sketch pad and the appropriate reaction conditions are selected using the appropriate menus (reagent, solvent, temperature, concentration, etc.). The reaction is then run. CAMEO evaluates the reaction in the mechanistic modules and suggested products are displayed on the product display screen. The use of physical parameters is an integral part of the mechanistic evaluation phase of reaction processing using CAMEO. They can be viewed directly by the chemist. Some of the available parameters are enthalpies of formation, bond dissociation energies, pK_a values, molecular formulae and weights, and Taft Es values. Further information is available via the 3D module, including molecular volumes and areas, log P values, dipole moments, bond lengths, bond angles, and dihedral angles. Some of these parameters require interfacing with the MOPAC program. Reaction products are displayed to the user in the order they are generated, and relative yield estimates are made on the grounds of enthalpies of reaction, activation energies, and rate constants. The results of the reaction evaluation are stored in the form of a tree with the reactant molecule(s) at the treetop and the products as nodes on the tree. The tree can be stored and reviewed. Products can be resubmitted for further evaluation so that the effects of extended reaction times or more severe reaction conditions can be examined. For multistep reactions, reactive intermediates are also stored on the tree and can be reviewed. In this way, the mechanistic course of a reaction can be examined.

Crabtree and El-Halwagi synthesize optimal "environmentally acceptable reactions" by identifying cost-effective, environmentally benign overall reactions that yield a certain product. A systematic tool is developed to screen reaction alternatives without enumerating them. This is formulated as a mixed-integer, nonlinear optimization program that examines overall reactions occurring in a single reactor to specify a particular product. It maximizes the economic potential of the reaction, subject to a series of stoichiometric, thermodynamic, and environmental constraints. The synthesis procedure can extract the most economical environmentally acceptable reactions from an infinite number of reactions, given a set of potential reactants and by-products. Since environmentally acceptable reaction synthesis is a screening tool, additional laboratory investigation, path synthesis, kinetics, and reactor design, etc. may need to be done, but location of the best environmentally acceptable reaction is an excellent starting point to plan experimental work efficiently.

Blurock introduces REACTION, a software system for the manipulation, management, and generation of molecular and reaction information. It automatically generates a detailed mechanism, given a small set of reaction classes. The output is in such a form that it can be used as input to a standard numerical package for solving the associated differential equations. He shows a concrete case involving a simplified mechanism of combustion (around 750 K). Detailed mechanisms are automatically generated for 30 different hydrocarbons. Numeric analysis of the mechanisms involves an interface with a numerical analysis package that solves the corresponding set of differential equations. Plots of concentration vs. time at a given temperature can be made, and symbolic analysis of the mechanisms characterizes their structure so that comparisons can be made. One such analysis is the production of parameters derived from the mechanism. One set of parameters count which and how often each individual reaction is used; another set of parameters is based on which and how often entire paths of reaction (from the initial molecules to the final end products) are used. The manipulation and analyses of paths are a preliminary stage to the more complex problem of finding an automatic means to generate reduced mechanisms, i.e., mechanisms in which the number of reactions are condensed with little or no loss of information.

Chemical reactions can be viewed in two directions: the synthetic direction, corresponding to laboratory operations, and the retrosynthetic direction, going backward from a target molecule to starting materials by way of retroreactions or transforms. Retrosynthesis is an important technique for designing synthesis routes to a target molecule. Retrosynthetic analysis lends itself well to a description of algorithms and thence to computer programs. Thus, the most important computer programs for synthesis design operate in a retrosynthetic fashion. A prime example is LHASA. In addition, other programs have developed that may assist in different stages of the synthesis planning process. An example is the computer program CHIRON, identifying specific chiral fragments in complex target molecules and suggesting eligible starting materials ("chirons") for their synthesis. An alternative tool is offered by the WODCA system, which allows an efficient lookup of starting materials with the same skeleton and generalized functionality as the query structure.

The course of a chemical reaction and its products can be predicted for given starting materials. The EROS system has been in development for over 15 years. In version EROS 6.0, the knowledge base and the problem-solving techniques are clearly separated. The knowledge base consists of methods for calculating important electronic and energy effects in organic molecules, as well as rules concerning the course and evaluation of elementary chemical processes.

Chen and Gasteiger illustrate knowledge discovery in reaction databases in landscaping organic reactions by a self-organizing neural network. They state that chemists derive their knowledge about chemical reactions by inductive learning from observations of individual chemical reactions. Predictions of the products of chemical reactions are made by analogy. Since there are large reaction databases available, this process can be automated; and they show a new method based on a Kohonen neural network and physicochemical variables for describing reaction centers. Results with two reaction data sets show how a set of chemical reactions with the same reaction center can automatically be classified, clearly revealing different levels of similarities of the reactions under investigation. The relative positions of reactions and clusters in the 2D Kohonen map offer extra chemical information. A third reaction data set is used to show how a trained Kohonen network can be used to predict reaction types for organic reactions.

There is much current interest in computer-based methods for selecting structurally diverse subsets for chemical databases, e.g., for inclusion in the biological screening program for the construction of combinatorial libraries. Recent work on dissimilarity-based compound selection seeks to identify a maximally similar subset of the molecules in a particular database, because the identification of the most dissimilar subset requires the use of a combinatorial algorithm that considers all possible subsets of a given data set. This is reviewed in a paper by Willett.

Paterson et al. wrote an article entitled "Neighborhood Behavior: A Useful Concept for Validation of Molecular Diversity Descriptors," which states that, when searching for a lead, testing molecules that are too similar is wasteful, but when investigating a lead, testing molecules that are similar to the lead is efficient. They studied which molecular descriptors should be similar and how much similarity is enough. A plot of differences in the values vs. differences in biological activities for a set of related molecules will exhibit a characteristic trapezoidal distribution enhancement, revealing a "neighborhood behavior" for the descriptor. Applying this to 20 data sets allows 11 molecular diversity descriptors to be ranked by their validity for compound library design. In order of increasing frequency of usefulness, these are random numbers = log P = MR = strain energy < connectivity indexes < 2D fingerprints (whole molecule) atom pairs = autocorrelation indexes = < steric CoMFA fields = 2D fingerprints (side chain only) = H-bonding CoMFA fields. Medicinal chemists have always sought to discover and patent "activity islands" and to identify the "peaks" in those islands as potential future drug candidates. In the long run, the appreciation that not all dimensions of molecular diversity are equally relevant in defining those activity islands and the ability to identify objectively and rank the relevant dimensions may be of the greatest importance in molecular discovery research.

A new version of a computer program called POLYRATE (6.5) for the calculation of chemical rates for polyatomics uses methods that are variational or transition state theory and multidimensional semiclassical adiabatic and large-curvature approximations for tunneling and nonclassical reflection. Rate constants may be calculated for canonical or microcanonical ensembles for specific vibrational states of selected modes with translational, rotational, and other vibrational modes treated thermally. Bimolecular and unimolecular reactions and gas-phase, solid-state, and gas–solid interface reactions are all included. Potential energy surfaces may be global analytic functions or implicit functions defined by interpolation from input energies, gradients and force constants (Hessian matrices) at selected points on a reaction path. The data needed for dynamics calculations may also be calculated from a global potential energy surface with more accurate calculations at stationary points. The program calculates reaction points by the Euler, Euler stabilization, or Page–McIver methods. Variational transition points are optimized from among a one-parameter sequence of generalized transition states orthogonal to the reaction path. Tunneling probabilities are calculated by numerical quadrature, using either the centrifugal-dominant-small-curvature approximation, the large-curvature-version-3 approximation, and/or optimized multidimensional tunneling approximations. In the large-curvature case, the tunneling probabilities may be summed over final vibrational states for exoergic reactions or initial vibrational states for endoergic reactions.

Reaction Patterns visualizes, interprets, and analyzes chemical reactivity and reactions. It combines kinetic models and the analysis of quantum mechanical energy surfaces to give an insight into fundamental chemical processes. It is available from Molecular Simulations, Inc.

REACCS (Reaction ACCess System) is a mainframe application used to build, maintain, and search proprietary and commercial databases of chemical reactions and related data and is available from MDL, Information Systems, Inc.

React is a kinetics program available from the CCL archives. It is available in compact form for the Macintosh, DOS, and OS/2 platforms. The program takes reaction mechanisms in the form:

```
5e-8 O2-+Cs+—>Cs+O2

1e-12 Cs++e—>CS

3.24e-3 Cs—>Cs++e-

.4O2—>O2+e-

1.76e-16 O2+Cs—>CsO2

1.24e-30O2+O2+e- —>O2+O2-

1.4e-16O2+e- —>O2-
```

then builds the differential equation set and finds a numerical solution. Specific reactions can be flagged to have their rate constants adjusted using a least-squares criterion and experimental measurements of time-concentration profiles can be obtained. Sample files and a brief manual are included in both text (ASCII) and MS Word versions. The program comes with no promises, warranties, or error messages, so it must be used at one's own risk. React was developed by Dr. Michael Whitbeck (whitbeckl@popcorn.llnl.gov) Chemistry and Material Science Division, Lawrence Livermore National Laboratory.

There are many process simulation programs and, to be brief, only two are mentioned here. One is called ASPEN. ASPEN stands for Advanced System for Process Engineering and is a flexible process model development system that can be used to simulate the performance and profitability of almost any process system. The flexibility of ASPEN lies in its input language and computer code as well as its ability to handle solids, streams, and to include user-developed models. It is a system of programs that are used for modeling, simulating, and analyzing chemical processes. These processes are represented by mathematical models that consist of a system of equations to be solved. To do the process analysis, the user specifies the interconnection and the operating conditions for the process equipment, and ASPEN solves for the unknown variables, given certain known quantities. ASPEN develops models of any type of process where there is a flow of materials and energy from one processing unit to the next and has modeled processes in chemical and petrochemical industries, petroleum refining, oil and gas processing, synthetic fuels, power generation, metals and minerals, pulp and paper, food, pharmaceuticals, and biotechnology. ASPEN has the capacity to perform material and energy balances, to compute sizing requirements, and to determine profitability of the process. It uses DIPPR for its needed physical property data. Another process simulation program comes from Simulation Sciences, Inc. which has developed programs to model both steady-state and dynamic operations effectively, as well as batch processes. Core applications include design and operational analysis; on-line modeling, control, and optimization; and plant and enterprise-wide economic models. These programs help companies develop better process designs, predict plant performance more accurately, and increase plant safety and efficiency.

Sometimes it is necessary to create programs to assist not only with environmental calculations but also with decisions in the office and even out in the field when assessing a site environmentally. The author has submitted over a dozen such proposals (only three of which are described here), chiefly to the Air Division of EPA, to help in projects that required knowledge of air quality and Superfund (CERCLA) assessment. He was a member of BTAG (Biological Technical Assistance Group) before which the project managers of hundreds of Superfund sites reported.

1. One such method set up matrices, one of which contained the measurement of components on site in the liquid phase believed to be hazardous. The other matrix contained numbers that convert these liquid measurements to gas concentrations, and various regulatory risk-concentrations for the compounds in question. The result of multiplying these two matrices then yielded the inhalation risk, and NAITCHPA, UNITRISK, AIR RISK, and RAL for these compounds. The result tells us whether or not this site represents a problem, how bad it is, and which compound should be concentrated on.
2. Suppose we are on a site where air measurements indicate a problem and somewhere there is a pool of hazardous substances but we do not know where it is. We can use a Markovian process to find the pool by essentially reversing case 1 using matrix equations, inverse matrices, a set of transition matrices, and triangulation.
3. A spreadsheet was set up with formulas for converting all the liquid measurements in the field to other quantities using Henry's law constants and other physical parameters in each column for almost 200 molecular species in each row and the physical results in each column. In this way, results of calculations and management decisions on the sites could be more immediate. This is discussed in the Appendix.

1.32 DIPPR

The DIPPR file contains textual information and pure component numeric physical property data for commercially important chemicals and substances. The data included in the file are compiled and evaluated by projects of the Design Institute for Physical Property Data (DIPPR) of the American Institute of Chemical Engineers (AIChE). The numeric data in the DIPPR file consist of single-value property constants and temperature-dependent properties. Subjects covered are:

Accuracy of data
Chemical engineering and chemistry
Electrical properties
Issue and revision dates
Optical properties
Physical properties
Property constants and property data
References and notes
Regression equations and coefficients
Source of data
Substance identifying information
Substance states
Temperature-dependent property values
Thermal properties
Thermodynamics
Environmental properties

It is reliable and quite complete in the subject classes. More will be said about it further on in the book.

1.33 NIST

The Physical and Chemical Properties Division is one of five divisions in the Chemical Science and Technology Laboratory (CSTL) of the National Institute of Standards and Technology (NIST). It provides a reference laboratory for thermophysical and thermochemical properties of gases, liquids, and solids — both pure materials and mixtures — the rates and mechanisms of chemical reactions in the gas and liquid phases, and physical processes and systems (separations, temperature refrigeration and heat transfer, and flow). It also publicizes the scope, reliability, accessibility, and economics of computational thermochemistry. More will be said about NIST later.

1.34 Equations of State of Real Gases, Liquids, and Solids

Although the van der Waals equation of state R is 0.08205 l-atm K^{-1} mol^{-1},

$$P_{VW}(T, V_m, a, b) = RT/(V_m - b) - a/V_m^2 \qquad (1.34.1)$$

Although the last equation has been used, it is not the best. A compressibility equation from Hougen and Watson and derived from first principles is better.

Another equation of state for solids and liquids is

$$V(p, T, V_o, p_o, T_o, k_o, a_o, B_o) = V_o[1 - k_o(p - p_o) + a_o(T - T_o) + B_o(T - T_o)^2] \quad (1.34.2)$$

1.35 The Enigmatic Substance

No one knows why water is able to carry so much heat. If the heat capacity of water were half of its actual value, the temperature fluctuations in our environment would be more extreme, and this would have important implications for life itself. The structure of water on the molecular level, specifically in the top one or two layers of liquid water where it meets water vapor in the air, is not known. No calculations can account quantitatively for the surface tension of liquid water. Supercomputer calculations now model water clusters, which are small groupings of water molecules that differ from bulk water in many ways. Most of the molecules in a small cluster are on the surface, where they have fewer interactions with other water molecules than do the molecules in bulk water. Quantum mechanical calculations are now being applied to the study of water clusters, and these can be used to test the usefulness of the water-interaction models used in simulating proteins and DNA. The calculations are aimed at accurately mapping the potential energy surfaces for different-sized water clusters. The calculations have also explored how the phase transition from solid to liquid water differs in water clusters from the behavior of bulk water.

1.36 Kinetic Theory of Gases

There are a number of approaches to obtaining properties from theory such as the theory of corresponding states, statistical mechanics, etc. (these will be discussed later), but one of the first to be used was kinetic theory. Using kinetic theory and Newton's law, it is easy to derive the pressure of a gas as

$$P = Nmv^2/3V \quad (1.36.1)$$

$$PV = nRT \quad (1.36.2)$$

and

$$E_k = 3/2RT \quad (1.36.3)$$

where v^2 and E_k are averaged, $n = N/N_A$ (n_A is the number of particles, and NA is Avogrado's number), $k = R/N_A$ with R the gas constant and $k = 1.38(10)^{-23}$ J/K. (The Boltzmann constant). It was after Maxwell attacked the problem that the probability distribution as a function of speed was derived as

$$f(v) = 4\pi(m/2pkT)^{3/2}v^2 \exp(-mv^2/2kT) \quad (1.36.4)$$

It was Boltzmann who later proved that the thermodynamic entropy S of a system (at a given energy E) was related to the number of microscopic states available to it by $S = k \log W$, k being Boltzmann's constant. He was then able to establish that for any system large or small in thermal equilibrium at temperature T, the probability of being in a particular state at energy E is proportional to $\exp(-E/kT)$.

Once we have stated that the total average energy per molecule is 1.5kT at temperature T, the specific heat per molecule is 1.5k and since $k = R/N_A$, the specific heat per mole comes out at 1.5R, which is confirmed for a monotomic gas. However, for a diatomic gas it can be 2.5R. This is because such a gas has more degrees of freedom. These problems were not resolved until the advent of quantum mechanics.

Some other formulas derived from kinetic theory include

$$\lambda = 1/\pi(\sigma)^2 n \quad (1.36.5)$$

where sigma is the effective diameter, n, molecules/cc;

$$n = 1/3(m\lambda)nc \quad (1.36.6a)$$

where c is the mean gas velocity;

$$K = nc\lambda(cv)/c \quad (1.36.6b)$$

where c_v is the specific heat at constant volume.

The ratio of thermal conductivity should yield $K/n = C_v/M$, or $(K/n)(M/C_v) = 1$. Actually, it lies between 1.4 and 2.5, partially due to repulsions that tend to reduce n.

The coefficient of self diffusion is derived as

$$D = c/(3\pi n(\sigma)^2) \quad (1.36.7)$$

There are equations derived from the kinetic theory with more finesse and also from models that are less perfect and more realistic.

1.37 Properties of Liquids

It is possible to obtain approximate descriptions of liquids. The empirical approaches, like Pade approximates, or empirical mixing of different approximations remain most common for calculating their thermodynamic properties.

The virial expansion is a polynomial in density p = N/V. Since the density cannot exceed the value of the closed packing (0.74), the equation should not diverge. The Pade approximant is

$$P(p) = pkT(1 + B(T)(1 + c_2p + c_2p^2 + \cdots + c_np^n)/(1 + d_1p + d_2p^2 + \cdots + d_mp^m) \qquad (1.37.1)$$

The expansion at small p should coincide with virial expansion. If N = n + m virial coefficients, (n,m)-Pade approximates can be constructed. (3,3) approximant describes the simulation (P,V,T) data for hard spheres a lot better than the corresponding virial coefficient with six virial coefficients. This approach is empirical and ambiguous. An alternative description is needed.

N-particle distribution functions are used for liquids. The Gibbs distribution, the probability that n particular particles of the N-particle system will be found at the positions r_1, r_2,...,r_n. This is corrected by the n-particle correlation function. The most important correlation function is the pair correlation function. Thus,

$$pg^{(2)}(r_1,r_2)dr = pg(r)dr \qquad (1.37.2)$$

This can be developed further to the structure factor and to define scattering. The total energy, pressure, pressure equation (equation of state), virial coefficients, chemical potential, and compressibility can be derived from the correlation function concept. The Kirkwood approximation is important for descriptions of colloids and ions in solution. The BBGKY equations (Bogolyubov, Born, Green, Kirkwood, Yvon) hierarchy relates the correlation functions of two adjacent orders and the intermolecular interactions potential. The BBGKY infinite hierarchy of integral equations can be closed at the second order with the Kirkwood superposition approximation. The Percus–Yevick approximation assumes a model of hard spheres

$$v(r) = \begin{cases} \text{infinity}, & r < \sigma \\ 0, & r > \sigma \end{cases}$$

and if an analytical solution is derived from the pressure equation

$$PV/NkT = (1 + 2n + 3n^2)/(1 - n)^2, \qquad (1.37.3)$$

$$n = 1/5(\pi)(\sigma)3p) \qquad (1.37.4)$$

and from the compressibility equation

$$PV/NkT = (1 + n + n2)/(1 - n)^3 \qquad (1.37.5)$$

The above two results are different! The theory is thermodynamically inconsistent.

We can suppose the energy in the correlation model is the sum of two terms.

We can then derive expressions for the free energy using the perturbation theory and using the hard-sphere system as a reference and the attractive part of the molecular interaction v(r) as a perturbation. We derive the van der Waals equation of state

$$p/kT = p/(1 - pb) - ap^2/kT \qquad (1.37.6)$$

There is great difficulty in deriving physical properties for liquids as compared with gases or solids because of the lack of adequate theory for the liquid state. Nevertheless, data for the physical properties of liquids do exist, largely because of the good programs of experimental measurements. Y. Marcus in his book, *Introduction to Liquid State Chemistry*, states that his classification follows that of the intermolecular forces, with allowance for the molecular complexity (atomic vs. molecular), shape (central vs. non-central forces) and extent of interaction (single particles vs. associated species).

When reviewing the properties of liquids, many of the bulk properties follow a classification of bulk properties. Recognized classes are simple fluids, nonpolar molecular liquids, polar molecular liquids, hydrogen-bonded molecular liquids, ionic liquids, and liquid metals.

There is little difference between the simple liquids (liquefied inert gases), which are monatomic, and spherically symmetrical (or practically so) nonpolar fluids, such as methane, neopentane, nitrogen, or carbon dioxide. They differ from other nonpolar liquids with nonspherical molecules (chain-like such as hexane or plate-like such as benzene) or large spherical molecules which are highly polarizable (such as carbon tetrachloride) where the interaction forces are not central. Because of the short range of the forces, interactions occur mainly between adjacent segments of neighboring molecules, rather than between the molecules as entities.

Strong intermolecular forces leading to self-association of liquids may be of several kinds: dipole-dipole interactions may lead to pair-wise association, but rarely to higher aggregates. Hydrogen bonding may lead to dimers primarily (e.g., carboxylic acids), to chain = like oligomers (e.g., liquid hydrogen fluoride, alcohols, N-alkylamides) or to three-dimensional networks (e.g., water, formamide). Donor-acceptor interactions may operate between the ions of an ionic melt, and lead to autocomplexation into discrete complex ions (e.g., mercury bromide) or to extensive networks (e.g., boron oxide).

1.38 The Solid State

The equipartition theorem is used to help calculate thermodynamics of such systems. The simple model is of a crystal with masses m on springs with coefficients k. An harmonic approximation is assumed and partition functions are used to derive the thermodynamic functions for free energy, entropy, energy, and heat capacity, respectively,

$$A = k_BT \ln X - mT \ln T \quad (1.38.1)$$

$$S = k_B \ln X + mk_B \ln T + mk_B \quad (1.38.2)$$

$$E = mk_BT \quad (1.38.3)$$

$$C = mk_B \quad (1.38.4)$$

where m is the number of degrees of freedom.

The equipartition theorem does not always work! It predicts different results for gases whether we allow vibration or not. It predicts heat capacity, 3kBN for solids. This is right, but at high temperatures. Vibrations are a quantum phenomenom, and we need to include quantum mechanics to describe them properly.

The Einstein model allows vibrations

$$E = hw(n + 1/2), \quad n = 0,1,2... \quad (1.38.5)$$

where the system has n phonons. It can have many phonons (Bose–Einstein statistics). We derive a partition function

$$Q = \exp(\Theta_E/2T)/(1 - \exp(-\Theta_E/T)) \quad (1.38.6)$$

where Θ_E is the Einstein temperature equal to hw/kB. Then the thermodynamic quantities for free energy, entropy, energy, and heat capacity are, respectively

$$A = k_B\Theta_E/2 + k_B T \ln r \quad (1.38.7)$$

$$S = -k_B \ln r + k_B\Theta_B \exp(-\Theta_E/T)/Tr \quad (1.38.8)$$

$$E = k_B\Theta_E/2 + k_B\Theta_E \exp(-\Theta_E/T)/r \quad (1.38.9)$$

$$C = k_B(\Theta_E/T)2 \exp(-\Theta_E/T)/(r^2) \quad (1.38.10)$$

where $r = (1 - \exp(-\Theta_E/T)$. (1.38.11)

The Einstein Model of Crystals

Assuming that each atom has a certain equilibrium position in the lattice and that vibrations are harmonic with frequency w, we have 3N harmonic oscillators and the heat capacity is

$$C = 3Nk_B(\Theta_E/T)^2 \exp(-\Theta_E/T)/r^2 \quad (1.38.12)$$

This yields too fast a value at low temperatures because it does not show a cooperative effect in movement.

Debye Model of Crystals

Collective motion for 3N oscillators is included. One uses modes as a sum of waves. The Debye model has some phonons excited even at low temperatures. Dispersion equations are applicable.

Z = Vibrations per unit volume
c = velocity of propagation of the motion
$U = \int^{vm}_{v=0} 4\pi V v^2/c^2 dv$
$U = 9N_0\int^{v=vm}_{v=0}(v/v_m)^3 hdv/\exp(hv/kT)-1)$

Now, if $x = hv/kT$ and $\theta = hv/k$

$C_v = 9k(T/\theta)^3\int^{\theta/T}_0 x^4dx/(\exp(x)-1)(1-\exp(-x))$
$U = 9N_0kT(T/\theta)^3\int^x_0 x^2dx$ at high T, x small
$C_v = 3R$

Now differentiate U with respect to T at constant V

$U = 9N_0kT(T/\theta)^3\int_0^{infinity}x^3/(\exp(x)-1)dx$ at low T, integration

$C_v = aT^3$ where a is a constant for a given element.

At low temperatures for u << 1, the integral is constant and C is proportional to T^3.

1.39 Outstanding Programs

There are a number of people advancing the techniques of process simulation. Among these are Ciric, who has been working in the areas of process stability analysis, process synthesis for dynamic stability, reactive distillation, optimization, and waste minimization. Other names that can be mentioned are Grossman and Floudas.

1.40 THERM

A set of subprograms collectively called THERM written by Ritter and Boselli have proved very useful in the two programs whose descriptions follow (the NASA LSENS and the DOE Chemkin). THERM supplies the thermochemical data needed by these two programs based on the data files in Benson's book *Thermochemical Kinetics*. The data in THERM are given for 330 groups. They are divided into three types of groups — BD (bond dissociation groups), CDOT (radical groups), and regular groups (all other groups with no unbonded electrons). The last groups consist of HC (hydrocarbon groups), CYCH (ring corrections for hydrocarbon, oxygen, and nitrogen-containing ring systems), INT (interaction groups/ substituent effects), CLC (chlorine and halogen containing groups), HCO (oxygen containing groups), and HCN (nitrogen containing groups). Recipes guide the user into the various subprograms. ENTER/ ESTMATE SPECIES asks for species ID, number of groups in the species, number of different groups to be entered, elemental formula, groups contained in the species, and the species symmetry number. (The code for the groups in the species takes a while to get used to.) The result when stored in Therm.lst is in the form heat of formation, entropy, and heat capacity at 300, 400, 500, 800, 1000, and 1500 K. The whole Therm.lst file for many species can be converted to the Therm.dat file, which is exactly what is needed in the NASA format for LSENS and Chemkin. This is done with amazing speed by the subprogram Thermfit. Incidentally, literature values for H, S, and Cp can be converted to the NASA format with Thermfit as well. The Therm.dat file is also needed in Thermrxn which is a subprogram where a chemist writes a reaction which is immediately converted to a table of enthalpies, entropies, free energies, equilibrium constants, ratio of Arrhenius factors at a series of temperatures and in units of one's option. All the information found by THERM can be sorted and stored in specific files that are easily recalled.

1.41 LSENS

LSENS, from the NASA Lewis Research Center, was developed for solving complex, homogeneous, gas-phase, chemical kinetics problems. It was motivated by the interest in developing detailed chemical reaction mechanisms for complex reactions, such as the combustion of fuels and pollutant formation and destruction. Mathematical descriptions of chemical kinetics problems constitute sets of coupled, nonlinear, first-order ordinary differential equations (ODEs). The number of ODEs can be very large because of the numerous chemical species involved in the reaction mechanism. Further complicating the situation are the many simultaneous reactions needed to describe the chemical kinetics of practical fuels. For example, the mechanism describing the oxidation of the simplest hydrocarbon fuel, methane, involves over 25 species participating in nearly 100 elementary reaction steps. Validating a chemical reaction mechanism requires repetitive solutions of the governing ODEs for a variety of reaction conditions. Consequently, there is a need for fast and reliable numerical solution techniques for chemical kinetics problems. In addition to solving the ODEs describing chemical kinetics, it is often necessary to know what effects variations in either initial condition values or chemical reaction parameters have on the solution. Such a need arises in the development of reaction mechanisms from experimental data. The rate coefficients are often not known with great precision, and, in general, the experimental data are not sufficiently detailed to estimate the rate coefficient parameters accurately. The development of a reaction mechanism is facilitated by a systematic sensitivity analysis which provides the relationships between the predictions of a kinetics model and the input parameters of the problem.

1.42 Chemkin

Complex chemically reacting flow simulations are commonly employed to develop a quantitative understanding and to optimize reaction conditions in systems such as combustion, catalysis, chemical vapor deposition, and plasma processing. They all share the need for accurate, detailed descriptions of the chemical kinetics occurring in the gas phase or on reactive surfaces. The Chemkin suite of codes broadly consists of three packages for dealing with gas-phase reaction kinetics, heterogeneous reaction kinetics, and species transport properties. The Chemkin software was developed to aid the incorporation of complex gas-phase reaction mechanisms into numerical simulations. Currently, there are a number of numerical codes based on Chemkin which solve chemically reacting flows. The Chemkin interface allows the user to specify the necessary input through a high-level symbolic interpreter, which parses the information and passes it to a Chemkin application code. To specify the needed information, the user writes an input file declaring the chemical elements in the problem, the name of each chemical species, thermochemical information about each chemical species, a list of chemical reactions (written in the same fashion a chemist would write them), and rate constant information, in the form of modified Arrhenius coefficients. The thermochemical information is entered in a very compact form as a series of coefficients describing the species entropy (S), enthalpy (H), and heat capacity (Cp) as a function of temperature. The thermochemical database is in a form compatible with the widely used NASA chemical equilibrium code. Because all of the information about the reaction mechanism is parsed and summarized by the chemical interpreter, if users desire to modify the reaction mechanism by adding species or deleting a reaction, for instance, they only change the interpreter input file and the Chemkin application code does not have to be altered. The modular approach of separating the description of the chemistry from the setup and solution of the reacting flow problem allows the software designer great flexibility in writing chemical mechanism–independent code. Moreover, the same mechanism can be used in different chemically reacting flow codes without alteration.

1.43 Kintecus

After encountering James Ianni's work on Kintecus on the Internet, the author arranged an interview with him at Drexel University, where is a graduate student in metallurgical engineering. The program models the reactions of chemical, biological, nuclear, and atmospheric processes. It is extremely fast and can model over 4000 reactions in less than 8 megabytes of RAM running in pure high-speed 32-bit under DOS. It has full output of normalized sensitivity coefficients that are selectable at any specified time. They are used in accurate mechanism reduction, determining which reactions are the main sources and sinks, which reactions require accurate rate constants, and which ones can have guessed rate constants. The program can use concentration profiles of any wave pattern for any species or laser profile for any hv. A powerful parser with mass and charge balance checker is present for those reactions that the OCR or operator entered incorrectly, yet the model is yielding incorrect results or divergent results. The operator can also create an optional name file containing common names for species and their mass representation. The latter can be used for biological and nuclear reactions. It is also possible to have fractional coefficients for species. It can quickly and easily hold one or more concentrations of any species at a constant level. It has support for photochemical reactions involving hv and Loseschmidt's number and can model reactions from femtoseconds to years. It automatically generates the spreadsheet file using the reaction spreadsheet file. It can do reactions in a continuous stirred tank reactor (CSTR) with multiple inlets and outlets. It can compute all internal Jacobians analytically, which is very useful for simulating very large kinetic mechanisms (more than 1000).

1.44 The CKS Package

This program (Chemical Kinetics Simulator) from IBM is a powerful, accurate, and easy-to-use package for experimental chemists, chemical engineers, and material scientists. Some of its simulation feature include:

- Reaction mechanisms in conventional, user-defined notation;
- Single-value or Arrhenius rate constants;
- User-configurable time, energy, and concentration units;
- Independently selectable temperature, volume, and pressure conditions;
- Capability for isothermal, adiabatic, and programmed temperature reactions;
- Ability for pressure and volume to be held constant or allowed to vary;
- Equilibrium detection and emulation function that can dramatically improve simulator performance;
- No predefined limit on the number of reaction steps or species in a mechanism (except for limitations imposed by Microsoft Windows); simulations can run up to 4 billion data points.

1.45 Groundwater Modeling and Analysis Software

Below are computer programs developed by the Groundwater Remediation Project of the National Water Research Institute of Environment Canada.

- BTEX — A computer plotting program that creates multivariate plots of hydrocarbon analyses of groundwater samples. The concentration of eight different compounds or groups of compounds (toluene, trimethylbenzene, benzene, cumene, sum xylenes, sum ethyl toluenes, ethylbenzene, and styrene) are each plotted on a separate axis to create a "fingerprint" of the analyzed sample.
- DPCT — DPCT simulates contaminant transport in a vertical cross section, accounting for advection, dispersion, diffusion, decay, and cation exchange for a single component, as well as a variety hydrostratigraphic and boundary conditions. The transport of contaminants uses a large number of reference particles, and the dispersive movement of the contaminant is simulated using a random-walk approach to move the reference particles.
- EXPRES — A combination of a knowledge-based system, geographic and pesticide databases, pesticide assessment models, and a graphically based user–system interface. EXPRES will enable a user to undertake a variety of assessments of the potential for groundwater contamination due to the application of pesticide.
- PC-PHREEQEP — A program that permits accurate calculations of mass transfer among aqueous and solid phases in very saline water and brines. To do this the Pitzer equation is used to calculate activity coefficients in high ionic waters.
- PIRS (Pesticide Information and Retrieval System) — A program that consists of an extensive database containing environmental and classification information for approximately 175 pesticides, combined with a user-friendly means of accessing this database. Searches can be made by pesticide name, synonym, or CAS registration number, and the program will retrieve and display numerous values for 10 environmental/chemical properties of the pesticide.
- SUTRAINV — This is an inverse analysis implementation of the SUTRA model of groundwater flow and transport. It links the SUTRA model to a robust method of function optimization.
- TCINV — TCINV is an automated type-curve-matching (inverse analysis) implementation of TYPCURV and RADT models of hydraulic and tracer tests conducted in fractured rock. The component models are equally applicable to groundwater flow and transport in porous media.
- TRIPLOT — This is a computer program that produces trilinear plots of the relative abundance of the major ions (Na, K, Ca, Mg, Cl, SO_4, CO_3, HCO_3) of groundwater analyses. This plotting technique is also referred to as the "Piper trilinear diagram."
- TYPCURV — TYPCURV generates type curves for the analysis of pumping and pulse interference tests in porous and fractured aquifers.

1.46 Process Synthesis

Friedler and Fan have discovered a method for process synthesis. It is an extremely versatile, innovative and highly efficient method that has been developed to synthesize process systems based on both graph theory and combinatorial techniques. Its purpose is to cope with the specificities of a process system. The method depicts the structure of any process system by a unique bipartite graph, or P-graph, wherein both the syntactic and semantic contents of the system are captured. An axiom system underlying the method has been established to define exactly the combinatorial feasible process structures. The method is capable of rigorously generating the maximal structure comprising every feasible possible structure or flow sheet for manufacturing desired products from given raw materials provided that all plausible operating units are given and the corresponding intermediates are known. The method is also capable of generating the optimal and some near-optimal structures or flow sheets from the maximal structure in terms of either a linear or nonlinear cost function. The task is extremely difficult or impossible to perform by any available process synthesis method. Naturally, the optimal and near-optimal flow sheets can be automatically forwarded to an available simulation program for detailed analysis, evaluation, and final selection. Such effective integration between synthesis and analysis is rendered by adhering to the combinatorial techniques in establishing the method. The maximal structure may be construed as the rigorously constructed superstructure with minimal complexity. The superstructure as traditionally generated in the MINLP (mixed integer nonlinear programming) or MILP (mixed integer linear programming) approach, has never been mathematically defined; therefore, it is impossible to derive it algorithmically.

The method has been implemented on personal computers with Microsoft Windows because the search space is drastically reduced by a set of axioms forming the foundation of the method and also because the procedure is vastly sped up by the accelerated branch and bound algorithm incorporated in the method. To date, a substantial number of process systems have been successfully synthesized, some of which are industrial scale containing more than 30 pieces of processing equipment, i.e., operating units. Nevertheless, the times required to complete the syntheses never exceeded several minutes on the personal computers; in fact, they are often in the order of a couple of minutes or less. Unlike other process synthesis methods, the need for supercomputers, mainframe computers, or even high-capacity workstations is indeed remote when the present method is applied to commercial settings. Intensive and exhaustive efforts are ongoing to solidify the mathematical and logical foundation, extend the capabilities, and improve the efficiency of the present method. Some of these efforts are being carried out in close collaboration with Friedler and Fan and others are being undertaken independently. In addition, the method has been applied to diverse processes or situations such as separation processes, azeotropic distillation, processes with integrated waste treatment, processes with minimum or no waste discharges, wastewater treatment processes, chemical reactions in networks of reactors, biochemical processes, time-staged development of industrial complexes or plants, and retrofitting existing processes. Many of these applications have been successfully completed.

Integration is apparent in discussions of computational software. The World Wide Web is one way of providing transparency for users of computational chemistry applications. Another way is the removal of seams between software applications along the drug-discovery spectrum from genomics and bioinformatics to chemoinformatics and drug development. It also occurs in the coupling of computational chemistry and process engineering. The integration is leading computational chemistry and molecular modeling to the desktops of more and more chemists. More cooperative relationships are occurring between computational chemistry software development companies and their customers, and there is an increasing development of technology. Web technology plays an increasing role. One example is that BASF and Silicon Graphics has applied Web technology to provide thermodynamics data to process engineers through the BASF intranet. An engineer has only to click on a Netscape browser Web page and furnish a 3D geometry estimate for the molecules in a reaction. The system calculates reaction thermodynamics data for the ideal gas phase — for example, enthalpy, entropy, or Gibbs free energy — by calculating total quantum chemical energies. The application has a limitation in that it is not feasible for large molecules of more than 100 atoms or very flexible molecules. By using MedChem Explorer, a molecule in a list resulting from a search can be clicked into the viewer for closer examination. The chemist can obtain property data for the molecule with another module; 200 properties, such as dipole moment, log P (octanol partition function), and molar refractivity are available while the chemist is doing other work. A Web-related development at ACD pertains to a Spectral Laboratory Information Management System (SLIMS), which has software programs for predicting properties. An early one included a sketcher called ChemSketch and programs that calculate boiling point, log P, 13C NMR spectra, and 1H NMR spectra. The company has more than 40 software products. In developing software called Jaguar, the company spent time getting numbers accurate to a few tenths of a kilocalorie in quantum mechanics and also much accuracy in solvation calculations. MOE is moving into bioinformatics. It is also going to make a major push in materials, Labute says — with crystal builders, for example — and he expects that the next MOE release will have a comprehensive suite of materials applications.

1.47 Modeling Air Quality

Brock Nicholson in the North Carolina Department of Environment, Health and Natural Resources used the Urban Airshed Model 4 (UAM-4) and an early Environmental Decision Support System (EDSS) on a Cray supercomputer to try out different emission control strategies. The computer power needed to run such photochemical grid models is massive. Modeling a control strategy with Model-3 has been used to simulate SO_2 movement across parts of the U.S. Model-3 runs on a Cray T3D, another powerful computer. The system will incorporate models that use regional to urban scales. It will be used to simulate pollutants in water as well. The Clean Air Act mandates that certain ozone nonattainment areas use an EPA-sanctioned photochemical grid model (the *de facto* standard is Urban Airshed Model-4, UAM-4) to demonstrate that proposed emission control strategies will enable attainment. Before the EPA sanctions the use of Models-3, the system will undergo 6 to 9 months of scientific verification. The Regional Oxidant Model (ROM) and the Urban Airshed Model-V (UAM-V) have also been used to develop regional strategies, but only the UAM-V uses a regional to urban grid that can model pollutants on the wide range of scales necessary to examine ozone locally and regionally.

More and better monitoring is needed. There is a need to check biogenic emissions such as those from trees. There will always be uncertainty in emissions, meteorology, model formation, etc. and people should know how much confidence there is in a model.

1.48 Chemical Kinetics and Photochemical Data for Use in Stratospheric Modeling

This report is available from JPL as Evaluation Number 12 (JPL Publication 97-4) in hard copy or downloaded from the Internet.

Parts of the research described on this Web page and material contained in the report were carried out at the Jet Propulsion Laboratory, under contract with NASA.

1.49 Measuring Air Quality

The PSI (Pollution Standards Index) enables the public to determine whether air pollution levels in a particular location are good, moderate, unhealthful, or worse. The EPA uses the PSI to measure five major pollutants for which it has established National Ambient Air Quality Standards under the Clean Air Act. The pollutants are particulate matter (soot, dust particles), sulfur dioxide, carbon monoxide, nitrogen dioxide, and ozone. For each of these pollutants, the EPA established air quality standards protecting against adverse health effects that can occur within short periods of time (a few hours or a day). The allowable concentration of SO_2 in the air of a community is 0.14 ppm. For ozone, the hourly average concentration permitted has been 0.12 ppm, but it is to be reduced.

The PSI converts these standard readings to 100 (on a scale of 0 to 500). A PSI level in excess of 100 means that a pollutant is in the unhealthful range on a given day; a PSI level at or below 100 means that a pollutant reading is in the satisfactory range.

1.50 Modeling in Leeds

At Leeds University there is excellent work on atmospheric modeling, and Nicola Carslaw, Philip Jacobs, and Mike Pilling have done modeling of OH and HO_2 in the marine boundary layer. The OH radical, though possessing a lifetime of only 1s, is very important in atmospheric kinetics. The Leeds group attempted to reproduce the OH and HO_2 concentrations measured by FAGE (fluorescence assay by gaseous expansion) instruments. The models that reproduce measurements of OH can then tell us much about the chemistry of the boundary layer. The OH radical removes many of the pollutants released as a result of human activities as well as a significant number of biogenic species. Equations show how the OH radical is produced in the troposphere and the OH and HO_2 radicals, with their short lifetime, react quickly to changes in local ambient conditions (solar flux, concentration of NOx and NMHC), but are unaffected by transport.

In this case, it is sufficient to use a zero-dimensional model with no spatial resolution to describe the fast chemical processes. In order to make the model meaningful, frequent measurements of ancillary data in the same air mass are required, and this was the case for recent campaigns at a place called Mace Head. At the Mace Head Atmospheric Research Station many atmospheric measurements were made of OH and HO_2 concentrations by the FAGE method. The concentration of about 30 NMHC (nonmethane hydrocarbons) were also made by the Leeds group in addition to peroxy radicals, NO, NO_2, O_3, HCHO, peroxides, CO, CH_4, NO_3, j(O1D), j(NO_2), halogen oxides, HONO, CO_2, CFCs, HNO_3, SO_2, NH_3, CH_3I, NOx, PAN, aerosol data, meteorological data, and black carbon. Results of both campaigns show a reasonable degree of agreement between the model and measurements for OH, HO_2, and peroxy radicals, although the agreement tends to be better on more-polluted days.

The Atmospheric Research Group in the School of Chemistry at Leeds has several projects, as follows.

Atmospheric hydrocarbons
Spatiotemporal urban air pollutant concentrations
FAGE
Atmospheric modeling
Smog chamber studies

NOx

The latest NO and NO_2 measurements in the U.K. are discussed. These species are responsible for ozone formation in the atmosphere (together with hydrocarbons) and also acid deposition.

O_3

High levels of ozone are generally observed under sunny, summertime conditions in locations where the air mass has previously collected emissions of hydrocarbons and nitrogen oxides. Because of the time required for chemical processing, ozone formation tends to be downwind of pollution centers (i.e., it is usual for the highest ozone levels to be in suburban or rural locations).

CO

The highest concentrations are found close to combustion sources. The atmospheric lifetime is about 1 month with respect to further oxidation to carbon dioxide.

OH

The dominant reaction is with CO and the majority of the remainder is with methane. In the presence of oxygen at concentrations found in the troposphere, the hydrogen atom and the methyl radical react rapidly to form hydroxy and methoxy radicals, respectively.

Studies of aerosol formation from biogenic organic carbon (AEROBIC) have proceeded from the oxidation of natural hydrocarbons in forest areas of Mediterranean countries by a detailed study of aerosol compositions, and of the gas phases in considerable products and in photochemical species.

1.51 Envirofacts Warehouse and Maps on Demand

The Envirofacts Warehouse allows retrieval of environmental information from EPA databases on Superfund sites, drinking water, toxic and air releases, hazardous waste, water discharge permits, grants information, facility index system, latitude/longitude information, master chemical integrator, and the EPA Spatial Data Library. Online queries to retrieve data or generate maps of environmental information by choosing from several mapping applications are available through the EPA Maps on Demand. Database information and spatial data are now (1998) available in convenient special formats. The Envirofacts Warehouse provides query forms that retrieve information from the various databases. The Envirofacts Query generates facility-based reports using information from the environmental databases. Maps on Demand provides several mapping applications which access the EPA Spatial Data Library. One can map the results of queries using Query Mapper or SiteInfo or choose another application to map the information.

Online query forms allow one to generate integrated reports from multiple databases or to search for more-detailed information by querying the databases individually. The databases include latitude and longitude coordinates, including accuracy and description (LRT); an index of facilities regulated by EPA with cross-references to program systems (FINDS); permit, emissions, compliance data on air pollution point sources (AIRS/AFS); information about facilities that discharge pollutants into the waterways (PCS); information about Superfund sites from discovery to listing on National Priorities List (CERCLIS); annual releases of toxic chemicals to air, land, and water (TRI); information about hazardous waste handlers (RCRIS); information used to track, award, administer and monitor grants (CIGS); information on water system violations of the EPA drinking water regulations (SDWIS), and an index of EPA regulated/monitored chemicals with cross-references to program systems (EMCI).

Project fact sheets will include identification information, data quality information, distribution information, metadata reference information, customer service standards, corporate linkage procedures, photo linkage, procedures, model of the Envirofacts database structure, table and column information, law reference, model, and query. Maps On Demand includes facility identification initiative, siteinfo, zipinfo, countyinfo, basininfo, facility density mapper, query mapper product pickup area, EPA GIS national product pickup area, and product pickup area.

1.52 Modeling Global Warming Remediation

Let us suppose that the accumulation of CO_2 alone has been responsible for global warming. We wish then to decrease the concentration of CO_2 in the atmosphere and will attempt to do this by efficient bimolecular collision reactions, not producing or involving toxic species. We must remember that the normal concentration of CO_2 in the atmosphere is quite small and that the temperature rise due to small increases in CO_2 concentration can be quite significant. Using GRI and the associated thermodynamic and kinetic data for all reactions involving CO_2 at 300 K, we will attempt to screen all reactions containing CO_2. Although the temperature may not be 300 K and although tabular data is available from 300 to 3000 K, we will use just this one value for simplicity in the screening shown in the following table.

The enhancement factors, such as those given in the table for the reaction involving $O + CO \rightarrow CO_2 + M$ mean that the reaction rate is increased by that factor when collisions occur between the species cited in the reaction equation and the third-body species in the atmospheric environment.

Now of the reactions listed in the table, the first three have rate constants that favor the reactions, with CO_2 on the left, to the right. The others have CO_2 on the right but do not favor the reaction to the left. The third reaction, however, merely uses CO_2 as the transfer mechanism for conversion of CH_2(s) to CH_2 and does not seem to reduce CO_2 at all. CH_2(s) in the first reaction may indicate solid hydrocarbon particles in the atmosphere and the products of the reaction, carbon monoxide and formaldehyde, are both toxic. The second reaction yields the undesirable carbon monoxide and the interesting HCO. It comes about from the reaction of "CH"with CO_2. It is possible that this indicates that such unsaturated and ring biogenic compounds as isoprene, pinene, limonene, etc., whose approximate stoichiometric formulas approximate CH, would be beneficial, but

Reaction	Δ-S	Δ-H	kf	kr	Keq
ch2(s)+co2->co+ch2o	3.2	−61.1	1.4E+13	9.15E−33	1.53E+45
ch+co2->hco+co	6.0	−65.1	1.07E12	1.97E-37	5.43E+48
ch2(s)+co2->ch2+co2	1.2	-9.0	7.0E+12	1.08E+6	6.47E+06
hnco+oh->nh2+co2	−3.8	−29.3	1.59E+7	4.48E−14	3.54E+20
hnco+o->nh+co2	−1.6	−40.1	1.96E+5	2.82E−24	6.95E+24
o+ch2co->ch2+co2	1.3	−48.4	1.82E+11	5.03E−25	3.62E+35
o2+co->o+co2	−6.7	−8.1	3.77E−23	1.42E−27	2.66E+04
ho2+co->oh+co2	−7.0	−61.2	9.63E−04	8.06E−47	1.20E+43
o+hco->h+co2	−13.6	−111.5	3.00E+13	1.56E−65	1.92E+78
h+h+co2->h2+co2	−23.6	−104.2	6.11E+15	4.38E−60	1.40E+75
o+co+M->co2+M	−34.7	−127.2	3.93E+12	1.32E−77	2.99E+89
nco+o2->no+co2	−3.1	−103.7	5.39E−03	6.90E−78	7.80E+74
nco+no->n2+co2	−9.0	−147.4	1.66E+13	6.97E−93	2.38E+105
Third-body efficiencies:	H_2 Enhanced by 2.0				
	O_2 Enhanced by 6.0				
	H_2O Enhanced by 6.0				
	CH_4 Enhanced by 2.0				
	CO Enhanced by 1.5				
	CO_2 Enhanced by 3.5				
	C_2H_6 Enhanced by 3.0				
	AR Enhanced by 0.5				

then one must be very careful of ozone increase. At least, this could be a start for the right compounds.

In attempting to reduce the amount of CO_2 in the environment, we must at the same time be careful not to raise the concentration of other species that have deleterious effects on health. One species that is bad for health is ozone and one must, at the same time, be watchfully concerned that all species that lead to tropospheric ozone increase are not increased in concentration. To do this, one must know the ozone formation mechanism as well.

Suffice it to say that in a run involving CO_2, two programs were used: Envirochemkin and Thermrxn (a subroutine of THERM) in a remediation attempt to eliminate or reduce benzene. Then, by chance in the calculations it was found that the reaction of CO_2 with benzene was quite feasible to form the product benzoic acid, a relatively innocuous compound that would be solid at ambient temperatures. This then is believed to be not far from the mark of considering that CO_2 will react with terpenes to remove CO_2 from the air. Terpenes are available in the atmosphere, for they are biogenic compounds and are generated from trees (some trees produce more of a specific variety of these ubiquitous terpenes) and these trees or other plant material can use CO_2 as a feedstock to produce more terpenes, thus creating a feedback loop or cycle to help eliminate the CO_2. This idea is also generated from the fact that terpenes have a structure similar to benzene in that they possess carbon-to-carbon aromatic bonds within ring structures. Incidentally, the physical state of these "terpene-CO_2-acids" will make the "sequestration" problem of the CO_2, now abundantly discussed in "remediation of global warming" concepts not necessary. It will also make the design of forests (kind, extent, and dispersion) a more prevalent topic, which has lent itself to an order–disorder treatment in work currently under way by this author.

1.53 A Computer Tool for Simulating Lipophilicity of Organic Molecules

Hansch and co-workers introduced the octanol/water system expressed as log P in 1962. Since then, log P has been used to simulate both transport and distribution in biological systems as well as interactions at receptor sites. Most of the available models have been developed from an insufficiently large training set to ensure acceptable application domains and reliability. Recently, a powerful log P model was developed and embedded into friendly software (Windows 3.1 or Windows 95) and called AUTOLOGP (Version 4.0). The model design has nonlinear modeling and a three-layer backpropagation neural network and is a superior tool for simulating lipophilicity. It also uses autocorrelation vectors to describe a large set of organic molecules from descriptors which are both easy to compute and meaningful in terms of physicochemistry. The model has been derived from more than 7000 molecules.

1.54 HPC, High-Performance Computing and Pollution from Combustion

During a meeting in Ohio in 1994 many papers were presented on the alliance between high-speed computing and combustion. Indeed, many of these posters had to be presented in poster sessions. The following is an abbreviated description of these papers, with the title, author's name(s), author's place of work, and a short synopsis of the work.

"A Lagrangian-Particle Numerical Method for Dense Particulate Flow"
Malcolm J. Andrews and Peter O'Rourke
Department of Mechanical Engineering, Texas A& M University
Group 3T, Los Alamos National Laboratory

A new more accurate numerical method, called MP-PIC has been developed for the calculation of dispersed multiphase flows in which the dispersed phase may be closely packed. The MO-PIC method should enable more accurate calculations of fluidized bed combustion, catalytic cracking, and other industrial processes involving dense particle flows with a large range of particle sizes and types.

"Applications of Envirochemkin"
Stan Bumble, Ph.D.
Consultant and author

A Chemkin derivative, this program was applied to pyrolysis of tobacco, tropospheric ozone generation from stationary (factories) and moving (cars) sources, industrial sources that emit pollution, stratospheric ozone depletion, new remedies for Superfund sites and dangerous spills, global warming, acid rain, reactions of anthropogenic VOC emissions and consequences, predictions of environmental quality in the near and distant future, etc.

"Detailed Modeling of Carbon Combustion Using Elementary and Semi-Global Heterogeneous Reaction Mechanisms"
H.K. Chelliah
Department of Mechanical, Aerospace and Nuclear Engineering
University of Virginia

A detailed description of chemical kinetic mechanisms to better understand the complex interaction between the heterogeneous and homogeneous kinetics to achieve clean and efficient combustion of coal.

"Pollutant Formation in Combustion — Prediction by Computer Modeling"
Michael Frenklach
Department of Materials Science and Engineering
The Pennsylvania State University

A review of combustion chemistry, development of detailed reaction mechanisms, methods of reducing the size of the chemical models, and currrent practices of combining the chemistry with fluid dynamics, illustrated mostly with examples from PAH and soot formation modeling undergoing study in their laboratories.

"Numerical Simulation of Multicomponent, Chemically Reacting Flows with Unstructured Mesh, Finite Element Approach"
J. Kang and F.-L. Chang
Engineering Mechanics Research Corporation

A computational fluid dynamics code, NISA/3D-FLUID, has been developed for the simulation of compressible flow with multicomponent chemical reactions. Chemical kinetics, thermodynamic and transport properties of chemical species/mixtures are assembled by incorporating a chemical kinetics package CHEMKIN and a transport property package TRANSPORT. The present approach provides a simple solution procedure, but yields reasonably good results with adequate convergence performance.

"Advances in High Performance Computing"
Charles Koelbel
Rice University

This paper discusses the state of the art in high-performance computing, particularly in the hardware and software areas. These include vector multiprocessors and massively parallel processors (MPP). Various parallel languages including task-parallel and PVM (parallel virtual machine) library and data-parallel languages highlight the advances in computing.

"Implementation of a Two-Dimensional Flame Model with Detailed Chemistry"
Gopal Patnail, D. Fyfe, and K. Kailasanath
Berkeley Research Associates Department
B. Moon, R. Bennett, A. Sussman, and J. Saltz
University of Maryland

This paper describes physical processes in hydrocarbon flames that are highly complex and that interact in a strongly nonlinear fashion. Numerical experimentation is used to isolate physical processes, study their interactions, and predict important properties such as flammability limits. The highly detailed models that include chemistry and diffusive processes obtain the correct flammability limits.

"A Parallel Algorithm for the Numerical Simulation of an Integro-Differential Forest Fire Propagation"
A. P. Macedo and A. C. P. Brasil, Jr.

Precise calculations of combustion kinetics of biomass burning is important for tropical countries. There is a need to know the amounts of gases released in the atmosphere and the mechanism of forest fires propagation. The models are integrodifferential for the fire propagation and include heat accumulation terms as well as radiative and convection head losses terms. The forest fuel mass decay and gases formation are given by Arrhenius terms. The heat accumulation term is given by a convolution integral that tries to quantify the radiative heat transmission. The wind is very important and is taken care of by advanced mathematics.

"Solution of Nonlinear Partial Differential Equations Using Parallel Inexact Newton Methods"
Michael Pernice and Lu Zhou
Utah Supercomputing Institute, University of Utah
Homer F. Walker
Utah State University

Models of combustion phenomena consist of nonlinear partial differential and integrodifferential equations that couple flow equations with chemical kinetics. Interactions between chemical and physical aspects of combustion occur over a wide range of spatial scales and require very high resolution simulations that are accurate enough to study phenomena of interest. Computational solution of practical problems tax the capabilities of today's most powerful computers. Recent work shows that techniques based on domain decomposition are potentially scalable to hundreds of processors and their extension to nonlinear problems is currently an active area of research.

"LSENS, A General Chemical Kinetics and Sensitivity Analysis Code for Gas-Phase Reactions"
Krishnan Radhakrrishnan, David L. Bittker and C. John Marek
Lewis Research Center
NASA

LSENS solves complex, homogeneous, gas-phase chemical kinetics for many problems, including nonisothermal situations, and contains sensitivity analysis for static problems. Sensitivity analysis eliminates unimportant reactions to increase integration speed. An implicit numerical integration method, which works efficiently for the extremes of very fast and very slow reaction, is used for solving the "stiff" differential equations that arise in chemical kinetics. LSENS is a flexible, convenient, accurate, and efficient solver for chemical reaction problems such as static systems; steady, one-dimensional, inviscid; reaction behind incident shock wave, including boundary layer correction; and perfectly stirred (highly backmixed) reactor. In addition, the chemical equilibrium state can be computed for the following assigned state: temperature and pressure, enthalpy and pressure, temperature and volume, and internal energy and volume. For static problems the program computes the sensitivity coefficients of the dependent variables and their temporal derivatives with respect to the initial values of the dependent variables and/or the three rate coefficient parameters of the chemical reactions. LSENS includes the checking of the legality and sufficiency of all input. Also, at the user's option, the code checks the reaction mechanism for the uniqueness and ensures that each reaction satisfies charge and atom balance requirements.

"The CHAD Computer Library for Automated Design"
Peter J. O'Rourke and Manjit S. Sahota
Group T-3, Los Alamos National Laboratory

Los Alamos is spearheading a multi-institutional development of a library of computer codes called CHAD (*C*omputational *H*ydrodynamics *M*odels for *A*utomotive *D*esign). CHAD effectively utilizes massively parallel computer architectures and the latest mesh generators for design issues. The increased computational power of massively parallel computers offers the promise of improved accuracy by permitting both increased spatial resolution and the use of more-refined submodels.

"Computation of Flow Rate, Temperature and Condensation in a Network of Vent Pipes from Natural Gas-Fired Appliances"
J. H. Saunders, D. B. Philips, J. C. Cross, R. D. Fischer, and A. L. Rutz
Battelle

An interconnecting network of vent pipes carry flue gases from gas-fired appliances in multistoried buildings. Flue gas is introduced to the vent by a forced flow with a fan, or by natural draft flow with buoyancy. The gas is maintained at negative gauge pressure from the buoyancy of the hot flue gases. Designers must ensure that the flue gas products always flow out to the ambient and cannot flow into the building by backflow through other appliances. A model of the mass and energy flows within the venting system is constructed. The program extends an earlier model, VENT-II, which models the venting from a two-appliance system, to consider large numbers of appliances.

"Parallel Computing for Combustion Simulation"
G. Skinner, R. Eigenmann, and D. Schneider
University of Illinois

The Chemkin software libraries are used for evaluating kinetic and transport properties. Because of the interest in using these libraries in real time and CAD environments, where time to solution is critical, their restructuring for parallel execution was undertaken. Premix is a one-dimensional premixed flame simulation that makes use of the Chemkin libraries. The governing tasks divide into evaluating kinetic properties and solving governing equations. However the equations might be solved, property evaluation accounts for most computing time and so represents much of the performance improvement to be gained from parallel computing.

"Reaction Kinetics for Pollution Applications"
Phillip J. Smith
University of Utah
Department of Chemical Engineering

Combustion problems encountered by industry often involve pollution prevention questions. Computational reacting flow simulations are attractive for evaluating engineering options to pollution abatement strategies. The largest challenge to the numerical simulation of reacting flow processes is the need to couple more and more chemical reaction pathways to the other first-order physical processes (radiative heat transfer, turbulent mixing, particulate dispersion, ash deposition, etc.). This presentation will discuss three strategies for pollution abatement: air staging in pulverized-coal flames, selective noncatalytic reduction of NOx and dry sorbent injection for SO_2 removal. The computational challenge, common to each strategy, is to define a minimal set of chemical reaction kinetic steps that adequately describe the controlling chemistry and are sufficiently coupled with the other important physical models to describe real systems.

"Computational Field Simulation for Complex Physics/Complex Geometry"
Bharat Soni and Hugh Thornburg
NSF Engineering Research Center (ERC) for Computational Field Simulation (CFS)
Mississippi State University

ERC for CFS provides U.S. industry with the capability for the computational simulation of large-scale, geometrically complex physics field problems for engineering and design applications. Thus, the academic and research and development efforts of this ERC are organized into six major cross-disciplinary thrusts for continued development of the following areas: solutions algorithms, grid generation, scientific visualization, computer architecture, system software, and integration. The role of high-performance computing in these multidisciplinary efforts is emphasized.

"The Use of Computational Thermodynamics in Energy Conversion"
D. B. Sundman
Max Planck Institute of Material Science
Powder Metallurgy Laboratory
Stuttgart, Germany

Bo Jansson and Malin Selleby
Royal Institute of Technology
Department of Material Science and Engineering
Stockholm, Sweden

Since energy conversion involves reaction between many different materials, it is important to control the formation of pollutants as well as low-melting oxide mixtures and salts. Also, the formation of corrosion products and interactions with linings must be controlled. Thermodynamics and kinetic properties are essential to solve these problems. Thermodynamic databases for liquid and solid mixtures as well as software for dealing with multiphase and multicomponent equilibria are essential tools for simulations of combustion and simulation as well as modeling of the flows and dynamic conditions. A software interface for accessing thermodynamic data has been developed for use in fluid flow simulations and some examples of its use are presented.

"Modeling Combustion Kinetics on High Performance Computers"
Charles K. Westbrook
Applied Physics Division
Lawrence Livermore National Laboratory

Chemical processes contribute to the overall rate of heat release as well as to the production of toxic emissions including oxides of nitrogen. Both represent critical problems that are central to the development of practical combustors. However, heat release and production of trace amounts of toxic species represent opposite demands on computer models for combustion chemistry. Heat release is relatively easy to simulate in a kinetic model. A reduced kinetic model will do provided it has been tested over the range of operating parameters. There are some methods for practical fuels, such as those used in automobile engines. For coal and other solid fuels, additional work must be done. For some problems in which greater chemical detail is necessary, the reaction mechanism can be made much more complex. Production of small amounts of minor intermediate compounds require such treatment. Unfortunately, these are the most toxic chemical species, including those covered by the Clean Air Act, and they are particularly important in practical problems and require detailed treatments.

"Thermal Decontamination of Soils — Dynamic Simulation of Rotary Kiln Incinerators on High Performance Computers"

Karsten Wilbrand
Technische Universitaet
Hamburg, Germany

A dynamic model was developed to understand rotary kiln behavior in the decontamination of soils that are polluted mostly with hazardous organic compounds. It includes transient heat and mass transfer. The rotary kiln incinerator is divided into computational volumes (wall, bed, and gas-phase region). Heat transfer by convection and radiation is calculated between gas and wall, gas and bed, and wall and bed. Vaporization of water, organic compounds, and heavy metals is considered in a simplified mass-transfer model.

"Simulation of the Formation of Multicomponent Particles in Flames by a Discrete-Sectional Model"

Chang-Yu Wu, Pratim Biswas, and Michael R. Zachariah
University of Cincinnati

Multicomponent particles in high-temperature environments include toxic aerosols in combustion systems and nanosized particles for novel materials applications. One needs to understand the particle formation and growth mechanisms to control emissions or produce materials with desirable properties. To understand the dynamics in short time steps, one has to model accurately the behavior of the particles in the cluster state. This can be done with a discrete-sectional model, which has the benefits of keeping the precise information of the small particles while saving computational time.

Part II

2.1 Introduction

The search for new compositions of matter has been an unending, pervasive activity for many centuries. These new states of matter were sought for many reasons — some having to do with their supposed possession of unique properties, some having to do with their precious quality or high economic value, some having to do with their healing value for some very fatal diseases, etc.

2.2 Preface to Synprops

During the 1950s and 1960s, when computers were first becoming useful, a program was written that revealed the structure of molecules which would have optimum properties relating to the atoms of which they were composed and how they were bonded together. A specific application was how such matter would furnish propulsion to solid rocket motors for missiles upon combustion. This and related hardware development was funded by the U.S. Air Force, and the public was unaware of the work because of its classified nature.

It was found that the program was valuable in the chemical process industries as well. It revealed species that were found to appear and cause problems in a chemical process; e.g., it revealed species that did not occur in the pilot plant, but did occur in the commercial plant. The program could help obviate such problems, and could find cheaper reactant or process materials and more profitable products for industrial products. The program could find better (economical and functional) materials and thus guide research and development toward better and more lucrative products.

However, the general nature of the work and the advent of personal computers made the program appropriate for desktop applications for many users where the method could be revised for general applications in industries, universities, hospitals, research laboratories, etc. Other software became available that could be used in conjunction with such a program and to facilitate its execution.

The program is now called SYNPROPS because it synthesizes the properties *you* want by computer. In the 1950s and 1960s, we were not conscious of pollution prevention or waste minimization as we are now. One of the major uses of the program is to design molecules to have a constellation of optimum physical properties and economics with minimum health risk problems. Thus, we are combining two very important concepts here: (1) the concept of maximum profit for the entrepreneur with (2) the concept of maximum safety for plant workers and the community.

These designed molecules can replace unsatisfactory molecules with better ones in industrial processes and can serve as target molecules for organic syntheses. The program also screens out those molecules that will contaminate the environment.

The optimization program determines the structures for molecules that can be either "taken off the shelf" or recommended for synthesis. The methods can be extended to the design of drugs with optimum biological activity and minimum side effects. Antibiotic and antineoplastic drugs have been examined. A matrix method has also been derived that shows how each chemical group affects physical properties and biological activities. A method is portrayed that shows optimum synthesis routes to desired optimum compounds.

The general method employs an array of numbers that were determined by regression on a great many experimental for properties of a great many compounds. The number and types of groups in each molecule are multiplied by numbers in five columns and the sum of these products then leads to five parameters for each molecule. The property of each molecule is then a linear combination of these five parameters with coefficients specific for each property.

There are many auxiliary structure aids within this large matrix. One can fix the number of groups in the target molecule and/or the number of atoms of each element of each compound and/or the nature of the carbon skeleton of the compound. Many other devices can be used.

One can set any number of constraints on any of the groups or properties to be greater than or smaller than or equal to fixed numbers. One can fix numbers to be integers, set a maximum or minimum on any variables, etc., and fix many constraints simultaneously.

A short-cut method was found by expressing the properties as a matrix equation and taking inverses. This method can be used for the cases where the desired properties are known and the structure in terms of parameters is desired. When the parameters are found by matrix algebra they are inserted in the large array and solved for the types and number of groups. It is a backward-chaining type of method.

This book has found the parameters for some 30 groups and set them up in a matrix. By matrix multiplication with the property parameters it then sets up the specific property in a column next to the group the property represents. The columns are then sorted, graphed in a radar-type plot, and printed. The properties are plotted as a function of the groups. If these graphs are printed on transparencies, they can be rotated to show a constellation of properties desired and the structure of the molecules that correspond.

The book also has summaries of computer synthesis of molecules, encoding of optimum molecular structure of molecules for optimum properties, the relationship of the coding to information theory and infosystems, and attempts to lay the groundwork for infinite three-dimensional search engines for computers for a "tree of search" for all structures of matter and their properties.

This brief preface to this work introduces you to the body of work contained herein. It is anticipated that when you embrace this work your costs will be reduced and your profits will rise.

This book contains dozens of graphs of physical properties, Risk-Based Concentrations and biological activities vs. either chemical groups or specific compounds chosen among the most toxic encountered in CERCLA (Superfund). The particular graph chosen is the logarithmic spiral, also known as the spiral of Bernoulli, as found in the nautilus, a genus of cephalopod mollusks of the South Pacific and Indian Oceans with a spiral chambered shell that is pearly on the inside. The history and importance of the spiral is both ancient and associated with famous names such as Hilbert, Wiener, Kolmogoroff, Von Neumann, etc. One must also remember that there are spirals or helices of genetics that make us what we are and, to some extent, what we will be.

Leonard of Pisa (1170–1250), better known as Fibonacci, introduced the *Fibonacci sequence of numbers*

0, 1, 1, 2, 3, 5, 8, 13, 21, 34, 55, 89, 144, ...

in which any member is the sum of the two immediately preceding members. This illustrates the exponential growth characteristic of any uncontrolled animal population.

There is a beautiful circle of ideas connecting various mathematical objects here. The Fibonacci numbers grow at a rate measured by a certain mathematical constant called the Golden Mean. The Golden Mean is thought by many people to describe the most aesthetically pleasing rectangle. Starting with such a rectangle, you can partition it so that it circumscribes a *logarithmic spiral*, which in turn models the shape of a chambered nautilus.

The logarithmic spiral was here chosen because it satisfies a practical principle. One can consider a number of species and, for each, plot the dependent variable and sort the magnitude of these dependent variables from high to low value or from low to high value, as the case may be. Then the peripheral circle of each graph for each property of these compounds or groups can be rotated about an axis through the center of the circle. Now if each graph is made into a transparency and each graph or transparency is positioned exactly on top of the other transparencies, about their common axis, which corresponds to the particular value of each property desired, then the collection of transparencies yields the structure of the compound or molecule with the constellation of properties desired. This is read off the circumference of each accumulated circle, and the totality of the accumulated groups are those from which the desired molecule is formed.

2.3 Summary

This book covers a wide variety of topics to point the way to synthesizing new molecules that will substitute for reactants, intermediaries, and products in industrial processes that will have better physical properties and environmental properties than the original. It does this by a spreadsheet program called SYNPROPS operating on a personal computer with optimization features. A large number of radar-type graphs are presented, one for each property, that visually sort the various groups in order of their contribution to the property and obviate the necessity for a computer to obtain answers for the structure of the optimum molecules for substitution or synthesis. There is also much background and discussion of the use of codes, information theory, advanced optimization techniques, and computer-assisted molecular design in this work. Matrix techniques are discussed that obtain optimal answers and help express preferences and decision making. There is a description of the Schwarz–Christoffel transformation and structure–activity relations as applied from complex variable mathematics. Recipes for Environmental Information Fields are discussed. Many applications to biologically active molecules without side effects, including antineoplastic drugs, are shown. Model compounds and many applications of the SYNPROPS optimization and substitution are demonstrated. The program Envirochemkin is explained. It solves for the chemical kinetics for 100 chemical reactions and 50 species at a time on a personal computer with concentration, and sorting of rate constants, at each instant of time, and many other output files. The inputs are temperature, pressure, mode of reaction (adiabatic, isothermal, etc.), designated intervals of time, and initial concentrations of reactants. The concept of the path probability method is described and there is also a plea for help from outside organizations to obtain good data for the inputs.

2.4 Early History

About 8 years ago, the author was hired as a consultant to the Dow Corning Company. This job was to derive physical properties for some 200 silicon compounds. Most of these were organic in nature and there were some, but not many, experimental sources, in addition to some compilations of valid data, such as DIPPR and a few other reference sources for strictly organic compounds. The author had 6 months to do this work and a portable Compaq computer to achieve the results. When he was finished, a spreadsheet program was used and there were three main categories recommended for the methods to estimate the needed properties. They were (1) the law of corresponding states, (2) group contributions; and (3) statistical thermodynamics.

2.5 The Law of Corresponding States

The law of corresponding states states that all gases behave alike when in corresponding states. This means that when any two out of three variables (pressure, temperature, volume) bear a given ratio to their respective critical values, then the gases are similar. In this regard one defines reduced pressure, reduced temperature, and reduced volume as follows:

$$p_R = p/p_c \tag{2.5.1}$$

$$T_R = T/T_c \tag{2.5.2}$$

$$v_R = v/v_c \tag{2.5.3}$$

where the subscript R denotes reduced and the subscript c denotes critical. The law of corresponding states states that any two gases at the same p_R and T_R will be at the same v_R and will also have the same value of the compressibility factor and thermodynamic properties. Critical properties are frequently found in the literature, but, if not, they can be estimated by a method such as that derived from Guldberg's rule with Lydersen's method. Here we define

$$\Theta = T_b/T_c \tag{2.53}$$

where

$$\Theta = 0.567 + (\Delta_T)^2 \tag{2.5.4}$$

Δ_T values are obtained from a list of contributions in a table. Then with this knowledge, T_b and T_c can be obtained. In a similar fashion Lydersens' groups or critical property increments are given in tabular form for ∇_p and ∇_v and then p_c and v_c can be obtained. Once values for T_c, p_c, and v_c are known, then the respective reduced T_R, P_R, and v_R can be computed, as well as a number of physical properties, since they are a function of the reduced values.

If the theory of corresponding states were rigorous, a single equation of state would suffice for all gases if expressed in reduced properties. Values of compressibility factors as a function of reduced temperature and reduced pressure are shown in a chart. This chart (Hougen and Watson) was derived as the average of data reported in the literature for seven gases. It would not agree exactly with each of the individual gases, as we must remember that the law of corresponding states is only an approximation, but the discrepancies are small enough so that for many purposes the chart may be taken as applicable to all pure gases. Indeed, Bumble has derived the compressibility chart from statistical mechanics utilizing only particles and the forces between them.

2.6 Group Contributions

Among the physical properties that can be derived from group contributions are critical properties, compressibility factors, acentric factors, normal boiling point, liquid molal volume at the normal boiling point, volume from bond distances and the van der Waals radii of atoms, heat of vaporization, average characteristic vibrational-bond wave numbers, ideal gas heat capacities, bond energies, standard heat of formation, heat of combustion, standard entropy of formation, standard free energy of formation, liquid heat capacities, the parachor, molar refraction, surface tension, viscosity, thermal conductivity, etc. Omitted in the above description are the many different equations that convert the tabular group contributions to the physical properties. It is recommended that the References be consulted for details. One of the most comprehensive uses of group properties is that of Benson's for thermodynamic properties. This will be mentioned again in the section on the computer program called THERM.

2.7 Statistical Thermodynamics

The thermodynamic functions for gaseous species may be calculated from spectroscopic constants. The properties are expressed as functions of the internal partition function Q; that is,

$$C_p^0 = T^2 d^2(\ln Q)/dT^2 + 2Td(\ln Q)/dT + 5/2 \quad (2.7.1)$$

$$(H_T^0 - H_0^0)/RT = (\ln Q)/dT + 5/2 \quad (2.7.2)$$

$$S_T^0/R = Td(\ln Q)/dT + \ln Q + 3/2 \ln M + 5/2 \ln T + S_c \quad (2.7.3)$$

$$-F_T^0 - H_0^0)/RT = S_T^0/R - (H_T^0 - H_0^0)/RT = \ln Q = 3/2 \ln M + 5/2 \ln T + S_o - 5/2 \quad (2.7.4)$$

where

$$S_{ST} = 5/2 + \ln[(2m_u kT_1/h^2)^{3/2} kT_1/p_0] \quad (2.7.5)$$

and Equation 2.7.5 is the Sackur–Tetrode constant.

The internal partition function Q in the above equations is given by

$$Q = L_{m=1} Q^m \quad (2.7.6)$$

where Q^m is the internal partition function for the mth electronic state and L is the number of electronic states. In monatomic molecules, internal energy consists of electronic energy only. The above equation then becomes

$$Q = L_{m=1} Q_{e=}^m L_{m=1} (2J_m + 1) e^{-m/kT} = L_{m=1} e^{-m/kT} \quad (2.7.7)$$

For diatomic and polyatomic molecules, Q^m in the above equations involves vibrational and rotational as well as electronic energy. The factored form below is used to calculate Q^m.

$$Q^m = Q_e^m Q v^m Q_R^m Q^m Q^m Q_W^m Q_c^m \quad (2.7.8)$$

or

$$\ln Q^m = \ln Q_c^m + \ln Q_v^m + \ln Q_R^m + \ln Q^m + \ln Q^m + \ln Q_W^m + \ln Q_c^m \quad (2.7.9)$$

The quantities Q_e^m, Q_V^m, and Q_R^m are the electronic, harmonic-oscillator, and classical-rotation contributions to the partition function, respectively. The remaining quantities are the following correction factors: rotational stretching (centrifugal distortion): Q^m; low-temperature rigid rotation: Q^m; Fermi resonance: Q_W^m; and both anharmonicity and vibration-rotation interaction: Q_c^m.

A number of species have internal rotation. Energy levels for internal rotation may be calculated from potential functions as below:

$$V = 1/2\, V_n(1 - \cos n) \quad (2.7.10)$$

where V is the potential, V_n is n-fold barrier, n an integer from 1 to 6, and the phase angle. These energy levels may be used to calculate the internal rotation contributions to the thermodynamic functions. Such computer programs are available.

Ideal Gas Thermodynamic Data for Combustion and Air-Pollution Use for a huge number of species was updated in March 1997. The database author is Alexander Burcat. It can be downloaded from the Internet. The latest print version was by Alexander Burcat and Bonnie McBride "1995 Ideal Gas Thermodynamic Data for Combustion and Air-Pollution Use" Technion Aerospace Engineering (TAE) Report 732, January, 1995. The origins of calculations are M. B. Colket, Chemkin Database, *JANAF Tables*, NASA Lewis Research Center, Thermodynamic Research Center (formerly American Petroleum Research Center), *Russian Tables* (TSIV), and Technion-Israel Institute of Technology. All the input that was used to calculate each species is given with the constants (in NASA format) for each species, including all vibration frequencies, moments of inertia, etc. It is truly a momentous achievement.

2.8 Hierarchical Analysis of Molecular Spectra and Intramolecular Dynamics

There are two models of intramolecular dynamics that are used to make sense of IVR (intermolecular vibrational redistribution). One preferred by spectroscopists, and the more common of the two, pictures the redistribution as occurring through coupled manifolds of states, and this is sometimes called the tier model. A second model preferred by the kineticists, is that the IVR is fast, except for the intramolecular bottlenecks. These bottlenecks might be due to isomerization barriers, or perhaps more bottlenecks in space. More bottlenecks may occur, for example, when the dynamics takes a molecule into a region of phase space where the vibrational frequencies of motion are out of resonance. In theories of unimolecular reaction, it is generally assumed that the IVR is complete and occurs on a timescale fast compared with the rate of reaction. One way in which the unimolecular reaction rate theories may break down is if the rate of IVR competes with the rate of reaction or if the IVR is not complete. Therefore, the study of IVR provides information concerning the applicability of unimolecular reaction rate theory, such as RRKM theory.

The hierarchical analysis of molecular spectra can be employed as an aid in understanding the meaning of a given spectrum, because it points to these features and provides a means to study trends. When a model of a spectrum generated from the eigenstates of the system is available, the hierarchical analysis can yield detailed information concerning assignments of various features of a spectrum and IVR pathways and lifetimes. The utility of this method can be appreciated from the two models discussed in the previous paragraph. When there are bottlenecks to energy transfer, such as weakly coupled tiers of states, the signature of these is a hierarchical structure in a spectrum which can be discerned from the hierarchical analysis.

The motivation for the generation of a hierarchical tree can be appreciated by studying the spectroccopy. Smaller peaks split off larger peaks as resolution is increased, suggesting a genealogy for the lines of a spectrum. By keeping track of this process, a hierarchical tree can be generated including the stick spectrum. The tree in this figure summarizes all the information in the branching as a function of resolution. The nodes that descend from the top of the tree down to a width of 0.7 represent the appearance of the large peaks observed for a middle portion, and the nodes that appear near 0.25 are indicative of the splitting of the small peaks from the large ones.

The hierarchical analysis can be used to understand intramolecular energy transfer pathways. These pathways can be investigated by examining smoothed states following a path down the tree. The reference shows a smooth state generated from a portion of the tree in each of five cases. The first panel on the bottom right shows an eigenstate. The upper right corners of the first five panels have a portion of the tree with a dot on each indicating which portion is used to resolve a group of states which are summed to form the smoothed states. These plots show one pathway for flow out for the overtone state presented in the upper left. Wave function density first moves perpendicularly away from the direction of the main density, indicating energy flow out of the overtone. Then density moves into a 2:1 resonance zone.

2.9 Vapor Pressure Used for the Design of Equilibrium Stage Processes

Vapor pressures can be obtained from groups indirectly. The law of corresponding states can be used to obtain vapor pressures using group contributions. This is possible by using group contributions to obtain critical properties T_c and P_c and also T_b. Then reduced properties T_r and P_r and T_{br} can be obtained. Two methods can then be found to find vapor pressure. One is the reduced Kirchhoff vapor-pressure equation and the other is the Riedel–Planck–Miller vapor-pressure correlation. The first uses an equation derived from

$$\log P_{vpr} = (B/T_c)[(1/T_r) - 1] \qquad (2.9.1)$$

where

$$B/T_c = -h \text{ and } h = T_{br}[\log P_c/(1 - T_{br})] \qquad (2.9.2)$$

which becomes

$$\log Pvpr = h(1 - 1/T_r) \qquad (2.9.3)$$

The second method uses the equation

$$\log Pvp = -(G/T_r)[1 - T_r^2 + g(1 - T_r)^3] \qquad (2.9.4)$$

where

$$G = 0.2471 + 0.4525h \quad 10 < Pvp < 1500 \text{ mmHg} \tag{2.9.5}$$

$$= 0.2271 + 0.4525h \quad Pvp > 1500 \text{ mmHg} \tag{2.9.6}$$

where h is the same as above and

$$g = [h/G - (1 + T_{br})]/(1 - T_{br})^2 \tag{2.9.7}$$

Once the vapor pressures are known, it is possible to calculate K (y/x) factors and relative volatilities ($\alpha_{1\text{-}2} = K_1/K_2 = y_1x_2/x_1y_2$) in order to design distillation towers and other mass transfer equipment and operations.

2.10 Handbook Optimization

The method that will be presented here can predict synthesis design accurately with optimum physical properties, risk-based concentrations, and biological activity and with accurate product ratios from a plethora of starting materials in a process with concomitant and appropriate reaction conditions. The method involves "spinning" the graphs in this handbook (Figures 1 to 40) to the appropriate properties desired, reading the corresponding chemical groups from the graphs, and joining the groups previously present in a data file for computer runs of a program called Envirochemkin at various reaction conditions. If a molecule is new, a program called THERM can derive its necessary thermodynamic functions. The results yield product ratios, conversion, etc. Alternatively, the molecule derived from the handbook graphs can be included in a reaction subprogram of THERM to predict equilibrium constants and ratios of Arrhenius constants as a function of temperature. This latter method is much more approximate as it is not in a milieu of reactions, but it does show the possibility of success. These methods will be illustrated below.

A new approach considered here produces a graph for many physical properties, solubilities in various solvents, risk-based concentrations, and biological activities for 36 chemical groups. Many examples are shown. The type of graph chosen is the radar graph with an imaginary line for each group (or x-axis value) extending from the center to the edge, like the spokes of a wheel; y-axis values (the properties) are graphed on the spokes. Growth trends show as outward spirals. (Figures 3 to 40).

2.11 Cramer's Data and the Birth of Synprops

Cramer's data is in the table of group properties (Tables 1 and 2). Results so obtained were from extensive regressions on experimental data from handbooks and were tested and statistically analyzed. The data were used to predict physical properties for compounds other than those used to derive the data. In this work, optimization procedures are combined with the Cramer data (in an extended spreadsheet), and applied for pollution prevention and process optimization. In addition, risk-based concentration tables from Smith et al. are included as constraints to ensure that the resulting composite structures are environmentally benign.

During the course of many years, scientists have recognized the relationship between chemical structure and activity. Pioneering work was been done by Hammett in the 1930s, Taft in the 1950s, and Hansch in the 1960s. Brown also recognized the relation between steric effects and both properties and reactions. QSAR methodologies were developed and used in the areas of drug, pesticide, and herbicide research. In the 1970s, spurred by the increasing number of chemicals being released to the environment, QSAR methods began to be applied to environmental technology.

Meanwhile, the hardware and software for personal computers have been developing very rapidly. Thus, the treatment of many molecules through their composite groups and the connection with their properties becomes an exercise of obtaining good data to work with. A Compaq 486 Presario personal computer with a Quattro Pro (version 5.0) program (20) was available. In the "Tools" part of the program is an Optimizer program, which was used in this work. The technology of the modern personal computer was matched with the power of mathematics to obtain the following results. The values of the parameters B, C, D, E, and F for 36 compounds were obtained from Table 1 and used to obtain physical properties and risk-based concentrations (shown in Table 3).

2.12 Physical Properties from Groups

It has also been known that a wide range of properties can be derived using The principle of corresponding states which used polynomial equations in reduced temperature and pressure. In order to obtain the critical properties needed for the reduced temperature and reduced pressure, the critical constants are derived from the parameters for the groups of which the molecules are composed.

Thus, the treatment of many molecules through their composite groups and the connection with their properties becomes an exercise of obtaining good data to work with. This is particularly difficult for drug and ecological properties that are not in the public domain.

Cramer's method consisted of applying regressions to data from handbooks, such as the *Handbook of Chemistry and Physics*, to fit the physical properties of molecules with the groups comprising their structures. The results considered about 35 groups and were used in the linear-constitutive model and a similar number of groups (but of a different nature) were used in the hierarchical additive-constitutive model. Statistically, a good fit was found and the prediction capabilities for new compounds were found to be excellent.

In all, 21 physical properties were fitted to the structures. The properties (together with their dimensions) were log activity coefficient and log partition coefficient (both dimensionless), molar refractivity (cm^3/mol), boiling point (°C), molar volume (cm^3/mol), heat of vaporization (kcal/mol), magnetic susceptibility (cgs molar), critical temperature (°C.), van der Waals $A^{1/2}$ (l $atm^{1/2}$/mol), van der Waals B (l/mol), Log dielectric constant (dimensionless), Solubility parameter (cal/cm^3), critical pressure (atm), surface tension (dyn/cm), thermal conductivity [$10^4 \times$ ($cals^{-1}$ $cm^{-2}$$(cal/cm)^{-1}$], log viscosity (dimensionless), isothermal (m^2/mol $\times 10^{10}$), dipole moment (Debye units), melting point (°C), and molecular weight (g/mol). Later the equations for molar volume (Bondi scheme) and molar refractivity (Vogel scheme) were included, as were equations for the log concentration X/water, where X was ether, cyclohexane, chloroform, oils, benzene, and ethyl alcohol, respectively. Risk-based concentrations and biological activity equations were also included. The units of the molar volume by the Bondi technique is 22 cm^3/mol and the other newer equations have dimensionless units.

The hierarchical model (Table 2), shows the parameters for the groups in five columns. This was set up in a spreadsheet and the structure of each molecule was inserted as the number of each of the groups that comprised the molecule. The sum of each column was then called B, C, D, E, and F after the parameters in each column multiplie the number of appropriate groups. In Table 1 column B contains the variables, which are the number of each of the groups denoted in column A, which can be manually set to find the values of the parameters B, C, D, E, and F, or determined automatically by the optimizer program. Column N and O essentially repeat column A and B, respectively, except near the bottom where there are equations to determine the number of gram-atoms of each chemical element for the molecule whose groups are currently displayed in column B. The top and bottom of column O and all of column Q have embedded in them formulas for physical properties, activities or risk-based concentrations in the general linear combination equation

$$P_{ij} = a_i + b_iB_j + c_iC_j + d_iD_j + e_iE_j + f_iF_j \qquad (2.12.1)$$

The i subscripts stand for different properties and the j subscripts indicate different molecules. The values for B, C, D, E, and F are found in cells D111, F111, H111, J111, and L111, respectively, and are linear equations in terms of all the group entries in column B.

It is seen that the spreadsheet (Tables 1 and 2) is like the blueprint of a molecule whose structure is the composite of the numbers in column B and whose properties are given in columns O and Q. The quantities B ... F are the conversion factors of the numbers in column B to the properties in columns O and Q. In this manner they are analogous to the genes (five in this case) in living systems. Values for B, C, D, E, and F are shown for 36 of the most hazardous compounds found on Superfund sites in Table 3.

Three linear graphs are presented that show how the parameters B, C, and D vary with the molecular groups. Also presented are graphs of how the parameters B, C, D, E, and F vary with the groups on spiral or special radar graphs. This is collated for all the parameters and all the groups on one spiral graph. Also, the values for all the hazardous compound is shown on a linear graph. A regression fits the plot of the parameter B vs. the groups on a spiral plot. A good fit was obtained for the parameters C, D, E, and F as well.

The linear model spreadsheet is shown in Table 1. It is similar to another table called the hierarchical model except that it uses groups that are different. The hierarchical model spreadsheet is shown in Table 2.

2.13 Toxic Properties from Groups

The equation derived was

$$-LN(X) = a + bB + cC + dD + eE = fF \qquad (2.13.1)$$

which can also be written as

$$X = \exp(-a) \cdot \exp(-bB) \cdot \exp(-cC) \cdot \exp(-dD) \cdot \exp(-eE) \cdot \exp(-fF) \qquad (2.13.2)$$

where X is MCL (mg/l), or tap water (μg/l), or ambient air (μg/m^3), or commercial/industrial soil (mg/kg), or residential soil (mg/kg).

Graphs for the risk-based concentrations for tap water, air, commercial soil, residential soil, and MCL for the hazardous compounds from Superfund sites are shown.

2.14 The Optimizer Program

The Quattro Pro Program (version 5.0 or 7.0) contains the Optimizer program under the Tools menu. This has been used to optimize structure in terms of a plethora of recipes of desired physical and toxicological properties. Such a program can be used for substitution for original process chemicals that may be toxic pollutants in the environment and also for drugs in medicine that need more efficacy and fewer side effects. These studies can be made while ensuring minimum cost. In order to do this the computer is instructed what the constraints are (= or ≧ or ≦) in the equations, what the variables are, what the constants are, and which variables are constrained to be integers. Conditions are also set up to constrain the number and types of bonds, if desired.

When the Optimizer is called up, a template appears, Table 4, in which you are to name the solution cell, whether you want a maximum or minimum, or none (neither), a target value and also for assigning the variable cells in the spreadsheet. Finally, the constraints are added. These may look like Q1..Q1 Integer, Q2..Q2 ← 5, etc. You may add, change, or delete any constraint. The difficulty is only in understanding what terms such as solution cell, variable cell, etc. mean. There is also an Options choice in the Optimizer Box. In it you can fix the maximum time, maximum iterations, and the precision and tolerance of the runs. It also allows choices for estimates: tangent or quadratic; derivatives: forward or central; and search methods: Newton and conjugate. It allows options for showing iteration results and assumptions of linear and automatic scaling. You can save the model, load the model, and have a report also. The system can use nonlinear as well as linear models. Before proceeding, it is well to set up your variable cells, constraint cells, and solution cells on your spreadsheet. This normally utilizes only a small part of your spreadsheet and the solution will appear within this small part of your spreadsheet that is set aside.

2.15 Matrix Mechanics Methods to Obtain Molecular Structures

A noncomputer program has been found to obtain the molecular structural parameters for species having a constellation of properties that are optimum or those most desired. Once the parameters (B, C, D, E, and F) are obtained they can be inserted in the SYNPROPS spreadsheet program to find the chemical groups of which the molecule is composed. To show this, a simple example is discussed below. Shown in the matrix for the properties boiling point, log viscosity, solubility parameter, melting point, log partition coefficient, and the heat of vaporization. This matrix is called A. Then a column matrix P is formed with the desired values of the physical properties in the same order as in the matrix A. The inverse of A (called A^{-1}), where A must be square, is formed (and it is very convenient to use the Tools/Advanced Math/Inverse operation of the spreadsheet routines for this) and the values of the parameters (B, C, D. E. F) are found by matrix multiplication ($A^{(-1)}P$). This is the structure matrix (S), and it is convenient to use the spreadsheet routine (Tools/Advanced Math/Multiply) for the matrix multiplication.

First, a matrix is constructed where each row is made up of the coefficients (a,b,c,d,e,f) for the linear equation of the parameters (B, C, D, E, F) to the properties.

To review: AS = P, $A^{(-1)}AS = A^{-1}P$. $S = A^{(-1)}P$. If you know values for six properties you desire = P and you have the property equations for these six properties = A, then take the inverse of A = $A^{(-1)}$, multiply by P and S = $A^{(-1)}P$. Now set S = the set: B, C, D, E, and F, and insert these in the optimization program and solve for the groups, leading to the structure.

An example is shown in Table 5 where the five properties T_c, P_c, T_b, Sol. Par., Dipole Moment, and Hv are selected. Two cases are chosen for the properties, one has high values for the six properties as shown and one has low values. The resulting values for B, C, D, E, and F are shown and the lead terms, which must come out to be unity. They are 1.002 and 0.999 as shown. When these values are inserted in SYNPROPS they yield 330 >C<, 20 C=–CH, 43 –F, 56–O– and 58 –COO– (a Bucky-type molecule) for the case with the high-valued properties. Another case gave 358 >C<, 290 –F, and 8 $-CF_3$. Another case was not completed because the lead terms were not close enough to unity, although this, together with the values of B, C, D, E, and F, has been adjusted on occasion to yield credible results.

2.16 Examples of SYNPROPS Optimization and Substitution

Some of the results for the linear model (using 21 groups) are indicated below:

1. Substitutes for Freon-13 can be CF_3Cl (a redundancy) or $CHBRFCH_3$.

	A						P
Bp	66.39	532.5	223.6	−365.4	−250.8	−794.6	99.9965
Log Visc	−0.023	2.44	1.26	−0.94	−8.1	5.06	−0.12626
Sol Par	8.97	5.22	12.1	−15	−12.9	−7.7	8.88631
Tm	77.2	295	194	−443	339	−406	85.3355
log Pc	1.604	3.65	−7.66	−5.74	−0.31	5.09	2.323085
Ht of Vap	8,197	14.92	9.615	−8.07	12.8	14.5	8.491805

A^{-1}

0.0006231	−0.361605	0.0496641	−0.003397	0.045857	0.075749
0.0012345	−0.031063	0.052957	−6.98E-05	0.0043579	0.0465844
−0.000724	0.0638264	0.0026857	0.0013682	−0.090351	0.0095093
0.0011136	−0.152961	−.031226	−.002209	−0.03663	0.0488307
−0.000944	0.0561439	−0.019593	−0.0010133	−0.009134	0.0158744
−0.000944	0.0551439	−0.01004	0.0007503	0.0010546	0.0129371

$S = A^{-1}P$
1.0000028
0.0455056
−0.061003
−0.017995
−0.009002
−0.004004

2. Substitutes for Freon-12 can be CF_2CL_2 (a redundancy) or CHF_2CL.
3. Substitutes for alanine can be $C(NH_2)_3CN$ or $CH(CONH_2)_2CN$ or $CH(CCF_3)BR$ or $CH(CF_3)CONH_2$.
4. A substitute for CH_3CL_3 can be CF_3I.
5. Substitutes for 1,1-dichloroethylene can be CH_2=CHOH and $CH_2=CHNO_2$.

If these substitute compounds do not fit exactly to the desired properties, they can serve as the starting point or as precursors to the desired compounds.

Skeleton compounds were used to find the best functional groups for each property. As examples, the linear model and 21 groups were used with the >C< skeleton (4 groups allowed) and the constraints:

1. T_c is a maximum: $C(\text{–NAPTH})_2(CONH_2)_2$;
2. Critical pressure smaller or equal to 60, boiling point greater or equal to 125, solubility parameter greater or equal to 15: $CF_2(OH)_2$;
3. Heat of vaporization a maximum: $C(CONH_2)_4$;
4. Heat of vaporization a minimum: CH_4;
5. Log Activity Coefficient greater or equal to 6, log partition coefficient smaller or equal to –2, critical pressure equal to 100: $C(CN)_2NO_2CONH_2$;
6. Minimum cost: CH_4;
7. Maximum cost: $C(NAPTH)_4$;
8. Maximum cost with critical temperature greater or equal to 600, critical pressure greater or equal to 100: $C(NAPTH)_2I(CONH_2)$;
9. Minimum cost with critical temperature greater or equal to 600, critical pressure equal to 60: $CH(OH)(CN)_2$.

Results for some of the runs made to ascertain which groups confer maximum and/or minimum properties to a substance follow, using the >C< skeleton. They show COOH for maximum magnetic susceptibility, minimum activity coefficient, maximum log partition coefficient, maximum heat of vaporization, maximum surface tension, and viscosity. NH_2 conferred minimum critical pressure, and maximum activity coefficient. C=O occurred for minimum dipole moment, minimum log partition coefficient, and minimum viscosity, NO_2 occurred for minimum critical temperature and minimum surface tension; CL appeared for maximum dielectric constant; $CONH_2$ appeared for minimum critical temperature; OH appeared for minimum boiling point; and F for minimum heat of vaporization.

An optimization leading to a most desired structure with noninteger values showed 8.67 hydrogen atoms, 1.88 cyclohexane groups, and 5.41 >C< groups. This is a string of >C< groups attached to each other with a proper number of cyclohexane rings and hydrogens attached. This was rounded off to eight hydrogens, two cyclohexane rings, and five >C< groups. Results show that a resulting molecule, cyclopentane with eight hydrogens and two cyclohexane groups appended, satisfies most of the desired physical properties very well.

The hierarchical model was used to find the best substitution model for methyl chloroform or 1,1,1-trichloroethane. It was $CH_3CH_{2.8}(NO_2)_{0.2}$, if the melting point, boiling point, and log of the ratio of the equilibrium concentration of methyl chloroform in octanol relative to its concentration in water are taken from the literature. The result was also obtained by constraining the molecule to be C–C surrounded by six bonds. This is a hydrocarbon in accord with practical results and that of Naser. In the same way, the TCE substitute (constrained to be C=C surrounded by four bonds) was $C{=}CI_{0.383}(CONH_2)_{1.138}(NO_2)_{2.480}$ and the PCE substitute was $C{=}CI_3COOH$.

The precision goal of the fit between predicted and actual risk-based concentrations was for adequate internal program control or constraint purposes. Comparisons between predicted and actual risk-based concentrations for air are shown in Figure 83. Tap water, soils, and MCL results are similar.

A SYNPROPS run that searched for a substitute for carbon tetrachloride, CCL_4, is shown in Table 6. Also tables for freons are shown in the freon tables for (CF_3CH_2F), R125 (CF_3CHF_2), HFC-338mccq ($CF_3CF_2CF_2CH_2F$), R32 (CH_2F_2), and hfc-245fa (CF_3CH_2CHF2) One can print out the intermediate results of a search, where the first result indicates that for the original compound, CCL_4, the second that for an intermediate SYNPROP result sheet on the way to an answer and the last the result that was the one the process found closest to the final answer. This last one cited for a substitute for CCL_4 was a compound with about nine –CH=CH– groups and about eighy $–CH{=}CH_2$ endcap groups indicating a highly olefinic molecule with the air risk concentration rising from about 1.7 to 94,000 with the solubility parameter remaining constant at 10.2.

A similar run with 1,1,1-trichloroethane was performed where the air risk concentration rose from 3.2 to 164, while the solubility parameter remained fairly constant, changing from 8.75 to 8.96. The molecule that was formed had two –CH=CH– groups and two $–CH{=}CH_2$ groups similar to the above but had to add a small amount of the naphthyl group. The molecule $C(NAPTH)_4$ had an air risk concentration of 26,000 and when the unlikely molecule, $C_{14}(NAPTH)_{30}$, was inserted in SYNPROPS, the air risk concentration was 2.6 E + 27 as predicted, indicating that this group needs a revision of data.

A compound such as $CCL_2{=}CCL_2$ can be formed from the molecule in the linear spreadsheet mode by taking 1 $>C{=}CH_2$ and –2 for –H and 4 –CL groups. Thus one can use negative numbers when the need arises. Notice that the air risk concentation here is 0.17 and the solubility parameter is 12.5.

2.17 Cubane

Since cubane (C_8H_8) was predicted by SYNPROPS in some of its optimization runs some mention is made of it here. It is a molecule consisting of eight carbon atoms arranged at the corners of a cube plus single hydrogen atoms sprouting symmetrically from each carbon. The C–C–C bond angle is therefore 90°, rather than 109.5° customary in other hydrocarbon molecules. Because of this, a great deal of strain energy is stored in the chemical bonds. Replacing the hydrogens with other chemical groups results in potentially more salubrious derivatives, some that may be effective in the fight against the AIDS virus, bone marrow cancer, and Parkinson's disease. Cubane exhibits a "plastic phase" (close to its melting point) in which the molecules start to swivel about one or more of their axes. The plastic phase of cubane is not cubic. A model of the intermolecular interactions correctly predicts this structure.

2.18 Biological QSARs Incorporated into SYNPROPS

Such QSARs that were derived elsewhere and incorporated into SYNPROPS include 1/S, BCF, and log K_{oc}. They are for S = solubility, BCF (the solubility in biochemical organisms such as fish over that dissolved in water), and K_{oc} (the adsorption of matter on carbon over the concentration in water), respectively. The bridge to the values of these properties from SYNPROPS is the log P or the log of the partition coefficient calculated automatically in SYNPROPS.

$$\text{Log P} = 1.604 + 3.65B - 7.66C - 5.74D - 0.31E + 5.09F \quad \text{from SYNPROPS} \tag{2.18.1}$$

$$\log \text{BCF} = 0.76 \log P - 0.23 \tag{2.18.2}$$

$$\log K_{oc} = 0.544 \log P + 1.377 \tag{2.18.3}$$

$$\log 1/S = 1.214 \log P - 0.85 + 0.0095(T_m - 25) \quad \text{when } T_m > 25C \tag{2.18.3}$$

The formulas above are all placed in cells in the SYNPROPS spreadsheet and the optimization program then does work with all the physical, environmental, and biological properties very well. As a matter of fact, the Quattro Pro used has changed from version 5.0 to 7.0 since the beginning of the writing of this report and the speed of the optimization program has increased very greatly.

2.19 Optimum Physical Properties with Minimum Toxic Properties: Hierarchical Model

Runs were made constraining MCL to be greater or equal to 1 mg/l, ambient air to be greater than 10 μg/l, tap water greater or equal to 10 μg/l, commercial/industrial soil to be greater or equal to 1000 mg/kg, and residential soil to be greater or equal to 1000 mg/kg. By constraining the number of chemical bonds to four the molecule that was created was methane. When the hydrogen-to-carbon bond was not permitted, CF_2I_2 resulted. When iodine was not permitted, CF_4 resulted. When fluorine was not permitted, CI_4 resulted. In another run, a prior selection was made of a ring compound with six aromatic carbon bonds and five carbon-to-hydrogen bonds. The program selected the COOH group, which results in benzoic acid. This procedure leads to phenylenediamine when two vacancies on the aromatic ring are to be filled. If amines were disallowed, then difluorobenzene is formed, and when fluorine is not allowed dibromobenzene forms. If a double bond between two carbons is the skeleton, four COOH groups are substituted, and, if this is not allowed, four bromines are substituted. When TCDD was studied, the four chlorines were substituted by two aldehyde groups, one SH group and one COOH group. A working graph of this is shown.

2.20 Model Compounds

Small model compounds were designed to learn more about the structure-optimizing process and to discover simple new structures with optimum properties. As an example, the risk-based tap water concentration was maximized, drug 2 was constrained to be above some arbitrary high activity value (5000), all variables were constrained to be greater than zero (positive), and the SUM INT (total number of bonds) was constrained to be equal to three. The optimizing program chose the polymer ($-C(COOH)_2-O-$), i.e., Delrin with the hydrogen atoms replaced with carboxyl groups. Another interesting substance chosen was (CN)n, which is a two- or three-dimensional lattice where every carbon atom is bonded to four nitrogen atoms and every nitrogen atom is bonded to three carbon atoms. This situation comes about because the nitrogen group (–N—) is favored often as is the tetrahedral carbon group and the oxygen group (–O–). Such results depend strongly on the compounds regressed and the care with which this is done. It is later proposed that more classes of molecules be regressed for their quantitative structure vs. physical, risk-based concentration, and biological activities relations and that an organization such as NIST can help with this. It is hoped that "hidden" data, possessed by private organizations, will become publicly available.

2.21 Preferences and Decision-Making Processes

About 30 groups, one at a time, were inserted in the SYNPROPS matrix to obtain the group properties. These properties, for each group, were then laid down as rows, one row for each property. When this was done for all the rows as properties and groups as columns, a 35 × 30 matrix was obtained where the columns were unique for each group and the rows for each property. Each element in the matrix was the number for the property of each group. Since specific matrix operations were to be performed, the matrix had to be square. Therefore, the rows of the less necessary properties were deleted to make the matrix 30 × 30. Let us call this matrix R. Then, a 30 × 1 matrix was constructed in which the elements represented the number of each individual group. The matrix was called Ng. Finally a 30 × 1 matrix called Pd was formed, which contained the desired properties. Then RNg = Pd and $R^{(-1)}RNg = R^{(-1)}Pd$, and $Ng = R^{(-1)}Pd$. Since the elements of Pd are known and the inverse of R, $R^{(-1)}$, can be easily obtained, the matrix Ng contains all the groups required in the molecule for all the properties to be satisfied. In order for this to work, R must be nonsingular and square. Elements of the matrices are shown. These include the values for a, b, c, d, and e and the constant term for each of the 30 properties. Also shown are the properties for each of 34 groups which are sorted from highest to lowest values.

2.22 "Bucky Balls"

In the course of applying the matrix method in several cases when conditions and constraints were imposed, a curious phenomenon occurred. The resulting molecules contained a number of >C< groups. In the cases invoked, they also had a number of fluorine atoms attached to some of the carbons. It was then considered likely that the derivative compounds belonged to the newer class of organic compounds named after Buckminster Fuller.

2.23 Trends in the Parameters and Properties

Several plots were made to see if the statistical model used does correlate with the theoretical precepts of physical chemistry. The plots of B, C, D, E, and F vs. the chemical group number are shown. These are for the most common toxic chemicals found on Superfund sites (about 36 compounds).

There it is shown that the magnitude of the parameters B, C, and D are much larger than E and F. Thus, it is shown that the parameters E and F could have been dropped to realize approximate answers if desired.

A plot of log(X/H_2O) vs. compound number for X = benzene, ethanol, oil, cyclohexane, chloroform, and carbon tetrachloride showed that the shapes of the curves including the maxima and minima are remarkably similar. However, when the log gamma is plotted on the same paper, it does not follow the trends. Yet, when the log partition coefficient is plotted on the same graph, it does follow the trend quite well. This is because the partition coefficient belongs to the same function X/H_2O where X is 1-octanol, whereas the activity coefficient is another concept entirely. The graph for the critical temperature and the normal boiling point again follows the same trend as does the graph for the heat of vaporization, surface tension, molar refraction, and thermal conductivity. This is corroborated as these properties can be estimated using reduced properties from the law of corresponding states. It is also noticeable how closely the solubility parameter follows the plot of the heat of vaporization. This is because the solubility parameter is defined in an equation linearly dependent on the heat of vaporization. A plot of group solubility for log(ethyl alcohol/water), log(oil/water), log(chloroform/water), and log(carbon tetrachloride/water) is shown in Figure 79 and one can study the trends shown there.

2.24 Virtual Molecular Analysis

Referring to previous equations, the parameters A (activity) and S (structure) are both real functions yielding P (property) of a molecule expressed in the complex plane. Thus, we set A equal to the set of numbers a, b, c, d, e, and f, and we set S equal to the set of numbers 1, B, C, D, E, and F in such a manner that a corresponds to unity, b to B, c to C, d to D, e to E, and f to F. The numbers are plotted in an Argand diagram with the A set along the x axis and the S set along the y axis. We thus set the S numbers along the imaginary axis and the A numbers along the real axis. Physically, the property is taken as orthogonal to the structure which has similarity to cases that occur in studies and applications of physics. Thus, P is the constellation of six points (1,a), (B,b), (C,c), (D,d), (E, e), and (F,f). If we connect the points, then the area under the curve is the integral of AdS which is proportional to the property for the structure i and the activity j for the ith particular molecule. This gives a geometric method for comparing properties of molecules. It does not matter what the property is. It could be a physical property or a toxic property. One graph shows such a plot for the tap water risk concentration for chlordane, dioxin-TCDD, ortho-dichlorobenzene, and glycine and another shows the air toxicity for the same four compounds, and in addition that for vinyl chloride. Examination shows that the relative toxicities are proportional to the area under the curves. The next two graphs show the tap water and air risk concentrations for the last 15 of the 36 compounds found on Superfund sites and Table 5 shows the numbers used.

Values of the parameters of B, C, D, E, and F have been plotted vs. the most toxic compounds found on Superfund sites in a radar-type plot and it is found that the "petals" are longest for the B and C parameters for the most toxic compounds among them. Another plot of the parameters B, C, D, E, and F has been made for the groups used in SYNPROPS and again the petals are longest for the B and C parameters.

2.25 Applications to Biologically Active Molecules without Side Effects

Many drugs that are beneficial for fighting disease are also harmful; i.e., they have side effects. As a matter of fact, many parts of these molecules are similar to the hazardous, toxic species that are found in hazardous waste sites, and they have been studied in environmental work using the very programs described here with data cited in the References. It is necessary that part of the molecule should be toxic to organisms causing disease, yet beneficial to the host. A proper balance must be obtained. The programs studied here are ideal for finding such molecules. Caution should be exercised here, as the results are based on limited sets of data and would be more reliable with extensive sets of data.

2.26 Hypothetical Example

The Free–Wilson model is used to treat the biological activities of 18 derivatives in a hypothetical example. An alternative method used was the hierarchical method of Cramer joined with the optimization model on the same data. Although the fit shown in Figure 58 is of course better for the regression of the first method than the second, it must be remembered that the second required only five variables to 12 of the former. Also, the second method was used to optimize and extend the variables beyond the scope (groups) of the original data which the second method was suited to do. When the regression found for the fit of the log of the activity (or 1/C) vs. the five parameters (B, C, D, E, and F) was inserted in the spreadsheet (as drug 1), this could then be optimized for the maximum activity nominated by the groups from the spreadsheet. When the number of groups was constrained to be three, they turned out to be OH groups with the activity as 14.879. Notice that the OH groups were not considered as a candidate in the experiments or observations and that the maximum observed activity had been 6.132. Observed vs. predicted activities are shown in Figure 57.

2.27 Biological Activities of Derivatives of 6-Deoxytetracycline

Only the three added groups were also considered for the second hierarchical model selected. The molecule considered tetracycline with additions at X, Y, and R positions. When the three amino groups ($-NH_2$) were added, the activity was 717.2. When only one $-NH_2$ was allowed, two –I groups were added but the activity dropped to 591.2. When no $-NH_2$ was allowed, three –I groups appeared and the activity dropped to 528.1. When two $-NH_2$ groups were allowed, one –H appeared and the activity was 600.7. When $-C-C-NH_2$ was considered with two –I groups, the activity was 472.8. Three $>N(CH_3)_2$ groups yielded 225.0 However, three $-C(NH_2)_3$ groups led to an activity of 1031.0. Observed vs. predicted values are shown in an additional graph.

A perusal of references then showed the interesting group $-NH-C(=NH)-NH_2$ found on such drugs as streptomycin, streptomycin B, streptomycozid, and deoxydihydrostreptomycin. This group is not one of those in the hierarchical model. However, one can add and subtract groups together with the numerical values of their parameters and doing so for the groups $3NH_2$ + 6NH–3-N— + 3=N– yields the group $-NH-$ $C(=NH)-NH_2$. Trying three such groups on a tetracycline skeleton using the hierarchical model shows that this derivative has an *in vitro* activity against *Staphylococcus aureus* of 1163.2. Observed vs. predicted activities are shown in Figure 58.

Some advantages of using the optimization technique coupled with the hierarchical model of Cramer are that

1. Less variables are required; it was found that frequently not enough observations (experimental data) were found in the literature to run a regression by former methods,
2. The method here allows optimizations to be made; and
3. The method here allows groups other than those used in the experimental observations to be considered in the observations, as the property-parameter equation carries with it the information about the shape, size, etc. of the groups, which may determine the activity of the drug.

2.28 Another Hypothetical Case

The third case studied also was a hypothetical one. In this case only eight observations on eight derivatives were made. The parameters B, C, D, E, and F were derived for these derivatives, regressed vs. the observed activities, and inserted into the spreadsheet as drug 3. The optimization, constrained to three substituents, again chose OH groups with the activity equal to 13.9, compared with 6.0 as the highest observed in the literature. Observed vs. predicted activities are shown in Figure 59.

2.29 Hypoglycemic Activities of Substituted Piperidinesulfamylsemicarbazides

A fourth case involved 10 derivatives of piperidine sulfamylsemicarbazide using substituents in three positions, one of which had provision for two substituents or a cyclic one. Carrying out the same procedure as before, optimization led to four amino groups with a maximum activity found of 489.9. Again, the amino group was not one of those among the selected candidates for optimization. Observed vs. predicted activities are shown in Figure 60.

2.30 Antagonism of *N,N*-dimethyl-2-Bromophenyethylamines to Adrenaline

A fifth study involved 22 derivatives of *N,N*-dimethyl-2-bromophenethylamines, where two substituents were allowed. The activity was measured as antagonistic to adrenaline. The resulting optimum compound was that with two $CONH_2$ groups, and the activity was 11.2. The $CONH_2$ group was not among the candidates previously observed. Observed vs. predicted activities are shown.

2.31 Inhibitory Potencies of Monamine Oxidase Cyclopropylamine

The sixth case involved inhibitors of monamine oxidase. Three substituents were allowed, and 18 derivatives were prepared and their activities observed. The optimization yielded three hydroxyl groups with an activity of 8.73. Experiments on the 18 derivatives had a highest activity of 7.56, and hydroxyl groups had not been considered. Observed vs. predicted activities are shown.

Some of the errors that do occur concern the identification of various groups (or variables) in the hierarchical model (or the linear model). As an example, the hierarchical model could not differentiate between two H substituents at R1, R2 and one

$(CH_2)_6$ substituent at R3 and one H substituent and one CH_3 substituent at R1, R2 and one $(CH_2)_5$ substituent at R3. Both were regarded as six C–C and 14 H in the model, yielding the same result for the parameters B, C, D, E, and F. Yet these two compounds are different. Also in 5f, the substituent $N{=}NC_6H_5$ used the group =N–; however, the group was derived by regression of the data on C=N– and not N=N–. These errors can be corrected by extending the original regression data to more compounds from the literature and also extending the variables to include more of the specific groups used.

2.32 Antineoplastic Drugs

Drugs that are active against various types of cancers and tumors include: (a) DNA-damaging drugs, (b) antimetabolites, (c) antibiotics, and (d) hormonal agents and biological response modifiers. In class a are Leukeran and Alkeran. In class b is Adrucil. In class c are Idamycin, Mutamycin, and Novantone. In class d are Lysodren, Eulexin, and Cytraden. They all have side effects. It is the object of this work to increase the biological activity while decreasing the side effects.

It is interesting to note how similar Alkeran and Leukeran are in structure. The former is identical to the latter except that it has an amino group replacing a hydrogen atom in a position beta to the carboxyl group, and Alkeran also has one fewer carbon atom in the chain containing the carboxyl group. The groups for Alkeran were placed in Table 1 so that the values of the parameters B, C, D, E, and F could be obtained for this compound. This also led to the stoichiometric formula and the quantity SUM INT which is the sum of the number of groups in Table 1 from B74 to B123. Then the optimization was run with the constraints fixed for B, C, D, E, F, SUM INT, x, y, z, w, r in (CxHyOzNwCLr) as was found. The values for B74 to B123 were set equal to unity and the run (about 5 to 10 min) made at a fixed number of fixed iterations and tolerance, etc. The result was Leukeran, except that the tertiary amine became a secondary amine with a chain of four carbon atoms instead of two chains of two carbon atoms, as shown as Compound A.

Another run replaced the two chlorines with sulfur atoms while keeping everything else the same. The result showed that the paraffinic carbon bonds (C–C) dropped by two while the aromatic C–C bonds increased by two. A run was then made where the parameter B was maximized with everything else remaining the same. Here, the iodo bond was added (with the two chloro bonds still present), no aromatic C–C bonds occurred, the molecular weight went up, as did all physical properties, most of which should be favorable to the therapy. This particular maximization of B was tried because, in past work, it was shown that the larger B becomes, the more "biological" the molecule becomes.

Another technique tried was to maintain the skeleton of a known drug and allow the computer to find the peripheral groups that raised the biological activity values of drug 1, 3, 4, 5, and 6 and also the risk concentration values for tap water, air, and the MCL. This reduced the toxicity or side effects and raised the biological activity at the same time. This was performed on antineoplastic drugs shown in Figure 42. Thus for Alkeran the skeleton became

$$(XCH_2CH_2)_2NC_6H_4CH_2CHYZ$$

where the computer program found X, Y, and Z to be three carboxyl (COOH) groups and one nitrile group. Similarly, for Leukeran, one finds

$$(XCH_2CH_2)_2NC_6H_4CH_2CH_2CH_2Y$$

X and Z become two COOH groups and one nitrile group. Incidentally, it is possible that the nitrile group can hydolyze to a carboxyl group in the body, cutting down on side effects after its activity has been effective on the cancer cells and the mechanism induced. In the case of Lysodren

$$XC_6H_5CH(CHX_2)C_6H_5X$$

the four X groups become three COOH groups and one nitrile group. In the case of Eulexin

$$(X)_2CHCONHC_6H_3Y(Z)$$

the four groups X, X, Y, and Z are replaced with a nitro group and three carboxyl groups. In all cases, the sum of the total number of groups was held to that in the original compound so that the number of replacements was equal to the difference between that in the original compound and that in the skeleton. Also, the value of both the drug activities and the risk concentrations were constrained to be unity, an arbitrary value but above that of the original compounds. This was true except for drug 2 and the air risk concentration for air, which may have been arbitrary. There may have been errors in either the experimental data for the second drug or the regression analysis, which made it difficult to retain in the optimization procedure. The air values often did not reach unity but did reach about 0.6 frequently.

Computations made with the linear model gave different groups on the same basis as shown above. Thus, for Alkeran, NH_2, BR, and naphthyl groups were indicated. Leukeran was replaced with a molecule with the end groups $2NO_2$. and NH_2. Lysodren utilized $2NO_2$, NH_2, and CL and Eulexin utilized 2I, NH_2, and C_6H_5. The fact that the groups are different for two models is partially explained by the different group bases. At any rate, there are more options for the researcher by using both models intelligently. Four antineoplastic drugs vs. their toxicity in tap water are shown.

2.33 Jamming the Helix

It is now possible to custom-design small molecules that wedge themselves into specific stretches of the DNA double helix. These molecules could be used to block the expression of harmful genes and perhaps improve gene therapy. The DNA helix consists of four nucleotides adenine (A), tyrosine (T), guanine (G), and cytosine (C). In the ladderlike double helix, an A on one side always pairs with a T on the other, likewise C with G, to form A-T, T-A, C-G, and G-C rungs. Searches have been made for molecules that can be tailored to bind to any chosen sequence of base pair rungs. It is necessary also for these molecules to be able to distinguish T-A bonds from A-T bonds. Such molecules cannot identify and latch onto a specific sequence of rungs, and can squeeze into the minor groove of the double helix, a narrow track that winds around the twisting DNA molecule. The molecule is hairpin shaped. These molecules each target a different sequence. Such sequences would have looked identical to other standard molecules. For instance, one molecule binds 77 times more tightly to TGGACA than TGGTCA. By glomming onto a target sequence, these molecules prevent regulatory proteins from binding and turning genes on. It remains to be seen if the new molecules can get into cells to block harmful genes — a prerequisite for gene therapy.

2.34 Physical Organic Chemistry/ Steric Hindrance

In physical organic chemistry we learn that it is not advisable to plan two functional groups (such as hydroxyl groups or carboxyl groups) on the same carbon atom. We also learn that it is not advisable to have two large bulky groups on adjacent carbon atoms. It is even not advisable to have groups with particular shapes next to each other if they freely rotate. In reactions where such molecules are the products, the reaction may not go and such molecules in many cases cannot exist. How does the program SYNPROPS know this? The fact is it does not. However, operators know these facts. In performing SYNPROPS operators perform runs with optimization and constraints. They prevent such molecules from resulting by how they set up the constraints. As an example, the operator prevents two groups on the same carbon atoms by constraining the number of groups X to be smaller or equal to the number of carbon atoms and prevents the steric crowding of Y groups by constraining the number of Y groups to be smaller than the number of carbon atoms minus one, etc. It is to be remembered that the answer for SYNPROPS is simply the number of groups of each type and it is the duty of the person who runs the program to assemble the groups into the molecule intelligently.

2.35 Environmental Properties

There are other important properties that have not been mentioned so far in this report. Such a property is bioconcentration factor. However, this factor can be approximated by groups by regression with the literature for each organism encountered when such data is available. It also can be approximated by a formula used by Veith which has the log (partition coefficient) as the independent variable. Other properties, such as fate and transport, are accessible by other models and the SYNPROPS regression can also succeed. Models are available for estimating volatilization rate (such as that of Liss and Slater based on the two-film-theory by Whitman, which describes the exchange of a chemical between two adjacent media). At present, the data of DIPPR Project 911 consists of 56 different properties of 601 nonregulated chemicals compiled from four different sources: CAAA (1990), OSHA, EPA-RPM, and a yearly selection of chemicals of special interest to the individual project sponsor. This has been collected under six specific property blocks: oxygen demand parameters, partitioning parameters, general physical properties, vapor–liquid equilibria, fire and explosion parameters, and aquatic toxicity.

2.36 Computer-Assisted Toxicity Assessment

Assessing certain molecular properties before the molecule is even synthesized is perhaps the most beneficial application of computer-aided techniques in chemical design. The most desirable chemical may not necessarily be the one with maximum activity, but rather the one which has maximum activity-to-toxicity ratio. Two of the approaches that have been used to develop tools are the structure–toxicity relationship (QSTR) model approach and the expert system approach.

There are two fundamental approaches to prediction of chemical toxicity from molecular structure. They are

1. Expert system approach, and
2. Linear free energy relationship (LFER) approach

Since TOPKAT 3.0 (based on the LFER approach) was discussed elsewhere, we will here discuss the first approach, the expert system approach.

An expert system is a collection of rules derived from the existing subject knowledge and stored in computer memory. When a chemical is presented to such a system for assessment, the rules associated with the presented chemical are identified from the array of stored rules and a decision is displayed. An expert system is called a human expert system or an artificial expert system depending on whether the rules are framed by human experts or are machine-learned.

The rules in a human expert system may be based on statistically insignificant frequency of an observation, and thus may not be suitable for predictive purposes. Also, since the rules can describe the then current knowledge only, the human expert systems eliminate the opportunity of discovering novel associations between structure and toxicity. The most serious drawback of these systems is that the rules, sometimes in thousands, relate only to "positivity" assuming all chemicals which are not governed by any of the slated rules as "negative." This generally leads to a large number of false positive predictions.

The artificial expert systems, on the other hand, carry out objective mining of the data to deduce rules. However, sometimes the rules from the artificial expert systems may represent false structure–toxicity associations, especially when the rules are built-in terms of mere presence or absence of certain groups of atoms. A pattern expressed in terms of binary digits cannot adequately encode the electronic structure information, let alone the influence of structural modification on this information. Since a toxic response is a result of the transport of a molecule from the point of exposure across biological membranes to the site of action and the electronic interaction with macromolecules, the rules constructed in terms of mere presence or absence of certain groups may lead to false associations, especially when the rules do not account for any synergetic effects of multiple groups.

2.37 Web Elements

There is a very good Web site called Web Elements that contains physical properties for all the elements in the periodic chart. Data for each element are divided into categories, each of which is further divided into subcategories. The categories and their subcategories are listed in a menu panel to the left. If one is not sure in which category the information sought belongs, then a general index is also available. For example, the data for praseodymium were called up. Graphic representations, with tabular properties, values, and units were easily found and the Web site is well developed.

2.38 Physical Property Data Needs in Clean Manufacturing Design

Clean manufacturing and environmental data are needed by industry, the EPA, and CenCITT. Thus, it is necessary to evaluate relevant data and provide access for specific CenCITT projects and all participating investigators. Existing modeling tools must be evaluated and new tools developed to fill data gaps. The DIPPR 911 database must be incorporated into process simulators or modules developed under CenCITT.

The proposed project is in the focus area of pollution prevention design tools. This project will impact on improving process performance predictions, fate assessment, and risk assessment. Also, for selecting a treatment technology the properties of a waste are required. Physical properties are also necessary to develop complete material and energy balances for integrated processes.

A compilation of environmental properties exists for over 1500 compounds (e.g., partitioning properties, group contribution breakdowns, and other data of environmental interest). This database has been put together to support the development of Software to Estimate Physical Properties (StEPP) by a companion CenCITT project under Dr. Mihelcic of MTU.

2.39 Molecular Descriptors

Most of the QSAR equations described up to now have been of two types. One type has been that where a simple equation has been proposed, a simple linear equation with one independent variable which may be determined from experimental data or from one other property, such as log (partition coefficient). The other has been the kind that uses humongous validated data fitted to six independent variables, such as that in SYNPROPS. However, we must remember that Kier and Hall in their work on connectivity had methods for actually describing molecular structure and the results were used in their equations for QSARs with good results. Now the use of molecular descriptors has become popular. S. Cerrini has done work where he calculates quantitative descriptors of molecular shape and molecular symmetry and size from mathematics and first principles. In Chem-X, molecular descriptors are stored as data in fields within the database. Fields are created automatically when either the substituent database or the AutoQSAR facility are used. Fields for other data, including data from external data files, must be created before the data is entered. The substituent database contains traditional 2D QSAR descriptors which are assigned to substituent groups in a series of structurally related compounds. The AutoQSAR facility allows the automatic calculation of a wide range of properties, based on the 3D structure of the molecules. It is also possible to calculate the 3D properties for fragments of molecules, where this is appropriate. C2.QSAR+, G-WHIM, and Hologram QSAR (HQSAR) also work with molecular descriptors. Software computer programs also list FRAGMENT (generator of substructural descriptors), FRAGPROP (fragment-based physicochemical descriptors), HMO (quantum-mechanical descriptors based on the Hueckel method), KAPPA (Kier's indexes of molecular shape), CONNECT (Kier–Hall connectivity indexes), BASAK (topological indexes based on informational content of graph), BALABAN (topological indexes based on generalized distance matrix), and ELEM (descriptors based on molecular gross formula). PRODRG is a program for generating molecular topologies and unique molecular descriptors from the coordinates of small molecules. This is converted to molecular descriptor strings. From this descriptor, a 3D structure can be regenerated using energy minimization. High information descriptors are molecular shape (shape graphs and topology), molecular vibration (EVA), electronic energy levels, molecular composite fields, and flexible molecular alignment using multiple fields.

"Molecular Descriptor Based Physical Property Estimation Method for Pure Organic Compounds"

Cheng Wang presented the above paper at the Spring 1998 AIChE meeting. He pointed out that the group contribution method is one where the property of a compound is estimated as a summation of the contributions of simple molecular groups that occur in the molecular structure. Its major advantages are its simplicity and its general applicability. However, it suffers drawbacks:

1. There is no theoretical basis for the identification of the basis groups.
2. The method ignores the interaction between molecular groups.
3. The effectiveness of the method is limited by the availability of the requisite group contribution constants.
4. The method is not able to capture the proximity effects and isomer effects and does not account for the hydrogen bonding effect.

In order to improve the estimation methodology for physical properties, it is known that the electronic charge density and its associated gradient vector field defines the spatial properties and the morphological features of atoms in the molecules. Thus, the space of the molecular charge distribution is partitioned into disjointed regions, each of which contains an individual atom or group of atoms. Each atom and functional group makes an additive contribution to the value of an observable property of the molecule. Since the surrounding molecular environment does exert influence on the individual functional groups, which in turn changes the topological feature of its charge distribution, the perfect additive scheme employed by group contribution is an unattainable limit. To overcome this limitation, an atomic level formulation is constructed to provide a better description of the structural interaction effects in molecules.

The atomic level formulation employs descriptors that are defined by computational results from electronic structure calculation of the molecule, such as molecular surface area, atomic partial charge, etc.

The major advantages of the molecular descriptors method include

1. The method accounts for all of the structural information provided by the molecule and considers explicitly all of the interactions between molecular groups. It is able to capture the proximity effects, isomer differences and hydrogen bonding effects.
2. Once the coefficients of the model are determined by using a large selective set of data, the only information needed to predict physical properties for a new substance is its molecular structure and atomic partial charge, all intrinsic to the substance.

Therefore, this method leads to pure predictive techniques with great flexibility. However, drawbacks are

1. From the computational point of view, this approach is not as simple as group contribution methods.
2. Unlike group contribution methods, this method cannot handle polymer molecules.
3. This method breaks down when considering small gaseous molecules, such as O_2, O_3, CO_2, etc., since the molecular interaction in these cases is too weak and the descriptor model tends to emphasize more of the interaction effects.

2.40 QSAR Using Interpretative Neural Networks

The applications of the techniques used to achieve predictive solutions with the neural network program AUTONET show that it is possible to extract information using a neural network from relatively small data sets and from networks that are not statistically predictive. The AUTONET method uses a series of multiple, short training neural networks to provide local minima as solutions. The information content is extracted from the coefficients of the hidden weights associated with the input descriptors with the overall solution provided by a consensus of solutions to the local minima. In spite of the short training periods, the neural network memorized random data very quickly. A characteristic profile of an overtrained network was identified as

1. A low training error on the learning set and a high testing error on the test set;
2. Generally (>75%) less than 100 epochs (one complete cycle through the data) in each network; and
3. The hidden weight coefficients similar to each other and similar to the coefficients of known random noise descriptors.

The addition of three descriptors of random numbers allow for a level from which to judge background noise and chance correction. The focus of AUTONET is not to achieve a predictive solution. The scope of the utility of neural networks in QSAR is to gain information in the absence of a predictive solution. The difficulty with neural networks is often their lack of interpretation, and the AUTONET approach addresses this through the visual display of the hidden unit weights and rapidly conveys useful and informative results to the user.

2.41 Indo-U.S. Workshop on Mathematical Chemistry

During the First Indo-U.S. Workshop on mathematical chemistry in January 1998, many papers were presented. About half of them are summarized below.

2.42 "Drug Design by Numerical Modeling of Molecules via Graphs" by Alexandru T. Balaban

Constitutional formulas of organic compounds can be easily modeled by graphs, whose vertices represent atoms and whose edges represent bonds. Heteroatoms and multiple bonding can also be modeled by vertex-colored multigraphs. By means of numerical local vertex invariants (LOVIs) and edge invariants, another step can be taken in the mathematical molecular modeling: these invariants, which can be either integers or real numbers, are combined according to certain algorithms into global numerical descriptors for molecules, usually called topological indexes (TIs). These TIs reflect the topological shape of branching, their cyclicity, and their compactness. Thus, they can serve directly in QSAR or QSPR studies, alone or in combination with other descriptors. Ideally, nonequivalent vertices and TIs should be distinct (nondegenerate) for different molecules. In practice, these goals are difficult to obtain, but there are efforts in this direction. A more recent aim of molecular modeling is to incorporate stereochemical features into the numerical descriptors, a task that can be accomplished successfully for certain types of diasterioisomerism, but enantiomerism raises more difficult problems.

2.43 "Graph Invariants, Molecular Similarity and QSAR" by S. C. Basak, B. D. Gute, and D. Grunwald

A recent trend in chemical graph theory and mathematical chemistry is the characterization of molecular structure and prediction of properties using graph invariants. Numerical graph invariants or TIs are quantitative descriptors of molecular topology. TIs encode information about various aspects of molecular architecture including size, shape, and branching of cyclicity. We have used TIs and atom pairs (APs) in the quantification of intermolecular similarity. Analogs of chemicals selected by such methods have been used for the estimation of properties of chemicals. A hierarchical approach has been developed in the use of TIs, geometric parameters, and quantum chemical indexes in the QSAR studies of various congeneric and structurally diverse sets of chemicals. Graph invariants are used in quantitative molecular similarity analysis (QMSA), analog selection, computational toxicology, and QSARs.

2.44 "The Importance of Hydrogen Bonding Parameters in Quantitative Structure–Activity Relationships (QSAR)" by J. C. Dearden

Hydrogen bonding is an important factor in many aspects of the interactions of xenobiotics with organisms, controlling such features as solubility, partitioning, and receptor binding. However, it is difficult to quantify, and consequently has been used in QSAR mostly as an indicator variable to specify the presence or absence of hydrogen bonding. Attempts at partially and completely quantifying hydrogen bonding contributions have been made, and there is now available a computer program that calculates, simply from a knowledge of chemical structure, several numerical indexes of hydrogen bonding.

2.45 "Application of 3-Dimensional QSAR in Drug Design" by I. Ghosh

Recent developments in molecular modeling methods and its application to the field of drug design emphasize the potential of 3D QSAR and its predictive power. Rational drug design using these tools has emerged as one of the promising fields in medicinal chemistry. The advantage of this method over classical QSAR is the introduction of parameters related to the interaction between the ligand and the receptor. Analyzing the interactions of pharmacore from selective compounds and correlating to their bioactivity has given more predictive power to QSAR-generated models. These parameters could be derived from steric and electrostatic properties of the pharmacore and may be independent of the exact chemical structure.

2.46 "Fujita Ban Analysis of Some Anticancer Glutamine Analogs" by Tarun Jha

In the study, 28 analogs of 5-*N*-substituted-2-(2′,3′,4′,5′-tetrasubstituted benzene sulfonyl)-L-glutamines were synthesized, biologically evaluated as anticancer agents by QSAR methods, and studied through *de novo* models and Hansch methods. Rational design through Fugita Ban analysis shows appreciable correction and proves that the method is equally good as earlier analysis and generation of new leads of such types of glutamines as potential anticancer agents have indicated.

2.47 "The Utility of Hierarchical Model Development in Examining the Structural Basis of Properties" by B. D. Gute, G. D. Grunwald, and S. C. Basak

There has been increased interest in theoretical structural descriptors for modeling activities of compounds because they can be calculated directly from molecular structure and also because of the lack of experimental data. We have used a hierarchical approach for developing QSARs using nonempirical descriptors for a wide range of physicochemical and biological properties for several reasons:

1. To limit the number of independent variables used in model construction to minimize the chance of spurious correlations;
2. To examine the contribution to modeling of distinct sets of indexes, e.g., topostructural, topochemical, geometric indexes, and quantum chemical indexes.

2.48 "Holographic Electronic Density Shape Theorem and Its Role in Drug Design and Toxicological Risk Assessment" by P. G. Mezey

The "holographic" property of molecular electron densities provides a strong foundation for the local, quantum chemical shape analysis of various functional groups, pharmacophores, and other molecular entities. Proofs are given for relevant molecular shape theorem, holographic density shape theorem, and its role in quantum chemical, quantitative shape–activity relations (AshAR). This can be extended to *ab initio* quality electron densities of macromolecules such as proteins based on the transferability and additivity of local, fuzzy density fragments, and the associated local density matrices within the framework of the ADMA (adjustable density matrix assembler) approach. Besides new results on chemical bonding and the development of macromolecular force methods, the new methodologies are also applicable to QSAR studies in computer-aided drug discovery and in toxicological risk assessment.

2.49 "Molecular Supergraph and Subgraph-Based Approaches in QSAR" by I. I. Baskin, E. V. Radchenko, and N. S. Zefirov

One determines the subgraph counts in the molecular graphs representing the training set structures and the use of them as descriptors of molecular structure. The vertices in the subgraphs are classified according to the properties of the corresponding atoms: the valence state, the number of hydrogens at the atom, the formal charge, the bond types, etc. This allows models for predicting physicochemical properties and various types of biological activity. Another approach constructs a supergraph from the topological alignment of the training set structures. By superimposing each molecular graph on the molecular supergraph, the local descriptors (attributed to atoms and bonds) are in a unified manner. Then the structure–activity relationships can be obtained and used for design of new structures. The parameters used as local descriptors are atomic charges, electronegativities, electrotopological states, atomic radii, the occupancy of the position, lipophilicity, H-bonding parameters, bond types, etc. PLS and artificial neural network models based on the approaches here provide better predictive performance than CoMFA and other methods.

2.49a "Inverse Problem in QSAR and Structure Generation" by V. A. Palyulin

The evaluation of structure–activity relationships is frequently done in QSAR studies, but the design of new structures is hardly done at all. The inverse problem in QSAR consists of the generation of new potentially active structures. Several approaches exist. In the case where topological descriptors exist, exhaustive generation of structures with a definite range of a given property (corresponding to a definite range of descriptor values according to the model) is possible. The inverse problem is then solved for first-order connectivity indexes, for κ-indexes of molecular shape, for the centric Balaban index, and for informational indexes. In more complex cases for the models including arbitrary descriptors, the following approach is used. New structures are generated taking into account the features of the training set structures (used for structure–activity model construction) and other restrictions to avoid the combinatorial explosion. Then the structure–activity models are applied as a filter for the generated structures. The set of filters (models) can be used to select the structures having the desired set of predicted activities/properties.

2.50 "The Learned Symmetry Conception in Revealing Quantitative Structure–Activity Relationships with Artificial Neural Networks" by T. V. Mukhina, N. M. Halberstam, V. A. Palyulin, and N. S. Zefirov

In the classical approach to revealing quantitative structure–activity relationships, we use substituent parameters (such as Hammett's and Taft's E_s constants, the lipophilicity p constants, the STERIMOL parameters L, B_1, B_2, etc.) in constructing QSAR equations for congeneric series of compounds. A problem is when several substituent positions are topologically equivalent (for example, two ortho or meta positions in pyridines). A typically used approach for this case consists in using symmetry-adapted functions of substituent parameters, such as the sum of s constants over all topologically equivalent positions, as independent parameters in constructing QSAR equations. The limitation of this approach is that simple sets of symmetry-adapted parameters cannot necessarily discriminate between different compounds and hence predict different values of activity for them.

We propose to (1) expand training sets of compounds several times by adding copies of the same compounds with the same values of activities but with permuted assignment of equivalent substitution positions (for example, 2-methyl-5-ethylpyridine in a data set should be supplemented with 5-methyl-2-ethylpyridine, but both are actually the same compound) and (2) use artificial neural networks for revealing QSARs. In this case neural networks learn to construct nonlinear structure–activity functions with necessary symmetry properties (such functions must be nonlinear in order to meet aforementioned requirements) and to reveal structure–activity relationships simultaneously.

2.51 "Prediction of Chemical Properties and Visualization of Chemical Reactivity by Computer-Aided Chemistry" by D. A. Gallagher

More than 100 properties can be predicted from just a sketch of the molecular structure: UV–visible spectrum, IR spectrum, dipole moment, ionization potential, various thermodynamic and kinetic data, polymer glass transition temperature, log P, susceptibility to nucleophilic/electophilic/radical attack, etc. Computer-aided chemistry can also be used for QSARs and to predict biological properties such as antibacterial activity and carcinogenicity. Visualization of chemical reactivity gives insights into reaction mechanisms, catalysis, and chemical processes unavailable elsewhere. Chemists can now experiment with new catalysts or reaction conditions on a personal computer. Some applications of computer-aided chemistry are listed below:

1. Prediction of environmental fate of pesticides;
2. Design of more ozone-friendly refrigerants;
3. Accelerating the search for better antibacterials;
4. Prediction of rodent carcinogenicity without animal testing;
5. Confirmation that a new street drug binds to amphetamine receptor;
6. Design of new multimode chemical transducers;
7. Elucidation of catalysis mechanisms;
8. Improvement of isomer yield in process and synthetic chemistry;

9. Unraveling thermal degradation mechanisms of polymers;
10. Endorsement of the weatherability of a new polyester.

2.52 "On the Topological Meaning of Physicochemical Properties" by J. G. Alvarez

The hypothesis of this paper is that TDs are able to describe specific physicochemical properties. They are essential magnitudes and are able to describe molecular structure. They can also directly design new compounds with predetermined properties. It is possible to design new lead drugs ignoring the explicit mechanism of action, but only by topological similarity with other active compounds. Experimental evidence of this ability is demonstrated in the design or selection of new leads, such as analgesics, antibacterials, bronchodilators, hypoglycemics, antivirals, antineoplastics, and antimalarials, as well as in the selection of new plaguicides and organic conductors.

2.53 "Computer-Assisted Toxicity Assessment — The TOPKAT Paradigm" by V. K. Gombar

Computational toxicology (CT) tools begin with the drawing of a chemical structure, and then computer-assisted CT systems can paint the toxicity profile, even before it is synthesized. Such a system can screen large combinatorial and virtual libraries for toxicity without any animal sacrifice at a tremendous rate. A CT system can guide in prioritizing testing schedules, thus saving time and money. Providing educated guesses in setting doses for toxicology experiments also seems to be an intelligent application of CT systems. TOPKAT is a CT system based on the principles of LFERs. A collection of robust and cross-validated parsimonious QSARs and associated databases, TOPKAT can provide assessment of sex-species specific carcinogenicity, mutagenicity, developmental toxicity, chronic adverse effect levels, LD_{50}, LC_{50}, EC_{50}. Unique to TOPKAT are algorithms to evaluate whether or not a submitted chemical is within the optimum prediction space of the QSTR model and methods to assign confidence to the predicted toxicity.

2.54 Biomolecular Computer

It is biomolecular devices based on principles of information processing inherent in biological entities dealing with "soft" problems of high computational complexity that could be significantly more powerful than von Neumann computers in solving many problems. Thus we consider biomolecular means as those that should be elaborated for solving problems of high computational complexity.

The information processing inherent in biomolecular and simple biological systems embrace

1. Giant parallelism, vastly exceeding that possible in contemporary digital semiconductor devices;
2. Molecular recognition processes ensuring the directed information transfer in a biomolecular system and excluding therefore the random search of variants;
3. The ability to self-assemble that utilizes shape-based pattern recognition mechanisms;
4. Complicated nonlinear mechanisms of information processing that provide a biomolecular system with ample opportunity to perform complex logical operations equivalent to a large number of binary operations;
5. The hereditary information code that provides biological systems with stability in the process of reproduction and at the same time allows for modification of this information in the process of evolution, that is, in the process of adaptation to changing external stimuli.

The conception of a DNA processor was initiated by Adleman in his paper "Molecular Computation of Solutions to Combinatorial Problems." Adleman showed how to use the unique possiblities of information storage by DNA molecules (in combination with biochemical techniques for transforming this information) to solve the Hamiltonian path problem.

In connection with this, some features of DNA processor should be mentioned. The DNA processor is not really a molecular processor but rather a chemical reaction information-processing device. Thus, molecular mechanisms that control the function of the DNA processor are mechanisms of chemical reactions between different molecular forms of oligonucleotides. They are based only indirectly on molecular information-processing mechanisms. In this sense, the Adleman approach is similar to known approaches and is fundamentally different from the Aviram approach, which used single molecules as switching devices. Because the DNA processor is a chemical reaction device based on giant multicomponent reaction media, some restriction to its practical possibilities should be considered.

The variants of biomolecular device elaboration considered herein are naturally only crude schemes intended to outline possible ways of future development. These variants represent a prospective ground for further investigations. They should be studied in detail at the level of real (hardware) molecular systems, taking into consideration the possibilities of their practical implementation.

2.55 Synthesis of Networks from Given Parameter Matrices

Recalling that P = a + bB + cC + dD + eE + fF, we see that there are five signs in the equation for the ith property of the jth molecule. This sign could be positive or negative. Thus, it is possible that the permutation and combination of these five signs could lead to an indication of all the molecular properties possible. In practice, we find

+ +	
log E	+ + − − −
Dipole moment	+ + − − −
Tb	+ + − − −
Solubility parameter	+ + − − −
Tc	+ + − − −
Hv	+ + − − +
log viscosity	+ + − − +
drug 3	+ + − − +
Tm	+ + − + −
Thermal conductivity	+ + + − −
drug 1	+ + − + −

+ −	
log benzene/H_2O	+ − − − −
log CCL_4/H_2O	+ − − − −
log Partition coefficient	+ − − − +
log Et/H_2O	+ − − − +
log oil/H_2O	+ − − + −
log cyclohexane/H_2O	+ − − + −
log chloroform/H_2O	+ − − − −
log CCL_4/H_2O	+ − − − −
$A^{1/2}$	+ − − + −
Molecular weight	+ − − + −
Volume (Bondi)	+ − + + +
V (exp)	+ − + − −
drug 5	+ − + − +
Molar refraction	+ − + + +

− +	
Pc	− + − − −
drug 6	− + − + −

− −	
Isocom.	− − + + +

Now,

$$\Delta P = \Delta a + B\Delta b + C\Delta c + D\Delta d + E\Delta e + F\Delta f + b\Delta B + c\Delta C + d\Delta D + e\Delta E + f\Delta F \qquad (2.55.1)$$

is the total derivative for the differential of the property, wherein the last five terms of the first line of the formula above is zero when the property remains constant and the molecules vary and the second line above drops out when the molecule remains constant and each differential there goes to zero. The above is also the error function in the property due to errors in each of the structural factors and each of the property parameters.

A network can be realized when its parameter matrix is given. The simplest matrix to construct is a sign matrix. We facilitate the correlation of algebraic sign patterns with geometric tree configurations. In the figure below are six tree configurations for n = 5. We can imagine the branches as being water pipes. If, in the starlike tree, we squirt water into pipe 1 in the reference direction it will obviously flow in positive reference directions in all the four other pipes. Hence in the first row of [G], all elements are plus. If we squirt water in the reference direction through pipe 2, it flows positively through pipe 2 and negatively through the other three. Thus, in the second row of [G], the first two elements are plus and the rest are minus. This method establishes the sign pattern for all the trees. The resulting sign patterns are called "sign matrices." These matrices are shown below:

$$S_a = \left|\begin{array}{ccccc} + & + & + & + & + \\ & + & - & - & - \\ & & + & - & - \\ & & & + & - \\ & & & & + \end{array}\right| \quad S_b = \left|\begin{array}{ccccc} + & + & + & + & + \\ & + & - & - & - \\ & & + & - & - \\ & & + & + & + \\ & & & & + \end{array}\right| \quad S_c = \left|\begin{array}{ccccc} + & + & + & + & + \\ & + & - & - & - \\ & & + & + & + \\ & & & + & + \\ & & & & + \end{array}\right|$$

$$S_d = \left|\begin{array}{ccccc} + & + & + & + & + \\ & + & + & + & + \\ & & + & - & - \\ & & & + & + \\ & & & & + \end{array}\right| \quad S_e = \left|\begin{array}{ccccc} + & + & + & + & + \\ & + & - & - & - \\ & & + & + & + \\ & & & + & - \\ & & & & + \end{array}\right| \quad S_f = \left|\begin{array}{ccccc} + & + & + & + & + \\ & + & + & + & + \\ & & + & + & + \\ & & & + & + \\ & & & & + \end{array}\right|$$

Since these matrices are symmetrical it is necessary only to record signs above the principal diagonal. The procedure for constructing a tree from a given sign matrix is similar to a method developed for determining the graph pertinent to a given cutset matrix.

There is a combinatorial problem of determining the number of possible trees for a given graph and of drawing all their configurations. However, as a by-product of the main topic there is an interesting algebraic method for determining the number of possible trees for any network graph. Knowing their number, it is possible to draw them by trial. Except for row and column interchanges, each sign matrix uniquely specifies a geometric tree pattern and vice versa, and there is a simple way of constructing the tree from a given sign matrix. The linear tree is the only one for which all signs in the [G] matrix are positive. It is the order and definition of the trees that can tell us the nature of the relation of structure to property and also the relationship among the defined properties themselves, as well as the nature of the properties that are as yet unknown.

2.56 Boolean Algebra, Logic Theory, Combinatorial Circuits, and SYNPROPS

A combinatorial circuit depends only on the present input and not on previous inputs or the state of the system. Combinatorial circuits can be constructed using solid-state devices, called gates, which are capable of switching voltage levels (bits). There are AND, OR, and NOT gates. In a digital computer, there are only two possibilities, 0 and 1, for the smallest, indivisible object. All programs and data are ultimately reducible to components of bits. The behavior of a combinatorial circuit can be represented by a logic table, which for a combinatorial circuit lists all possible combinations of input values for a combinatorial circuit together with the resulting outputs. The constraints in any particular SYNPROPS run can be expressed using Boolean algebra as a logic table and/or a combinatorial circuit. Thus, any particular study using SYNPROPS can be done as a complete study using such devices with symbols such as 0 or 1, AND, OR, or NOT gates in a logical fashion to arrive at intermediate or final results.

2.57 Switching Circuits and SYNPROPS

In a similar fashion, a switching circuit can be on or off (open or closed). This also can be represented by the symbols 0 or 1 and by logic tables. There can be combinatorial circuits in parallel or series. This procedure is also applicable to SYNPROPS.

2.58 Sequential Circuits and SYNPROPS

In the combinatorial circuits considered before, the output depended only on the input. Those circuits had no memory. Circuits in which the output depends not only on the input, but also on the state of the system at the time the input is introduced and in which the state of the system is determined by previous processing do have memory. These circuits are called sequential circuits and are important in computer design. This is particularly important for in SYNPROPS one has to select starting values of the groups that can be all zeros, all ones, a mixture of zero or ones, or any mixture of real numbers, and this can affect the course of the iterative run. Also, one can stop SYNPROPS during the course of its run and change the constraints.

2.59 Petri Nets and SYNPROPS

Petri nets are graph models of concurrent processing and can be a method for studying concurrent processing. A petri net is a bipartite graph where the two classes of vertices are called places and transitions. In modeling, the places represent conditions, the transitions represent events, and the presence of at least one token in a place (condition) indicates that that condition is met. In a petri net, if an edge is directed from place p to transition t, we say p is in an input place for transition t. An output place is defined similarly. If every input place for a transition t has at least one token, we say that t is enabled. A firing of an enabled transition removes one token from each input place and adds one token to each output place. Not only do petri nets have relations to SYNPROPS, but also to chemical reactions and flow sheet synthesis methods such as SYNPHONY.

2.60 Hologram Quantitative Structure–Activity Relationship (HQSAR)

In the Introduction, the development of QSARs was described. They can be broadly divided into two generic classes, fragment-based and 3D QSAR. The models provided a good fit to the training dataset but could be improved to gain predictive power and to indicate where changes could be made to improve properties of interest. 3D QSAR techniques allowed receptor site binding into the statistical model, and the CoMFA technique allowed for the biological activity from steric and electrostatic fields that a ligand presents to a receptor. The process is very time-consuming. However, HQSAR easily generates QSAR models of high statistical quality and predictive value and is rapid. It proposes that if the structure of a molecule is encoded within its 2D fingerprint and that structure is the key determinant of all molecular properties (including biological activity), then it should be possible to predict the activity of a molecule from its fingerprint. This is validated by the 2D similarity searching of chemical databases where the similarity measure is the Tanimoto coefficient between fingerprints. HQSAR uses an extended form of fingerprint known as a molecular hologram, which encodes more information, including branched and cyclic fragments as well as stereochemistry than the traditional 2D fingerprint. The key difference, however, is that the molecular hologram contains all possible molecular fragments within a molecule, including overlapping fragments, and maintains a count of the number of times each unique fragment occurs. This process of incorporating information about each fragment, and each of its subfragments, implicitly encodes 3D structural information.

2.61 Integration of Software and Programs and Their Uses and the Movement of Computational Chemistry Further into the Mainstream

As evolution of technology continues, the difference between software tools and their application promises to increase and broaden the use of computational chemistry. Integration can use the World Wide Web technology as a means for providing transparency for users of computational chemistry applications. It may remove boundaries between software applications along the drug-discovery spectrum from genomics and bioinformatics to cheminformatics and drug development. It confers a new modularity designed into a computational chemistry software application that enables speedy inclusion of new features. It may also refer to the coupling of computational chemistry and process engineering.

Web-based technology available for deployment of software tools is expanding and gathering strength. Activity relating to drug discovery is continuing to sweep the software development effort along, and new attention is being paid to the science and technology underlying the software. These efforts result in a developing level of integration that is leading computational chemistry and molecular modeling inexorably to the desktops of more and more chemists. Cooperative relationships between software companies and also between software companies and their customers have led to an array of developing technologies including computational chemistry, molecular modeling, simulation, and chemical information management and delivery. A factor common to the development of software for both computational chemistry and information management is the blossoming use of Web technology for disseminating applications and information.

2.62 Physical Properties of Inorganics, Plastics, Metals, Semiconductors, Ceramics, Composites, and Other Materials

Four basic elements and the relations among them are basic. They are

1. The properties or phenomena that make a material interesting or useful;
2. Performance, the measure of usefulness of the material in actual conditions of application;
3. Structure and composition, which includes the arrangement of as well as the types of atoms that determine properties and performance; and
4. Synthesis and processing, by which the particular arrangements of atoms are achieved.

The creative use of computers can be used to solve research and engineering problems. Materials science and engineering can be actively exploited by using computers in several areas, from the calculation of electronic-based structures, through simulation of nonequilibrium processes, to real-time monitoring and control of processing. Progress can be enhanced through increased emphasis on computer modeling and analysis of materials phenomena and properties based on the underlying physical and chemical principles. Three factors are leading to an explosion of activity, opportunities, and results. The first is increasing speed, capacity, and accessibility of computers and the concomitant decreasing cost of computing. The second is the growing complexity of materials research and manufacturing, and the third is the need in industry to speed the introduction of new designs and new processes into production and to improve production processes and products. This can strengthen fundamental understanding of materials science and engineering and integrate understanding with applications. The ability to analyze composition and structure at increasingly smaller levels is surely one of the great engines of progress of modern materials science and engineering. Of equal importance to materials science and engineering progress today is the ability to control structure and composition in new ways and at new levels of precision.

Numerical simulations, modeling, and calculation from first principles are increasingly used by scientists and engineers as a means to shorten the time to develop understanding and applications of materials. It is almost inevitable that computers will be used increasingly to simulate the properties of new assemblies of atoms or to simulate a new process to shape structural materials. This needs to be encouraged at the level of practice and teaching.

The rapid development of fast and versatile computers provides opportunities for improving materials processing technologies. Online computational control of process parameters can improve product quality and performance, increase efficiency, and reduce costs. Processing technologies such as steelmaking, crystal growth, fabrication of integrated circuits, near-net-shape forming, and rapid solidification can benefit. Materials processing involves unusual and extreme temperature, pressure, strain rate, flow velocity, or other material or system parameters. Phenomena that occur during materials processing are generally complex and nonlinear; they must be modeled with precision in order for the numerical results to be meaningful. Improved process control requires a detailed knowledge of material and system parameters during processing, and improved real-time sensing capabilities are required to monitor parameters such as temperature, composition, and flow speed at many locations in the processing system. Advanced numerical schemes for optimizing materials processing may ultimately be able to include not just the process parameters themselves, but also criteria relating to performance and life prediction. Fully integrated simulations of this kind will require better understanding of the relations between processing and performance.

Three different areas of activity that divide analysis and modeling in materials research can be characterized by length scales at which the properties of materials are being considered. Condensed-matter physicists and quantum chemists deal with microscopic length scales, where atomic structure plays an explicit role. At intermediate length scales, continuum models are appropriate. These include fracture mechanics and microstructural pattern formation in alloys. Then there is work at macroscopic scales, where the bulk properties of materials are used as inputs to models of manufacturing processes and performance.

Classical statistical mechanics has described how huge numbers of atoms collectively determine the bulk properties of materials. The use of pair-wise forces has led to an understanding of phase transitions, the crystalline ordering of solids, the compositional ordering of alloys, the magnetization of ferromagnets, and the computation of phase diagrams for multicomponent metallic alloys. Many of the most fundamentally challenging and technologically important problems, however, have to do with systems that are far from equilibrium.

The distribution of chemical constituents in metallic alloys, or multicomponent ceramic or polymeric materials, can be far from what would be the case if they were to relax by annealing or indefinitely long aging. Predicting relaxation rates and structural changes on the way to equilibrium is important to manufacturability or the performance of such materials. In casting, welding, extrusion, melt spinning, etc., complex patterns are being made to emerge from simple structures, meaning the natural trend toward equilibrium is being reversed.

The advent of more powerful computers and quantum theory with the theory of the cohesive energy of metals make it possible to describe irregular configurations such as crystalline deformations near defects, surfaces, or grain boundaries. Perhaps studies of metastable or strongly disordered states will follow.

The total energy of the many-body system of interacting electrons and nuclei needs to be found and in particular the configuration that minimizes the energy. Now large arrays of atoms (100) in a supercell geometry have their total energy minimized with their electronic and ionic degrees of freedom relaxing toward equilibrium simultaneously. This is called "simulated annealing." It is mathematical optimization theory borrowed from statistical mechanics performed on today's computers.

Rather than calculations for long-range stresses and phenomena, there is a need for combining atomistic calculations — the interface structure — with continuum theories — elasticity, plastic deformation, and so on that predicts interfacial bonding and electronic and magnetic structure of interfaces, which is the case now for special electronic properties.

There is also a need, for example, in epitaxial growth using molecular beam or chemical vapor deposition techniques, to model the kinetics of atoms interacting with a substrate during growth.

In continuum models one must deal with density, temperature, strain, and magnetization. Diffusion equations are to be used to describe transport of heat, composition, or chemical reagents; hydrodynamic equations to describe motion of fluids; and elasticity theory to relate strains in solids to applied stresses. Computers have shown many surprises in the results of continuum calculations.

A molten alloy solidifies when quenched and chemical constituents segregate. It happens when equilibrium thermodynamics predicts stable and homogeneous solutions. This occurs because the liquid and solid phases fall out of equilibrium with each other and chemical constituents rearrange by being driven across the solidification front. The last liquid to solidify can be compositionally very different from that solidified first. An intricate pattern of cellular or dendritic structures results, on the scale of tens or hundreds of microns.

The processes of forming these microstructural patterns and their control are important in determining the grain structure. They affect the mechanical structure and affect the motion of the dislocations. They affect behavior under heating, deformation, aging, or exposure to corrosive environments. The strength of a weld depends on the microstructure in the resolidified material and on microstructural changes in the heat-affected zone. The suitability of a semiconductor for use in electronic devices depends on suppression of microstructural solute segregation.

Usually, solution of the problem involved diffusion equations with simple boundary equations. However, the boundaries move and microstructural patterns are caused by morphological instabilities of these boundaries. Recently, it has been found that the patterns generated by the process are controlled by a delicate interplay between a basic diffusional instability and a number of ostensibly much weaker effects, such as surface tension, crystalline anisotropy, interfacial attachment kinetics, and ambient noise. The instabilities produce intrinsically nonlinear behavior and the weak controlling effects are singular perturbations. Now by the use of mathematical analysis and computers, it is possible to compute the growth rates and geometries of dendritic or cellular solidification patterns.

Fracture mechanics may be even more complicated than solidification in its physical ingredients and applications. The linear theory of elasticity in solids is more computationally complex than theories of heat transport or solute diffusion. Real cracks in real solids need theories of nonlinear elasticity, plastic deformation, and viscoelastic dissipative mechanisms. One needs to consider defects, inclusions, grain boundaries, and the overall shape of the object. Perhaps a quantum theoretic description of atomic bonding is needed.

Fracture mechanics, although complex, has much impact on materials technology because progress has been made without directly confronting mathematical problems as difficult as free boundary problems encountered in solidification. Applications of digital computers to numerical calculation of elastic stresses in irregularly shaped objects subject to forces applied at their fixed boundaries were quite early. These calculations with estimates of elastic limits and yield stresses are important in the design of structural components such as pressure vessels and airplane wings. When brought down to micromechanical length scales, such techniques are used in fracture mechanics to compute the stress intensity at the tip of a crack and, from this, to determine whether the crack will grow.

Microcracks form spontaneously in stressed or heterogeneous materials and lead to crack propagation. They also involve nonequilibrium phenomena; irreversible, dissipative processes play essential roles. They are also free boundary problems and their solutions require detailed analyses of the actual shapes of voids, crack tips, zones of plastic deformation, etc.

Many problems can be incorporated into computer simulations carried out during the initial stages of the design process. One could thus examine possible processing paths to optimize metallurgical microstructures for different properties in different regions of a component. For example, in a turbine disk, the microstructure could be optimized for creep strength at the rim and for low cycle fatigue and ultimate tensile strength in the bore. Such detailed simulations can also address technical feasibility and economics. The models that relate microstructural properties to processing paths can also be used to examine manufacturing options and to optimize ease of maintenance.

Computational computer modeling of the complex processes involved in synthesis, fabrication, deformation, degradation, and failure of materials is very important to performance. The more powerful computer, with new theoretical developments, makes it possible to develop quantitative methods for solving major problems. Accurate codes may become more available to aid in microstructural design of complex materials, such as composites, or for making predictions of lifetimes of structural materials in service. A periodic system reflects periodic variations in properties as a function of atomic number. Only those characteristics that are related to the actual arrangement of the electrons can vary periodically with atomic numbers. The vast number of the chemical and physical properties are dependent upon electronic configurations, and only comparatively few, such as X-ray spectra are dependent upon numbers of electrons alone. Among the properties that are periodic in character are atomic volume, atomic radius, ionic radius, ionization potential, electron affinity, electronegativity, standard oxidation potential of the free element, oxidation state or number, ion mobility, melting point, boiling point, compressibility, optical spectrum, magnetic behavior, heat of formation of a given compound type, parachor, hardness, and refractive index. Predictions as to the magnitude of periodic variations are impossible to make exactly, but it is possible to approach them rather closely.

Hsueh and Ching consider any periodic property determining the periodicity and the other an amplitude factor causing numerical change in the property within a given family of elements. The periodicity factor is a function of valency or outermost electronic configuration, and the amplitude factor is a function of energy state and atomic radius. The periodicity function may be either a maximum at the center of a period or a minimum at the center. Periodic properties of the increasing class are atomic frequency, melting point, boiling point, etc., and those of the decreasing class are atomic volume, atomic radius, atomic parachor, etc. Also, the amplitude function is either parallel or crossing combination, and the 16 periodic families may simultaneously increase or decrease or they may change in reverse order for positive or negative elements. Properties with the parallel combination are atomic volume, atomic radius, ionic radius, and ionization potentials, and those with crossing combination are melting point, boiling point and hardness.

Hsueh and Ching derived the following equation from periodicity and amplitude functions:

$$P = k\ Z^a/n^b[\int VdZ]^c \qquad (2.62.1)$$

where the property P is related to the atomic number Z of the element in question in terms of the valence V, a function of the periodic factor c, the principle quantum number n, and the two parameters a and b, which are constants for a given family of elements but different for different families. Curves plotted from theoretical values so calculated against atomic number agree closely with similar ones drawn from measured values.

Some of the properties that should be considered in this class of materials are

- Mechanical
  - Strain
  - Strength
  - Yield
  - Tensile
  - Ductility
  - Elongation
  - Area reduction
  - Toughness
  - Hardness
  - Modulus of elasticity
- Thermal Characteristics
  - Temperature
  - Heat Content
  - Heat Capacity
  - Specific Heat
  - Thermal Expansion
  - Linear Expansion Cofficient
  - Volume Expansion Coefficient
  - Thermal Conductivity
  - Thermal Diffusivity
- Electrical
  - Conductivity
  - Resistivity

In order to give a representation of the kind of data needed for different industries, here are some examples:

- Ductile Iron Data for Design Engineers
  - Physical properties
  - Density
  - Thermal expansion
  - Thermal conductivity
  - Specific heat
  - Electrical resistivity
  - Magnetic properties
  - Wear resistance
  - Corrosion resistance
- Tensile Properties
  - Modulus of elasticity
  - Poisson's ratio
  - Proportional limit
  - Yield strength
  - Tensile limit
  - Elongation
  - Relations between tensile properties
  - Hardness
  - Tensile properties vs. Hardness
  - Effect of microstructure on tensile properties
  - Effect of graphite shape
  - Effect of nodule count
  - Effect of graphite volume
  - Effect of carbide content
  - Effect of matrix
  - Low-temperature tensile properties
  - High temperature tensile properties
  - Effect of temperature on design stresses
  - Growth and oxidation
  - Effect of environment of tensile properties
  - Fatigue strength
  - Fatigue limit
  - Effect of nodule shape and size
  - Effect of metal cleanliness
  - Effect of matrix
  - Effect of environment
  - Effect of surface conditions
  - Shot peening
  - Surface rolling
  - Surface heat treatment
  - Designing for fatigue applications
  - Thermal fatigue
  - Fracture behavior

- Fracture testing
- Impact properties effect of microstructure
- Effect of composition — carbon
- Effect of composition — silicon
- Effect of composition — manganese, nickel, phosphorus
- Effect of heat treatment
- Modified charpy test results
- Dynamic tear testing
- Temper embrittlement
- Galvanizing embrittlement

Physical Properties for Fibers
- Specific gravity
- Sticking temperature
- Melting temperature
- Ultimate elongation

Other Mechanical Properties
- Modulus of rigidity
- Compressive properties
- Torsional properties
- Damping capacity
- Damping mechanism
- Effects of microstructure

Physical Properties of Liquids
- Melting point
- Boiling point
- Density
- Specific latent heat of vaporization
- Specific heat capacity
- Cubic expansivity
- Viscosity

Selected Physical Properties of Solid Materials
- Melting point
- Density
- Specific latent heat of fusion
- Specific heat capacity
- Linear expansivity
- Thermal conductivity

A list of physical properties includes

- Physical properties
- State
- Absorption
- Internal properties
- Absorptivity
- Anisotropy
- Capillarity
- Compacity
- Consistency
- Density
- Diffusivity
- Dryness
- Explosiveness
- Imperviousness
- Isotropy
- Permeability
- Physical heterogeneity
- Physical homogeneity
- Porosity
- Specific gravity
- Specific surfaces
- Turbidity
- Volatility
- Wettability
- Etc.

Other chemical and physical properties found in libraries for some substances:

- Activation energies
- Activity coefficients
- Boiling point
- Bond dissociation energy
- Bond lengths/bond angles
- Conductance, electrical
- Conductivity, thermal
- Conversion factors
- Critical constants
- Crystal structures
- Density
- Dielectric constant
- Dipole moment
- Dissociation constant
- Electrical conductivity
- Electrode potentials
- Electron affinity
- Enthalpy
- Enthalpy of formation
- Equilibrium constants
- Extinction coefficient
- Free energy
- Freezing point
- Fundamental physical constants
- G-factors
- Gibbs free energy
- Heat capacity
- Heat of combustion
- Heat of dilution
- Heat of dissociation
- Heat of formation
- Heat of fusion
- Heat of ionization
- Heat of mixing
- Heat of neutralization
- Heat of polymerization
- Heat of reaction
- Heat of solution
- Heat of sublimation
- Heat of transition
- Heat of vaporization
- Henry's law constant
- Internuclear distance
- Ionization constant
- Ionization heat

- Ionization potential
- Kinetic data
- Luminescence
- Melting point
- Molar absorptivity
- Optical rotation
- Partition coefficients
- Physical constants
- Rate coefficients
- Refractive index
- Solubility
- Specific gravity
- Specific heat
- Specific rotation
- Stability constants
- Surface constant
- Thermal conductivity
- Thermal expansion
- Thermodynamic properties
- Vapor pressure
- Virial coefficient
- Viscosity
- Physical Properties of Plastics
    - Specific gravity
    - Refractive index
    - Transparency
    - Water absorption
    - Coefficient of friction
    - Flammability
- Mechanical Properties of Plastics
    - Tensile
        - Strength
        - Elongation
        - Modulus
        - Flexstrength
        - Modulus of elasticity in flex.
    - Compressive strength
    - Impact strength
        - Izod (Notched)
        - Gardner
    - Hardness
        - Rockwell
        - Shore
        - Barcol
    - Bending fatigue strength (Woehler)
    - Abrasion resistance
    - Coefficient of static friction
        - Against self
        - Against steel
- Electrical Properties of Plastics
    - Volume resistance (dry)
    - Dielectric strength
        - Short-time
        - Step-by-step
    - Dielectric constant (dry)
    - Dissipation factor (dry)
    - Arc resistance
- Thermal Properties of Plastics
    - Thermal conductivity
    - Coefficient of thermal expansion
    - Specific heat
    - Softening point
    - Heat deflection
    - Brittleness
    - Maximum temperature for continuous use

The design technology of plastics including prediction of performance and endurance is quite different from that of metals and other more traditional materials. The tests for the engineering properties of plastics are different, and so is the interpretation of data produced by the traditional tests, such as tensile strength, stress–strain modulus, and Izod impact. These standard properties are not suitable for material selection or design of parts because they do not account adequately for the effect of time under load, response to sustained temperature exposure, or relate to performance of plastics for the complexity of the applied stress in the designed use. Fundamental engineering properties required to overcome these limitations include strength and rigidity as a function of time, temperature, and stress (creep rupture and creep), as well as realistic measures of strength at high rates of loading (impact) and repeated loading (fatigue).

2.63 Membrane Research and Technology at McMaster University

Existing membranes are being studied for the separation of organics and inorganics from water and the design of membrane systems using reverse osmosis/nanofiltration and pervaporation membranes, particularly for environmentally important applications. Such results are analyzed using mathematical models for transport in polymeric membranes. By using traditional separation processes, such as distillation, design of membrane-based systems is founded on a combination of experimental testing and experience, rather than mathematical calculations. This is due to the lack of a model that can truly predict membrane performance for each new separation problem.

The models being developed are based on fundamental transport equations with a potential function that represents the attractive/repulsive interactions between solute and membrane. Since these interactions can be described from independently developed theory, a model can predict the performance from a minimum of experimental data.

After a model is developed, the equations have to be integrated to describe a large-scale membrane separation system. The approach focused upon is extremely flexible in terms of accounting for changes in feed channel conditions, physical properties, mass transfer coefficients, and pressure drop and allows the independent choice of a membrane transport model. Yet, this model is simple to implement. Research continues to extend this approach to more-complicated problems including mixed solute systems.

Also, membranes for fuel cells/electrolysis that are ion-containing membranes have had their ionic transport mechanism investigated by computer simulation.

2.64 Environmentally Benign Chemical Synthesis and Processing

The NSF has sponsored a program to support research projects in chemical engineering and chemistry aimed at reducing pollution at its source. Active participation by scientists and engineers in the chemical, pharmaceutical, agrichemical, electronics, and energy industries dependent on chemistry-based manufacturing is required. Chemicals that go up the stacks and down the drains are both a serious detriment to the environment and a waste of valuable natural resources. By-product production and pollution at the manufacturing source can be minimized. This joint initiative joins researchers in designing environmentally benign synthesis and processing methodologies which will undergird the clean, efficient, and profitable industries of tomorrow in the production of chemicals, pharmaceuticals, plastics and fibers, electronics, agriculture, and energy. The joint initiative of the NSF and the CCR (Council for Chemical Research, Inc.) is intended to finance long-range research supporting new synthetic methodologies, novel process designs, and process improvements that reduce the potential for environmental release in the first place. Projects are sought that can lead to feedstock substitutions, alternative synthetic and separation procedures, more specific and efficient catalysts, product reformulations, and process modifications that would minimize by-product formation and reduce waste production at the source.

Some appropriate problems and areas of research are illustrated by the following examples:

- New chemistries and methodologies for on-demand, on-site production and consumption of toxic intermediates for use in manufacturing;
- New, more highly selective catalysts to increase product yield and reduce byproduct formation in chemical manufacturing;
- Low-energy separation technologies for feedstock purification and recycling;
- Improved membranes and membrane/molecular sieve technologies that integrate transport and reaction to enhance specificity;
- Alternative chemical syntheses that bypass toxic feedstocks and solvents such as chlorocarbons and nitrocarbons;
- New processing methods that eliminate production of hard-to-trap micron-sized aerosols;
- Alternative chemical syntheses that eliminate or combine process steps, resulting in a net reduction of pollutants;
- Design of novel low-temperature or other energy-efficient methods for chemical synthesis and processing.

2.65 Research in Electrochemical Synthesis

Another program sponsored by the NSF was electrochemical synthesis defined as the production of chemical products or materials using electricity as the driving or controlling factor. Understanding of the following must be improved: electrode processes including mass transport; dynamics of electron transfer; chemical transformations induced by electrical charge or electric fields; generic principles for design, scaling, and operation of electrochemical reactors; and the identification of electrically induced chemical transformations that may form the basis of electrosynthetic processes. An additional objective is to facilitate integration of this knowledge into industrial design and process methodology and to lower the barriers that separate electrochemical technology from more traditional chemical technologies.

This initiative supports research leading to opportunities for the use of electrochemical synthesis broadly defined, including

- Synthesis of high-value-added products by electrochemical methods;
- Electrochemical production of commodity chemicals offering advantages in cost or availability of raw materials, energy utilization, or pollutant reduction;
- Production of chemicals with concomitant generation of marketable amounts of electricity or other energy;
- Use of electrochemical techniques to destroy by-products or to convert unwanted by-products or other waste materials into useful products; such byproducts include the products of combustion;

Novel electro-organic syntheses, including polyfunctional molecules and new feedstocks;
Coupling of organometallic chemistry to electrosynthesis;
Electrochemistry of small molecules, such as sulfur dioxide, carbon monoxide, nitrogen oxides;
Modification of electrode surfaces to improve selectivity or catalytic efficiency;
Kinetics and mechanisms of processes at electrode surfaces;
Electrosynthesis of composite or layered materials, including those used in electronic applications, by metal plating, electrodeposition of polymers, or intercalation reactions;
Electrochemical processes or techniques for recycling of process streams to achieve zero effluent;
Design or synthesis of more stable and more selective electrocatalysts and membranes;
Engineering principles for scale-up of electrochemical processes, including design and integration;
Electrowinning of light metals.

2.66 Design of Molecules with Desired Properties by Combinatorial Analysis

Suppose that a set of groups to be considered and the intervals of values of the desired properties of the molecule to be designed are given. Then, the desired properties constitute constraints on the integer variables assigned to the groups. The feasible region defined by these constraints is determined by an algorithm involving a branching strategy. The algorithm generates those collections of the groups that can constitute structurally feasible molecules satisfying the constraints on the given properties. The molecular structures can be generated for any collection of the functional groups.

The proposed combinatorial approach considers only the feasible partial problems and solutions in the procedure, thereby resulting in a substantial reduction in search space. Available methods exist in two classes. One composes structures exhaustively, randomly, or heuristically by resorting to expert systems, from a given set of groups; the resultant molecule is examined to determine if it is endowed with the specified properties. This "generate-and-test" strategy is usually capable of taking into account only a small subset of feasible molecular structures of the compound of interest. It yields promising results in some applications, but the chance of reaching the target structure by this strategy can be small for any complex problem, e.g., that involving a large number of groups. In the second class, a mathematical programming method is applied to a problem in which the objective function expresses the "distance" to the target. The results of this assessment may be precarious since the method for estimating the properties of the structure generated, e.g., group contributions, is not sufficiently precise. The work here is a combinatorial approach for generating all feasible molecular structures, determined by group contributions, in the given intervals. The final selection of the best structure or structures is to be performed by further analysis of these candidate structures with available techniques.

2.67 Mathematical Background I

Given:

a. Set G of n groups of which a molecular structure can be composed,
b. The lower bounds, p_j terms, and the upper bounds, P_j terms of the properties to be satisfied, where j = 1, 2, ..., m;
c. Upper limit L_i (i = 1, 2, ...,n) for the number of appearances of group i in a molecular structure to be determined; and
d. Function f_k (k = 1, 2, ..., m) representing the value of property k which is estimated by the group contribution method as

$$f_k(x_1, x_2, \ldots, x_n)$$

In the above expression, x_1, x_2, ..., x_n are, respectively, the number of groups #1, #2, ..., and #n contained in the molecular structure or compound.

The problem can now be formulated as follows:

Suppose that f_k(k = 1, 2, ... m_i) is an invertible function on the linear combinations of coefficients a_{ki} (i = 1, 2, ..., n), on $\Sigma a_{ki}x_i$. Furthermore, assume that function f_k(k = m_i + 1, m_i + 2, ..., m_2) has a sharp linear outer approximation, i.e., there are coefficients a_{ki} and a_{ki}' such that

$$\Sigma a_{ki}x_i \geq f_k(x_1, x_2, \ldots x_n) \leq \Sigma a_{ki}'x_i \quad (k = m_i + 1,\ m_i + 2,\ \ldots,\ m_2) \qquad (2.67.1)$$

We are to search for all the molecular structures formed from given groups, #1, #2,... and #n, whose numbers are x_1, x_2, ... x_n, respectively, under the condition that the so-called property constraints given below are satisfied.

$$p_j \leq f_j(x_1, x_2, \dots x_n) \leq P_j \quad (j = 1,2\dots, m)$$

Throughout this book, the constraints imposed by the molecular structure on feasible spatial configurations are relaxed, and the molecular structures are expressed by simple connected graphs whose vertices and edges represent, respectively, the functional groups from set the G and the associated bonds. Thus, the set of such connected graphs needs to be generated from the set of functional groups G that satisfies the property constraints admitting multiple appearances of the functional groups.

In the conventional generate-and-test approach, all or some of the connected graphs, i.e., structurally feasible graphs, are generated from the available functional groups and are tested against the property constraints. This usually yields an unnecessarily large number of graphs. To illustrate the inefficiency of this approach, let the structurally feasible graphs be partitioned according to the set of functional groups of which they are composed; in other words, two graphs are in the same partition if they contain the same groups with identical multiplicities. Naturally, all the elements in one partition are either feasible or infeasible under the property constraints. Moreover, the graph generation algorithm of this approach may produce all elements of the partition, even if an element of this partition has been found to be infeasible earlier under the property constraints; obviously this is highly inefficient.

2.68 Algorithmic Generation of Feasible Partitions

The feasible partitions can easily be generated for the problem defined in the preceding section by a tree search algorithm similar to the branch-and-bound framework.

A novel approach is proposed in the present work which is substantially different from the generate-and-test approach. It first identifies the feasible partitions satisfying the property constraints as well as the structural constraints; this is followed by the generation of the different molecular structures for each of the resultant partitions. The proposed approach is more effective than the generate-and-test approach because each partition need be considered only once, and the algorithm for generating molecular structures is performed only for a feasible partition. In addition, the approach can be conveniently implemented by means of a tree search.

Suppose that the values of variables $x_1, x_2, \dots x_k$ (k n – 1) are fixed a priori as $l_1, l_2, \dots, l_k$ at an intermediate phase of the procedure; then, the problem is branched to L_{k+1} partial problems for $l_{k+1} = 0, 1, 2, \dots, L_{k+1}$ according to the following two cases.

Case 1. $k \leq n - 2$:

$$p_j' \leq \Sigma a_{ji} l_i + \Sigma a_{ji} x_i \leq P_j' \quad (j = 1, 2, \dots, m_i) \tag{2.68.1}$$

$$p_j \leq \Sigma a_{ji}' l_i + \Sigma a_{ji}' x_i \quad (j = m_i + 1, m_i + 2, \dots, m_2) \tag{2.68.2}$$

$$\Sigma a_{ji} l_i + \Sigma a_{ji} x_i \leq P_j \quad (j = m_i + 1, m_i + 2, \dots, m_2) \tag{2.68.3}$$

$$0 \leq x_i \leq M_i \quad (i = k + 2, k + 3, \dots, n)$$

$$\Sigma x_i - \Sigma x_i \geq -2$$

where the values of the x variable are extended from integers to real values; p_j' and P_j' denote $f^{-1}(p_j')$ and $f^{-1}(P_j')$ $(j = 1, 2, \dots, m_1)$ respectively; constraint ( ) expresses a necessary condition to have a connected graph in a partition; and i_1 indicates the indexes of branching functional groups, while i_2 indicates the indexes of terminator groups.

The feasibilities of partial problems generated above must be tested; for example, this can be done by the first phase of the simplex algorithm. If any of the problems pass the test, it needs to be branched further.

Case 2. $k = n - 1$:

A test must be performed by simple substitution to determine if the constraints and condition below are satisfied for $l_1, l_2, \dots, l_n$.

$$p_j \leq f_j(l_1, l_2, \dots, l_n) \leq P_j \quad (j = 1, 2, \dots, m) \tag{2.68.4}$$

$$\Sigma l_i - \Sigma l_i \tag{2.68.5}$$

is an even number not less than –2.

Condition 1. If the partition given by $l_1, l_2, \dots, l_n$ includes functional groups with different types of bonds (e.g., single and double bonds), then there must be a group included in the partition, which has at least two different types of bonds, and each type belongs to at least one group of the partition containing other types of bonds.

After the feasible partitions are generated, the feasible molecular structures can be generated by an available computer program or a combinatorial algorithm. The present procedure is most advantageous when applied to problems involving large numbers of constraints on the predicted properties, especially if most of them are linear or can be sharply bounded by linear functions.

2.69 Codes

Codes have been considered in the book by Pierce. Figures 43 to 48 refer to code formation, generation, and choice. The groups have been sorted according to the magnitude of the property considered in each graph. They are then plotted according to this magnitude in descending order along the circumference of the circle. Since there are 36 groups, it was expedient to limit each quadrant to 9 groups. Those with the highest values (i.e., in the first quadrant) were then abstracted and placed in a code (1,3,4,5,6,7). The code simply used unity if a group was in the first quadrant. A semicolon was placed after every five groups for ease of reading the code. Thus, if 15 groups were used in the sequence

NO2/C=O/NH2/COOH/CH3:
OH/CONH2/NAPTH/H/>C<

the code would read

10011:11001

if NO_2, COOH, CH_3, OH, $CONH_2$, and >C< were present and the other groups were not present. The sequence of groups is arbitrary; i.e., it does not depend on an exact sequence within the first quadrant and the sequence once set is fixed. It will be seen that this code has similarities to the genetic code. Thus, it determines all the properties of the molecules. A shortcoming is that the code does not place or position each group in the order it occurs in the chain. Thus, it will be seen that in a couple of cases two properties will have identical codes. This is not correct because the positional placement of the groups is not exactly the same. However, the group-molecular-property dependence is approximately the same and this shortcoming was incorporated to simplify the procedure and can be corrected. We can now operate on the code to optimize physical, solubility, and biological activities and also to minimize risk concentrations, promoting pollution prevention. In such a way this code can be used to screen out or prohibit new molecules scrutinized for synthesis or production because they do not meet particular environmental requirements. It can also be used to convert molecules that do not meet environmental standards or other property desires.

The code adopted obeys the tenets of group theory. It is found that the position of numbers in the above code occupies positions in the sequence that corresponds to a pattern whereby the physical properties can be separated into subgroups or cosets which fit into groups of properties that have similar attributes or that arise from similar roots. For example, all the solubility properties have numbers in the 10th, 11th, 12th, 13th, 14th, 16th, and 20th positions and the numbers in these positions are largely identical. Similarly, the risk concentration properties have numbers in the 1st, 4th, 5th, 6th, 9th, 26th, and 28th positions. Other similar properties can also be separated into groups of similar codes. This corresponds to the fact that similar chemical groups give rise to similar physical properties as well as chemical properties. Figures 47 and 48 show codes for 15 groups that make each property highest.

In order to transform or synthesize molecular structures with minimum risk features as optimum properties, we would like to incorporate chemical groups with the code of the risk concentration variety (air, tap water, MCL, commercial soil, residential soil, etc.) and also include in that same code the part of the property codes desired. In effect we must cut and paste such codes to create the hybrid code of our desire. This code when mapped onto a particular molecule must also make sense as a practical molecule and not just a virtual molecule.

An example would be the use of the code method to find a molecule that has minimum risk concentration and minimum heat of vaporization. To do this we consider the codes from the table below. We find that the groups COOH, OH, and $CONH_2$ are common to the codes of min Hv and air-risk concentration. In this table the Hv value of 12.8 and the air risk concentration of 1E-6 does satisfy the requirement.

Groups	Hv	Air
OH, COOH, CONH2	12.8	1E-6
>C=CH2,OH,COOH	16.7	1.5E-7
>C=CH2,OH,CONH2	20.0	E-8
>C=CH2,COOH,CONH2	22.2	6.8E-9
>C=CH3,2OH	14.5	4.3E-7
>C=CH2,2COOH	18.2	5.1E-8
>C=CH2,2CONH2	25.6	9E-10

If we lay out the code for each physical property as a row and form a matrix where the columns are the chemical groups and the rows the properties (the group matrix or G matrix) and multiply each row by the desirability of each physical property (the desirability matrix or D), then we have G·D = P (the preference matrix). Let us suppose we only have three groups and three properties and the desirability of the first two properties is primary and the last property is secondary. Also let us assume that the order of the groups for the three properties is, respectively, 3,2,1; 1,2,3; and 2,3,1. Then we have

$$G \cdot D = P \quad \begin{matrix} 3 & 2 & 1 \\ 1 & 2 & 3 \\ 2 & 3 & 1 \end{matrix} \quad \begin{matrix} 1 & 0 & 0 \\ 0 & 1 & 0 \\ 0 & 0 & 2 \end{matrix} = \begin{matrix} 3 & 2 & 2 \\ 1 & 2 & 6 \\ 2 & 3 & 2 \end{matrix} \qquad (2.69.1)$$

The sum of the numbers in the columns of the P matrix are 6, 7, and 10, respectively. The lowest sum is the first column which means the first group is preferred for synthesis followed by the second followed by the third. This could just as easily be done for 10 or 20 or 30 or 40 groups and properties using spreadsheet methods. A more accurate method will be discussed later.

A polynomial code can be set up as follows. Call the term x^n a term for a group in the nth ranking or sort for a group that is not solely composed of carbon and hydrogen, and call y^m a term for a group in the mth ranking or sort for a group that is composed solely of hydrogen and carbon. Thus, the x terms are the groups bonded to a carbon backbone or skeleton and the y terms are those in the carbon skeleton that may need a group bonded to it. Furthermore, we have x_i, where i identifies the particular appending group of interest, and y_j, where j identifies the particular skeleton group of interest. Furthermore, we have $a_i x_i^n$ or $b_j y_j^m$, where the a coefficients represent the valence of the groups which will be negative for the y factors and positive for the x factors. Our aims in constructing a molecule will be to saturate the valences for carbon and the other elements and also to have the lowest numbers for n and m. In that way we can achieve the optimum molecule for a given property, be it the highest numerical number or the lowest, as the case may be. Remember that our numbering system proceeds as 0,1,2,3,4.... and thus we have

$$a_0 + a_1 x + a_2 x_2^2 + a_3 x_3^3 + \ldots \quad \text{and}$$
$$b_0 + b_1 y + b_2 y_2^2 + b_3 y_3^3 + \ldots$$

The above two polynomial codes, the multiplication of which gives terms for the bonding of a pendant group with a skeleton group and the preferred or optimized molecule, will have the exponents of x and y at a minimum. In many cases this will be a constant term; both exponents will be zero. A molecule can then be defined as the sum of terms as follows $(-1)^k\ a_i b_j x_i^n y_j^m$ and since the valence of the whole molecule must be zero, the sum of $(-1)^k\ a_i b_j = 0$.

To decode we may use digital signal analysis, the Fourier transform DFT (discrete Fourier transforms), FFT (fast Fourier transforms), combinatorial algorithms, radio engineering, system designs (trees), branch and bound techniques, etc. This will be explained below.

The branch of information theory that provides a mathematical basis for problems involving data compression is called rate distortion theory. We discuss here how this can be useful for the relation between molecular structure and properties. A useful way to introduce dependencies between the words of a source code is to confine them to a tree structure like that of Figure 43. In general the tree has branches stemming from each node with a string of b digits from the reproducing alphabet on each branch. The code words are formed by cascading the letters encountered along any path from the bottom to the top of the tree. For example, the code word corresponding to the path in the figure is 11011010. If the paths are 1 branch long, we say the tree has depth 1, or has 1 level. Figure 44 shows the circuit used to generate the words of the binary convolutional tree code of the figure in which l = 4 and b = 2. It consists of a three-stage tapped shift register and a mod 2 adder. This can be generalized to convolutional tree codes shown in Figure 45 with a branches per node and b letters per branch. This is a code word–generating circuit (source decoder) for a general tree code. The input is a sequence of letters from an alphabet of size a that specifies a path through the tree. Each input gives rise to b letters from the reproducing alphabet that are deposited on the tree branch designated by the portion of the path map sequence that has shifted into the register up to that time. The multiplicative weight vectors gi = (g1(1), ..., gi(n)), i = 1,...,b are called the generators of the code.

The "tree of choice" is shown in Figure 46. As we proceed from the root to the twigs, let 0 signify the left branch and 1 signify the right branch. Then, 011 means left, right, right and takes us to the octal digit 6. Three digits gives enough information to determine 1 among 16 digits and 20 digits to determine 1 among 1,048,576 alternatives. We can do this by assigning the required binary numbers to the alternatives in any order we wish.

In the traveling salesman problem, a minimum cost route is sought which passes through each node once, and only once, and returns to the starting node. The $(n - 1)!$ tours are partitioned into two distinct groups and a lower-bound cost is calculated for each group. The group with the larger lower-bound cost is discarded and the surviving group is again partitioned, and so on. The procedure continues until a single route remains. It is the optimal route.

2.70 Information Theory

Certain physical systems possess the ability to gather, store and utilize information. The information may pertain either to the state of the immediate environment of the system or to the inherent nature of the system itself, as in the case of genetic information. For the sake of convenience, we shall refer to systems with these abilities as infosystems. It is well known that information cannot be gathered without expending energy. The pool of expendable energy it possesses is known as the Gibbs free energy of the system. An efficient infosystem is one that expends as little of its free energy in the process of collecting the required information as possible. The mechanism operates according to a set of physical equations that are the exact mathematical analogs of the variational equations that determine points on a rate distortion function of an information source with respect to a fidelity criterion. Such a phenomenon is multiphase chemical equilibrium. Here chemical substances mix together and the system tends toward an equilibrium configuration characterized by the partitioning of the system into stable phases. Detailed multiphase chemical equilibrium models of the lung and kidney indicate that the enormous number of accept–reject decisions (reducing concentrations of many undesirable chemicals to a tolerable level) that the kidney has to make accounts for a significant fraction of its overall energy budget, and its efficiency is between 30 and 40%. Also, data compression by the sensory organs involves time constants that are quite short. Other organs, such as the lung and kidney, process data on a somewhat more protracted timescale.

2.71 Computer Synthesis of Molecules

Once more molecules preferable (due to their better physical, risk-based and/or biological properties) to the present ones under scrutiny are known, it is desirable to know the path (that conforms both to thermodynamics and kinetics) in order to transform the less desirable molecules into the more desirable ones. Reaction prediction systems or synthesis design routes now appear more frequently in the literature. It is necessary in many cases to determine which part of the target molecule is synthetically accessible and what starting materials and reaction conditions could make the desired reaction proceed according to the synthesis strategy; the chemist must judge possibilities and side reactions. A reaction knowledge base is frequently necessary and frequently the product ratio cannot be predicted. A set of such programs is available from LHASA UK in the School of Chemistry in the University of Leeds. These include LHASA and CAMEO for synthesis planning, DEREK for toxicological hazard assessment, toxicity prediction to find suggested precursors, and StAR for toxicological risk assessments. LHASA uses retrosynthetic analysis and can suggest synthetic routes based on user-defined starting materials. In "Molecular Computation of Solutions to Combinatorial Problems" by Adleman graph theory for a directed graph with a Hamiltonian path is used to encode DNA. Another (retro)synthesis program is CAOS (Computer Assisted Organic Synthesis). It is called igor2.Zip and can be found and downloaded from Belgium's LOSH (Laboratory for Organic Synthesis Hoornaert) hyperlink.

2.72 Computer-Assisted Molecular Design

Computer-assisted molecular design (CAMD), computer-aided drug design (CADD), molecular modeling, molecular graphics, computational chemistry, structure-based drug design, and computational biology describe a group of techniques for the drug discovery process. The following abridged listing neglects programs that involve recognition and usually noncovalent binding of small molecules (ligands) to specific sites on target biological macromolecules (receptors). QSARs predict biological activity from chemical structure or properties. Spreadsheet programs have revolutionized accounting and information storage. All-purpose CAMD software includes SYBYL, Chem-X, Insight/Discover, and Quanta/CHARMm. Mainly modular in nature, some of the features include sophisticated molecular graphics, molecule building, structure minimization, molecular dynamics, distance geometry, protein homology, modeling including polymer modeling, QSAR, and 3D database searching. Molecular dynamics (MD) is used to "relax" structures, relieve local strain, and refine models. It consists of solving Newton's equations of motion for each atom in a system as a function of time and energy functions describing interatomic forces. Free energy perturbation (FEP) is a molecular dynamics technique in which a thermodynamic cycle is used to calculate the difference in free energy between two states. Searching databases of 3D structures of known compounds, which have not yet been examined for activity of interest, is a powerful and recent method. Such nonproprietary 3D databases are commercially available from Cambridge Structural Centre, Chemical Abstracts Service, Molecular Design, Ltd., and Chemical Design, Ltd.

2.73 Advanced Optimization Techniques

Older programs include the Simplex Technique, Linear Programming, Search Techniques and Nonlinear Programming, Dynamic Programming, Multiobjective Decision Analysis, Integer Linear Programming, Combinatorial Programming, etc. Newer Programs that may be useful are Genetic Programming, Molecular Phylogeny Programming, Evolutionary Programming, Process Synthesis, Adaptive Control Programming, Automated CAD of Control Systems through AI, Computer-Generated Reaction Networks for Stability, branch and bound methods, computer programs for the calculation of chemical reaction rates for polyatomics, such as POLYRATE, degradation pathway maps for specific molecules, PUMA, a system that integrates biological sequence data from within a functional context that offers access to an overview of an organism through structure presentations of the data such as metabolic pathways, alignments, and phylogenetic trees, global optimization including annealing codes, etc. The A.I.Ch.E. Meeting in San Francisco in 1994 was rife with papers on the development of these programs. Of particular interest in this summary is the Path Probability Method. Notice how many of these programs are based on biological processes or natural processes now that the genetic code has been solved. They have been found to be practically efficient. All these programs are ready now, but some may cost a bit.

2.74 Global Optimization Methods

This section surveys general techniques applicable to a wide variety of combinatorial and continuous optimization problems. The techniques involved are

- Branch and bound
  - Mixed-integer programming
  - Interval methods
- Clustering methods
- Evolutionary algorithms
- Hybrid methods
- Simulated annealing
- Statistical methods
- Tabu search

Global optimization is the task of finding the absolutely best set of parameters to optimize an objective function. In general, there can be solutions that can be locally optimal but not globally optimal. Thus, global optimization problems are quite difficult to solve exactly; in the context of combinatorial problems, they are often NP-hard. Global optimization problems fall within the broader class of nonlinear programming (NLP). Some of the most important classes of global optimization problems are differential convex optimization, complementary problems, minimax problems, bilinear and biconvex programming, continuous global optimization, and quadratic programming.

Combinatorial problems have a linear or nonlinear function defined over a set of solutions that is finite but very large. These include network problems, scheduling, and transportation. If the function is piece-wise linear, the combinatorial problem can be solved exactly with a mixed-integer program method, which uses branch and bound. Heuristic methods like simulated annealing, tabu search, and genetic algorithms have also been used for approximate solutions.

General unconstrained problems have a nonlinear function over reals that is unconstrained (or have simple bound constaints). Partitioning strategies have been proposed for their exact solution. One must know how rapidly the function or an analytic formulation of the objective function (e.g., interval methods) can vary. Statistical methods can also partition to decompose the search space, but one must know how the objective function can be modeled. Simulated annealing, genetic algorithms, clustering methods and continuation methods, can solve these problems inexactly.

General constrained problems have a nonlinear function over reals that is constrained. These problems have not been as well used; however, many of the methods for unconstrained problems have been adapted to handle constraints.

Branch and bound is a general search method. The method starts by considering the original problem with the complete feasible region, which is called the root problem. A tree is generated of subproblems. If an optimal solution is found to a subproblem, it is a feasible solution to the full problem, but not necessarily globally optimal. The search proceeds until all nodes have been solved or pruned, or until some specified threshold is met between the best solution found and the lower bounds on all unsolved subproblems.

A mixed-integer program is the minimization or maximization of a linear function subject to linear constraints. If all the variables can be rational, this is a linear programming problem, which can be solved in polynomial time. In practice, linear programs can be solved efficiently for reasonably sized problems. However, when some or all of the variables must be integer, corresponding to pure-integer or mixed-integer programming, respectively, the problem becomes NP-complete (formally intractible).

Global optimization methods that use interval techniques provide rigorous guarantees that a global maximizer is found. Interval techniques are used to compute global information about functions over large regions (box-shaped), e.g., bounds on function values, Lipschitz constants, or higher derivatives. Most global optimization methods using interval techniques employ a branch and bound strategy. These algorithms decompose the search domain into a collection of boxes for which the lower bound on the objective function is calculated by an interval technique.

Clustering global optimization methods can be viewed as a modified form of the standard multistart procedure, which performs a local search from several points distributed over the entire search domain. A drawback is that when many starting points are used, the same local minimum may be identified several times, thereby leading to an inefficient global search. Clustering methods attempt to avoid this inefficiency by carefully selecting points at which the local search is initiated.

Evolutionary algorithms (EAs) are search methods that take their inspiration from natural selection and survival of the fittest in the biological world. EAs differ from more traditional optimization techniques in that they involve a search from a "population" of solutions, not from a single point. Each iteration of an EA involves a competitive selection that weeds out poor solutions. The solutions with high "fitness" are "recombined" with other solutions by swapping parts of a solution with another. Solutions are also "mutated" by making a small change to a single element of the solution. Recombination and mutation are used to generate new solutions that are biased toward regions of the space for which good solutions have already been seen.

Mixed-integer nonlinear programming (MINLP) is a hybrid method and refers to mathematical programming algorithms that can optimize both continuous and integer variables, in a context of nonlinearities in the objective and/or constraints. Engineering design problems often are MINLP problems, since they involve the selection of a configuration or topology as well as the design parameters of those components. MINLP problems are NP-complete and until recently have been considered extremely difficult. However, with current problem-structuring methods and computer technology, they are now solvable. Major algorithms for solving the MINLP problem can include branch and bound or other methods. The branch and bound method of solution is an extension of branch and bound for mixed-integer programming.

Simulated annealing is a generalization of a Monte Carlo method for examining the equations of state and frozen states of n-body systems. The concept is based on the manner in which liquids freeze or metals recrystallize in the process of annealing. In an annealing process a melt, initially at high temperature and disordered, is slowly cooled so that the system at any time is approximately in thermodynamic equilibrium. As cooling proceeds the system becomes more ordered and approaches a frozen state at T = 0. Hence the process can be thought of as an adiabatic approach to the lowest energy state. If the initial temperature of the system is too low or cooling is done insufficiently slowly, the system may become quenched, forming defects or freezing out in metastable states (ie. trapped in a local minimum energy state). By analogy the generalization of this Monte Carlo approach to combinatorial problems is straightforward.

Simulated annealing was designed for combinatorial optimization, usually implying that the decision variables are discrete. A variant of simulated annealing called tree annealing was developed to minimize continuous functions globally. These problems involve fitting parameters to noisy data, and often it is difficult to find an optimal set of parameters via conventional means.

The basic concept of tabu search is a metaheuristic superimposed on another heuristic. The overall approach is to avoid entrainment in cycles by forbidding or penalizing moves which take the solution, in the next iteration, to points in the solution space previously visited (hence, tabu).

2.75 Math Background II

This section merely serves to point out the Research Institute for Symbolic Computation (RISC-LINZ). This Austrian independent unit is in close contact with the departments of the Institute of Mathematics and the Institute of Computer Science at Johannes Kepler University in Linz. RISC-Linz is located in the Castle of Hagenberg and some 70 staff people are working at research and development projects. Many of the projects seem like pure mathematics but really have important connection to the projects mentioned here. As an example, Edward Blurock has developed computer-aided molecular synthesis. Here algorithms for the problem of synthesizing chemical molecules from information in initial molecules and chemical reactions are investigated. Several mathematical subproblems have to be solved, and the algorithms are embedded into a new software system for molecular synthesis. As a subproblem, the automated classification of reactions is studied. Some advanced techniques for hierarchical construction of expert systems have been developed. This work is mentioned elsewhere in this book. Blurock is also involved in a project called Symbolic Modeling in Chemistry, which solves problems related to chemical structures.

A remarkable man is head of the Department of Computer Science in Vesprem, Hungary. Ferenc Friedler has been mentioned three times before in this book for his work on process synthesis, design of molecules with desired properties by combinatorial analysis, and reaction pathway analysis by a network synthesis technique.

2.76 Mathcad Applications in Chemistry

This is a site (http://science.widener.edu/~svanbram/mathcad.html) on the Web that has many applications of Mathcad, particularly analytical and physical chemistry. It has links to other Web pages and among them is http//www.mathsoft.com/appsindex.html. The latter has 210 links to other sites in the following categories: mathematics and statistics, astronomy and navigation, electrical engineering, physics, chemistry, chemical engineering, industrial engineering and operations research, civil and mechanical engineering, business and economics, and linguistics. Among the applications are the following that could pertain to topics in this book:

1. Focusing in on Gibb's phenomena
2. Conformal mapping and Joukowski airfoils
3. The convolution integral
4. Bifurcation diagrams using the brute force method
5. The game of life by elementary programming
6. Plotting a logarithmic spiral
7. Plot of the Lorenz butterfly
8. The butterfly attractor
9. Neural networks in Mathcad
10. Calculation and Bode plots of transfer functions
11. Plotting the time response directly from a Laplace transfer function
12. A handy circuit for filter design
13. Passive low-pass filter by network synthesis
14. Passive Tchebyshev low-pass filter
15. Carbon dioxide chemical equilibria and applications
16. A problem involving carbon dioxide equilibria
17. Rate expression — differential method
18. Simulation of Chemical reactions by randomly selecting pairs of molecules
19. Simulation of two consecutive irreversible reactions
20. Estimation of vapor pressure from the Redlich/Kwong equation of state using Mathcad
21. Fitting a logistic growth model to colony data.

2.77 The Schwarz–Christoffel Transformation and Structure–Activity Relationships

Consider a physical situation where electric current flows along straight lines parallel to the x axis from right to left. Then the streamlines are y = constant and the equipotentials are x = constant. If the potential drop is k_1 V per unit distance with x = 0, then V = k_1x. Also, the stream function S is the amount of current crossed in traveling along the line x = constant starting at the edge of the field at y = 0. So S = k_1y and combining S and V into the complex quantity, F = V + iS, so

$$F = V + iS = k_1(x + iy) = k_1 z \qquad (2.77.1)$$

This is a relation that transforms or maps the upper half of the z plane (with parallel flow) onto the upper half of the F plane where there is a similar net of orthogonal lines of equipotential and lines of constant stream function. We have essentially performed a stretch or contraction in the ratio k_1.

This discussion, which dealt with the flow of current, applies equally well to the flow of a fluid or to the flow of heat or to electrostatic flux. In heat flow, we have temperature as the potential, isotherms as the equipotentials, and the quantity of heat as the stream-functions. In fluid flow, velocity potential replaces potential, and the amount of fluid represents the stream-function. For electrostatic flux, we have the potential with lines of force as the streamlines.

The mapping of the function e^z onto the w plane would be the transformation

$$w = e^z \qquad (2.77.2)$$

The mapping of the polygon in Figure 49 transforms the lines y = constant into rays 0 = constant and the mappings of x = constant yield the circle. Thus, the rectangular region a < x < b, c < y < d is mapped onto the region $e^a < r < e^b$ and c < 0 < d. These are sectors, bounded by portions of circles and rays.

We will endeavor to use this development here to represent a conformal mapping between the properties of molecules and their structure, and in order to do so we will employ the Schwarz–Christoffel transformation, which maps the x axis and the upper half of the z plane onto a simple closed polygon and its interior in the w plane.

The derivative of the function that is to map the x axis onto a polygon is

$$f(z) = A(z - x_1)_1^{-k}(z - x_2)_2^{-k} \ldots (z - x_{n-1})_n^{-k-1} \quad (2.77.3)$$

This mapping function can be written f(z) = F(z) + B, where B is a complex function. The resulting transformation is

$$w = A\,(s - x_1)^{-k1}(s - x_2)^{-k2} \ldots (s - x_{n-1})_n^{-k-1}ds + B \quad (2.77.4)$$

The equation above is such that the polygon in the w plane will be mapped onto the x axis of the z plane, and is known as the Schwarz–Christoffel transformation. The quantities k_1, k_2, ... k_n are all determined from the exterior angles at the vertices of the polygon, and must satisfy the equation

$$k_1 + k_2 + \cdots + k_n = 2 \quad (2.77.5)$$

It has been shown that the representation of molecular structure can be represented by three or less parameters rather than the full set (B, C, D, E, and F) with a small degree of error. So, below is shown the Schwarz–Christoffel transformation for the triangle with various angles and sides.

Example of Schwarz–Christoffel Transformation

The simplest example of a polygon is the triangle with angles api, bpi, and cpi, so that

$$a + b + c = 1 \quad (2.77.6)$$

Then the function determining the conformal representation of this w triangle on the half z plane is

$$W = \int dz/((z - r)^{1-a}(z - s)^{1-b}(z - t)^{1-c}) \quad (2.77.7)$$

$$dz/dw = (z - r)^{1-a}(z - s)^{1-b}(z - t)^{1-c} \quad (2.77.8)$$

If a = o, the only instance when the integral is a uniform function of w is when b = 1/2, c = 1/2, and the function is simply periodic. Then the w figure is a strip of the plane of finite breath, extending in one direction to infinity and terminated in the finite part of the plane by a straight line perpendicular to the direction of infinite extension.

If none of the quantities a, b, c is zero, then, since a + b + c = 0, the only cases when the integral gives z a uniform function of w are as follows:

I. a = 1/3, b = 1/3, c = 1/3; an equilateral triangle:
II. a = 1/2,b = 1/4, c = 1/4; an isosceles right-angled triangle:
III. a = 1/2,b = 1/3, c = 1/6; a right-angled triangle, with one angle equal to $(1/3)\pi$

z is a uniform doubly-periodic function of w and we find

I. 2i/z = P

where the invariants g_2, g_3 are 0, 4.

II. $z^2/(z^2 - 1) = P^2$

where the invariants are –4, 0.

III. $(1 - z)/(1 + z) = P^3$

where the invariants g_2, g_3 = 0, 4.

A further case is the isosceles triangle with an angle of 2/3 and a = 2/3, b = 1/6, and c = 1/6. Then

$$z^2/z^2 - 1 = P^3$$

where the invariants g_2 and g_3 are 0, 4.

It must be mentioned here that when the polygon is of a more complicated form with many unequal angles and many unequal sides, then the transformation can be quite difficult because of the difficulty of the integration.

In previous work, the properties of molecules (P) were found to be equal to

$$P = a + bB + cC + dD + eE + fF \quad (2.77.9)$$

where B, C, D, E, and F were dependent on the structural groups of the molecules and a, b, c, d, e, and f were determined by regression of the properties of the molecules cited in the handbooks. It was found, however, that toxicity–structure factors were more closely described by relations such as

$$-\mathrm{Ln}P = a + bB + cC + dD + eE + fF \quad (2.77.10)$$

It is the object of this study to identify the structure of molecules as a set of points in the complex plane. This set of points can be joined to form polygons whose sides and concomitant angles will define the mapping of molecular structure as identifying features of the parent molecules. It is a further object to identify the orthogonal geometric functions with the molecular structure (potential) geometry as the properties or stream functions. This will simplify the structure–activity relations and lead to correlations for the dependence between structure and properties. A Schwarz–Christoffel transform computer program exists for computers, it is called SCTRANS and is authored by Peter Wallenten (1988). There is a manual. The program calculates transforms for a given polygon and is used for building physics and ground heat applications.

2.78 Chaos

The complex plane of initial values is subdivided into two subsets. Every point in the complex plane is in the first set or outside. The first set collects points for which the iteration escapes. It is called the *escape* set. The iteration for all the other initial values remains in a bounded region forever, and we collect these points in the so-called *prisoner* set.

The electrical function for a very long wire is proportional to 1/r; thus the potential field is logarithmic and the base of the logarithm does not matter.

$$p(z) = \log |z| \quad (2.78.1)$$

If the prisoner set P_c is connected, then the escape set E_c carries a system of field lines and equipotentials and the Riemann mapping theorem, which allows the relationship of the potential of the unit disk to that of the connected prisoner set: equipotentials and field lines of P_0 and P_c can be brought into one-to-one correspondence. Moreover, the dynamics of the field lines of P_c can be played back to the dynamics of the field lines of the unit disk P_0.

Referring back to a previous equation, we see that

$$\log |z| = -(a + bB + cC + dD + eE + fF) \quad (2.78.2)$$

and it is here that we make the connection between the *Julia set* and the properties of matter. We shall see that by taking a subset of the Julia set and iterating a finite number of times, the result will be the complete Julia set. This says that the immensely complicated global structure of the Julia set is already contained in any arbitrary small section of it.

In the *Mandelbrot set*, the same shapes are repeated everywhere, yet each repetition is somewhat different. It is essentially a theme and variation. The complexity of the Mandelbrot set is in a different class than the Julia set. It has a solid interior without any structure, and on the other hand it is bordered by a very complex boundary with an infinity of different shapes. The Mandelbrot set can also be used as a road map to all possible kinds of Julia sets. It is also a table of contents of the book of Julia sets.

We may liken the prevention of pollution in chemical systems to a system we wish to control, particularly in the design phase. In the theory of servomechanisms, we depict a system with an input, an output, and a feedback loop. The transfer function is the ratio of output to input, each expressed as a complex polynomial. There is a frequently used criterion (the Nyquist criterion) for stability of the system. This occurs when the locus of the feedback transfer function does not pass through or encircle the point (–1,0) in the Y plane. Then the system is stable. In the Mandelbrot system, however, if the locus falls inside the interior it may be pollution free. There is then an analogy between stability and the pollution-free state from our perspective.

2.79 Complex Systems

The symbiosis between science and the computer is now so successful that it is making it feasible for life itself to gain a new meaning. Life is a *process*, and it is the *form* of this process, not the matter, that is the essence of life. Life is fundamentally independent of the medium in which it takes place, and the implications of separating living complexity from its medium are stunning. Attempts to create living complexity within computers have already been taking place.

An appropriate language for depicting the complexity of the real world is chemistry. The full gamut of complexity can emerge from the simplest chemical circumstances, providing a powerful nonlinear paradigm for the patterns found in nature. Chemistry can solve mathematical problems. A self-propagating chemical reaction front can explore a labyrinth in a manner akin to the way a ripple of chemical activity can probe the networks of the brain. Computers can unravel the secrets of chemical complexity, ranging from the highly unorganized to the utterly chaotic.

To work with complexity, computers need to be combined with a description of chemistry that works at a higher level. This means a description in terms of macroscopic properties rather than a microscopic one. If we are to describe these emergent properties, we need concepts to handle the *global* properties of large collections of molecules, whether in terms of concentration or temperature changes, to avoid becoming overwhelmed by redundant detail. We need to be equipped with tools such as nonlinear dynamics and nonequilibrium thermodynamics, together with computers.

Nonequilibrium thermodynamics divides into two parts: the linear version describes behavior close to equilibrium, whereas the nonlinear version is concerned with behavior far from the equilibrium state. No simple laws state how thermodynamic evolution will occur far from equilibrium, other than it must involve the dissipation of energy, where dissipation can be thought of as synonymous with increasing entropy. As a result of this dissipation taking place far from equilibrium, genuine complexity can emerge. A richer language than thermodynamics is needed when we are to come to grips with real-world phenomena. This is supplied by nonlinear dynamics.

2.80 Elusive Heat, Chaotic Patterns

Experiments have replicated a classic form of heat transfer known as Rayleigh–Bernard (R–B) convection for decades. Now the CRAY Y_MP and C-90 have simulated R–B convection. This has revealed unpredicted chaotic patterns in R-B convection, unveiled new R-B convection structures, and paved the way for mathematically accurate, 3D renditions of chaotic behavior. This is significant because for an easy-to-replicate phenomenon, it provides insight into understanding how heat energy moves through a flow system.

Supercomputing has helped reveal novel unpredicted states. Recently, the CRAY C90 has been used to monitor the onset of R–B from a nonconvective state. The resulting 2D images reveal a kaleidoscope pattern of local spirals. The spirals not only rotate, they also move around in the fluid, annihilate each other, grow at the expense of one another and fluctuate in size and number. Gone is the regularity exhibited in the parallel rolls, with the fluid moving in continuous circular waves. Velocities and temperatures vary throughout the fluid. Some portions of it rise and fall faster than others, and everything about the system changes from one moment to the next. It is a crock pot of symmetry turned bubbly cauldron. Since these local spirals are erratic in time-changing position and size, there is an irregular behavior in space as well. Chaos is a realm beyond order, whose disorder has a recognizable pattern over time. We are now using supercomputing to create 3D models of the chaotic patterns, which will provide further understanding of both convective behavior and chaos.

2.81 Minimax Group Selections for Optimum Properties

Max	Property	Min
-cycloh/-naphth/-c6h5/ -ccl3/-ch=ch2	logccl4/h2o	-conh2/-oh/-cooh/ -nh2/-c=o
-conh2/-cooh/-oh/ -cn/-no2	Solubility parameter	>c</>ch-/-nh-/ -ch=ch2/-ch2-
#-napth/-conh2/-I/ -c6h5/-cooh	Tc	>c</>ch-/-nh-/-ch2-/-ch=ch2
-conh2/-cooh/-oh/ -nh2/-cn	Log activity coefficient	>c</>ch-/-ch2-/ -ch3/-h
-cycloh/-napth/-c6h5/ -ccl3/-ch=ch2	Log partition coefficient	-conh2/-c=o/-nh2/ -oh/-o-
-conh2/-napth/-cooh/ cycloh/-ccl3	Log viscosity	>c</-nh-/>ch-/ -0-/-s-
-conh2/-cn/-no2/-cooh/ -napth/	Thermal conductivity	>c</>ch-/-ch2/ -ch=ch-/-nh-
-conh2/-cn/-I/-no2/-oh	Pc	>c</>ch-/-nh-/-ch=ch-/cycloh

2.82 Envirochemkin

Once the nature of the desired or substituted product or intermediate or reactant is known, we wish to describe how it and the other species change with time, while obeying thermodynamic laws. In order to do this we use another program called Envirochemkin.

Envirochemkin is derived from a program called Chemkin. Chemkin is a package of FORTRAN programs which are designed to facilitate a chemist's interaction with the computer in modeling chemical kinetics. The modeling process requires that chemists formulate an applicable reaction mechanism (with rate constants) and that they formulate and solve an appropriate system of governing equations.

The reaction mechanism may involve any number of chemical reactions that concern the selected named species. The reactions may be reversible or irreversible, they may be three-body reactions with an arbitrary third body, including the effects of enhanced third-body efficiencies, and they may involve photon radiation as either a reactant or product.

The program was used by Bumble for air pollution, water pollution, biogenic pollution, stationary sources, moving sources, remedies for Superfund sites, environmental forensic engineering, the stratospheric ozone problem, the tropospheric ozone problem, smog, combustion problems, global warming, and many other problems. It was found to function well with room-temperature reactions, working well with free radicals, etc.

In order to describe Envirochemkin, a simplified case is shown involving the cracking of ethane, to convert it to a less toxic and more profitable species. First, create the reaction file called ethane.dat.

Then create the input file: ethane.sam.

The output Spec.out is shown in Table 20, from which we can plot 2D and 3D graphs.

In the ethane.dat file, we first type the word ELEMENTS, then all the chemical elements in the problem, then END, then the word SPECIES, then all the chemical formulas, then the word END, then all the chemical equations and next to each three constants a, b, and c from the rate constant for the equation that comes from the literature: $k = aT^{-b} \exp(-c/RT)$. Finally, at the end of the reactions, which may number 100 in some problems, we type END. The program can solve for 50 unknowns (species) and 100 differential equations, and such problems are often run.

In the ethane.sam file we first type 0 for the isothermal problems, where T and P are constant, then the temperature in degrees K, and the pressure in atm next to it. Other modes for running problems are 1 for constant H and P, 2 for constant U and V, 3 for T varies with time with V constant, and 4 for T varies with time with P constant. Below the numbers 3 and 4 are the coefficients for $dT/dt = c1 \exp(-c2T) + c3 + c4T + c5T^2$, displayed as 3000. 1000. 0.0 0.0 0.0. Below that we put 0.0d-6, then the residence time in microseconds in the form shown (which is 100000 usec. or 0.1 sec.) and then the interval in microseconds between times for calculation and display. Then all the chemical formulas for the species and below each one the initial concentration in mole fractions (shown as 1.0 or 0.0) and a three-digit code consisting of 0 or 1. The ones indicate that a sensitivity analysis calculation is wanted and is placed in the second position or that data is needed to make the plotting of results simple.

The spec.out file presents the mole fraction of each chemical species as a function of time, temperature, and pressure as indicated. The mole fraction of each species is presented in a matrix in the same position as the chemical formulas at the top.

The program is originally in FORTRAN and will not run if there is the slightest error. Now if the program refuses to run, type intp.out and it will indicate what the errors are, so you can correct them and run the program again. Intp.out (Table 21) reveals your errors with an uncanny sort of artificial intelligence that will appear at the appropriate equation or below the last statement.

In order to run, the thermodynamic data for each species is needed and is contained in either the file sandia.dat or chemlib.exe\thermdat. The data that is used is

$$Cpi/R = a1 + a2iT + a3iT^2 + a4iT^3 + a5iT^4 \quad (2.82.1)$$

$$Hi/RT = a1i + a2i/2T + a3i/3aT^2 + a4i/4T^3 + a5i/5T^4 + a6i/T \quad (2.82.2)$$

$$Si/R = a1ilnT + a2iT + a3i/2T^2 + a4i/3T^3 + a5i/4T^4 + a7i \quad (2.82.3)$$

There are seven constants for each species (a1...a7) and each species is fitted over two temperature ranges, so there are 14 constants for each species in all. The NASA format for the species C_4H_6 would then look as shown in Table 22.

Other information embedded in the NASA code are name, formula, date of creation, physical state, temperature range of validity, and temperature at which the two temperature ranges fit smoothly together.

Now to run the program, type in the order shown below:

```
c:\ckin\intp
ethane.dat
C:\ckin\sandia.dat
```

Then wait a few moments and type

```
ckin\ckin
[enter] [enter]
ethane.sam [program will run]
```

After the first, second and third line, press enter. After the fourth line press, enter twice.

If the thermodynamic data is in chemlib.exe\thermdat substitute that for sandia.dat.

In every run the file rates.out will be created. Its format is shown in the tables.

The above is repeated for every instant of time for which there is a calculation in the forward and reverse direction, and then the equations are ranked according to the speed of their reaction. Note that when the sign is negative in the fifth column the reaction proceeds in the reverse direction. Also note that these data are very important in determining the mechanism of the reaction.

Another file, sense.out, is created when the code in ethane.sam indicates that it is desired for up to five species.

There is often a great deal of uncertainty in the rate constants for some reaction mechanisms. It is, therefore, desirable to have an ability to quantify the effect of an uncertain parameter on the solution to a problem. A sensitivity analysis is a technique which is used to help achieve this end. Applying sensitivity analysis to a chemical rate mechanism requires partial derivatives of the production rates of the species with respect to parameters in the rate constants for the chemical reactions. This file, Table 24, shows the partial derivatives of how the increase or decrease of each species changes the speed or velocity of each reaction shown for every interval of time and, like the rates.out file, is very important in determining the mechanism and optimizing the reactions.

2.83 Additions to Sandia.dat (the Original Thermochemical File)

In order to run problems in Envirochemkin a thermochemical file is needed containing the thermodynamic functions for all the species contained in the problem. For each such species there are 14 numbers for the standard heat of formation and entropy in addition to the enthalpy, entropy, and heat capacity as a function of temperature. The original such file was called Sandia.dat, and a later file was called Chemlib.exe/thermdat. Each of these files contained a huge number of species. Sandia.dat was furnished with Chemkin and was used in many combustion-type problems. In order to extend the range of problems, additional species were added to Sandia.dat so that certain other problems could be run. The thermodynamic values for the new species were obtained by several means including literature values converted to the NASA format using the Thermfit routine in THERM, commonsense methods ("this is to that as that is to that"), THERM methods based on Benson's group properties, crude methods honed by later sensitivity analysis, etc. Thus, Chemkin became Envirochemkin and some of the additional problems that were run follow:

1. The toxicity of solutions of methyl tertiary butyl ether;
2. Nucleotide chemistry and the mechanism of benzopyrene toxicity;
3. Atmospheric oxidation and ozone formation;
4. Petroleum refinery unit compositions and pollution;
5. Maleic anhydride process chemistry;
6. Chlorohydrocarbon composition reactions and pollution formation;
7. The harm of smoking, nicotine, and second-hand smoke;
8. An analysis of the Bhopal incident and how it could have been prevented, etc. A list of such species is listed below.

$MeCOO_2NO_2$
$C_2H_3O_3$
$C_{22}H_{14}$
MTBE(L)
C_4H_9O
$C_2H_3NO_5$
C_4H_9OOH
TBN
$C_5H_{10}O_2$
PYRTHYM
O(1D)
CYBUTTHY2
PBTHYM2
$C_{25}H_{16}O_4N$
$C_{20}H_{12}O_2$
BENZOTHYM
$C_{20}H_{12}O$
$C_{20}H_{10}O$
THYRAD2
$C_{20}H_{11}$
$C_{20}H_{11}O$
$C_{20}H_{11}O_2$
$C_5H_6N_2O_4$
$C_4H_7N_3O_3$
$NH(CHO)_2$
NH_2CHO
NH_2CONH_2
THYCYTDIM
DITHYMINE
THYMINE
ADDUCT1
CYTOSINE
$C_{20}H_{12}$
$C_{14}H_{10}$
$C_{18}H_{12}$
PYRIMIDIN
ANILINE2
PAB2
$C_{10}H_8$
AZULENE
C_5H_8
C_5H_7
C_5H_7O
$C_5H_7O_2$
C_4H_7O
C_2H_5
$C_5H_8O_2$
SO_2*
EG
STYRENE
$C_2H_6S_2$
C_2H_6S
$S^*C_4H_9$
$N^*C_4H_9$
$P^*C_4H_9$
BUTENE2

C_4H_7
OCTANE
TMPEN*
C_6H_6(L)
C_7H_8(L)
$C_2H_6O_2$
CF_2CL_2
$C_{10}H_{14}O$
C_2H_5O
$C_2H_5O_2$
C_2H_5SH
$C_4H_{10}O$
PXYLENE
MXYLENE
OXYLENE
NAPTHAL
METBUETH
PHENOL
TNBEN*
OCTENE*
ETBENZENE
ISOHEXANE
HEPTANE
HEPTENE*
ISOOCTANE
IC_4H_{10}
IC_4H_{10}
$1C_4H_8$
TOLUENE
$C_6H_5CH_3$
CH_4S
C_4H_{10}
CUMENE
$C_{10}H_8$
NICOTINC
C_3H_6O
MALEIMIGG
MALEOACID
CL^-
$(NH_4)_2SO_4$
NH_4PLUS
SO_{-4}
$CH_2CHCOOH$
ACRYLAMID
MALEIMIDE
MALEONITR
CH_3COCH_3
CH_4O_2*
NICOTINE
C_4H_5NA
$C_5H_{11}NA$
$C_5H_{11}N$
CH_2CHCN
C_4H_9N
PYRIDINE1
NICOTINEA
C_4H_5N
CH_3NCO
$C_4H_8O_4$
$C_2H_3O_4$
C_4H_6O
$C_4H_7O_2$
$C_5H_8NO_3$
C_4H_6O
C_5H_8O*
$C_3H_4O_2$
C_5H_8O
$C_5H_8N_2O_6$
C_5H_8*
C_5H_9O
C_4H_6O
$C_2H_8N_2$
$C_{12}H_8O$
HNO_4
CH_3CL
CH_2CCL_2
CH_2CHCL
CH_2CLCH_2CL
CH_3CHCL_2
CHCLCHCL
CL_2
CL_2CCCL_2
CH_2CCHCL
HCL
O*
HSO_2
$C_2H_3O_3$
CH_3NO_4
CH_3NO_2
CH_3O_2
CH_3NO_3
CH_2O_2
N_2O_5
CH_3O_3
H_2SO_4
H_2SO_4(L)
C_4H_5
C_5H_5
C_6H_5
#$C_{12}H_{10}$
CL
CLO

2.84 Gasteiger

The work of Prof. J. Gasteiger of Computer-Chemie-Centrum Erlangen is presented here as a well-organized example of our domain of interest. It consists of the following:

Research topics:

- Program development
- Programs (index)

- Methods
- In house databases
- Collaboration and support
- Detailed research projects

Development of computer programs for the following:

- Reaction prediction
- Synthesis design
- Acquisition of knowledge on chemical reactions
- Interpretation and simulation of mass spectra
- Analysis and simulation of infrared spectra
- Analysis of modeling of biological activity
- Finding new lead structures
- Generation of 3d molecular models
- Assessing molecular similarity
- Prediction of physical, chemical, and biological properties
- Assessment of molecular diversity
- Databases and algorithms
- Electronic publishing
- Fast and efficient generation of 3d molecular models
- Analysis of molecular surfaces
- Perception of molecular similarity with a genetic algorithm
- Integrated information management in chemistry

Solution of these tasks achieved by the following:

- Statistical methods
- Pattern recognition methods
- Neural networks
- Genetic algorithms
- Graphics programming

Databases — Development of knowledge-based systems strongly builds on information and knowledge derived from computerized databases:

- SpecInfo and MassLib (Chemical Concepts) — 99,000 13C NMR spectra; 44,000 mass spectra; 20,000 infrared spectra
- ChemInform RX reaction database (FIZ Chemie, MDL Information Systems) — 500,000 reactions
- MACCS Drug Data Report-3D (MDL Information Systems) — 78,000 structures with biological activity
- ACD (Available Chemical Directory) (MDL Information Systems) — 1777.000 commercially available starting materials
- Acros Organics Catalog (Acros Organics) — 8000 commercially starting materials
- NMR-Sharc — 4000 *ab initio* NMR calculations
- VCH Trivialname Catalog (FIZ Chimie Berlin) — 20,000 structures
- Flukalog (Fluka Chiemie AG) — 16,000 commercially available compounds

Index of programs and program packages:

- CORINA (Generation of Three-Dimensional Molecular Models)
- EROS (Elaboration of Reactions for Organic Synthesis)
  Description of the program
  Application in: reaction prediction
- FM3D
- FRANZ (Fragmentation and Rearrangement Analyzer)
  Description of the program
  Application in: interpretation and simulation of mass spectra
- MAGIC (Mouse Aided Graphical Input for Chemists)
- MASSIMO (Mass Spectra Simulator
  Description of the program
  Application in: interpretation and simulation of mass spectra
- MEDTOOL
- PETRA (Parameter Estimation for Treatment of Reactivity Applications)
- VEGAS
- WODCA (Workbench for the Organization of Data for Chemical Application in Synthesis Design)

2.85 Reaction Prediction

The course of a chemical reaction and its products are predicted for given starting materials. For over 15 years they have been developing EROS. In the present version, EROS 6.0, the knowledge base and the problem-solving techniques are clearly separated. The knowledge base consists of methods for calculating important electronic and energy effects in organic molecules, as well as rules concerning the course and evaluation of elementary chemical processes. At present, the following organic reaction types are being investigated:

Electrophilic aromatic substitution
Diels–Alder reaction
Amide hydrolysis
General hydrolysis

2.86 Acquisition of Knowledge on Reactions

Reaction databases provide a rich source of information on organic reactions. Machine learning methods (classification and generalization of reactions) are used to combine individual reactions into reaction types. In this way, the important characteristics — and the driving forces — of a class of reactions can be shown. Repetition of the classification and generalization steps leads to an increasing level of abstraction and thus to a hierarchy of reaction types.

The methods are part of the HORACE system (Hierarchical Organization of Reactions by Attribute and Conditions Education). In a different approach, a set of reactions is sent into a Kohonen neural network. This leads to 2D landscapes of organic reactions that allow one to identify the different reaction types, the scope and special cases of each reaction type, as well as the more unusual reactions. The automatic classification of reactions can be used for knowledge acquisition for reaction prediction, for synthesis design, and for more efficient search in reaction databases.

2.87 Interpretation and Simulation of Mass Spectra

Mass spectra play an important part in deducing the structure of organic compounds. In order to analyze mass spectra more efficiently we have developed the system FRANZ. Using the structure of an organic compound, its mass spectrum, and a set of reaction types, the fragmentation scheme for the given structure is derived. By using intensities of the experimental mass spectrum each reaction is assigned a reaction probability and stored in a reaction database. This database can then be used to generate classifying fragmentation rules that can evaluate the reactivity of a mass spectroscopic process with statistical methods or neural networks. To simulate mass spectra the system MASSIMO applies these reaction rules to a structure, calculates the fragmentation scheme and reaction probabilities, and converts them into a simulated mass spectrum.

2.88 Analysis and Simulation of Infrared Spectra

The goal of this project is the development of empirical methods for the rapid simulation of infrared spectra to be used in the structure elucidation process. This task critically hinges on the way structures are coded. A molecular transform has been developed that encodes the 3D structure of a molecule by a constant (fixed) number of variables. Furthermore, this encoding scheme allows the consideration of different atom properties, such as mass, partial atomic charge, and atomic polarizability. The relationship between structure and infrared spectra is stored in counterpropagation neural networks.

2.89 Synthesis Design

The objective is to develop methods for defining appropriate, easily obtainable starting methods for the synthesis of a desired product. In addition, the individual reaction steps of the desired synthesis plan are derived. At present, our work consists of deriving methods for the definition of structural similarities between the target structure and available starting materials, developing methods for finding strategic bonds in a given target, and development of rating functions to assign merit values to starting materials. The methods are integrated into the WODCA system.

2.90 Generation of Three-Dimensional Molecular Models

A program CORINA has been developed that automatically generates 3D atomic coordinates from the constitution of a molecule as expressed by a connection table augmented by stereoscriptors. This system is a comfortable means for the first and automatic conversion of large databases containing 2D molecular structures to the 3D coordinates or for the convenient generation of input structures for the quantum mechanical, semiempirical, or molecular mechanics calculations, and for any other molecular modeling method requiring 3D atomic coordinates. CORINA is applied to the entire range of organic chemistry. All structures that can be expressed in a valence bond notation can be processed, including cyclic, polycyclic, macrocyclic, and polymacrocyclic ring systems, as well as radicals, reactive intermediates, or organometallic systems. Atoms with up to six neighbors are allowed. The input structures can be processed in a variety of standard formats such as MDL MOLFILE, SYBYL File Formats, SMILES, or an internal file format. CORINA has been applied to a set of 126,705 structures provided in the form of connection tables by the National Cancer Institute (U.S.). Of these, 126,089 structures (99.5%) could be successfully converted into 3D coordinates. These 3D structures have been made publicly available (ftp helix.nih.gov then:cd nicidata/3D file: ncioprn3d .mol.Z). CORINA can be accessed on the World Wide Web.

2.91 Prediction of Physical, Chemical, and Biological Properties

A series of empirical methods were developed to calculate fundamental electronic and energy effects in organic compounds.

Charge distribution
Inductive effect
Resonance effect
Polarizability
Bond dissociation energy
Heats of formation

These methods have been integrated into the PETRA. Physical, chemical, and biological properties can be calculated and correlated with the obtained parameters. Statistical and parameter recognition methods are used to determine relations between structure and the examined property.

2.92 Integrating Information Management in Chemistry

A system was developed that

- Handles, organizes, and manipulates chemical information of any kind, including *ad hoc* defined types;
- Provides computational and data analysis modules as well as data sets on a worldwide net of algorithm databases; modules can be built upon in a hierarchical fashion and are reusable entities;
- Defines complex computational sequences for molecules by reusable modular iconic operative units following principles of visual programming; interaction of objects leads to computational actions;
- Carries through data generation, analysis, I/O, and display through a distributed network of client/server applications.

2.93 Analysis and Modeling of Biological Activity

The molecular electrostatic potential gives detailed information for studies on chemical reactivity or pharmacological activity of a compound. A Kohonen network is used to perform a nonlinear projection of the 3D molecular surface into a 2D feature map. In this way the entire electrostatic potential on the molecular surface can be shown on one single picture. This allows the perception of similarity of substrates that bind to the same receptor. In a similar manner other molecular surface properties can be represented and investigated with the program KMAP (Kohonen MAP generator).

2.94 Perception of Molecular Similarity with a Genetic Algorithm

The maximum common substructure (MCSS) is a similarity measure that allows the study of a variety of chemical problems. By knowing relationships between structure and biological activity, appropriate starting materials for the target of a synthesis and biosynthetic pathways can be investigated proceeding from the MCSS or the 3D MCSS. The determination of the largest substructure contained twice in a single molecule allows one to derive synthesis precursors that can be used several times to make a synthesis short and convergent. The determination of an MCSS is a computationally expensive task that is solved by a genetic algorithm, and this is an optimization that imitates the adaptation methods of nature.

2.95 Assessment of Molecular Diversity

A major goal of combinatorial synthesis is to obtain compounds that have a wide structural variety while simultaneously covering the structural space as completely as possible. Biological activity is intimately tied to molecular surface properties such as the molecular electrostatic potential. Molecular surface properties can be encoded by autocorrelation vectors, and molecular structures thus coded can be mapped by a Kohonen neural network. These maps show both molecular similarity and diversity.

2.96 Laboratory of Mathematical Chemistry and Computer Design

In the field of computer-assisted synthesis, the following programs have been developed:

- STORM, aimed at the hierarchic planning of key steps in multistep synthesis;
- ICAR for the investigation of multistep carbocationic rearrangements;
- GREH for the classification, enumeration and prediction of new ring transformations in heterocycles.

In the field of QSARs, a computer program BIBIGON is being developed for revealing new structure–property relationships (as well as the prediction of compounds with a given activity/property value) on the basis of structural descriptors, represented by atom chains of certain types.

2.97 NASA, GRI-MECH

Version 2.11 expands GRI-MECH 1.2 by including nitrogen chemistry relevant to natural gas chemistry and reburning. It contains 277 elementary chemical reactions of 49 species. Version 2.11 contains the chemistry identical to that of version 1.2 but adds 102 reactions of 17 additional nitrogen-containing species. Optimizing the nitrogen chemistry relevant to natural gas flames and reburning is qualitatively a very different challenge than optimizing the carbon–hydrogen–oxygen chemistry. We thus regard GRI-MECH 2.11 as only a preliminary starting point with less degree of confidence than can be applied to the version 1.2 carbon–hydrogen–oxygen chemistry. It is believed, however, that it is an improvement for the experiments modeled over previous attempts to describe NOx formation and removal in natural gas flames.

2.98 THERM

THERM is a thermodynamic program that prepares the thermodynamic functions for Envirochemkin and calculates equilibrium constants and other thermodynamic values for reactions together with ratios of Arrhenius constants in tabular form as functions of temperature. Methods and parameters were taken from Sidney Benson's book *Thermochemical Kinetics*. The user's manual is from Dr. Ritter, now at Villanova University and from Dr. Boselli at the New Jersey Institute of Technology. THERM can be used for insertion of thermodynamic data in Sandia.dat or Chemlib.exe/thermdat "libraries" of Envirochemkin, using literature values in the file c:\ckin\thermo.lst converted to the c:\ckin\thermo .dat file within THERM or, when lacking literature values, using the applicable codes of the 380 chemical group codes for the molecule concerned to insert values into c:\ckin\thermo.lst, etc.

UNITS:KCAL
THERMO LISTING

SPECIES	HF (298)	S (298)	CP (300)	CP (400)	CP (500)	CP (600)	CP (800)	P (1000)
CUMENE	0.94	92.87	36.47	48.00	57.90	66.20	78.60	87.30
ETBENZENE	7.12	86.15	30.88	40.76	49.35	56.44	67.15	74.77
TOLUENE	11.95	76.64	24.94	33.48	40.98	47.20	56.61	63.32

and so on. The list consists of hundreds of compounds. The numbers can be obtained from Stull, Westrum, and Sinke, *The Chemical Thermodynamics of Organic Compounds*, Wiley, 1969, or the JANAF Tables, etc.

A NASA subroutine called c:\ckin\thermfit fits each species from the c:\ckin\thermo.lst table to the 14 constants previously described and forms the c:\ckin\thermo.dat table at incredible speed, and this is copied into c:\ckin\sandia.dat or c:\ckin\chemlib.exe\thermdat as needed.

Another useful program in THERM is called c:\ckin\thermrxn. In this program you type the equation of a reaction, all of whose species are entered in thermo.dat.

CUMENE+H2=BENZENE+PROPANE

and with incredible speed a table is printed out

```
Rx    cumene+H2=benzene+propane
Hf {kcal/mol} -14.890    .000   19.8000   -25.300
S {cal/mol K} 53.130   31.200   64.200    64.500

dHr {kcal/mol} (298K) = 9.39  dHr avg. (298., 1500. K) = 35.83
dU (dE) {kcal/mol}" = 9.39    dUr avg(298., 1500. K) = 35.83
dSr {cal/mol K}" = 44.37      dSr avg (298., 1500. K) = 75.24
dGr {kcal/mol}" = -3.84       dGr avg (298., 1500) = -31.82
Kc " = 6.520E+02              Kc avg (298., 1500) = 5.433E+07
```

T (K)	dH (Kcal/mol)	dU (Kcal/mol)	dS (cal/mol)	Kc	dG (Kcal/mol)
300.00	9.418E+00	9.418E+00	4.446E+01	7.190E+02	-3.921E+00
400.00	1.137E+01	1.137E+01	5.001E+01	5.252E+04	-8.639E+00
500.00	1.405E+01	1.405E+01	5.598E+01	1.236E+06	-1.394E+01

and so on from 300 to 2000 K. There are many user-friendly variations of the table that can be displayed, extending temperature ranges, changing temperature increments, etc. One variation is to change the quantity computed and displayed in the column for Kc to Af/Ar, changing the equilibrium constant to the ratio of Arrhenius constants, forward to reverse, so that in the top half of the table the line Kc would read

```
Af/Ar " = 4.987E+09     AF/Ar avg (298., 1500K. K)
                          = 2.788E+16
```

and in the bottom half the Kc column would be changed to

```
(Af/Ar)

5.229E+09
8.523E+10
1,718E+12
```

This table is valuable in deciding on mechanisms for remedies to environmental problems. If the equilibriums are favorable and the ratio of Arrhenius constants for rate constants are favorable, the reaction(s) are worth a try in the Envirochemkin program for more-detailed computation with many competing reactions.

Another routine in THERM calculates the thermodynamic values as in thermo.lst from groups. Over 380 groups are considered, including free radicals, etc. Values are taken from Benson's book and represented by a code. Knowledge of physical chemistry and some statistical mechanics are helpful, but the user's manual and instructions are valuable. This program is useful when values are not available in the literature. Tables from the program or Benson's book can also be used for calculating results manually.

2.99 Reaction Files

Once a program such as Chemkin becomes more widely used, it requires reaction files for each problem that it runs. On an ordinary personal computer one can run problems with 50 chemical species and 100 chemical equations or reactions. It becomes necessary with some very important problems to have more species and more reactions. The author did participate with the IEEE and its computer and supercomputer groups in a white paper to the U.S. government for funds for supercomputers with just such a justification. The author has run some programs on personal computers with less than the full complement of variables out of necessity but was able to justify the answers because of his background in science and mathematics.

At first, the data for Chemkin runs was supplied by NASA and DOE together with the literature and some professors at key universities. Now that the Chemkin style of computation has gained wide credibility, other organizations have joined in to provide the data in excellent format and such data have become more widely available, particularly to people on the Internet. A sample of such data, as derived from the Internet, is included here for a system of 50 species with the chemical elements O, H, C, N, and AR. However, the molecules containing carbon are limited to those containing no more than two carbon atoms per molecule. Even with this restriction, the number of equations or the number of reactions is almost 300 equations! However, a great service has been offered in that the constants, etc., have been checked out to be the latest and best available. The data can be obtained from http://www.me.berkely.edu/gri_mech/version21/files21/grimech211.dat and can be used with GRI-MECH thermodynamics which is the same as that described before. Some additional information in the file will be described here.

2.100 Rate Coefficient Format

A crude estimation of rate constants is derived from a simple application of statistical mechanics from the physical constitution of the molecules and is given in Table 25. However, a more stringent approach is shown below.

The rate constant parameter for bimolecular reactions are set up in the form

$$K = At^m \exp(-E/RT) \qquad (2.100.1)$$

with concentration units mol/cm^3. The A units are cm^3/mol/s, T is in K, and E is in cal/mol/K. The low pressure limit rate constants and the rate constants of unimolecular reactions that are always near their low pressure limits for combustion conditions have been assigned "enhanced efficiencies" in Chemkin format. From the model concentrations [M_i] and the efficiencies {f_i}, an effective concentration is computed through the formula:

$$[M] = \text{sum}\{f\_i[M\_i]\} \qquad (2.100.2)$$

where f_i = 1 for all species except those with enhanced efficiencies listed on the last line of the reaction input.

The Lindemann fall-off form treatment applies if no specific fall-off parameters are given. At pressures intermediate to the high and low pressure limits, the rate constant is given by the Lindemann formula:

$$k = k\_inf/(1 + k\_inf/k\_o[M]) \qquad (2.100.3)$$

In cases where no high pressure limit rate constant parameters are given (i.e., the collider M as a reactant is not in parenthesis), the reaction is in the low pressure limit.

The fall-off parameter F_cent for a unimolecular reaction is calculated from the values of a, b, and c and d by the formula of Troe

$$F\_cent = (1 - a)\exp(-T/b) + a\exp(-T/c) + \exp(-d/T) \qquad (2.100.4)$$

which gives the temperature dependence of F_cent, the factor by which the rate constant of a given unimolecular reaction at temperature T and reduced pressure P_r = k_o[M]/k_inf of 1.0 is less than the value of k_inf/2, which it would have if unimolecular reactions behaved according to the Lindemann formula.

The broadening factor F, which is 1 for the Lindemann case where no parameters for F_cent, are is

$$\text{Log F} = \log \text{F\_cent}/(1 + [(\log \text{P\_r} + \text{C})/(\text{N} - 0.14\{\log \text{P\_r} + \text{C}\})]^2 \qquad (2.100.5)$$

with N = 0,75 – 1.27 log F_cent and C = –0.4 – 0.67 log F_cent.

2.101 Modeling Microbial Chemotaxis in a Diffusion Gradient Chamber

M. T. Widman, D. Emerson, and C. C. Chiu have found that the diffusion gradient chamber (DGC) has proved a useful experimental tool for studying population-level microbial growth and chemotaxis. A mathematical model capable of reproducing the population-level patterns formed as a result of cellular growth and chemotaxis in the DGC has been developed. The model consists of coupled partial differential balance equations for cells, chemoattractants, and a nutrient that are solved simultaneously by the alternating direction implicit method. Modeling simulated results were compared with population-level migration patterns of *Escherichia coli* growing on glycerol and responding to a gradient of the chemoattractant aspartate for two different initial conditions. To reproduce the experimental results accurately, a second chemoattractant equation was necessary. The second chemoattractant has been identified as oxygen by directly measuring oxygen gradients in the DGC. Important trends observed experimentally and reproduced by the model include the formation of a chemotactic wave, a reduction in the wave velocity as it encounters higher chemoattractant concentrations, and chemotaxis in response to two different chemoattractants simultaneously. The model was also used to study the relative magnitude of cell fluxes due to random motility and chemotaxis, and the suppression of chemotaxis due to receptor saturation.

2.102 The Path Probability Method

The path integral from quantum mechanics and theoretical physics is used to plan the best chemical groups for a given constrained stoichiometry. This is then utilized to ascertain whether or not there is a reaction scheme or mechanism to produce a molecular species composed of the assigned chemical groups. The program used to find the set of groups is SYNPROPS, and the program Envirochemkin is used to find whether the reactions are feasible under a set of conditions (P, T. t. mode and mechanism) used in Envirochemkin. The program THERM is used to obtain thermodynamic functions for the molecule if they are not known or contained in the Envirochemkin file. It is to be noticed that the variable emphasis in this undertaking is chemical groups (of which there are 380 choices as in the program THERM) rather than chemical species as originally used in SYNPROPS in which there were 32 in the linear model and 33 in the hierarchical model, many of which were duplicated in both models.

Each of the groups in THERM has thermodynamic functions associated with it, namely, heat of formation at 298 K, entropy at 298 K, and heat capacities at constant pressure at 300, 400, 500, 600, 800, 1000, and 1500 K. Therefore, the free energy can be found at any temperature (F = H – TS), and thus the free energy difference between the species and its precursor or descendant or that for any reaction between precursor and descendant can be obtained if the free energies of the precursor or descendant species are known as well. The data for 380 groups include that for free radicals, etc., so activated complexes can be included in the scheme and thus a tree can be drawn for the progress of the reactions from the initial reaction to the final reaction with the probability of occupancy of the different levels of the tree as well as the various participants in the progress of the reaction. This probability can be assumed to be proportional to the equilibrium values of the species, which is proportional to the value of the quantities exp(–F/RT) which is obtained from the change in free energy of the reactions involved. The process can thus be used to arrive at the mechanism of the overall reaction and each of its constituent parts.

Another way to obtain mechanisms of reaction is from rate distortion theory. Here a tree can be constructed to decode messages that are used in communication theory but will now be used to ascertain and depict needed mechanisms.

The path probability method of irreversible statistical mechanics has been applied to pollution prevention and waste minimization. The most probable path in time taken by a system is derived by maximizing the path probability after adding a space axis to equilibrium statistical mechanics. The path probability formulation is based on the Markoffian character of the process, and this depends on the choice of variables used in describing the system. In a cooperative process, all cooperating degrees of freedom must be taken into account. The cluster-variation method in which a finite size of a cluster is used to represent the whole system violates the Markoffian requirement of the process. The formulas for the most probable path can be interpreted based on a superposition approximation, which is a shortcut to the expressions for the most probable path without going through the path probability formulation and its maximization each time that will greatly increase the maneuverability of the technique.

It is interesting to note that Feynman's space–time approach can be used to write down quickly results rigorously derived from quantum field theory (or second-quantized Dirac theory). Matrix elements derived using quantum field theory can be obtained much more quickly using the space–time approach of Feynman. Feynman's approach (based on the particle wave theory of Dirac) is simple and intuitive. It visualizes the formulas that we derive rigorously from field theory correctly. The Feynman graphs and rules have had a profound effect on a number of areas of physics, including quantum electrodynamics, high energy (elementary particle) physics, nuclear many-body problems, superconductivity, hard-sphere Bose gases, polaron problems, etc. Although we do not use the technique here, there is a relation between the path probability method and Feynman's approach.

Consider a system of N atoms, each of which has two energy levels, g and e (for ground and excited). During a short time interval t, states of atoms may change by exchanging energies with a heat bath of temperature T. When we look at the system, its configuration may change in time, as shown on the left where a system of two-level atoms changes in time (e and g stand for the excited and the ground states, respectively). At the right is a configuration of an assembly of one-dimensional Ising model (x and o are a plus and minus spin, respectively).

The correspondence between these cases suggests that the irreversible problem on the left can be treated in analogy with the equilibrium problem when the time axis is treated as the fourth space axis. Thus, the probability function **P** to be written for the case on the left is expected to be constructed in analogy with the state probability function **P** or the free energy **F** for the case on the right.

Let us consider a tree, upside down with the root at the top. As we proceed from top to bottom we have choices about which compounds form from the original compound.

$$C_0^0$$
a^c
$$C_1^1 \; C_1^1$$
d^e f^g
$$C_2^1 \; C_2^2 \; C_2^3 \; C_2^4$$

Thus, the first compound, which might be PAN or some other pollutant we wish to eliminate, is made to react with additives so that it forms the first tier of compounds, indicated by the subscript 1, and the two compounds in the first tier react with additives to form the four compounds in the second tier. The question is which is the most probable path that will be followed, a → d, a → e, or c → f, etc?

To decide this, one may utilize the path probability method of Kikuchi together with some of the author's own publications on order–disorder theory. The probability that a given path is followed is proportional to the reaction rates that take place. In the reaction A + B = C + D, this can be approximated by knowing the structure of the molecules A and B and the activated complex A–B. Table 25 is included to approximate the range of such rate constants. Also needed is the energy of the reaction that can be

t −Δt t t +Δ t	k − 1 k k + 1
atom1 —>e—>g—>g—>e—>g—>e—>	system 1-x—o—o—x—o—x
atom 2 —>g—>g—>e—>e—>e—>g—>	system 2-o—o—x—x—x—o-
— — —	— — —
— — —	— — —
atom N—>g—>g—>e—>g—>e—>e—>	system N -o—x—o—o—x—x-

obtained from the program THERM. This is usually printed out in a table and may be the enthalpy or the free energy of reaction under reaction conditions. The equation for the probability of reaction comes from the order–disorder theory which is an equilibrium method because the kinetic theory reduces to equilibrium expressions as stated in Kikuchi's paper. Finally, we need the concentration (and conditions) of the pollutant, and this is available either from Envirochemkin calculations or actual measurements in plant processes.

Thus we have:

$$\begin{aligned} P4 &= \text{(conc. of pollutant)(path probability)} \\ &\quad \text{(rate constant)(energy factor)} \\ &= \text{pollution prevention path probability} \end{aligned} \qquad (2.102.1)$$

The concentration of pollutant is derived from the Envirochemkin program or measurements, the rate constant comes from the structure of the reactants and activated complex or the value of A_f/A_r from the output of the Thermrxn program, the (free) energy factor also comes from the subroutine Thermrxn of the THERM program, and the path probability comes from the formalism developed here. Initially, the path probability can be set equal to unity and the three factors can be checked: the rate constant from the estimated rate table, the ratio of Arrhenius factors from THERM, and the equilibrium constant from THERM. If the magnitude of these three constants are positive and the magnitude sufficiently large, then the proposed reaction, concentrations, conditions, and/or additional reactants can be tried in Envirochemkin for a final computation in a proper molecular environment. Thus, one can control the environment and have clean production.

Order–disorder theory, also called cooperative phenomena or chaos, has been used to find critical conditions for systems with interactions between particles. This can be very helpful to find discontinuities in chemical potentials and phase equilibrium, which can affect the reaction kinetics and equilibrium of systems that we are studying here.

2.103 Multistep Reaction Mechanisms and the Most Probable Path

One can study the basic figure (d) signifying the rhombus where the superscript d will be left out in this discussion. The first number in the subscript of the rhombus figure signifies the degree of its particular occupancy. The other numbers in the subscript signify the distribution of occupancy in that particular subfigure of the basic figure. Now in going from the completely empty figure p_0 to the completely occupied rhombus p_4, six paths are permissible:

$$p_0 \rightarrow p_{12} \rightarrow p_{212} \rightarrow p_{31} \rightarrow p_4 \text{ Path a}$$

$$p_0 \rightarrow p_{11} \rightarrow p_{212} \rightarrow p_{31} \rightarrow p_4 \text{ Path b}$$

$$p_0 \rightarrow p_{11} \rightarrow p_{211} \rightarrow p_{31} \rightarrow p_4 \text{ Path c}$$

$$p_0 \rightarrow p_{12} \rightarrow p_{22} \rightarrow p_{32} \rightarrow p_4 \text{ Path d}$$

$$p_0 \rightarrow p_{11} \rightarrow p_{212} \rightarrow p_{32} \rightarrow p_4 \text{ Path e}$$

$$p_0 \rightarrow p_{12} \rightarrow p_{212} \rightarrow p_{32} \rightarrow p_4 \text{ Path f}$$

Now in these paths two kinds of bonds can form, the shorter stronger bond (c) and the longer weaker diagonal bond (c′). In proceeding from p_0 to p_4, the stronger c bond will tend to be formed first and figures containing such bonds are favored, followed by figures containing the longer, weaker diagonal bonds with time. However, on proceeding from p_4 to p_0, the opposite will occur as the kinetics is not truly reversible.

This has been applied to the attempt to bring the concentration of the greenhouse gas, CO_2, in the atmosphere down to concentrations where its effect will not be as deleterious in the phenomenon known as global warming. Here the sites in the basic figure represent a degree of forestation, and empty sites represent deforestation. When the concentration of such forests is high, we have canopies present. The forests can absorb the CO_2 and can also lose some CO_2, but we begin to see from the foregoing discussion that hysteresis is present and that the forest can capture, store, destroy, or degrade a good portion of the CO_2. This will be especially true with anthropomorphic mechanical devices to aid in the speed and equilibrium of the process. This model has also been applied to other biochemical and photosynthetic processes with good results.

2.104 Hysteresis

Viscosity is a general term for all those properties of matter in which the resistance that a body offers to any change depends upon the rate at which the change is effected. The existence of viscous resistances involves a dissipation of the energy of the substances, the kinetic energy of the molar motion being transformed, as is generally supposed, into kinetic energy of molecular agitation.

Lord Kelvin pointed out that the decay of vibrations could be accounted for by supposing that, even for the very small strains involved in vibratory motions, the effects of elastic afterworking and plasticity are not wholly absent. These effects, as well as viscous resistances of the ordinary type, are included in the hysteresis phenomena. All of them show that the state of the body depends at any instant on its previous states as well as on the external conditions at the instant. Hysteresis always implies irreversibility in the sequence of states through which a body passes, and is generally traced to the molecular structure of matter.

Noise has been found in hysteresis loops. Even though they look smooth, hysteresis loops often consist of many small jumps. These jumps can be thought of as the jerky motion of a domain boundary, or as an avalanche of many local spins or domains. Many hysteric systems make screeching noises as they respond to their external load. One is not hearing the individual, microscopic domains transforming: the crackling noises are large avalanches of domains. The little domains can be pushed over the edge by the external field or by the connection with their neighbors; as the field increases, one domain transforms, triggering some of its neighbors, who kick over some of their neighbors.... The noise in hysteresis is the popping sound made by this cascade of avalanches.

Self-similar behavior is governed by certain universal critical exponents; universal here means that different systems will have the same exponents. These exponents as a function of dimension have been predicted using the "renormalization group."

The parameter R measures the randomness relative to the strength of the connection between domains. If the randomness is large compared with the coupling, the hysteresis overall is smooth, with only small jumps; if the randomness is small, one infinite avalanche transforms most of the system. A system returns to the original curve that it has left in hysteresis. This return-point memory effect is known as "wiping out," since the memory of the excursion is wiped from the behavior of the material as soon as the field is raised past the previous extreme level. A critical point separates smooth hysteresis loops from loops with a macroscopic jump or burst. At large disorder, every domain flips independently; at disorder small compared to the coupling between domains, a single domain flip can send its neighbors flipping until most of the system transforms in one burst.

Avalanches come in a wide range of sizes. The change in the shape of the hysteresis loop is important. If the system is very clean, the first domain to be pushed over by the external field can push over its neighbors, leading to an infinite avalanche, which turns most of the domains. If the system is very dirty, the domains will probably flip only a few at a time: the avalanches get stopped by the dirt.

2.105 Energy and Environmental Research Enhancement Program

Artificial Neural Networks: A New Tool for the Petroleum Industry

This project explores the application of artificial neural networks (ANN) to petroleum engineering–related problems. A new and powerful tool, ANN can help engineers solve problems that conventional tools (analytical and numerical methods) are unable to solve. ANN has proved efficient in solving problems. This results in higher efficiency in the production of fossil fuels, better management of oil and gas fields, and considerable savings compared with conventional methods. The main objective here is to develop a methodology for hydrocarbon reservoir heterogeneity characterization using ANN.

Previously it had been found that using ANN for virtual measurement of formation permeability is a feasible methodology and that ANN can provide accurate permeability values from well log data. This will eliminate the need for coring all the wells in a field in order to get a detailed picture of flow characteristics.

This method is now being applied to virtual measurement of other formation parameters, such as effective porosity, irreducible water saturation, and residual oil and residual gas saturation.

Design and Development of an Expert System for Energy Analysis and Diagnostics: Energy Conservation and Efficiency in Manufacturing Industries

In today's environment of depleting energy resources, this project provides a way to conserve and use industrial energy more efficiently, thus increasing the competitive edge of industries by lowering their energy costs while, at the same time, addressing our national energy concerns. The main goals of the project were to identify ways and means to reduce industrial energy consumption by analyzing industrial operations and support systems. In specific terms, the objective of the project was to design and develop computer-based software systems that can aid the industrial user in self assessments of energy use.

Earlier, we researched industrial energy conservation and efficiency, we studied industrial operation and support systems in order to develop energy-efficiency action plans, and we developed computer-based computer systems to address energy conservation and efficiency with respect to specific industrial equipment and operational practices.

2.106 The Game of Life

"Cellular automaton" extends the analogy of treating objects as simple computers, each obeying its own set of laws in order to provide a way of viewing whole populations of interacting "cells." By building appropriate rules into cellular automation, we simulate all kinds of complex behavior. The properties include *self-organization.* Successive states cause a pattern to emerge. *Life-like behavior.* Wolfram and others show that even simple linear automata behave in ways reminiscent of complex biological behavior. The fate of any initial configuration of a cellular automaton is either to die out, to become stable or cycle with fixed period, to grow indefinitely at a fixed speed, to grow and contract irregularly, or to exhibit *"thermal" behavior.* Here, models that force a change of state for few configurations tend to "freeze" into fixed patterns, whereas models that change the state of the cell in most configurations tend to behave in a more "gaseous" way. That is, fixed patterns do not emerge.

As an example, consider a system whose states are of the form (s,t), where t denotes "time since fire burned out the cell" and s takes one of the values: "bare earth" (E), "grass" (G), "woodland" (W), or closed forest (F). After state transition rules are defined, we start with an arbitrary "landscape." Consisting of a 2D grid of cells in random states, the rules quickly produce patterns. "Forest zones" quickly develop, even though the model contains no assumptions about the environment or site preferences of the vegetation. Although this may be a mere caricature of true forest succession, it is only a short jump to management models taking (as an example) satellite data as inputs.

2.107 Game Theory and Decision Theory

In 1944 von Neumann and Morgenstern published *Theory of Games and Economic Behavior.* A game has a goal or end state for which the players strive by selecting courses of action permissible under the rules. The games we consider here are competitive. A game is competititive when there is an end state (winning) such that each player deserves it, but not all can obtain it. Each player has a set of possible choices. To select one is to move. A play is a sequence of choices that brings the game to an end-state. In many games, the attainment of the goal is accomplished by a payment of some kind, usually money. A strategy is a player's predetermined method for making his or her choices during the game. Hence, a strategy is a set of decision rules. A payoff matrix is a table that specifies how payments should be made at the completion of the game. A zero-sum game is one in which the payments, upon completion of the game, are equal to zero. The above does not try to describe how the game is to be played. Rather, it is concerned with the procedure and principles by which plays should be selected. It is a decision theory applicable to competitive situations.

The author's role as an "expert witness" in a number of environmental cases has prompted the inclusion of this section here. These cases have included the evaluation of proposed remediation steps where environmental damage has been done, the cause and extent of environmental damage from toxic releases from operating plants, and the culpability and forensic damage for illegal dumping of hazardous waste. In each of these cases the proceedings take place as delineated above for "game theory." Also, the matter of physical properties of the materials concerned can be very important in the proceedings. There is a "decision maker" in each of the cases or someone who represents the decision maker. This person is frequently the CEO of the errant industrial company concerned. His or her agenda of important corporate plans is kept very much in tact until, finally, there is a breakthrough, and the proceedings slide steadily from there. This occurs when the opposing side finds an agreeable place on the corporate agenda.

2.108 Process Design

Process design uses molecular properties extensively and it is a very important part of such work. El-Halwagi uses the concept of integrated process design or process integration which is a holistic approach to design that emphasizes the unity of the whole process. He states that powerful tools now exist for treating industrial processes and sites as integrated systems. These are used together with a problem-solving philosophy that involves addressing the big picture first, using fundamental principles and dealing with details only after the major structural decisions are made. In further work, two approaches are developed: graphical and algorithmic. In the graphical approach, a new representation is developed to provide a global tracking for the various species of interest. The graphical approach provides a global understanding of optimum flow, separation, and conversion of mass throughout the process. It also provides a conceptual flow sheet that has the least number of processing stages. In the algorithmic approach, the problem is formulated as an optimization program and solved to identify the optimum flow sheet configuration along with the optimum operating conditions.

A systematic tool is developed to screen reaction alternatives without enumerating them. This task of synthesizing "environmentally acceptable reactions" is a mixed-integer, nonlinear optimization program that examines overall reactions occurring in a single reactor to produce a specified product. It is designed to maximize the economic potential of the reaction subject to a series of stoichiometric, thermodynamic, and environmental constraints. It is a screening tool so additional laboratory investigation, path synthesis, kinetics, and reactor design may be needed, but it is an excellent starting point to plan experimental work.

2.109 Recipes for Environmental Information Fields

In some recently published books, Rupert Sheldrake has stated that there is a "field" in the environment that determines the attributes or growth and morphology of developing biological organisms. This field can amount to "intelligence" or a set of "information" that determines the future development of organisms. Thus, once something has occurred, it is embedded in the universe, making it more likely that this development can occur again. While running the programs in this work, certain phenomena that occurred seemed to corroborate such thoughts. It is here visualized that certain molecules first appeared on "Earth" because there was a "recipe" of physical and toxicological properties that were needed, in a similar manner to the way the molecules in the present programs are formed with the properties as constraints. It has also been thought appropriate to replace toxic or hazardous materials that have been cleaned from sites with their benign "clones" to conserve the information network of the planet and, in effect, "trade credits." Stuart Kauffman states that systems at the boundary of chaos ("rugged landscapes") have the proper structure to interact with and internally represent other entities of their environment. Adaptive evolution achieves the kind of complex systems that are able to adapt.

It has been attempted to map the structure for each molecule in the complex plane as a potential and its orthogonal streamlines as properties, or vice versa. The earlier-discussed Schwarz–Christoffel transformation and conformal mapping can help with these mnemonics. The computer could then follow the chemical synthesis process, in detail, from reactants to products.

2.110 Conclusions and Recommendations

SYNPROPS synthesizes molecules of optimum structure for the most-desired physical properties and minimum toxic properties. Its constellation of properties can replace molecules with environmentally hazardous properties with those with environmentally benign properties and better physical properties at lower cost. The program works well, but it is proposed that more classes of molecules be regressed for their quantitative structure–physical properties and structure–toxicity relations and that more "hidden" data become publicly available. The program is now ready for extension into other fields, such as medicine.

The work of L. T. Fan and F. Friedler et al. is very significant. It is now in a program called Synphony from Epson International. It revises a flow sheet for Plants into ones free of hazardous waste with the profit motive of the plant owners carefully considered.

2.111 The Need for Good Data

The information that can be obtained from the programs above can only be as good as the data that were originally used for the SYNPROPS data, and the thermodynamics and rate constants for the Envirochemkin programs. In this work the best procedures were used to obtain such data from the literature and from programs such as THERM. However, there is a need to do concentrated work involving many people to obtain good validated data and to make such data available to many scientists so that such work as this can become even more reliable and accurate. Such work that was done during the days of the beginning of the space endeavor and the missile defense efforts carried out by the government and the cooperative efforts of many industrial companies would be appropriate. An organization such as NIST should supply needed information to the scientific and engineering community free of charge, supported financially by the government in conjunction with industrial companies. We all stand to benefit. One can readily see that with such compendiums of knowledge there would come new states of matter to fight disease and create a better environment to benefit greatly this country and the world in times to come.

Part III

3.1 Application of Order–Disorder Theory to Cancer

The sensitivity of the smallest events as input to a system can lead to the most massive of consequences. The term "butterfly effect" captures the idea that through nonlinear systems or chaos, "the beating of a butterfly's wings in Brazil could spark off a tornado in Texas." This is a sensitive dependence on initial conditions. Such events also occur in solid-state physics and can be seen dramatically in a physics laboratory where the proper doping of GE or SI drastically alters a measured physical property known as the Hall effect.

Computer modeling has indeed succeeded in changing many phenomena from an art to a science. In the following pages a model is studied on a computer. It is then applied to the disease — cancer.

The adsorption of molecules (on cells) and their interaction lead to physical changes of the adsorbate. These physical changes can be responsible for the attributes of cancer and can suggest models of cause, therapy, and cure for the disease.

This study of the disease of cancer uses terminology such as adsorption rather than docking and the order–disorder theory of the adsorbed molecules as a stage for their reaction that leads to carcinogenic properties and their cause and prevention.

A mathematical procedure combined with statistical mechanics leads to a group theoretic approximation that describes conditions resulting in critical regions for the adsorbate. This then suggests some changes in methods that can reduce or prevent the growth of the disease and also verifies some of the present methods.

Methods are suggested that can then change the physical, chemical, and environmental properties of the adsorbate and then abate or even cure the disease. Avoidance of the critical region is a key ingredient for arresting the growth of the cancer.

Molecular replacement, photochemistry, heat, signal transduction, etc. are some of the techniques suggested as beneficial. Newly existing computer programs, matrix techniques, data from the Unifac model, exaltation data, etc. are also suggested to provide candidate molecular surrogates and methods in oncology.

A previous publication (Bumble and Honig, 1960) has applied order–disorder theory to gas adsorption. Since much of the mathematical groundwork was given in that paper, it will not be repeated here. Other approaches have also been applied by others to ferromagnetism, binary alloys, lattice models of fluids and solids, etc. A very brief review of the paper shows that the general treatment of order–disorder theory leads to solutions that can be approached to any desired degree of approximation using a systematic and well-delineated approach. In this description a lattice consists of a regular array of repetitive sites. This lattice is decomposed into subfigures, a limited number of sites in close proximity (such as the point, the bond, the triangle, or the rhombus), abstracted from the lattice. The largest subfigure is termed the basic figure.

Figure 90 shows how the rhombus is nested within a star configuration. Some of the lattice sites may be occupied and each occupation state of the lattice is reflected in an appropriate configuration of the various figure assemblies. Figure 91 shows the rhombus as a basic figure in its various occupied states. One assigns to each configuration of a given subfigure an energy and a probability of occurrence. The energy and entropy of these assemblies may then be computed. The free energy then leads, after minimization, to the chemical potential for the phase in equilibrium with the adsorbed phase. The resulting expression can be called an isotherm equation.

Plots of the pressure of the gas phase vs. the occupancy of the adsorbent will lead to loops if the temperature is below the critical temperature. The interaction is denoted by c = exp(−w/kT), where w is the interaction energy between nearest neighbors in the lattice. The plots show the plots of various subfigures as the value of c increases. It is seen that the triangle and the rhombus approximations are relatively sluggish in comparison with the bond approximation in looping or reaching the critical point. The critical region is really that which is first reached when the temperature or the interaction energy first indicates a flat region of the graph or one that first becomes parallel to the abscissa. The critical region shows that there are two compositions or theta (occupancy) values with the same pressure. Physically this can be interpreted as a gas and liquid in equilibrium or a change of phase.

The method used is one based on a principle introduced by R. Kikuchi and developed further by J. Hijmans and J. de Boer. In these systematic procedures, the lattice is decomposed into certain simple figures, and the various possible modes of distribution of particles are considered. The free energy of the system is shown to be a linear combination of the free energy of these simple units. At equilibrium, the chemical potentials of the physical phases in equilibrium are set equal to each other. One or more equations result from this which relate the physical state variable of one phase (such as gas pressure) with the degree of occupancy of the other. The distribution numbers for the configurations of the different figures are interrelated by geometric consistency relations within the lattice, and by normalization requirements. The set of all equilibrium, consistency, and normalizing relations leads to the elimination of all unnecessary variables and finally provides an equation that relates the gas pressure to the surface coverage only in terms of parameters containing the interaction energies and temperature and lattice coordination number. The work presented here sacrifices some accuracy for simplicity. Thus, the equations can be incorporated into a spreadsheet and manipulated most readily for results.

Results

1. In order to inhibit the growth of malignant cell and tissue, this study suggests that the carcinogenic tissue be washed or flushed with a liquid that can insert or replace molecules or ions with ones that do not interact with their neighbors as strongly as the original ones.
2. Also, the surrogate molecules or ions should have a radius more conducive to blocking the number of deleterious interactions by changing the coordination number of the destructive carcinogenic molecules.
3. The signal communication or transduction between cells must be ameliorated for those that are beneficial and destroyed for those that are harmful. This can be expedited by studying the pathways, using the methods shown here, that are conducive to good health.
4. Critical regions must be avoided at all costs. The transition between states of matter resulting in the liquefication of cells is a fatal phenomenon.
5. The results indicate that the effects noted in the literature regarding the application of heat or withdrawal of heat to tumors are in accord with the model given here and it would be well for medical and surgery specialists to study the ramifications of heat transfer to tumor size and how to lessen the pain involved in the process.
6. Photochemistry, photodynamics, and laser therapy are elucidated by the model, and clinical observations are again predicted by this model.
7. A matrix technique is used to divide the molecular aspects from those of the interaction aspects and in so doing surrogate candidates can be found for these individual aspects to reduce or destroy carcinogenic tissue.
8. Biology does not render itself into simple order. It is governed by a systematic disorder and requires mathematics of disorder or chaos to reflect reality. This model is strictly nonlinear.
9. Electomagnetic fields can be set up from cells or molecules with dipoles aligned in the + − + − + − manner and can act to send signals or create a morphogenetic field.
10. The UNIFAC model for solutions has surface areas and volumes for chemical groups and interaction energies for many chemical groups that can be optimized to provide the best candidate molecular species to prevent or abate cancer.
11. Three computer programs exist to
    a. Provide kinetics for reactions involving chemical species involved in cancer (Envirochemkin),
    b. Provide thermodynamic functions for complex species involved in cancer (THERM), and
    c. Provide and optimize properties for chemical species involved in cancer (SYNPROPS).
12. The pathways between the subfigures of the basic figure become more numerous as the basic figure becomes larger and next-nearest neighbors appear and are taken into account. The calculated probabilities of these paths are close in value, yield logical values, and provide signal possibilities that can be important to growth.

13. The refraction exaltation (difference between the experimental refraction and the calculated refraction) is due to a conjugated system of double bonds. Frequently, such molecules are higher polyaromatic hydrocarbons, etc. and are carcinogenic.
14. Signals are transmitted by an "antenna" as charges oscillate back and forth. According to field theory, these charges produce an electromagnetic wave. The wave reaches a receiving antenna and sets the charges in that antenna into oscillation, with results that are detected in the receiver. The paper presented below shows some ways that carcinogenic materials are deposited on a substrate of cells and are prepared for chemical reactions that form the basis for signal transmittal.
15. When the force between molecules is of a considerable magnitude and repulsive, then the probability for an occupied rhombus can become not only more sigmoidal, but also unstable, whether the cause is mathematical or real.
16. The expression for q as a function of p/po, resembles a Fermi–Dirac distribution function and a plot shows a very sharp step function at body temperature from q equals 0 to unity. This leads to a model from solid-state physics where there are bands of energy within living creatures such as the valence band and the conduction band. When cancer cells can reach the conduction band, then metastasis is prevalent.
17. The model studied here concentrates on the order–disorder of the substrate. The molecules that then come in contact with a relatively stationary substrate can undergo chemical reactions with those in the substrate. It is the chemical kinetics of such reactions, where the substrate is near or in the critical region, that can be one of the important causes of cancer.
18. Figure 104 elucidates the above. It utilizes the Michaelis–Menten equation as a model. This defines the quanitative relationship between the enzyme reaction rate and the substrate concentration [S] if both Vm and Km are known. Substitution of q for [S] then shows that when the attractive forces in the substrate are large, the reaction rate is much above that appropriate for the Langmuir isotherm, whereas when the forces are very repulsive, the reaction rate can fall much below.

In the chaotic regime, the dynamics are very sensitive to initial conditions. The transition from the ordered to the chaotic regime constitutes a phase transition, which occurs as a variety of parameters are changed. The transition region, on the edge between order and chaos, is the complex regime.

Many cellular and biochemical processes exhibit a response that changes an S-shaped, or sigmoidal curve, as a function of altered levels of some regular input. For example, hemoglobin is a tetrameric protein, with each monomer binding oxygen. The binding behavior of the four monomers exhibits positive cooperativity. At first, the binding increases the affinity, but then it saturates.

It is also true that in a sigmoidal system, where the maximum slope is less than vertical, coupled systems are often idealized as on–off systems. Over iterations of such a system, feedback signals through a sigmoidal function tend to sharpen to an all-or-nothing response.

The internal homogeneity of a system depends on the kind of lattice involved and increases as the number of neighbors of each point in the lattice increases. It is higher for a triangular lattice (Z = 6) than for a square lattice (Z = 4).

Parallel-processing systems in the phase boundary interface region (between order and chaos) may be those best able to adapt and evolve. Also, natural selection may be the force that pulls complex adaptive systems into this boundary region so that a tool is available to examine the collaborative interaction between self-organization and selection.

Complex systems exhibit spontaneous order. Thus, it is possible that adaptive evolution achieves the kind of complex systems that are able to adapt.

Approximation of the Equations to Be Used in a Personal Computer

Basic figures are shown for the bond, triangle, and rhombus and the figure that can be assembled from them. We call the probability of empty sites p_0 and those occupied by 1 to 4 particles, p_1 to p_4, respectively. The superscripts a, b, c, and d refer to the point, bond, triangle, and rhombus basic figure, respectively. When the point becomes occupied we have $a_1 = p_1^{(a)}/p_0^{(a)}$. Likewise, a bond can become $b_1 = p_1^{(b)}/p_0^{(b)}$ or $b_2 = p_2^{(b)}/p_0^{(b)}$. A triangle can become c_1, c_2, or c_3 and a rhombus can become d_{11}, d_{12}, d_{22}, $d_{2,12}$, $d_{2,11}$, d_{31}, d_{32}, and d_4, similarly. Multiple numbers are needed in the rhombus because of equivalent sites. It turns out from the section above and from algebra and group theoretic concepts that the variables can be expressed as direct products.

The setup of the equations for each basic figure derived from the equilibrium, the consistency, and the normalizing relations is not repeated here but is described in the literature. A new method is shown that is appropriate for a personal computer. To do this it is convenient to define $K = (p/p_0(1 - \theta/\theta))^{1/Z}$, where p is the pressure, Z is the coordination number, and θ is the fraction of the surface covered.

$$b_1 = Ka_1 \quad (3.1.1)$$

$$b_2 = c(Ka_1)^2 \quad (3.1.2)$$

$$C_1 = K\tau \quad (3.1.3)$$

$$C_2 = (K\tau)^2(\sigma/\tau) = K^2\tau\sigma \quad (3.1.4)$$

$$C_3 = (KT)^3(\sigma/T)^3 = K^3\sigma^3 \quad (3.1.5)$$

$$d_{11} = K^2\tau' \quad (3.1.6)$$

$$d_{12} = K\tau' \quad (3.1.7)$$

$$d_{22} = (K\tau')^2c' = c'K^2(\tau')^2 \quad (3.1.8)$$

$$d_{2,12} = (K^2\tau')(K\tau')(\sigma'/\tau) = K^3(\tau')(\sigma') \quad (3.1.9)$$

$$d_{2,11} = (K^2\tau')^2c = cK^4(\tau')^2 \quad (3.1.10)$$

$$d_{31} = (K^2\tau')^2(K\tau')(\sigma'/\tau')^2c = cK^5(\tau')(\sigma')^2 \quad (3.1.11)$$

$$d_{32} = (K^2\tau')(K\tau')^2(\sigma'/\tau')^2c' = c'K_4(\tau')(\sigma')^2 \quad (3.1.12)$$

$$d_4 = (K^2\tau')^2(K\tau')^2(/\tau')^4cc' = c'cK^6(\sigma')^4 \quad (3.1.13)$$

and the functions σ, τ, σ′, and τ′ are simple algebraic functions of c, c′, and K.

One may now postulate that in the formation of the rhombus figure described by $m_{is}{}^d$ from the completely unoccupied figure, a factor x will occur for every occupied site on the diagonal, a factor x′ for every occupied site, on the long diagonal, a factor y for every external bond formed, a factor z for every long internal bond, and a factor y′ for every short internal bond. Each d_{is} is then a direct product of x, x′, y, and z and leads to the identification of $K\tau'$ with x, $K^2\tau'$ with x′, σ'/τ' with y, c′ with z, and c with y′. The equation

$$d_{is} = [x]^{g-h}[x']^h[y]^r[z]^t[y']^s \quad (3.1.14)$$

then holds where g is the total number of occupied sites in the subfigure, h is the number of particles on sites equivalent to the short internal bond, r is the number of external bonds, t is the number of internal long diagonal bonds, and s is the number of short diagonal bonds. The equation above can then be used not only for the rhombus figure but also for any other subfigure. It also facilitates the use of personal computer spreadsheet programs for calculating simply the effect of many variables upon this problem. It is, of course, more than a coincidence that the structure of the expression, shown above, corresponds to the form of the grand canonical ensemble of statistical mechanics. In the above equation

$$\sigma = ((c/c')(K)^2 - 1) + ((1 - (c/c')(K)^2 + (4cK/c')(K - 1)(c' - cK))^{0.5}/2K(c' - cK) \quad (3.1.15)$$

and

$$\tau = (K(1 + K\sigma') - 1)/(C'K) \quad (3.1.16)$$

If c′ − 1 = 0 and K − 1< 1, the above two equations can be greatly simplified to yield

$$\sigma = c(K - 1)/(c'(c' - cK)) \quad (3.1.17)$$

and

$$\tau = (K - 1)(c'^2 + cK)(K - c'))/c'^2(K(c' - cK)) \quad (3.1.18)$$

Now $K = \{(p/po)(1 - \theta/\theta)\}^{1/Z}$ and we can solve for K for each basic figure, i.e.,

for the point: $a_1 - K = 0$

the bond: $cK\,a_1 + K(1 - a_1) - 1 = 0$ (3.1.19)

for the triangle: $K^4 - a_1K^3 + a_2K^2 - a_3K + a_4 = 0$,

where

$$a_1 = ((2 - 5\theta)c + (2 - 3\theta))/(c(1 - 2\theta)),$$
$$a_2 = (c + 5)/c,$$
$$a_3 = ((3 - 5\theta)c + 1 - 3\theta)/c^2(1 - 2\theta), \text{ and}$$
$$a_4 = 1/c^2, \text{ etc.} \quad (3.1.20)$$

In fact, higher approximations to the lattice can be seen to be linear polynomials in K with the coefficients as functions of c, c′, and θ. Since polynomials higher than the quartics cannot be solved analytically, they must be solved numerically.

Now it can be seen that solutions for K in these polynomial equations will lead to p/po or the chemical potential or the free energy, etc. in terms of θ (the occupancy of the sites). As an example of such plots, shown in Figures 93 and 94 are plots for the isotherm of p/po vs. for c = 2.5 and c = 3 for a gas adsorbed on a solid adsorbent with an idealized triangular array of sites. It can be seen that when c rises in value, the isotherm becomes parallel to the abscissa and then loops. The temperature or value of the interaction energy (or the combination of both) where this first happens is where the critical region first occurs.

Attractive lateral interactions arise from van der Waals–London dispersion forces, where the energy of attraction falls off as the sixth power of the distance between adsorbed molecules. Hence, inclusion of next-nearest neighbor interaction terms will lead to some significant changes in the isotherms. Repulsive energies due to parallel alignment of molecules with permanent dipoles vary inversely with the cube of the distance between dipoles and will be of even greater importance. The nature of the biological molecules present in this application and the particular lattice chosen make this quite important here.

Plots of p or d vs. θ show first-order transitions for p and d depending on the interaction parameters c and c′. Normally when the parameters for c′ are of the van der Waals–London type they occur at very high values of θ, but when they are repulsive and vary with the cube of the distance between charge centers the peaks shift toward intermediate values of θ and they are always quite wide in breath.

Discussion

Now it can be seen that solution for K in these polynomial equations will lead to p/po or the chemical potential or the free energy, etc., in terms of θ (the occupancy of the sites). Examples of such plots are shown for the isotherm of p/po vs. θ for c = exp(–w/kt.) = 2, c = 2.5, and c = 3 for a gas adsorbed on a solid adsorbent with an idealized triangular array of sites. It can be seen that when c rises in value, the isotherm becomes parallel to the abscissa and then loops. The temperature or value of the interaction energy (or the combination of both) where this first is where the critical region first occurs. It can be seen that this occurs for the bond approximation when c > 2.5 and that the critical region for the triangle or rhombus occurs when c is approximately 3. Figure 95 shows the isotherm when the force is repulsive, and there will be a discussion about this later. Now the critical region is also of importance because it shows the region where there is a phase change. When there is a loop or when two values of surface coverage, θ, are equal, then there are two phases, one of which is a liquid so that the cancer has changed the cells to liquid. Figure 96 shows a plot of the equation of state: u/kT vs. a, where u is the chemical potential. Here the chemical potential for the rhombus remains constant over most of the range but the bond and triangle rise at high a and the bond shows a critical region at c = 2.77, and Z (the coordination number) is equal to 10, the most probable number for the liquid phase. Figure 96 shows shows the activity of two components for c = 2.77 and Z = 12, according to the Guggenheim *quasi-chemical equilibrium* model. It resembles the activity and activity coefficient curves for two components in the liquid phase. Figure 98 also shows the free energy F/kT vs. θ and it reveals minima for all the values of c plotted (2 to 3) and Z (6 to 12). There is a plot of the probability of the basic figure occupied vs. θ and it shows that the curves for the bond and triangle are similar to the curves for the kinetics of the reactions $A \rightarrow B \rightarrow C$ with the falling concentration of A, the rise and then fall of B, and the rise of C. Figure 104 also also shows similarity in the probability of the occupation of the rhombus as the basic figure. It is seen that some of the subfigures of the rhombus rise and then fall in the low region of θ, but tend to converge at higher values of θ. The plots of d_1 to d_4 vs. θ for the cases of c = 2.77 to and c = 3 show that their values ascend sharply for the highly occupied subfigures of the rhombus at high values of θ. That this should be so is obvious when c′ – cK approaches zero as this expression is the denominator of both σ and τ (used below). This means that K = c′\c, and since K gets smaller as θ gets larger, the critical value will more likely be at higher occupation. Also, since c′ is the next-nearest neighbor interaction, it first occurs in the rhombus approximation, and this critical phenomenon may have been missed in previous approximations that only took into account figures smaller than the rhombus. It is also true that keeping the pressure low and the coordination number low will help diminish the probability of this singularity. Conversion probabilities d for the rhombus when c = 0.36 and c′ = 0.6 are shown. Here it is shown that probabilities can also become great when the forces are repulsive and shift to lower values of θ. Figure 103 also shows the case where c = 0.36 (strong repulsion) at 1 and 310 K. This follows the equation $\theta = 1/(K^Z/p/po + 1)$ or $\theta = 1/(e^{(W-u)/kT} + 1)$ which resembles the Fermi–Dirac distribution function. This led to a model where metastasis is the conduction of cancer cells in the conduction band as in solid-state physics. Figure 102 shows plots of the Michaelis–Menten equation: v = Vmax[S]/(Km + [S]) where [S] is replaced by θ for three cases: the Langmuir isotherm and the rhombus approximation with c = 3 and c = 0.36. This shows that order-disorder phenomena in the substrate will raise the reaction rate or lower the reaction rate when there are attractive or repulsive interactions, respectively.

Thus, we see that a carcinogenic process proceeds as follows. A carcinogenic agent (which may include viruses, radiation, mechanical simulation, etc.) is adsorbed onto primary sites of cell proteins or protein complexes. As this agent accumulates, it produces an inductive effect (which may be due to short-range or longer-range electronic effects due to the structure of the molecules or the environment produced by neighboring molecules) and this leads to an unstable ensemble. The unique ensemble can be brought about by an order–disorder transition and the resulting disorder affects not only the layers of molecules above the adsorbate or "docked" species but eventually the organs of the complete organism and the complete organism itself. This yields a metastable state which represents the neoplastic state. It is one of more rapid proliferation and is no longer under the normal regulating influence of bioregulants. Most of the experimental findings on the action of carcinogens and the nature of neoplastic cells are consistent with the above schemes and tend to verify their general applicability.

There are many other graphs that were made. Figure 104 is an interesting one in that it shows the reaction $A \rightarrow B \rightarrow C$ that occurs in the section on chemical kinetics. The correlation between this graph and the following one is interesting. Time is absent from the succeeding graph. Yet it shows the progress from p_{0c} to p_{1c} to p_{2c} to p_{3c}, the progress of the occupancy of the triangle. The spiral graph (Figure 105) shows p_{0c} spiraling in to the center from the outer right and p_{3c} spiraling out from the center towards the circumference on the left, whereas p_{1c} and p_{2c} just spin around in the center.

Figures 106-113 show graphs of the probabilities of the subfigures of the rhombus as θ varies from 0 to unity and c ranges from 0.36, 0.9, 2, 2.5, 2.77, and 3. The coordination number was six in all cases, and the figure was was blown up with θ varying only from 0 to 0.1.

A plot of the toxicity of eight toxic chemicals for six organisms is shown in Figure 115. It appeared in the author's chapter in *Clean Production* in 1996. It bears a similarity to the graphs in this book. Here, of course, the graphs do not have the linear quality. The earlier graph was a mathematical simplification, and it too would not have been so linear.

Figure 116 is similar to earlier graphs except it is in terms of mole fraction rather than θ. Figure 117 multiplies the factor p/p_0 by $(1 - \exp(-kt))$ to introduce time into the expression. It is, of course, similar to the former graphs, but the values of the probability are diminished. Figures 118 and 119 are similar except one is for activity and one is for chemical potential. Figure 120 shows repulsion in the rhombus figure for c equal to 0.9 and c equal to 0.36. Notice that it takes a higher pressure to reach the same occupancy when c = 0.36 than when c = 0.9. This is what you would expect for the greater repulsion. In fact, when the repulsive forces are great there would be a breakup into "islands." Figures 121-123 show models that were used for study of cancer models for the bond and triangle when c = 2.777, 3, and 100. Figure 124 shows a plot of free energy over kT. Here a search was made for the minimum of the curve as c and Z varied. Figure 125 shows probability vs. θ for the bond and triangle at c = 2.77 and z = 12. Figure 126 shows the value of K_r versus I for the rhombus, where c = 3 and $K = ((p/p^0)(1 - \theta)/\theta))^{1/Z}$, as used in the original main frame computer calculations from which all the variables here can be derived.

3.2 UNIQUAC

The two-liquid model and the concept of local compositions are used in the semi-theoretic development of these equations. The model considers both considerations of differences in sizes and shapes of the molecules (configurational or combinatorial part) and also a contribution due to energetic interactions between them (residual part). The equations of activity coefficients of binary mixtures are given below. In addition to group interaction parameters, group volume parameters and surface parameters are involved. The interaction parameter $D\mu_{ij}$ is usually taken as independent of temperature. The coordination number, Z, has been correlated in terms of the temperature ($Z = 35.2 - 0.1272T + 0.000147T^2$). Examples of the use of the equations in the system of water + acetonitrile + ethyl acetate and water + 1-hexanol + methane (forming three liquid phases), the quaternary system heptane + 2-amino ethanol + benzene + heptane and the limits of the metastability of the ternary system morpholine + methanol + heptane are given in the literature. Values for the parameters of many substances are given in Prausnitz, the DECHEMA VLE Data Collection. Also they can be obtained from Reid and Sherwood in addition to special factors from Anderson and Prausnitz for water and alcohols. Some similarities can be noticed between the earlier part of this book on applied order–disorder theory and the quasi-chemical and regular or strictly regular treatment of solutions.

Mathematical Equations

$$\ln \gamma_1 = \ln \gamma_1^C + \ln \gamma_1^R \qquad (3.2.1)$$

$$\ln \gamma_1^C = \ln \vartheta_1/x_1 + Z/2q_1 \ln \theta_1/\vartheta_1 + \vartheta_2 (l_1 - (r_1/r_2)l_2) \qquad (3.2.2)$$

$$\ln \gamma_1^R = -q_1 \ln (_1 + \theta_{21}) + \vartheta_2 q_2(\tau_{21}/(\vartheta 1 + \vartheta 2\tau 21) - \tau_{12}/(i1\tau 2_{+\vartheta 2})) \qquad (3.2.3)$$

$$\ln \gamma_2 = \ln \gamma_2^C + \ln \gamma_2^R \qquad (3.2.4)$$

$$\ln \gamma_2^C = \ln_2/x_{2+z}/2q_2 \quad \ln \theta_2/\vartheta_2 + \vartheta_1 (l_2 - (r_2/r_1)l_1) \tag{3.2.5}$$

$$\ln \gamma_2^R = -q_2 \ln (\theta_{12} + \theta_2) + \theta_1 q_2(\tau_{12}/(\theta_{12+}\theta_2) - t_{12}/(\theta_1 + \theta 12\theta_2) \tag{3.2.6}$$

$$l_i = Z/2(r_i - q_i) - (r_i - 1) \tag{3.2.7}$$

$$\theta_i = q_i x_i/\Sigma_j q_j x_j \quad \text{area fraction of component i} \tag{3.2.8}$$

$$phi_i = r_i x_i/\Sigma r_j x_j \quad \text{volume fraction of component i} \tag{3.2.9}$$

$$\tau_{ji} = \exp((\mu_{ji} - \mu_{ii})/RT) \tag{3.2.10}$$

where

- q_i = area parameter of component i
- r_i = volume parameter of component i
- μ_{ij} = parameter of interaction between components i and j; $\mu_{ij} = \mu_{ji}$
- Z = coordination number
- g_1^C = combinatorial part of activity coefficient of component i
- $g_i C$ = residual part of activity coefficient of component i

3.3 Schwarz–Christoffel Copies of SYNPROPS Parameters, Filters, and Hyperspace

In an earlier part of this book, the Schwarz–Christoffel transformation was discussed, showing how the parameters of the molecular structure could be drawn in the complex plane as a polygon and then how the polygon could be mapped into another polygon using the Schwarz–Christoffel transformation. This discussion will be continued. Also, since a molecule is thought of as atoms and/or groups joined together and these groups each have at least one frequency of vibration when joined to the molecular assembly, the molecule is possessed of a characteristic set of frequencies. Thus, a circuit consisting of filters (electrical) would change the set of characteristic frequencies of the molecule producing a new set of parameters and, thus, a new molecule. This will be followed by transformations, using the equality of the chemical potential through hyperspace from two to three to four dimensions. Geometric figures for these changes will be denoted.

3.4 Reducing Risk by Controlling the Environment

We will now discuss how an analog computer can express toxicity for six mammals and eight chemical species. General control theory is discussed and the system transfer function is shown to be similar to a simple analogue toxicity equation. The general kinetic equations of Lotka are of this nature. Electrical network equations can be solved for $LC_{50/100}$ for humans and animals in a more complex system by the network systems model of the environment. By analogy then, the system can be controlled by feedback control or any of a dozen other methods to ameliorate the overall $LC_{50/100}$ of the ecological population, at any site, by reducing the emissions of specific chemicals at a site whose ecological nature is known.

While studying risk and toxicology of small mammals, as a member of the U.S. Environmental Protection Agency Region III Biological Technical Assistance Group, the author discovered that the use of simple analogue techniques led to an acceptable description of the ecological impact for several species on Superfund sites. This occurred in spite of the dearth of ecological data in the literature. Encouraged by this, further general electrical and mechanical analogues revealed specific exponential and algebraic equations, with parameters that could be fitted to the sparse laboratory data found and a regression equation was also found that fit the experimental data well and bore strong similarity to the simple analogue equation found. When the analogue method was joined with network theory, control theory, system theory, chemical kinetics, operational (e.g., Laplace transform) methods, order–disorder techniques, system techniques used in medicine, and catastrophe theory, then powerful techniques for planning the control of pollution, as well as describing its impact and the exact nature of its character to humans as well as animals became evident.

3.5 Mathematical Ecotoxicity

Below is a description of the mathematical character of ecotoxicity and its control with several descriptions of applications. The simple electrical analogue first used is

$$I(t) = (1 - \exp(-(R/L)t)) \tag{3.5.1}$$

where E is e.m.f., R is resistance, L is inductance, and t is time. An analogue was used by setting R equivalent to k, L equivalent to m, and E equivalent to p′, where k is the toxicological reference concentration, m is the mass, and p′ is the physiological factor for the organism in question. This equation is also found in the description of mechanical or viscoelastic behavior. Also, $I(t) = LC_{50/100}$. Then,

$$LC_{50/100} = p'/k(1 - \exp - (kmt)) = p'/k(f(t)) \tag{3.5.2}$$

In practice, when t is very large, the function of t (1 – exp(k/mt)) is equivalent to unity and an additive constant, C, is added so that

$$LC_{50/100} = p'k + C \quad (3.5.3)$$

A toxicologist at Region III of EPA provided information on many of the more common mammals and such data include the physiological factor, so that the factor p' is readily available, namely, the physiological rate of the animal compared with humans. As an example, it has the following data for small mammals:

	Rat	Mouse	Hamster
Normal respiration	85	160	83
Respiration range	65–110	80–240	35–130

Figure 115 shows how well the data fits the model for small mammals: rats, cats, mice, etc., for the scant applicable data for LC_{50}.

The reason there is so much spread in the laboratory experimental data for animals in measuring LC_{50} is because their respiration rate is not controlled or measured during the process. If the respiration rate of a particular animal can vary between 80 and 240 breaths per minute, the LC_{50} can vary by a factor of three during this process.

Equation 3.5.2 is very similar to equations in inductive and capacitive circuits and it is constructive to proceed by analogy with such circuits to obtain the link for p_j when $LC_{50ij/100}$ could not be obtained from the literature. Thus, for the circuit,

$$Lq + Rq + q/C = E_o \exp(iwt) \quad (3.5.4)$$

and proceeding by analogy with the "ecophysical" system,

$$q \leftrightarrow \text{concentration} = C_i$$

$$E \leftrightarrow p_j$$

$$R \leftrightarrow k_i$$

$$L \leftrightarrow m_j$$

$$1/C \leftrightarrow p_i/C_i$$

$$v_i \leftrightarrow r_j$$

where the "ecovariables" are the same as before with the addition of m_j as the mass of the ecosystem receptor and r_j as the respiration rate (breaths per minute). Then by analogy,

$$m_j d^2C_i/dt^2 + k_i dC_i/dt + p_j = r_o \exp(i2\pi vt) \quad (3.5.5)$$

(The solution of this equation proceeds by analogy with the solution to the Schrodinger equation from quantum mechanics and is of the form: A sin(2πvt) + B cos(2πvt)).

The resonant frequency of this equation is

$$2\pi v_o = w_o = p_j/k_iC_j \quad (3.5.6)$$

from which

$$p_j = 2\pi v_o k_i C \quad (3.5.7)$$

This result portends to yield pj when LC_{50} data is not available. The missing bridge is v_o, the number of breaths per minute for the organism.

For vinyl chloride at a concentration of 3 μg/m³ (yields an inhalation risk of 1E-4, the action level for a Superfund site for humans), k_j = 0.03 μg/m₃, v_o = 26 breaths/min (resting to moderate activity) yields p_j = 14.7. Then for a small mammal, we have been using p_j = 10 and normalizing the number for humans, 14.7 → 1, the breaths/min for a mouse should be 147 breaths/min which from the table above is reasonable.

Each chemical has a different value of k, which is a risk-based concentration. Tables have been prepared by R. L. Smith, Ph.D., Senior Toxicologist at EPA, Region III, on a quarterly basis and include values for tap water (μg/l), ambient air (mg/m³), commercial industrial soil (mg/kg), and residential soil (mg/kg); all are for humans and MCL (mg/l) has also been used. When $LC_{50/100}$ = y is plotted vs. 1/k = x, then the experimental data fall on a straight line with slope p′ and intercept C. The eight chemicals used to plot Figure 115 were vinyl chloride, xylene, acrolein, formaldehyde, benzene, styrene, epichlorohydrin, and chlorobenzene, since some good published data were found for the six organisms. The data were extended to the horse with data from the University of Pennsylvania Veterinary School and found to fit logically on the simple graphs.

The more rigorous derivation leads to an expression with hyperbolic trigonometric functions as well as exponential functions and is nonlinear, bending toward the abscissa at higher values of 1/k. It is this expression that should be used for very toxic substances, such as "dioxin."

The analogy referred to above can also be extended from that of the field theory to the circuit theory:

$$E \rightarrow V_f$$

$$H \rightarrow I_r$$

$$e' \rightarrow Ci$$

$$u' \rightarrow L$$

$$e'w \rightarrow G$$

$$u'w \rightarrow R$$

We can change over from equations of the field theory to those of the circuit theory, and vice versa, by replacing symbols or by analogy. Since equations of the field theory were shown to be analogous to those of the circuit theory and risk factors for humans and mammals were shown to be analogous to circuits, then risk factors for humans and animals must be analogous to field theory. In other words, pollution can be shown to be analogous to a wave effect.

In the publication, *Dielectrics and Waves*, Arthur R. Von Hippel, shows that the frequency characteristics of two substances: water and *pyranol* at room temperature can be represented by simple three-element circuits with fixed values of the resistance and capacitors, and this has led to the use of "filters" applied to SYNPROPS in order to optimize molecular structure and properties shown below.

3.6 Control Theory

Then, describing control as Initial Conditions,

$$\text{input } (u) \rightarrow \text{SYSTEM} \rightarrow \text{output } (y)$$

and

$$P(D) \cdot Y(t) = Q(D) \cdot U(t) \qquad (3.6.1)$$

and calling

$$g(s) = Q(s)/P(s) \qquad (3.6.2)$$

the system transfer function,

$$u(s) \rightarrow g(s) \rightarrow y(s)$$

where s is a complex variable, D = d/dt, and P(D) are polynomials. It is then true that the parallel, series, and feedback configurations can be represented as

$$g(s) = g_1(s) + g_2(s) + \cdots g_m(s) \qquad (3.6.3\ )$$

$$g(s) = g_1(s)g_2(s) + \cdots g_m(s) \qquad (3.6.4\ )$$

$$g(s) = g_1(s)/(1 + g_1(s) + g_2(s)) \qquad (3.6.5\ )$$

and Figures, and , respectively.

Parallel Control

$$g_1(s)$$
$$\nearrow \quad \searrow$$
$$u(s) \rightarrow g_2(s) \rightarrow y(s)$$
$$\searrow \quad \nearrow$$
$$g_m(s)$$

Series Control

$$u(s) \rightarrow g_m(s) \rightarrow g_2(s) \rightarrow g_1(s) \rightarrow y(s)$$

Feedback Control

$$u(s) \rightarrow g_1(s) \rightarrow y(s)$$
$$\uparrow\!\!— \ g2(s) \leftarrow\!\!|$$

3.7 The Transfer Function

$$G(s) = Q(s)/P(s) = a_o + a_1 s + — A_1/(s - s_1) + — + a_2/(s - s_2) + A^{p2}/(s - s_{p2})^{2+—} A^{pr}/(s - s_2)^r \qquad (3.7.1)$$

The inverse transform is then

$$g(t) = a_o u_1(t) + a_1 u_2(t) + — + A_1 \exp(s_1 t) + — + A_{pr}/(r - 1)! t^{r-1} \exp(spt)(-1)^{r-1} \qquad (3.7.2)$$

where u is the unit step function. In this connection, it is noted in a text on ecotoxicity (Duffus and Worth) that a function such as

$$LC_{50/100} = a \exp(kt) + a \exp(kt) + — \quad (3.7.3)$$

fits the data for selected chemicals and animals very well.

In Bumble and Honig, as well as in Bumble, applications of order–disorder theory to various physical phenomena are described. A lattice approach and states of occupation of the vertices and neighbor interactions were proposed. It was found that the ratio of probabilities of the unoccupied basic figure could be expressed as the ratio of polynomials and, if the lateral interactions were very small, as a series in one variable which would rapidly converge when the variable was very small.

$$G(s) = (a_o p_m + a_1 p_{m-1} + — a_m)/ (b_o p_n + b_1 p_{n-1} + — b_n) \quad (3.7.4)$$

This leads to simpler results in many cases. It also indicates that there may be a critical value for the transfer function of value in controlling pollution cases by analogy to order–disorder theory. The grand canonical ensemble of a space–time lattice in three dimensions in its various states of occupation can then be thought of as a model for the toxic components traveling into the biological system. A Boolean, or an on–off idealization, captures the main features of many continuous dynamic systems. Many cellular and biochemical processes exhibit a response which follows an S-shaped, or sigmoidal curve, as a function of altered levels of some molecular input. This is because there is a greater interaction or positive cooperation at the beginning of a process than near the end, where the response saturates. This can be seen from order–disorder theory referenced earlier under Bumble. Even a function whose maximum slope is less than vertical (e.g., coupled systems governed by such sigmoidal functions) is often properly idealized by on–off systems. Over iterations, feedback of signals through a sigmoidal function tends to sharpen to an all-or-nothing response. A biological rate equation dependent on the Hill function law can lead to

$$x = (k_1/k_{-1})\ (1 - \exp(-(k_{-1})(t))) \quad (3.7.5)$$

where k_1 is the rate constant of the constant in the forward reaction and k_{-1} that in the reverse direction. This regulation endows biological systems with the possibility to choose between two or more distinct states of regime. This, characterized by multiple steady states, is epigenetic and is ensured by feedback loops.

3.8 Networks and Ecosystems

Lotka proposes general kinetic equations that lead to similar results for the last equations above. If X is mass, an excess x_i of each mass X over corresponding equilibrium values is

$$x_i = X_i - C_i \quad (3.8.1)$$

$$X_i = F_i(X_1, X_2, — X_n, P, Q) \quad (3.8.2)$$

where F is equal to growth of any component, dependent on the others and the parameters P and Q. Then

$$dx_i/dt = f(x_1, x_2, — x_n) \quad (3.8.3)$$

and using Taylor's theorem, we obtain

$$dx_1/dt = a_{11}x_1 + a_{12}x_2 + — a_{1n}x_n + a_{111}x_1^2 + a_{112}x1x_2 + a_{122}x_2^2 + — \quad (3.8.4)$$

$$dx_2/dt = a_{21}x_1 + a_{22}x_2 + — a_{2n}x_n + a_{211}x_1^2 + a_{212}x_1x_2 + a222x_2^2 + —$$

$$\ldots \quad \ldots \quad \ldots$$

$$dx_n/dt = a_{n1}x_1 + a_{n2}x_2 +— a_{nn}x_n + a_{n11}x_1^2 + a_{n12}x_1x_2 + a_{n22}x_2^2 + —$$

A general solution is

$$x_1 = G11 \exp(m_1t) + G_{12} \exp(m_2) + — G_{1n} \exp(m_nt) + G_{111} \exp(2m_1t) + — \quad (3.8.5)$$

$$x_2 = G21 \exp(m_1t) + G_{22} \exp(m_2t) + — G_{2n} \exp(m_nt) + G_{221} \exp(2m_1t) + —$$

$$\ldots \quad \ldots \quad \ldots$$

$$x_3 = G_{n1} \exp(m_1t) + G_{n2} \exp(m_2t) + — G_{nn} \exp(m_nt) + G_{n11} \exp(2m_1t) + —$$

Now the G terms are constants (n arbitrary) and m_1 — m_n are n roots of the equation for m.

$$\begin{vmatrix} a_{11} - m & a_{12} — an \\ a_{21} & a_{22} - m — a_2n \\ a_{n1} & a_{n2}—a_{nn} — m \end{vmatrix}$$

If some of G are positive and some negative, then oscillations may occur. If the roots m are complex, exp(a + ib)t = exp(at)(cos bt + i sin bt) and there are damped oscillations about equilibrium.

Consider a general type of network made up of independent circuits. Each circuit contains an e.m.f., resistances, inductances, and capacitances so that current will flow in each circuit. Letting $z_{jk}(p)$ be the operator acting on I_k in the jth equation

$$z_{jk} = L_{jk}p + R_{jk} + 1/(C_{jk}p) \qquad (3.8.6)$$

$$z_{11}(p)I_1 + z_{12}(p)I_2 + — + z_{1m}(p)I_m = p(L_{11}I_{10} + L_{12}I_{20} + — + L_{1m}I_{m0}) + E_{11} \qquad (3.8.7)$$

$$z_{21}(p)I_1 + z_{22}(p)I_2 + — + z_{2m}(p)I_m = p(L_{21}I_{10} + L_{22}I_{20} + — + L_{2m}I_{m0}) + E_{22}$$

$$\cdots \quad \cdots \quad \cdots \quad \cdots$$

$$z_{m1}(p)I_1 + z_{m2}I_2 + — + z_{mm}(p) = p(L_{m1}I_{10} + L_{m2}I_{20} + — + L_{mm}I_{m0}) + E_{mm}$$

At t = 0, I_k = 0, (k = 1,2, — m) and the right parentheses are all zero, and all E on the right are zero except E_{11} = E(t), as a single e.m.f. will produce the same effect as if all others were operating. Then,

$$I_k = (M_1k(p)/\Delta k(p))E(t)$$

where Δ p is the determinant, $M_{1k}(p)$ is the cofactor $(-1)^{k-1}$ times the minor of $z_{ik}(p)$.

$$\begin{vmatrix} z_{11}(p) & z_{12}(p) & \text{———} & z_{1m}(p) \\ z_{21}(p) & z_{22}(p) & \text{———} & z_{2m}(p) \\ — & — & & — \\ z_{m1}(p) & z_{m2}(p) & \text{———} & z_{mm}(p) \end{vmatrix}$$

If E(t) is a function of t, the current in the circuit k is

$$I_k(t) = E(0)A_1k(t) + \int_0^t A_1k(t - z)E'(z)dz \qquad (3.8.8)$$

where A_1k is the transfer indicial admittance and E′(z) denotes dE(z)/dz.

Consider the circuit: The descriptive equations are

$$LdA/dt + RA = 1 \qquad (3.8.9)$$

$$A(t) = 1/R(1 - \exp(-Rt/L)) \qquad (3.8.10)$$

Then

$$1 = 1/R\int_0^t (1 - \exp(R(t - z)/L))E_o\omega \cos \omega z dz$$

$$1 = (E_o/R^2 + L^2\omega^2)(R \sin \omega t - L\omega \cos \omega t + L\omega \exp(-Rt/L)) \qquad (3.8.11)$$

Solution of the equation,

$$LdI/dt + RI = E_o \sin \omega t \qquad (3.8.12)$$

would also lead to a result for I(0) = 0.

It is now seen that if we designate R = k, L = m, E = p′, etc., by analogy, we can solve the equations above for an ecological or an anthropomorphic system and we will obtain $LC_{50/100}$ for humans and other mammals in a more complex system than shown above by the network systems model of the environment.

3.9 Feedback Loops and the Environment

There have been applications in medicine that are similar to the analogue results, such as the equation relating the rate of hemoglobin production P(t) at time t as a function of plasma erythropoietin concentration [E(t)], which has an equation P(t) = 1 – b exp c[E(t)].

This equation is similar to the above. A deterrent to the development and use of mathematical models is that ambiguous and ill-defined relationships will be difficult to convert to quantitative equations. This can screen out speculation. Also, models of respiratory control do correspond as examples.

The polynomials of the functions form minimums that are equilibrium states and are a surface (manifold) in n + k dimensional space, where n is the dimension of the state and k is the dimension of the control state. The behavior of the system is described by a trajectory on the manifold. The canonical polynomial form

$$f(x) = x^m + u_1x^{m-2} + u_2x^{m-3} + — u(n - 2)x \quad (3.9.1)$$

represents the universal unfolding of singularities. Regions higher than the solution represent catastrophic separation and are seen as jumps to another branch.

3.10 Optimizing Environmental Quality

Once the analog has been made between the physical and biological cases and a mathematical fit for the ecotoxic function is present, it can be ascertained whether or not the toxic limit is exceeded for humans and every other animal that may be present in the area of assessment. The limit is here defined as the k factor above or the risk-based concentrations. If the limit is exceeded, then remediation is necessary, control steps can be used for the system and the control design can alleviate the environmental impact on humans and the ecology by bringing down the concentration of the hazardous materials to below their particular risk-based concentrations. These steps can be made prior to the final design of the plant, or afterwards as a retrofit, if necessary and permissible.

Various kinds of control, such as cascade, feedforward, adaptive, proportional, integral or reset, derivative or rate, bang bang, epidemic, impulsive, singular, and profit are examples, and various combinations of these controls are used today. It is realized that the control systems can be both characterized and optimized by mathematical techniques to yield optimum feedback systems for humans and the ecology. There are two types of feedback control systems. The first type is called a regulator, whose function is to keep the output or controlled variable constant in the face of load variations, parameter changes, etc. The second type is the servomechanism, whose inputs are time varying. The function of a servomechanism is to provide a one-to-one correspondence between input and output. The system speed of response is another important consideration in feedback control. Information about response time can be gained from the frequency-response characteristic of the system.

Stability of the system is important. Thus, if any perturbation of the system occurs, the effect on the system should be transitory, with the system ultimately coming to rest. Nyquist has made use of contour mapping methods in the analysis of stability, and this has been used for many years as an important technique in the analysis of stability of linear servomechanisms and regulating systems. The use of contour integration was extended to systems with simple time delays. It is procedures such as these that can lead to the prevention of such pollution catastrophes as did occur at Bhopal.

Now that the system is characterized, analyzed, analogized, and solved mathematically, it can be optimized using calculations of variations, filter, control and predictor methods, dynamic programming methods, automatic control and the Pontryagin principle, linear and nonlinear prediction theory, in addition to the newer methods. Then the system can be converted back to the original ecosystem to plan or direct the prevention or minimization of pollution to the ecosystem. Sensitivity analysis can also be used to determine the specific influence of the parameters on the system process, so that steps can be taken to bring it into synchronization with the goals of reducing risk optimally.

3.11 General Control Equations Expressing Toxicity as a Function of Chemical Structure

A general equation has been found to fit the toxicity of a wide variety of chemical structures to a degree wide enough to be incorporated in a computer control program. The equation has the following form:

$$-LN(Tox) = a + bB + cC + dD + eE + fF \quad (3.11.1)$$

where B, C, D, E, and F are various parameters for the molecular structure described by the groups of which the molecule is composed. The parameters a, b, c, d, e, and f are coefficients for fitting the equation to data and sets have been obtained for the risk-based concentrations for tap water, ambient air, commercial soil, residential soil, and MCL. Such equations are now incorporated in a program that optimizes the synthesis of molecules for dozens of physical properties as well as assuring their most environmentally benign properties. It is found that certain groups, such as $>CCL_2$ and $-CCL_3$, orthochlorines on an aromatic ring, conjugated hydrocarbons, etc., lead to increased toxicity. The results show that the integrated structure (gestalt) of the molecule rather than the localized structure becomes more important in toxicity than physical phenomena and yet show very useful signposts that are needed by regulatory agencies, synthetic chemists, chemical companies, drug companies, etc. in the effort to obtain better environmental quality by replacing hazardous materials with benign materials (with even better physical properties) and by, perhaps, forbidding production of a new venture with hazardous materials. Now Equation 3.11.1 can be rewritten

$$Tox = \exp^{-a} \exp^{-bB} \exp^{-cC} \exp^{-dD} \exp^{-eE} \exp^{-fF} \quad (3.11.2)$$

where $\exp^{-a} = K$, a constant. That this bears a correspondence to some part of statistical mechanics may be no coincidence. The number and kinds of groups of which the molecule is composed combine to form complexions producing the stage for the statistical toxic response of an individual. The probability for doing so can be factored to individual probabilities in the same way that the energies can be factored into translational, vibrational, rotational modes, etc. in deriving thermodynamic properties in statistical mechanics. The work has been tested successfully and is based on the work shown earlier in this book. It bears a similarity to the Hammett function where $\log(k/k_o) = (\rho)(\sigma)$. Here k and k_o are the rate constants for the reaction of substituted and unsubstituted compounds (k/k_o can also be thought of as an equilibrium constant), σ is characteristic only of the substituent and is the ability of the group to attract or repell electrons by a combination of its inductive or resonance effect, and ρ is characteristic only of the reaction series at hand, a measure of the sensitivity of the type of reaction series to (ring) substitution. However, this equation, much used by chemists, has been applied to limited types of reaction and has a very limited number of parameters.

3.12 Computer Simulation, Modeling and Control of Environmental Quality

Programs such as Envirochemkin and THERM discussed earlier can help controllers bring systems or plants into the optimum mode for pollution prevention or minimization. Self-optimizing or adaptive control systems can be developed now. These consist of three parts: the definition of optimum conditions of operation (or performance), the comparison of the actual performance with the desired performance, and the adjustment of system parameters by closed-loop operation to drive the actual performance toward the desired performance. The first definition will be made through a regulatory agency requiring compliance; the latter two by a program such as Envirochemkin. Further developments that are now in force include learning systems as well as adaptive systems. The adaptive system modifies itself in the face of a new environment to optimize performance. A learning system is, however, designed to recognize familiar features and patterns in a situation and then, from its past experience or learned behavior, to react in an optimum manner. Thus, the former emphasizes reacting to a new situation and the latter emphasizes remembering and recognizing old situations. Both attributes are contained in the "mechanism" of Envirochemkin.

Envirochemkin can also use the artificial intelligence technique of "backward chaining" to control chemical processes to prevent pollution while maximizing profit during computation. Backward chaining is a method whereby the distance between the nth step and the goal is reduced while the distance between the n – 1th step and the nth step is reduced and so on down to the current state. To do this, time is considered as negative in the computation and the computations are made backward in time to see what former conditions should be in order to reach the present desired state of minimum pollution and maximum profit. This has been applied in forensic work, where people were sickened by hazardous material, not present when the analytical chemistry was performed at a later date. However, the computer kinetics detected the hazardous material during the reaction of the starting material. Then the amount of each starting species, the choice of each starting species, the process temperature and pressure, the mode of the process (adiabatic, isothermal, fixed temperature profile with time, etc.), and the associated chemical reaction equations (mechanism) are chosen such as to minimize pollution and maximize profit.

3.13 Forced Vibrations and the Spiral of Group Physical Properties

The equation for forced vibrations is

$$md^2x/dt^2 + 2qdx/dt + k^2x = A \cos wt \quad (3.13.1)$$

Here m, q, A, w are constants, $m > 0$, $q \geqq 0$, $k > 0$, $A \geqq 0$, $w > 0$. Now

$$(mD^2 + 2qD + k^2)x = A \exp(iwt) \quad (3.13.2)$$

The real part of A exp(iwt) = A cos wt and the real part of each solution of the second equation will satisfy the first equation, and the imaginary part will satisfy

$$(mD^2 + 2qD + k^2)x = A \sin wt \quad (3.13.3)$$

Now in the homogeneous case, A = 0:

$$(mD^2 + 2qD + k^2) = 0 \quad (3.13.4)$$

and the characteristic equation has roots s_1 and s_2 as

$$(-q_-^+(q^2 - k^2m)^{1/2})/m$$

There are three cases for these roots:

I. Overdamped case: $q^2 > k^2m$. The roots are real and negative; the real complementary function is

$$x_c = c_1 \exp s_1t + c_2 \exp s_2t, \text{ and}$$
$$x_c(t) \to 0 \text{ as } t \to + \text{infinity} \quad (3.13.5)$$

II. Underdamped case: $q^2 < k^2m$. The roots have the form a + –bi, where a is negative if q > 0, and a = 0 if q = 0. The real complementary function is

$$x_c = \exp(at)(c_1 \cos bt + c_2 \sin bt) \quad (3.13.6)$$

$$= C \exp((a + bi)t) + (\text{im } C)\exp((a - bi)t)$$

$$Re = [C_1 \exp((a + bi)t)]$$

$$C_1 = c_1 - ic_2 = 2C$$

The solutions are vibrations of decreasing amplitude approaching 0 as t Æ infinity, provided by q > 0.

III. Critical damping: $q^2 = k^2m$. The roots are real, negative, and equal. The real complementary function is

$$x_c = c_1 \exp s_1t + c_2t \exp s_1t \quad (3.13.7)$$

Again $x_c(t) \to 0$ as $t \to$ infinity.

Returning to case II, when q > 0, the solutions can be written in the form

$$x_c(t) = Re[C_1 \exp(at) \exp(bit)] \quad (a > 0) \quad (3.18.8)$$

The curve $z(t) = \exp(at)C_1 \exp(bit)$ is obtained from the curve $z = C_1 \exp(bit)$ by multiplying z, at each t, by exp(at). As t increases, exp(at) decreases, approaching 0 as $t \to +$ infinity. Hence z(t) follows a spiral approaching the origin.

The application of the above then is to arrive at a fit to the spiral graphs of physical properties vs. groups that are prevalent in this book and also at a model to help explain by analogy the physical process or phenomena that may explain the graphs. In order to do this, each group and its corresponding value of the parameters B, C, D, E, and F were placed in adjoining columns of the spreadsheet of Quattro Pro and then each numerical parameter was sorted in descending order with its associated group formula sorted as the secondary field. Then, the five sets of results were plotted yielding radar or spiral graphs. The regression was applied as: a + b exp(kx) to find the values of a and b for each graph that best fits each of the graphs with statistical and error parameters. The comparison is shown in Figures 73-77.

3.14 The Radial Wave Function of the Hydrogen Atom and the Spiral of Physical Properties

In quantum mechanics the wave equation in three dimensions for the hydrogen atom can be derived from the Schrodinger equation. This wave equation can then be separated into three functions: the translational motion of the atom, the angular motion of the electron, and the distance between the nucleus and the electron. The second of these is composed of two equations with two necessary angles: θ and ψ. The radial equation is

$$1/r^2d/dr(r^2dR/dr) - B/r^2R$$
$$+ 8\Pi^2u/h^2(E + Ze^2/r)R = 0 \quad (3.14.1)$$

This can be transformed into the associated Laguerre equation. Without showing the mathematical details, the solution of the radial hydrogen atom then becomes that of a number of states including the ground state characterized by quantum numbers n, l, and m, which are, respectively, that for total energy, angular momentum, and the magnetic moment of the particle. These quantum numbers have the values 1, 0, 0 for the ground state and the ground state, wave function is

$$\psi = (Z/a_0)^{3/2} \exp(-Zr/a_0) \qquad (3.14.2)$$

Here Z for hydrogen is unity and a_0 is the Bohr radius.

We must remember that probability in quantum mechanics is ψ^2, not ψ. This does not change the form of the equation.

3.15 The Particle in the Box and the Spiral of Physical Properties

Here we consider the quantum mechanics of the one-dimensional box. Again we consider the Schrodinger equation and in substituting for the energy we obtain

$$1/X(d2X/dx2) + (8\Pi^2 m/h^2)(E_x - V_x) = 0 \qquad (3.15.1)$$

The solution to this equation is

$$X = A \exp(qx) + B \exp(-qx) \qquad (3.15.2)$$

where $q = ((2m(V_x - E_x))^{1/2})/h$.

Here we are to visualize the container as a circle rather than a box. The line (or plane) at the bottom of the box is to be bent or constrained into a circle and the form of the equation is to be conserved. Within this circle we visualize the spiral of physical properties with the groups at the rim and the values as the radius of the spiral.

The examples of forced oscillators, or particles in a box, or the hydrogen atom all lead to one solution, among others, that gives credence to the form of the equation that has been found to fit the curves of the parameters for B, C, D, E, and F in addition to the physical and toxic properties for the groups. The form of this equation is simply

$$\text{Property or B or C, or D, or E, or } F = a + b \exp(-mk)$$

Here k is arrived at by placing the groups on the circumference of a circle equidistant from each other and with the groups (33 in number) completely filling the circumference. The values of the properties or the parameters are the values of the radius of the logarithmic spiral within the circle.

Consider the system:

$$x' = y$$

$$y' = -x$$

These produce figures in txy space and y vs. t and x vs. t space. In order to untangle the first impression of these figures (which can be guessed at from the equation $x'' = -x$ that some of the solutions are

$$x = u(t) = C \sin t$$

$$y = v(t) = C \cos t$$

For each choice of the constant C, the last equations above represent in txy space a circular spiral, moving forward in time, for different values of C; each solution starts at t = 0 with $x_o = 0$, $y_o = 0$.

All the values of the physical properties were sorted so that the highest values were at the top of the column and the lowest were at the bottom. Then counting what kind of groups were at the top position and which of the groups were at the bottom, it was found that the top approximately 50% of the groups were **CONH2,** 21% were **CYCLOHEXYL**, and 21% were **NAPHTHYL** groups. Likewise at the bottom of the column, 50% were **>C<** groups, but 27% were **CONH2** and 23% were **NAPHTHYL** groups. In a true global optimization of properties with groups, some of the properties will be ascending properties and some will be descending properties. In previous Chemkin runs for the beginnings of life from primordial molecular compositions proposed at that time, there was an abundance of NH_3, CH_4, etc., which did lead also to a composition of molecules with COOH, CNH_2, $CONH_2$, etc., similar to the mix found here.

At first, the values for the parameters were checked using a regression routine and then since the properties are linear functions of the parameters, it was natural that the properties could be approximated in that way as well.

3.16 Exponential Representation of Functions of a Variable

The function $r = C_1 \exp(-at)$ in polar coordinates produces logarithmic spirals approaching the singular point F at the origin. The rotation is clockwise, and the trajectory approaches the singular point, which is called a focal point. If $a < 0$, the preceding conclusions remain except that the rotation of R takes place in the opposite direction so that, instead of approaching the singular point F, it departs from it. Thus we have to distinguish between the stable focal points ($a > 0$) and the unstable ones ($a < 0$). The whole family of spirals either converges toward or diverges from F according to the sign of a. Another set of references to the spiral is a "source" or "sink" depending on the direction of R. In the clockwise direction it is a sink and in the counterclockwise direction it is a source.

In terms of matrices and exponents, one can describe the above as

$$\exp(tA_1) = \exp(ta) \begin{vmatrix} \cos bt & -\sin bt \\ \sin bt & \cos bt \end{vmatrix} \qquad (3.16.1)$$

Now, it can be seen that the trajectory of a point simultaneously turns at a speed b radians per unit time, under the influence of the rotation matrix above and approaches or leaves 0, under the influence of exp(ta) depending on whether $a < 0$ or $a > 0$. Then,

1. It is a spiral sink, if $a < 0$, when the trajectories spiral in to (0,0).
2. Center. If $a = 0$, the trajectories turn periodically on an ellipse centered at (0,0).
3. It is a spiral sink, if $a > 0$, when the trajectories spiral out to infinity.

The direction of rotation of the spiral trajectories depends on the sign of c, the lower left entry of the matrix above. If $c > 0$, the rotation is counterclockwise: if $c < 0$, the rotation is clockwise. Centers are exceptional, but they turn up often in differential equations describing conservative mechanical systems.

There are many different spirals. The one here is called the spiral of Bernoulli, or of Descartes, or of Gregory, or of Torricelli, depending upon whose drummer you march to, and it is also referred to as the logarithmic, the exponential, or the equiangular spiral. In a certain sense, probably because of the role it plays in the theory of linear differential equations and because of its seeming omnipresence in nature, this is the spiral *par excellence.*

The exponential exp(+jwt) is widely used to represent periodic time functions. Re[exp(jwt)] = cos wt and Im[exp(+jwt)] = sin wt and a phase shift is accomplished by multiplying exp(+jwt) by exp(+jphi) so that exp(+jwt)exp(+jJ) = exp(+j(wt+J)) and Re[exp(+j(wt+J))] = cos(wt+J) and Im[exp(+(wt+J))] = sin(wt+J) and sine and cosine waves can be described as follows: coswt = (exp(+jwt)+exp(-jwt))/2 and sinwt = (exp(+jwt)-exp(- jwt))/2. When the constant a in the first equation in this section is imaginary, the exponential time function is complex. Then the equation is described in a two-dimensional manner in terms of a vector of unit amplitude rotating at a constant angular velocity w, so that the locus is a circle of unit radius. The time function exp(jwt) may also be depicted three-dimensionally as a helix generated by a unit vector rotating at a constant velocity w and moving at a constant velocity along the time axis. Now in our application we have 33 physical properties, each one represented as a spiral rather than a circle. These distorted circles are connected one to another along the quasi-temporal axis, and each can be thought of as a gear which connects the chemical groups in such a way that the sorted properties are either ascending or descending depending on the particular property represented by each gear. There is a resemblance to the helix in living organisms where each gear or property represents a chromosome and the number of configurations of properties is huge and can represent many molecules (people) with many property variations (characteristics).

3.17 Wave Filters, Finite Differences and Benign-Property Pass Substances

The SYNPROPS spreadsheets have within them numbers that came from regressions on the properties and structures of a number of compounds and their experimental data in the literature. These numbers have been broken up into the groups of which such compounds are composed. It is also true that there are a number of procedures whereby such compounds can be analytically determined such as infrared spectroscopy. Here we find that the bonds in such molecules can be detected by frequencies or wavelengths characteristic of such bonds. We will now proceed by analogy to classify the numbers in SYNPROPS to synthesize substances with properties and structures with preferred values as deemed preferable.

The parameters B, C, D, E, and F are made up from the numbers in the SYNPROPS Table as follows:

$$B = \Sigma_i m_i n_i \qquad (3.17.1)$$

where m is the number of groups of type i and n is the constant number in the B column for group i.

Similar equations occur for C, D, E, and F with n replaced by o, p, q, and r, the respective constant numbers for those parameters. Now let us consider the network ladder shown below. Side AB is at zero potential, and the other side CD consists of a series of identical coils and capacitors in series.

The voltage of the point P_x where the xth rung of the ladder joins the side C is denoted by V_x and the three currents meeting in P_x by I_{1x}, I_{2x},I_{3x}, where

$$I_{1x} + I_{2x} + I_{3x} = 0 \qquad (3.17.2)$$

The impedance Z_1 of the combination L_1 and C_1 is given by $Z1 = L_1 iw + 1/(C_1 iw)$, that of the combination C_2 and L_2 by

$$Z_2 = L_2 iw + 1/(C_2 iw) \qquad (3.17.3)$$

Then

$$I_{1x} = (V_x - IZ_1,\ I_{2x}) = (V_{x+1} - V_x)/Z_1, \text{ and } I_{3x} = -V_x/Z_2 \qquad (3.17.4)$$

Substituting these expressions in the first equation, we obtain the following relation between the consecutive voltages V_{x-1}, V_x and V_{x+1}:

$$V_{x-1} - (2 + Z_1/Z_2)V_x + V_{x+1} = 0 \qquad (3.17.5)$$

Substituting

$$Q = 1 + Z_1/2Z_2 = 1 + 1\ (L_1 w^2 - (1/C_1))/2\ (L_2 w - (1/C_2)) \qquad (3.17.6)$$

In the above equations Z_1/Z_2 is real as Z_1 and Z_2 are pure imaginary numbers. Progressing waves can pass through the system only if $-1 < Q > 1$.

A. If CD contains coils only and the rungs of the ladder merely capacitors as shown in the first diagram, then

$$L_2 = 0 \text{ and } C_1 \text{ equals infinity and } Q = 1 - 1/2L_1C_2w_2 \qquad (3.17.7)$$

The device lets through waves of the frequency w if $Q > -1$. Thus , the device acts as a low-pass filter with the cut-off frequency

$$w_1 = 2/(L_1C_2)^{1/2} \qquad (3.17.8)$$

B. If BD contains only capacitors and the rungs are constructed of coils, then $L_1 = 0$, C_2 = infinity, and we obtain

$$Q = 1 - 1/(2C_1L_2w_2) \qquad (3.17.9)$$

Here Q is a large negative quantity for small values of w and it passes through $Q = -1$ for $w_{22} = 1/(4C_1L_2)$ and it converges to 1 if w → infinity. Therefore, the device cuts off the waves of low frequency up to $w_2 = 1/2(C_1L_2)^{1/2}$. The device is a high-pass filter.

C. If L_1 and L_2 are not zero and C_1 and C_2 are finite, we obtain waves between two limiting frequencies w_1 and w_2, which correspond to $Q = 1$ and $Q = -1$. It follows from the equation above that

$$w_1 = 1/(L_1C_1)^{1/2} \text{ and } w_2 = 1/(L_1C_1)^{1/2}((1 + 4C_1/C_2)/(1 + 4L_2/L_1))^{1/2} \qquad (3.17.10)$$

The device acts as a bandpass filter. If w_1 and w_2 are near together, only a narrow range of frequencies will pass through the filter. It is seen that this will be the case if C_1/C_2 and L_1/L_2, i.e., C_1L_1 and C_2L_2 are only slightly different.

The parameter Q as a function of the frequency w for the three cases: A, B, and C. In this way we see that it is possible to filter out unwanted groups and synthesize optimum molecules to produce those with the desired physical properties.

3.18 Multidimensional Recursive Filters via a Helix

If a wire is coiled into a helix whose surface is a cylinder, then the filter on the 1D space of the wire mimics a 2D filter on the cylindrical surface. The 2D convolution can be done with a 1D convolution program. 2D recursive filtering (also called 2D deconvolution or 2D polynomial division) can be performed. For many years our most powerful signal-analysis techniques have been in 1D space, while our most important applications have been in multidimensional space. The helical system makes a giant step toward overcoming this difficulty.

3.19 Convolution

Convolution is the operation we do on polynomial coefficients when we multiply polynomials. Deconvolution is similar for polynomial division. A basic idea of filtering, in any number of dimensions, is that you have some filter coefficients and some sampled data; you pass the filter over the data; at each location you find an output by cross multiplying the filter coefficients times the underlying data and summing the terms. Ordinarily, a plane of data is thought of as a collection of columns, side by side. Now, imagine the columns stored end to end, and then coiled around a cylinder. This is simply the helix. The values output from filtering can be computed in any order. So we can slide the filter coil over the data coil in any direction. The output order is irrelevant; and we could compute the results in parallel. We could, however, slide the filter over the data in the screwing order that a nut passes over a bolt. The screw order is the same order that would be used if we were to unwind the coils into 1D strips and then convolve them across one another. The same filter coefficients overlie the same data values if the 2D coils are unwound into 1D strips. The helix idea allows us to obtain the same convolution output in either of two ways, a 1D way or a 2D way.

3.20 Deconvolution

Deconvolution (polynomial division) can undo convolution (polynomial multiplication). The helix makes it possible to consider 1D convolution to be the same as 2D convolution. Thus, we can use 1D deconvolution to undo convolution, whether the convolution was 1D or 2D and have thus discovered how to undo 2D convolution and that 2D deconvolution on a helix is equivalent to 1D deconvolution. The helix enables us to do multidimensional deconvolution, which is recursive filtering. Recursive filter outputs cannot be computed in parallel, but must be computed sequentially as in one dimension, namely, in the order that the nut screws on the bolt.

In the computers of the 1970s, a track or disk would hold several thousand words; a sizable delay is incurred on the average before the first word of a desired page appears under the read–write mechanism. A paging drum or disk has a drum track i divided into a number of fixed–size sectors j, each corresponding to a page. A hardware sector pointer is synchronized with the rotating drum, and an IO request queue is maintained for each sector; each queue element contains a track number and read or write indicator. When the pointer reaches a sector with nonempty queue, the read or write operation at the head of the list is automatically initiated over the drum channel. The performance of this drum is theoretically much superior to one which schedules each operation on a first-come, first-served basis. The University of Michigan Multiprogramming System (MTS, 1967), running on the IBM 360/67, was one example of a system that uses such a paging drum.

3.21 Wavelets

Wavelets are mathematical functions that cut up data into different frequency components, and then study each component with a resolution matched to its scale. They have advantages over traditional Fourier methods in analyzing physical situations where the signal contains discontinuities and sharp spikes. The fundamental idea behind wavelets is to analyze according to scale. Wavelet algorithms process data at different scales or resolutions. If we look at a signal with a large "window," we would notice gross features; similarly, if we look at a signal with a small "window," we would notice small features. The result in wavelet analysis is to see both the forest and the trees, so to speak. An analyzing wavelet does temporal analysis and it is performed with a contracted high-frequency version of the prototype wavelet, while frequency analysis is performed with a dilated, low-frequency version of the same wavelet. Because the original signal or function can be represented in terms of wavelet expansion (with coefficients in a linear combination of the wavelet functions), data operations can be performed using just the corresponding wavelet coefficients. And if you further choose the best wavelets adapted to your data, or truncate the coefficients below a threshold, your data are sparsely represented. This coding makes wavelets an excellent tool in the field of data compression.

The fast Fourier transform (FFT) and the discrete wavelet transform (DWT) are both linear operations that generate a data structure that contains segments of various lengths. Similarly, the inverse transform matrix for both FFT and DWT is the transpose of the original. Both transforms can be seen as a rotation in space to a different domain which for FFT contains basis functions that are sines and cosines, but for the wavelet transforms are wavelets, mother wavelets, or analyzing wavelets. Similarly, both transforms are localized in frequency, making mathematical tools such as power spectra useful at picking out frequencies and calculating power distributions. A dissimilarity is that individual wave functions are localized in space; Fourier sine and cosine functions are not. This localization feature, along with wavelet localization of frequency, makes many functions and operators using wavelets "sparse" when transformed into the wavelet domain. This sparseness, in turn, results in a number of useful applications such as data compression, detecting features in images, and removing noise from time series. An advantage of wavelet transforms is that the windows vary. To obtain detailed frequency analysis is to have short high-frquency basis functions and long low-frequency ones. This is what is obtained with wavelet transforms. Wavelet transforms have an infinite set of possible basis functions.

3.22 Time Series

There are two main goals of time series analysis: (1) identifying the nature of the phenomenon represented by the sequence of the observations and (2) forecasting (predicting future values of the time series variable). In time series analysis it is assumed that the data consist of a systematic pattern and random noise (error), which makes the pattern difficult to identify. Most time series analysis techniques involve some form of filtering out noise in order to make the pattern more salient. Most time series patterns can be described in terms of two basic classes of components: trend and seasonality. In trend analysis smoothing always involves some form of local averaging of data such that the nonsystematic components of individual observations cancel each other out. In a study, there appears to be (by time series analysis) geographic relationships between gross national product per capita and CO_2 per capita to the North and between population and deforestation to the South.

3.23 Molecular Information Theory and the Theory of Molecular Machines

Classical information theory has been used to study genetic systems. Shannon found a way to measure information, and this measure has been used to characterize the sequence conservation at nucleic-acid binding sites precisely. The resulting methods completely replace the use of "consensus sequence," and therefore provide better models for molecular biologists. Molecular interactions have been described by using the mathematics of information theory, which has led to the prediction of proteins that bind DNA. Through connections with the second law of thermodynamics and Maxwell's demon, this approach also has implications for the development of technology at the molecular level.

3.24 SynTree

SynTree is a set of programs for exploring organic chemical synthesis on a personal computer. It is designed to facilitate the discovery of possible synthetic routes to a given target structure by the repeated application of retrosynthetic transforms. The well-established chemical reactions are held in the database which may be customized. The capability for customization by adding transforms for important reactions is potentially very useful and is an essential feature of a program of this type. The application of the transforms to the target molecule generates precursors, together with the required reagents, in sequence, until one is selected and becomes the new target. Groups that interfere with the desired reaction or require protection are duly considered. There is a reasonable amount of common functional groups and synthetic reactions. However, important retrosynthetic steps can be obscured by a plethora of irrelevant suggestions. The complete synthetic tree is saved as the analysis proceeds. There was no superior competitor package (1991) available for use on personal computers and, with appropriate customization, SynTree might be a useful way of searching possible synthetic routes exhaustively.

3.25 Moirai: Modeling Biochemistry

Efforts to describe the structure and function of simple biochemical systems began with a sufficiently large-scale model that included detailed information on both the structure of the molecular parts and their functions in biochemical reactions. Representations for molecules and reaction biochemistry for use in databases of biochemical function were developed by capturing the "natural language" of biochemistry in a layered graph grammar *Klotho*, which permits interconversion among a family of equivalent representations for compounds and then operates on these with rules which express chemical and mechanistic aspects of the biochemical reaction. These representations are used to develop algorithms for important computations, which include logical and kinetic simulation of sequential biochemical changes to molecules (tracing the atoms through the pathway); automated pattern recognition and classification for molecules and biochemical reactions; and fundamental inquiries into the biochemical and dynamic structure of metabolism. The applications of this work may range from understanding fundamental cellular processes, to the prediction of probable metabolic routes of drugs and xenobiotics, to the design of organisms for increased yield of desirable compounds in industrial processes.

Many databases draw sharp intellectual and implementation among objects (or data) processes to query or manipulate the objects, and constraints which apply either to particular manipulations or to the system, as a whole, such as numerical simulation of a set of differential equations. In extreme cases, data are stored in one language, retrieved in another, and manipulated for "advanced" computations by a third. Two common consequences of this approach are that representations of these three informational dimensions tend to be mutually incompatible and that programming and execution of the code can be noticeably confused and inefficient. This phenomenon has been called impedance mismatch.

To avoid these problems, the database and its computations have been implemented in a symmetric manner, the terms are meant in two senses; representations of the three categories use the same method and representations of the three categories and use the same language.

Other languages and software are used as appropriate for particular parts of the computations, such as advanced mathematics or visual display, but overall coordination of the computations resides exclusively in Moirai, with results from other tools passing in and out of the central database.

3.26 Reaction Pathway Analysis by a Network Synthesis Technique

In the 1997 Annual Meeting Session 264 of the AIChE, Drs. Fan and Friedler presented a method for selecting candidate pathways or mechanisms of a reaction through synthesis of networks of elementary reactions. The difficulty is attributable to the combinatorial explosion of the number and structural complexity of networks which can be possibly generated from the elementary reactions and also to the highly involved nature of the problem itself. It is difficult to establish an algorithm to generate the networks either synthetically from precursors or starting reactants toward targets or final products or retrosynthetically from targets toward precursors. Nevertheless, this can be done and the resultant method is capable of rapidly yielding a complete network, i.e., the maximal network, for a given set of plausible elementary reactions based on the combinatorial properties of the reaction pathway; moreover, it exactly decomposes the maximal network into a set of potentially feasible subnetworks. All the subnetworks among these potentially feasible subnetworks containing intermediaries as products in addition to targets of the reaction of concern (as byproducts) are eliminated in the light of the definition of reaction pathway. In other words, all intermediates generated in one elementary reaction or one step must be consumed by other steps in any potentially feasible subnetwork. Among the remaining subnetworks, all that satisfy mass conservation or, more specifically, element conservation are finally selected as the feasible candidate networks or mechanisms. The number of these mechanisms is usually small, thereby greatly facilitating verification of an individual mechanism. The method proposed is firmly rooted in a set of combinatorial axioms and expressed in the terminology of process graph or P-graph. The method and its efficacy has been amply demonstrated.

3.27 TIMECHS, TIME-MECHANISMS, Pollution Prevention by Optimizing Kinetics and Mechanisms on a Spreadsheet

Chemkin (the root of Envirochemkin) has been a brilliant technological advance in the study of chemical reactions, computer science, and chemical kinetics. In a way, though, it has been too much of a good thing. That is because frequently there are 100 chemical reactions and 50 chemical species on a single run on a personal computer that need human analysis. It then becomes practically impossible "to see the forest for the trees." There is too much data to assimilate to find solutions to pollution prevention in the plant, globally, in the troposphere, in the atmosphere, etc. Now the problems are becoming even more complex with more then 50 species and one 100 reactions for each run with the advent of larger supercomputers. Granted the ability to run large and complex sets of problems is needed, because the problems themselves are complex and what is needed is completeness or as near to completeness as can be achieved for each run. Yet, the ability to spend more time in the study of *mechanisms* and the ability to *isolate* the important series of reactions called the mechanisms for study by the computer itself and the human analyst may lead to more actual solutions to process and environmental problems than confrontation with reams of data, whose analysis, meaning, and relation to problem solution themselves become projects of inestimable size. In an attempt to keep it simple, while still capable of solving important problems, the concept of TIMECHS was born.

In a little-known book, *Consecutive Chemical Reactions*, by N. M. Rodiguin and E. N. Rodiguina, operational calculus is used in the solution of complicated systems of linear differential equations. There is an extensive collection of the results of such analysis on a large number of general types of consecutive reaction systems. Furthermore, the Russian to English translation of this monograph presents the necessary application of the Laplace transformation to the solution of the equations for the essential kinetic systems.

Simple chemical reactions are grouped on the basis of their molecularity or their order. Molecularity is the number of molecules that interact simultaneously to undergo chemical transformation. Mono-, bi-, and termolecular reactions are known, but higher molecularities are practically unknown. The order is numerically equal to the sum of the powers to which the concentrations are raised in a given rate equation. There are zeroth-, first-, second-, and third-order reactions that are known. Higher orders have never been observed.

Complex chemical reactions are divided according to the way in which the component simple reactions are interconnected into consecutive parallel, reversible, conjugated, and mixed reactions. Consecutive chemical reactions are those in which the initial substance and all the intermediate products can react in one direction only. Parallel chemical reactions are those in which the initial substance reacts to produce two different substances simultaneously. Reversible reactions are those in which two substances entering a single simple consecutive chain reaction interact in both forward and backward directions. Reversible consecutive reactions combine some characteristics of consecutive irreversible and reversible reactions. Parallel consecutive reactions are mixed type also and have characteristics of both parallel and consecutive reactions.

The Laplace transform is defined as

$$L[f] = \int_0^\infty \exp(-st)f(d)dt \qquad (3.27.1)$$

It is used in the solution of initial-value problems for linear ordinary differential equations. The operator method is useful in solving linear differential equations, as well as for other purposes. Denoting the definition of the Laplace transform as

$$L\{f\} = \int_0^\infty \exp(-st)F(t)dt \qquad (3.27.2)$$

and using integration by parts, we obtain

$$L\{F'(t)\} = sf(s) - F(0) \qquad (3.27.3)$$

where $f(s) = L\{F(t)\}$.

Therefore, differentiation of the object function corresponds to the multiplication of the result function by its variable s and the addition of the constant $-F(0)$. The formula above gives the fundamental operational property of the Laplace transformation, the property that makes it possible to replace the operation of differentiation by a simple algebraic operation on the transform.

Through its use, the calculation procedure involved here can often be both simplified and shortened. An example follows. To solve the equation:

$$dx/dt + ax = 1 \qquad (3.27.4)$$

let P be d/dt and a a constant coefficient; x and t are variables. The equation above becomes

$$Px + ax = 1 \qquad (3.27.5)$$

Treating P as a constant, we solve the above equation as an ordinary algebraic equation with one unknown:

$$x = 1/(P + a) \qquad (3.27.6)$$

The final solution, i.e., determination of x as a function of t, is arrived at directly, by applying the appropriate equation from the tables of transforms and originals that are present in the literature.

$$x = (1/a)(1 - \exp(-at)) \qquad (3.27.7)$$

Here the value P is called an operator, the function of time $(1/a)(1 - \exp(-at))$ is an original, and the function $1/(P + a)$ is a transform.

Another example, solving

$$d^2/dx^2 + (a + b)dx/dt + abx = 1 \quad (3.27.8)$$

leads to

$$P^2x + (a + b)Px + abx = 1 \qquad (3.27.9)$$

$$x = 1/(P^2 + (a + b)P + ab) \qquad (3.27.10)$$

or

$$x = a/((P + a)(P + b)) \qquad (3.27.11)$$

This transform and the table of transforms leads to

$$x = 1/(ab) - (1/a(b - a))\exp(-at) - 1/(b(a - b))\exp(-bt) \qquad (3.27.12)$$

This method for solving linear differential equations is valid only for the specific initial conditions, namely, where f(t) and all its derivatives with respect to t become equal to zero when t = 0.

Now a few of the mechanisms that follow are depicted by name, description, and differential equation.

a. Single exponential decay — $A \xrightarrow{k_1} B$

$$dA/dt = -k_1[A] \quad dB/dt = k_1[A] \quad (3.27.13)$$

b. Double exponential decay — $A \xrightarrow{k_1} B,\ A \xrightarrow{k_2} C$

$$dA/dt = -k_1[A] - k_2[A],\quad dB/dt = k_1[A],\quad dC/dt = k_2[A] \quad (3.27.14)$$

c. Three exponential decay — $A \xrightarrow{k_1} B,\ A \xrightarrow{k_2} C,\ A \xrightarrow{k_3} D$

$$d[A]/dt = -k_1[A] - k_2[A] - k_3[A],\ d[B]/dt = k_1[A],\ d[C]/dt = k_2[A],\ d[D]/dt = k_3[A] \quad (3.27.15)$$

d. Consecutive reaction — $A \xrightarrow{k_1} B \xrightarrow{k_2} C$

$$d[A]/dt = -k_1[A],\ d[B]/dt = k_1[A] - k_2[B] d[C]/dt - k_2[B],\ PA - PA_o = -k_1A,\ PC1 = k1Co - k_1C_1,\ PC_2 = k_2C_1 - k_3C_2,\ Co = PCo^o/P + k_1,\ C_1 = k_1PCo^o/(P + k_1)(P + k_2),\ C_2 = k1k_2PC_o^o/(P + k_1)(P + k_2)(P + k_3) \quad (3.27.16)$$

e. Reversible — $A \underset{k_2}{\overset{k_1}{\rightleftarrows}} B$

$$d[A]/dt = k_1[A] + k_2[B],\ d[B]/dt = k_1[A] - k_2[B] \quad (3.27.17)$$

f. Double Reversible Consecutive Reactions — $A \underset{k_2}{\overset{k_1}{\rightleftarrows}} B \underset{k_4}{\overset{k_3}{\rightleftarrows}} C$

$$d[A]/dt = -k_1[A] + k_2[B] \quad (3.27.18)$$

$$d[B]/dt = k_1[A] - (k_2 + k_3)[B] + k_4[C],\quad d[C]/dt = k_3[B] - k_4[C] \quad (3.27.19)$$

g. Second Order Reaction — $A + B \xrightarrow{k_1} C$

$$d[A]/dt = -k_1[A][B],\ d[B]/dt = -k_1[A][B]\ d[C]/dt = k_1[A][B] \quad (3.27.20)$$

Now these are but seven out of many mechanisms. Later we may single out and discuss some more of these equations, such as the ones concerning hysteresis when we have a special interest or need for them. Remember that there are four steps in deriving the analytic equations for the concentration of each species in a mechanism. They are

1. Write directly the differential equations for the reaction under consideration.
2. Write P for the differentials d/dt and then solve for each concentration (C_0, C_1, C_2, etc.). Each concentration will then be an algebraic expression in terms of k_1, k_2, etc., P, and $C^o o$, which is the initial concentration of the initial substance.
3. Solve these equations for concentrations; the resulting concentration expressions are the transforms for the unknown functions.
4. Use the proper entries in the tables of transforms and originals to replace the transforms of the unknown concentration functions by their originals.

Spreadsheet programs have been written to do this. Item 1 can be done intuitively or as an expert program (see Section 2.102). The rate constants $k_1 \ldots k_n$ can be assigned different cell addresses, the concentrations $C_0 \ldots C_n$ can be assigned different column cell addresses, and time can be assigned as different cell addresses in an assigned column. The type of reaction can be ascertained by the exponents, signs of the rate constants, etc. and the transform to original conversion by the table-lookup method. A series of graphs and a sample of spreadsheet entries are shown.

Now the interesting thing about setting up the chemical kinetics in a spreadsheet format is that it is now possible to call into play the optimization subprogram that is found in Quattro Pro, versions 7, 5, etc., just as we did in the SYNPROPS program. Thus, we can maximize or minimize concentrations as a function of time. We can vary the rate constants, as parameters that affect the rate of complex reactions, and we can prescribe and limit certain parameters and study the effect on the outcome of the kinetics. Most important, we can search for minimums and maximums in concentrations as a function of time and thus find solutions for global, stationary, mobile, etc., the most important problems of the environment, the commercial problems, and many others that face us today. Together with other programs such as Chemkin, Envirochemkin, SYNPROPS, THERM, and available data such as that from NIST the ability and power for analysis, study, and perfection of methods to solve problems have been steadily increasing.

There are libraries of mechanisms such as the Mathematical Models of Biological Systems of Georgetown University Medical Center, the University of Pennsylvania, and Vanderbilt University which can be accessed on the Web at http://biomodel.georgetown.edu/model/model.html. Below is an overview of 29 chemical models contained in the last Web site.

1. Single exponential decay
2. Two exponential fit
3. Three exponential fit
4. Consecutive reaction
5. Reversible reaction
6. Reversible consecutive reaction (1)
7. Reversible consecutive reaction (2)
8. Second-order reaction
9. Second-order reversible reaction
10. Second-order consecutive reaction
11. Second-order consecutive reversible reaction (1)
12. Second-order consecutive reversible reaction (2)
13. Hysteresis reaction (1)
14. Hysteresis reaction (2)
15. Michaelis–Menten
16. Bisubstrate-ordered reaction
17. One substrate reversible reaction
18. Bisubstrate ping pong
19. Bisubstrate reaction
20. Trisubstrate reaction
21. Competitive Inhibition
22. Uncompetitive inhibition
23. Noncompetitive inhibition
24. Mixed-type reaction
25. Substrate activation
26. Simple bimolecular reaction
27. Mass transport limited reaction
28. Surface heterogeneity reaction
29. Double exponential decay

SAAM II is a powerful research tool to aid in the design of experiments and the analysis of data. It is a new compartmental and numerical modeling program developed to help researchers create models, design and simulate experiments, and analyze data more quickly, easily, and accurately. Compartmental models are constructed graphically by creating a visual representation of the model. Numerical models are built by entering algebraic equations directly. SAAM is especially powerful because of the way it integrates model development and testing features, creates systems of ordinary differential equations from the compartmental model structure, permits the simulation of complex experimental designs on the model, and solves the model and fits it to data using state-of-the-art mathematical and statistical techniques. This program has been used for kinetics studies.

3.28 Master Chemical Mechanism

At Leeds University in England, some excellent work is being done in kinetics and particularly atmospheric kinetics and mechanisms.

The master chemical mechanism (MCM) from Leeds University is believed to be the most comprehensive chemical mechanism in the world treating the degradation of 120 volatile organic compounds (VOCs) that are emitted into the atmosphere. The MCM has been constructed by considering the chemical kinetics data concerning rate coefficients, absorption cross sections, and reaction products, including recent evaluations and reviews. The MCM is an explicit mechanism and, as such, does not suffer from the limitations of a lumped scheme or of one containing surrogate species to represent the chemistry of many species. Rate constants that are known are determined by extrapolating data from previously measured systems using structure–activity relationships. The mechanism here includes the oxidation of 120 VOCs by OH and O_3, and the chemistry of the subsequent oxidation products. These steps continue until CO_2 and H_2O result as the final products of oxidation. This can lead to many reactions in the scheme of just one hydrocarbon: the isoprene scheme alone contains 200 reactions, due in part to the production of four possible peroxy radicals following the initial attack by OH. The model contains a large number of distinct, often generic photochemical reactions involving inorganic species, aldehydes, ketones, and organic hydroperoxides.

The Web site that is included provides a platform on which to make the MCM available to a wide scientific community. It describes the complete tropospheric oxidation of 120 VOC.

It is in excess of a 7000-line mechanism code. It has some text on the project and also relevant references. It would be impossible to publish in hard copy and also would be very unreadable. A small initial sample is presented.

Users can browse code on the Web site by selecting the VOC of interest to them, from a primary classification table. These then link to the individual VOC mechanism sheets.

All the coded species names are linked to a species file, which gives an interpretation of the true chemical stucture of the species.

3.29 Multicomponent Diffusion and Reaction in Three-Dimensional Networks — General Kinetics

During the last 10 years the design of catalyst particles and porous structures has made considerable progress. Due to the complicated interaction of diffusion and reaction in catalysts, more-detailed models of porous structures are needed. We have based our model on a three-dimensional network of interconnected cylindrical pores as pore model, although the treatment is applicable to alternative pore geometries, e.g., slit pores. The network assumed has predefined distributions of pore radii, connectivity, and porosity. Mass transport in the individual pores of the network is described by the dusty-gas model. In contrast to previous publications, the present network model can be applied to any common reaction kinetics. This becomes quite inevitable in order to make 3Dnetwork models applicable to practical problems in industry.

To solve the mass balances within the entire network, the mass balances for individual pores have to be solved simultaneously, since the mass balances are coupled by the boundary conditions at the nodes of the network. The system of differential equations has been solved by the finite-difference method. To solve the resulting large nonlinear system, a Schur complement method was employed. Due to a decoupling technique, the Schur complement method has relatively small computer storage requirements. The use of the algorithm is demonstrated for a complex reaction network.

Porous catalyst supports are in widespread use in the chemical industry. Reactants diffuse into the voids pace of the support and reactants and products formed diffuse out of the pellet. Multicomponent diffusion within the catalyst supports has yet to be described. Molecular diffusion and Knudsen and surface diffusion can all occur. On fractal surfaces anomalous diffusion will take place, and in microporous materials configurational diffusion may be the dominating mass transport mechanism. A quite useful approximation based on Boltzmann's equation, the dusty-gas model, has been developed. This model describes the molecular diffusion, Knudsen diffusion, viscous flux, and, in principle, the surface diffusion, with sufficient accuracy.

One should differentiate between distinct but interconnected micro- and macroporous networks of pores, to provide a realistic representation of a bimodal porous network. These networks are advantageous in several respects:

- The effect of connectivity of the pore space is taken into account.
- Any type of network can be employed.
- Any pore size distribution can be accommodated and any pore geometry, e.g., cylindrical or slit-like pore size.
- It is possible to model local heterogeneities, e.g., spatial variation in mean presize.
- Any distribution of catalytic active centers may be taken into account.
- Percolation phenomena can be described.

The basic feature of this work are

- Multicomponent diffusion has been modeled using a Maxwell–Stefan approach.
- Multicomponent diffusion and reaction occur in a random network of interconnected pores with a predefined average connectivity and distribution of pore radii, whereby any type of network can be employed. Pores with smooth or fractal walls and of any shape may be used as long as transport behavior within the pores is known.
- Percolation phenomena can be studied.
- Different distributions of active centers may be investigated.
- Any common type of reaction kinetics can be applied. This is the most important aspect of the present work. In this work only the isothermal case was considered. Recently, developed numerical techniques have led to a considerable reduction of computer memory requirements for such calculations.

An efficient algorithm to calculate multicomponent concentration profiles for general kinetics within 3D prenetworks has been developed. The algorithm may be employed for any type of a 3D network and any pore shape. The pore walls may be smooth or of a fractal nature. Percolation processes can also be described by including a cluster-counting algorithm. The present algorithm has been applied successfully to a complex example. It can be employed as a basic model for pore structure optimizations.

3.30 Programs for Other Physical Properties

CHETAH stands for Chemical Thermodynamic and Energy Release Program. It is widely used in the chemical industry for hazard evaluation. Its chief aim is to predict deflagration/detonation potential from molecular structure, but it can also be used to estimate heats of reaction, heat capacities, and entropies. It is available in an IBM personal computer–compatible version, so that the program is accessible to many users. It performs these calculations solely through a knowledge of the materials of interest through the implementation of Benson's method of group activity. It must be emphasized that all of the thermodynamic calculations performed by CHETAH are for species in the gas phase. Since most of chemical processing is carried out in the condensed phase, this may seem a weakness in the program. However, the deviations from the condensed values are either too small to affect the result significantly or can be estimated by other means. Also, the correlations used in the hazard evaluation section of the program are based on the experimental data for condensed phase species. CHETAH was never intended to replace physical testing of materials. It should be used only in an overall scheme involving both physical testing and other predictive tools.

3.31 Solvent Substitution

Solvent substitution data systems include

- Integrated Solvent Substitution Data System (ISSDS)
- Solvents Alternatives Guide (SAGE)
- Hazardous Solvent Substitution Data System (HSSDS)
- Joint Service Pollution Prevention Technical Library
- Department of Defense Ozone Depleting Chemical/Substance Information
- Solvent Handbook Database System (SHDS)
- Solvents Database (SOLV-DB)
- Materials Compatibility Database (MATCOMPAT)
- Related Documents

ISSDS — ISSDS facilitates access to solvent alternative information from multiple data systems through a single, easy-to-use command structure.

SAGE — SAGE is a logic tree that evaluates the user's current operating scenario and then identifies possible surface cleaning alternative solvent chemistries and processes that best suit the defined operating and material requirements. More information is given below.

HSSDS — HSSDS is an online, comprehensive system of information on alternatives to hazardous solvents and related subjects. The HSSDS contains product information, material safety data sheets, and other related information. More will be mentioned about HSSDS later.

Joint Service Pollution Prevention (P2) Technical Library — This is an electronic resource maintained by Naval Facilities Engineering Service Center (NFESC). It is subject to technology transfer with other organizations by direct access or by search facilities.

Department of Defense (DoD) Ozone Depleting Chemical/Substance Information — U.S. Air Force Ozone Depleting Chemicals (ODC) Information Exchange, courtesy of the Brooks Air Force Base, Human Systems Center (HSC), DoD Ozone Depleting Substance (ODS) MILSPEC Database. The central point of contact is U.S. Navy CFC & Halon Clearinghouse.

SHDS — SHDS is a database providing access to environmental and safety information on solvents used in maintenance facilities and paint strippers. DHDS contains information from laboratory testing.

Solvents Database — SOLV-DB will help you find a wide variety of data quickly and easily. It will help you make intelligent choices about which solvent will do the job, help you understand what you are dealing with, and help you find out how to handle the materials wisely. It will also help you determine your responsibilities under routine conditions and in the case of accidents.

3.32 SAGE

SAGE (version 2.1) recommends solvent replacements in cleaning and degreasing operations. The system leads the user through a question-and-answer session. The user's responses allow the system to develop a list of alternative chemistries and processes. The user reviews descriptions of the alternatives on the computer screen, and, if any of the alternatives are of interest, the user requests a detailed report for each. Safety/environmental concerns, industrial case studies, and state technical system programs in the user's geographic location that can provide more information are included in the report.

The program is from the EPA and follows its common menu form. It provides information on just one of the many physical property needs that can be supplied from SYNPROPS.

3.33 ISSDS

A multitude of solvent substitution data systems were created when ozone-depleting and hazardous solvents were banned. Each of these systems provided the user with only part of the solution. ISSDS provides a better way for those searching for information on solvents and cleaning processes to quickly retrieve information regarding solvents and process alternatives for a variety of cleaning agents and cleaning processes is called the Integrated Solvent Substitution Data System, or ISSDS. It is based on data distributed across the Internet.

ISSDS supports users who are experts in the field of solvent substitution, as well as those who have minimal exposure to this type of information. The novice ISSDS user performs solvent searches using the Guided Query. As the users become more educated, they move to the Assisted Query. When a user becomes familiar with ISSDS contents, they migrate to the Direct Query.

Currently ISSDS accesses information from all of the above data collections and more.

ISSDS is a subset of the Enviroene Web site. The latter is funded by Strategic Environmental Research and Development Program (SERDP) through the EPA, DOE, And DoD. The prototype of ISSDS is currently on the World Wide Web and can be accessed at the Environene Web site. Ongoing work is under way to refine the current interface and provide more data from other existing solvent databases. The URL for accessing ISSDS is http://es.inel.gov/issds/index.html.

3.34 Sources of Physical Property Data

CRC Handbook of Chemistry and Physics
Tables of Physical and Chemical Constants and Some Mathematical Functions
The Merck Index
JANAF Thermochemical Tables
Chemical Technicians' Ready Reference Handbook
Composite Index for CRC Handbooks (also a CD-ROM)
Chemical Abstracts CA File on STN
Kirk-Othmer Encyclopedia of Chemical Technology (also a CD-ROM)
The Search for Data in the Physical and Chemical Sciences
The CODATA Directory of Data Sources for Science and Technology
Handbooks and Tables in Science and Technology
Chapman & Hall Dictionaries
Handbook of Data on Common Organic Compounds
Properties of Organic Compounds (CD-ROM)
AIChE publications, includes *AIChE Journal, Chemical Engineering Progress, International Chemical Engineering*, and the *AIChE Symposium Series*
Bourton, K., *Chemical and Process Engineering Unit Operations*, 1967
Ray, M. S., *Chemical Process and Plant Design Bibliography*, 1959–1989, 1991
CHERUB: Chemical Engineering Reference User Bibliography, 1989–
Applied Science & Technology Index, 1958–
Chemical Abstracts (CAS), 1907–
Engineering Index, 1919–1988
*Ei Compendex*Plus*; 1980-Electronic databases; El Compendex *Plus is available via the Internet from Ei Compendex *Web at http://cpxweb.ei.org
Encyclopedia of Chemical Technology, 3rd ed., 1978–1984
Ulmann's Encyclopedia of Industrial Chemistry, 5th ed., 1985–
AIChE Manual for Predicting Chemical Process Design Data, 1984–
Properties of Gases and Liquids, 4th ed., 1987
Azeotropic Data, 2 volumes, 1994.
Experimental Results from the Design Institute for Physical Property Data: Phase Equilibria, 3 volumes, 1985–1989
International Critical Tables of Numerical Data: Physics, Chemistry and Technology, 7 volumes. 1926–1930
Jacobson, C. A., *Encyclopedia of Chemical Reactions*, 8 volumes, 1946
Journal of Chemical and Engineering Data, 1956–
Journal of Physical and Chemical Reference Data, 1972–
Kaye, G. W. C. and Laby, T. H., *Tables of Physical and Chemical Constants*; 16th ed., 1995
Lange's Handbook of Chemistry
Macmillan's Chemical and Physical Data, 1992
Perry's Chemical Engineers Handbook
Solubilities of Inorganic and Organic Compounds, 4 volumes, 1963
Yaws, C. L., *Physical Properties: A Guide to the Physical, Thermodynamic and Transport Property Data of Industrially Important Chemical Compounds*, 1977.
DECHEMA's Chemistry Data Series, Heat of Mixing Data Collection, 4 volumes, 1984–1991

Recommended Data of Selected Compounds and Binary Mixtures, 1987
Critical Data of Pure Substances, 2 volumes, 1986
Vapor-Liquid Equilibrium Data Collection, 16 volumes, 1977–1992
INFOCHEM and DIPPR data and data banks

Infochem has two proprietary physical properties databanks, one for fluids (Infodata) and one for condensed components (Infocond), and it is a distributor for the DIPPR data bank, produced by the AIChE. Tables of comparison between the data banks for constant properties and for temperature-dependent properties are listed. Infodata currently contains about 100 components, mainly light gases and hydrocarbons, especially useful to the oil and gas industry. Infocond includes about 140 solid components. The data are limited to molecular weight, normal melting point, normal boiling point, the enthalpy of formation in the condensed state at standard conditions, the standard entropy in the condensed state at 298.15 K and the heat capacity in the condensed state as a function of temperature. It is intended for use with the Infochem chemical and phase equilibrium module.

DIPPR contains data on 39 properties for around 1300 components. The properties cover 26 single-valued property constants and 13 properties which are functions of temperature. These latter are calculated from correlation coefficients.

Of particular importance to scientists and engineers are the vast resources of the federal organization NIST, particularly the Physical and Chemical Properties Division. In this division there are attempts at technology transfer in computational thermochemistry. It states as its purpose and approach the encouragement of communication among engineers, computational scientists, and experimental scientists, building upon their common interests in themochemistry. It states that global competition is shrinking the time lines for industrial process scale-up (from laboratory to production). This has resulted in an increased reliance on computerized modeling and simulations to replace expensive, time-consuming, and hazardous experiments. There has been a drop in the cost of computations, allowing the development of new predictive techniques that are practical in an industrial environment. "Unfortunately, the chemical engineering community is largely unaware of this impressive recent progress. This poor communication is a major hindrance to the conversion of the science into useful technology."

In cooperation with Dow, NIST organized a symposium. Four recommendations came out of this.

1. A critically evaluated, consensus database of thermochemical data is needed for methods development by scientists.
2. The scientific research should be broadened to include the accurate prediction of a wider spectrum of engineering properties such as kinetics, vapor–liquid equilibrium parameters, critical data, and the properties of molecules on surfaces.
3. Better dissemination of these techniques to the engineering and industrial community is needed, including both publicity and user-friendly software.
4. Corporate cultures should be changed to allow private (but nonproprietary) thermochemical data to be added to public resources such as databases and estimation software.

"In the future, we intend to focus on recommendations 1 and 3 and to encourage others to pursue recommendations 2 and 4."

The division also provides models for thermophysical and thermochemical properties of gases, liquids, and solids for pure materials and mixtures and for the rates and mechanisms of chemical reactions in the gas and liquid phases. It also develops and evaluates theoretical models, estimation methods and computational approaches. The division is also responsible for providing thermodynamic information necessary for the prediction of movement of heavy metals through groundwaters (aquifers) and surface waters and for prediction of remediation yields in treatment methodologies. In addition to providing accurate thermodynamic data for remediation, use in simulators or calculational models for the prediction of the rate of migration of contaminants through water will be monitored. This division will also provide industry with high-accuracy property data and models for alternative refrigerants and refrigerant mixtures. The chlorofluorocarbon and hydrochlorofluorocarbon refrigerants must be replaced with fluids that do not deplete stratospheric ozone. This requires knowledge of mixtures and transport properties. Knowledge of thermophysical properties is also very important to this work.

3.35 Computational Thermodynamics and the Physical Properties Database (PHYSPROP)

Computational thermodynamics (CT) makes use of computer software and thermodynamic databases to perform realistic calculations of thermodynamic properties of multicomponent systems. Consistent databases where each phase is described separately using models based on physical principles and parameters assessed from experimental data is a key feature. CT is a valuable tool for the properties of a wide range of commercial materials as well as for process development. It is used for metals, ceramics, geochemistry, electronic materials, aqueous solutions, etc. A recent book describing such applications is the *SGTE Casebook* published by The Institute of Metals in 1996. The Thermo-Calc system is used for these purposes at more that 100 universities and industries all over the world.

PHYSPROP contains chemical structures, names, and physical properties for over 13,000 chemicals. Physical properties are collected from a wide variety of sources, and include experimental, extrapolated, and estimated values for

- Melting point
- Boiling point
- Water solubility
- Vapor pressure
- Dissociation constant
- Octanol/water partition coefficient
- Henry's law constant

available in either a ChemBase or ISISBase format (MDL Information Systems, Inc.). These programs allow PHYSPROP to be searched by substructure, name fragment, or any of the physical properties.

3.36 Laboratory for Applied Thermodynamics and Phase Equilibria: Research; Computer Aided Design of Technical Fluids

Environmental, safety, and health restrictions impose limitations on the choice of fluids for separation and energy processes. Based on group contribution methods computers can assist in the design of the desired compounds. These compounds or mixtures of these compounds have to fulfill the requirements from an integrated process point of view. This program includes on the one hand the design of the components and on the other hand the experimental verification of the predicted design of components.

3.37 Computational Fluid Dynamics

Computational fluid dynamics (CFD) means the description of the properties of fluids by using physical equations and computational methods. CSC has wide experience in modeling fluid phenomena and has an extensive selection of commercial software for CFD and powerful supercomputers to carry out the computations.

The most important fluid-computing applications are CFX and FIDAP. Both are versatile and it is possible to model a wide variety of problems with them. A new version of ANSYS includes the FLOTRAN module for fluid mechanics. However, it is not as versatile as CFX or FIDAP in terms of physical models. In presenting the results of CFD it is possible to use Funcs visualization program developed in CSC. Also, a general-purpose program, Ensight, may be used.

3.38 Web Collections of Physical Property Sites

Thermodynamics Research Laboratory (Thermodynamic Data and Property Calculation Sites on the Web, http://www.uic.edu/~mansoori/Thermodynamic.Data.and.Property_html)

CHEMINFO, SIRCh: Physical Property Information, Link to Lecture Notes on this topic (http://www.indiana.edu/~cheminfo/ca_ppi.html

3.39 Computerized Emergency Response Systems — PLG, Inc.

PLG has developed a software system called MIDAS (Meteorological Information and Dose Assessment System) to support emergency action decisions. It receives data automatically from online meteorological and toxic material sensors and rapidly calculates the dilution due to atmospheric dispersion from which the off-site exposures to radiological or chemical releases are determined. Results are displayed in color graphics format showing areas where emergency action would be appropriate.

3.40 Tackling the Scale Problem

Turbulent diffusion is ubiquitous. We wish to know whether the mathematical laws believed to govern turbulent diffusion hold true. The main problem in testing them has been the immense range of scales involved in some of nature's turbulent systems. To test the universal laws over a turbulent diffusion expanse that occurs in nature requires handling a scale range of from 1 to 100,000, with the number of variables and the amount of computation increasing exponentially with each increase. By tapping the parallel processing power of a CRAY T3D and measuring the results against meteorological data, it has been shown that the universal laws hold true, and the landmark work offers potential for accurately modeling extremely large, complicated flow systems.

The problem with existing models is that the algorithms employed to follow the movement of two particles while simultaneously assessing turbulence data around them can only track the particles for a relatively short time. The information grows exponentially with each jump to a larger scale, because the calculations try to assess turbulence in all the eddies of each scale.

The T3D, however, can run millions of independent calculations and then pool the results. Snapshots are produced of many velocity fields. The resulting data are used to build a statistical framework that helps illuminate dispersion paths. They are interested in their distance from each other, how rapidly that grows, as the particles move away from each other. That reveals the characteristics of the turbulent system, which relates to the transport of tracers like pollutants, heat, or water droplets.

The modeling technique, Monte Carlo simulation, involves independent repetitions of the same experiment. If these repetitions are run simultaneously on different processors, then the code achieves perfect scaling. Doubling the number of processors halves the running time. Because the algorithm requires small memory and because of the large number of available T3D processors, we can run many experiments simultaneously, while sampling the particle position intermittently allows the program to run economically.

3.41 Process Design and Simulations

Aspen is a tool that can be used to develop models of any type of process for which there is a flow of materials and energy from one processing unit to the next. It has modeled processes in the chemical and petrochemical industries, petroleum refining, oil and gas processing, synthetic fuels, power generation, metals and minerals, pulp and paper, food, pharmaceuticals, and biotechnology. It was developed at the Department of Chemical Engineering and Energy Laboratory of the Massachusetts Institute of Technology under contract to the U.S. DOE. Its main purpose under that contract was the study of coal energy conversion.

Aspen is a set of programs that are useful for modeling, simulating, and analyzing chemical processes. These processes are represented by mathematical models, which consist of systems of equations to be solved. To accomplish the process analysis, the user specifies the interconnection and the operating conditions for process equipment. Given values of certain known quantities, Aspen solves for the unknown variables. Documentation is available and the ASPEN PLUS™ Physical Properties Manual is very important.

The AspenTech Smart Manufacturing Systems (SMS) provides model-centric solutions to vertical and horizontal integrated management systems. These embody the AspenTech technology in the area of modeling, simulation, design, advanced control, online optimization, information systems, production management, operator training, and planning and scheduling. This strategy is enabled by integrating the technology through a design-operate-manage continuous improvement paradigm.

The Consortium in Computer-Aided Process Design (CAPD) is an industrial body within the Department of Chemical Engineering at CMU that deals with the development of methodologies and computer tools for the process industries. Directed by Professors Biegler, Grossmann, and Westerberg, the work includes process synthesis, process optimization, process control, modeling and simulation, artificial intelligence, and scheduling and planning.

Unique software from Silicon Graphics/Cray Research allow virtual plant computational fluid dynamics analysis and complex simulations. The CFD analysis solution focuses on analyzing the fluid flows and associated physical phenomena occurring as fluids mix in a stirred tank or fluidized bed, providing new levels of insight that just were not possible through physical experimentation.

Floudas at Princeton has discussed the computational framework/tool MINOPT that allows for the efficient solution of mixed-integer, nonlinear optimization (MINLP) methodologies and their applications in process synthesis and design with algebraic and/or dynamic constraints. Such applications as the areas of energy recovery, synthesis of complex reactor networks, and nonideal azeotropic distillation systems demonstrate the capabilities of MINOPT.

Paul Matthias has stated that the inorganic chemical, metals, and minerals processing industries have derived less benefit from process modeling than the organic chemical and refining industries mainly because of the unique complexity of the processes and the lack of focused and flexible simulation solutions. He highlighted tools needed (i.e., thermodynamic and transport properties, chemical kinetics, unit operations), the new data and models that are needed, the ways models can be used in day-to-day operations, and, most important, the characteristics of the simulation solutions that will deliver business value in such industries.

Advances in CFD software have started to impact the design and analysis processes in the CPI. Watch for them.

The industrial perspective of applying new, mostly graphical tools for the synthesis and design of nonideal distillation systems reveals the sensitivity of design options to the choice of physical properties representation in a more transparent way than simulation, and they are very useful in conjunction with simulation.

Barton discusses three classes of dynamic optimization problems with discontinuities: path constrained problems, hybrid discrete/continuous problems, and mixed-integer dynamic optimization problems.

3.42 Chem Info, Khase, and Physical Property Needs

Infochem Computer Services Ltd. has a Web site where one can use the software products Multiflash Phase Equilibrium Package, Thermodynamic Models and Gas Hydrate Model, Chemreact Chemical Equilibrium Package, and Physical Property Databanks.

Multiflash is a multiphase equilibrium calculation package. It handles multiple models and uses different thermodynamic models. It can solve realistic problems where the phases may have differing properties that need to be modeled in different ways. It can carry out the full range of flashes: (T,P), (H, P), (H,T), (S,P), (S,T), (V,P), (V, T), (P, phase fraction), (T, phase fraction) and (U,V). It can work out which phases are present at equilibrium and starting value and can search for dew or bubble points. It uses convergence accelerators and Gibbs energy minimization.

Basic models used include standard implementations of ideal gas, RK, RKS, PR, ideal solution, Wilson, UNIQUAC, NRTL, BIP correlations for oil and gas systems for cubic equations of state. UNIFAC is present in both VLE and LLE variants. The Infochem version of the DIPPR data bank contains UNIFAC group structures for all eligible components. Models are used for the transport properties: viscosity, thermal conductivity, and surface tension.

Chemreact calculates chemical equilibria for a mixture based on initial composition, temperature, and pressure. Chemical equilibria, and the associated heat changes, can be calculated for either a liquid phase or, a gas phase, or both, plus a number of solid phases. It can be used for applications such as the effect of temperature and pressure on equilibrium, modeling behavior of reactors, checking catalyst efficiency and reactivity, comparing reaction routes, solid precipitation from a reacting mixture, equilibrium among isomers, and preferential reactivity.

Infochem has two proprietary physical properties data banks, one for fluids (Infodata) and one for condensed components (Infocond), and it is also a distributor for the DIPPR databank. Infodat has 100 components, mostly light gases and hydrocarbons, especially suitable for the oil and gas industry. Infocond has 140 solid compounds, and DIPPR has 39 properties for 1300 components.

KHACE Consulting uses up-to-date control system design for the process industries. Among its capabilities are

- Dynamic simulation of processes and process control systems
- Advanced process control feasibility and design studies
- Controllability, operability, performance studies, and troubleshooting
- MATLAB/Simulink process control application engineering and training
- SPC and neural networks process control applications and training
- Physical properties and thermodynamics tools for MATLAB and Simulink

MATLAB versions of established thermodynamics and physical properties tools have been developed and are available for both design calculations and rigorous Simulink modeling.

Their application areas include:

- Oil and gas production
- Petrochemicals
- Process, chemical, and control engineering
- Manufacturing
- Control system design/analysis
- Process control/monitoring
- Simulation
- Training/instructional tools
- Thermodynamics and physical properties

3.43 Thermo Database

Assessed thermodynamic databases consist of model parameters which describe as accurately as possible experimental data on binary, ternary, and higher-order systems. The models, based on physical principles, make it possible to extrapolate to commercially interesting multicomponent systems. Extrapolation from several assessed lower-order systems require that these systems are internally consistent, and this takes time and effort to achieve. At present, a number of commercially interesting as well as scientifically challenging materials have been collected into databases which are available and listed below. Most of these are available for a fee.

Databases available from the Thermo-Calc group:

- KP, Kauffman binary alloys database
- GEO, Saxena Geochemical database
- G35, Group III-V binary semiconductor database
- TC-AQ Therm-Calc aqueous database
- PURE, SGTE pure element database
- SSUB, SGTE substance database

SSOL, SGTE solution database
SALT, SGTE molten salt database
TC-Fe, Steel database, modified subset of SSOL
Fe-DATA, Steel database, developed by FCT and TT
SLAG, database from IRSID for Fe-containing slags
ION, database for some oxide, sulfide, and nitride systems

Databases available from ThermoTech Ltd.

TT-Ni, Ni-based superalloys (13 components)
TT-Ni11, Ni-based superalloys (11 components)
TT-Al, Al-based alloys (12 components)
TT-Ti, Ti-based alloys (15 components)
TT-Ti9, Ti-based alloys (9 components)

Databases available from AEA Technology, U.K.

NUCLM, AEA nuclear materials database
COR, AEA corium materials database

3.44 Diffusion Program

DICTRA is a general software program for the simulation of *DI*ffusion *C*ontrolled *TRA*nsformation in multicomponent systems. Any number of components may be treated (provided that necessary thermodynamic and kinetic data are available). DICTRA is particularly suitable for treating problems involving a moving boundary, e.g., Stefan problems. Boundary conditions may be flexible, enabling the user to treat problems of practical interest. Examples of problems treated are:

One-phase problems:
- Homogenization of alloys
- Carburizing and decarburizing of steel in austenitic state

Moving boundary problems
- Carbide dissolution during austenitizing of steel
- Solidification of Al alloys
- Growth of intermediate phases in Ti alloys
- Reactions in clad steels

Diffusion in dispersed systems
- Carburizing of superalloys
- Carburizing of high-speed steels

Cooperative growth — Growth of pearlite in alloyed steel

The program is based on a numerical solution of the multicomponent diffusion equations in the various regions of a material, assuming that thermodynamic equilibrium holds locally at all phase interfaces. DICTRA is interfaced with Thermo-Calc which handles all thermodynamic calculation needed by DICTRA.

Dictra is divided into four modules:

Solution of differential equations
Calculation of thermodynamic equilibrium
Solution of flux-balance equations
Displacement of phase interface positions and adjustment of grid points, etc.

In addition, there is a post-processor that enables the user to plot and list data from a simulation, e.g., concentration profiles, particle size as a function of time, etc., and a database module to handle thermodynamic and kinetic data.

3.45 Physical Properties of Mixtures

This can be the most difficult problem of all, especially when the mixture is very imperfect. As an example, the author once had to help in the design of a plant that contained a mixture of an olefin and an alcohol, both of low molecular weight, mixed with water. Now this mixture was estimated to be near its critical point and was designed to use up the heat of reaction that came from the oxidation of the olefin. However, at the critical point, the heat of vaporization of the mixture is zero. If this is so, the whole plant would blow up. Therefore, in order for this to work the mixture temperature must be far enough away from its critical point. To assure the success of the plant design, some experimental critical point mixture measurements were made and calculations were performed several different ways to assure success, which did occur.

Above a certain temperature a gas will not change to the liquid phase with increased pressure. Instead, it changes to a supercritical phase, which is neither gas nor liquid, but has properties of each. Carbon dioxide (CO_2) is the most commonly used supercritical fluid (SCF) in cleaning applications. In this state it is an excellent solvent and can penetrate into very small cracks, which makes it very good for cleaning complex parts with tight tolerances.

The process works well on liquid contaminants, including silicone oils and machining fluids. Efforts are under way to improve the removal of particles and salts, which currently are not removed well by SCFs. Some particle removal is being seen in liquid-phase CO_2. Nonmetallic materials such as gaskets and O-rings need to be tested for compatibility with the process. The process can be designed as a closed-loop system in which the contaminant is collected and the CO_2 is reused. The only waste is removed material.

Terpenes are organic solvents that are usually derived from natural sources such as pine trees or citrus fruit. They generally have strong characteristic odors. Specific terpenes used in cleaning are α-pinene, *d*-limonene, and turpentine, which is a mixture of terpenes. Terpenes are VOCs and are flammable or combustible. Terpenes are good solvents for rosin fluxes, fingerprints, heavy petroleum greases, and oils. They can be used as a component of semiaqueous cleaning solutions or by themselves. Terpenes are strong solvents and are not compatible with some elastomers. They can be used with immersion and ultrasonic systems, often working well at room temperatures. Because terpenes may produce explosive mists when sprayed, they should only be used in spray applications with proper safety precautions, such as inert gas blanketing. In semiaqueous systems, the terpene can often be separated and reused. Terpenes are considered biodegradable, but should not be sent to the drain. One method of disposal is by incineration at an approved facility.

There are numerous articles in the literature for the physical properties of mixtures, such as

- Isobaric vapor–liquid equilibrium of binary mixtures of 1-propanol and chlorobenzene and 2-propanol and chlorobenzene;
- Vapor–liquid equilibrium for the binary systems of 2-methyl-1-propanol with some halohydrocarbons at 40.0 and 101.3 kPa;
- Vapor–liquid equilibria and saturation densities in the ternary system helium–nitrogen–methane;
- Isobaric vapor–liquid equilibra of the methanol, methyl acetate, and methy acrylate system at atmospheric pressure;
- The critical properties of alkanoic acids using a low residence time flow method;
- Excess thermodynamic properties of binary mixture of ethyl acetate with benzene, ethanol, and 2,2,2-trifluoroethan-1-ol at 298.15 K;
- Vapor pressures and excess functions of 1,4-dimethylpiperazine and *n*-heptane, or cyclohexane, measurement and prediction.

Another case where the author was called in to help was where we obtained methods, predictions, and a report for over 200 compounds of an element other than carbon. We had just a few measurements. We used UNIQUAC, which was new at the time, and matched measurements through with groups as parameters. We used DIPPR, proceeded by analogy with organic compounds, used the law of corresponding states, etc., etc., and finally filled in a spreadsheet with some novel programs using Macro programming.

3.46 Physical Properties of Natural Substances

Many physical properties of natural substances can be found on the Web. As an example, at the site http://www.zonge.com/ferro/prope_10.htm can be found tables of physical properties of ore materials. They are arranged by crystal class and alphabetically by space group (Schoenflies symbol). Columns of data include mineral name, composition, space group, resistivity, dielectric permittivity, charge separation phenomena, Curie temperature, magnetic properties, optical properties, etc. Sometimes physical properties of solid materials are included among manufactured materials.

Selected Physical Properties of Some Solid Materials (http://www.york.ac.uk/depts/chem/course/studhand/solids.html) includes melting points, density, specific latent heat of fusion, specific heat capacity, linear expansivity, and thermal conductivity. Physical Properties of Liquids includes melting point, boiling point, density, specific heat of vaporization, specific heat capacity, cubic expansivity, and viscosity and can be found near the same Web site as above. Physical Properties of Fibers (http://www.radioworks.com/nrope1.html) includes fiber type, specific gravity, sticking temperature, melting temperature, ultimate elongation, and recommended working loads. Physical Properties of Ice at 0 C contains 21 different physical properties and can be found at http://skua.gps.caltech.edu/hermann/ice.table2.html. The site http://www.ductile.org/didata/tenprop.html contains ductile iron data for design engineers. There are 42 tensile properties among its engineering data with six other mechanical properties and eight additional properties classified as physical properties.

The cosmochemistry group at the University of Arkansas in the Department of Chemistry and Biochemistry uses techniques of thermoluminescence, cathodoluminescence, instrument neutron activation analysis, electron microprobe analysis, and laboratory measurement studies of physical properties and processes. In gemnology the physical properties are crystal structure, optical properties, pleochroism (dichroism, trichroism), electrical effects, cleavage, fracture, and hardness.

3.47 Computer Modeling of Polymers

An important new area of research is the prediction of polymer properties using computerized molecular modeling techniques. Computer graphics workstations perform Monte Carlo and molecular dynamics calculations, and results are visualized in three dimensions. We are also concerned with the interactions of molecules at surfaces and interfaces, which encompasses such subjects as molecular epitaxial growth, polymer crystal growth, and alignment of liquid crystal molecules by polymer surfaces. An understanding of the polymeric aligning layer used in all crystal displays will lead to improvements in speed and contrast, ultimately paving the way for large-scale flat-screen television sets. Although polymer crystal growth studies have been active for many years, yet, even now new discoveries are being made that challenge the accepted view of how polymer crystals form. It has been suggested that crystallization occurs via an intermediate mesophase. Computer simulations are being used as a tool to examine the importance of such mesophases and to explore the conditions required for their formation. An understanding of such mesophases will have important consequences for polymer processing.

3.48 Properties of Semiconductor Materials

Modern semiconductor devices and integrated circuits are based on the unique energy band, carrier transport, and optical properties of semiconductor materials. The physical properties can be chosen and optimized for operation of semiconductor devices. Emphasis is on the quantum mechanical foundations of the properties of solids, energy bandgap engineering, semiclassical transport theory, semiconductor statistics, carrier scattering, electromagneto transport effects, high-field ballistic transport, Boltzmann transport equation, quantum mechanical transitions, optical absorption, and radiative and nonradiative recombination.

3.48a Metals and Insulators

A metal has loose electrons and they flow from one place to another. They conduct and electrons in copper form a current. Insulators have all the electrons tightly bound to the atoms, and they are not free to make electricity. A metal can not change to an insulator, and vice versa, which is one reason metal–ammonia solutions are so interesting. Michael Klein who directs the Laboratory for Research on the Structure of Matter at the University of Pennsylvania is pursuing a project at the Pittsburgh Supercomputing Center, investigating the properties of metal–ammonia solutions. The results provide, for the first time, a detailed picture of the transition between insulating and metallic states in these fascinating solutions.

3.49 Theoretical Condensed Matter Physics, Polymer Physics, and Statistical Mechanics

Acceleration Algorithms

Computers are a major source of information in all areas of physics. Microscopic mechanisms of systems undergoing phase transitions can really only be obtained by computer simulation. At phase transitions, large numbers of particles act coherently. Algorithms have been designed that move large correlated regions so that physical processes can be associated with large regions of coherence. These are called acceleration algorithms and have removed the critical slowing down associated with critical points. These algorithms will be adapted to first-order transitions and the modeling of nonequilibrium processes.

Kinetics of Phase Transitions

We need to understand the evolution of systems far from equilibrium. How a metastable state decays into equilibrium is important in alloy structure, cloud formation, membrane structure, and decay of the false universe in the early universe. The proper theoretical tools to describe evolution far from equilibrium are not yet in existence.

Recently, massively parallel computation has made it possible to formulate computer models of systems far from equilibrium and to obtain information on a microscopic level. The second helped formulate condensed matter physics problems as problems in cluster growth, percolation, and fractals. Cluster dynamics and structure, field theory, and renormalization group, as well as the Connection Machine, a massively parallel computer, have now been applied to the problem of systems far from equilibrium. Fractal structures in the early stage mechanisms of nucleation and spinodal decomposition, the effect of conservation laws on the evolution of systems out of equilibrium, and the presence and influence of fractal structures in nonlinear dynamics are being studied.

Strongly Interacting Electron Systems

High-temperature superconductivity, fullerene superconductivity, heavy fermion systems, persistent currents in mesoscopic rings, and the spectroscopy of quantum dots are some of the physical problems where strongly interacting electron systems are prevalent. To study this mean-field theory, exact diagonalizations of small systems and numerical group analysis are being used.

The interplay between electronic and phononic effects in fullerenes is under study. As the electronic timescale is comparable with the phonon time scale in these materials, the phonons have the time to relax into a distorted configuration on a fullerene, according to the Jahn–Teller effect. A novel approach to superconductivity in these materials based on the semiclassical approximation is being pursued.

Kinetics of Stochastic Processes

Under study are the effects of trapping and adsorption on the statistical properties of random walks. The most useful applications of these results may be the kinetics of chemical reactions. Such applications have helped with the understanding of the dynamics of disparate classes of reactions, such as aggregation, catalysis, recombination, and trapping. Frequently, the temporal evolution of the reaction can be described in terms of a system of interacting random walks. It has been seen that spatial fluctuations in the densities of the reactants can play a profound role in influencing the rates of chemical reactions. Thus in competitive reactions between two equivalent species, an initially homogeneous system can evolve into a continuously coarsening mosaic of single species domains and the rate of this reaction is profoundly affected by this spatial organization. Such studies will be helpful for understanding reversible and oscillatory systems, prey–predator population biology models, and biological morphogenesis.

Theories of Diffusion-Limited Aggregation (DLA)

DLA has attracted attention because of the many applications to the morphology of growing interfaces and because of fundamental issues associated with disorderly growth processes. DLA growth is completely characterized by assigning to each perimeter site i the number Pi, the probability that site i is the next to grow. Pi may form a multifractal set — that which cannot be characterized by a single exponent, but rather require an infinite hierarchy of exponents. Since the hottest tips of a DLA aggregate grow much faster than the deep fjords, the scaling of Pi, with respect to the aggregate size, depends on the particular site. Analogies to percolation systems are made by assigning different sites in the DLA model to particular kinds of percolation bonds. Additionally, conditions for viscous-fingering experiment yield patterns that obey the same scale symmetry as DLA have been found, leading to new questions about the physics underlying the viscous-fingering phenomenon.

Quantum Phase Transitions

Research for the understanding of the quantum hall effect has been in progress. Theory has predicted universal values for the critical conductivities at the plateau transitions in agreement with experiments on macroscopic samples. Our understanding of mesoscopic conductance fluctuations in simple disordered metals has been well established. Research concentrates on the poorly understood quantum Hall systems when a strong magnetic field is present, with particular emphasis on the critical regime of the metal–insulator transitions. This work is expected to provide, for the first time, a microscopic theory of conductance fluctuations in the critical regime, which will then be contrasted to the theory of universal conductance fluctuations in simple disordered metals.

Nonlinear Excitations on Surfaces

Investigations of solitons and their realtionship to surface physics are being carried out. Theoretical derivations of Hamiltonian descriptions of collective variables for the sine-Gordon, double–sine-Gordon, and four systems were performed. The Hamiltonian approach has been applied to double-sine soliton collisions, discrete soliton lattices including phonon radiation, and soliton Brownian motion. Present problems include soliton impurity scattering, kink pinning and depinning, exact equilibrium statistical mechanics of the double-sine-Gordon system, and kink transport phenomena. Studies in surface physics also include applications of dynamic and statistical mechanics models to the problem of structural and vibrational properties of atomic layers absorbed on a solid surface. Studies on biological molecules include energy transport on DNA and the derivation transition on DNA.

3.50 SGTE

SGTE is a consortium of European laboratories working together to develop high-quality thermodynamic databases for a wide variety of inorganic and metallurgical systems. It enables users to do calculations of complex chemical equilibria efficiently and reliably. It is a time- and cost-saving basis for guiding materials processing of many different types and provides crucial process-related information regarding the nature, amounts, and distribution of environmentally hazardous substances produced during different processing stages. Further developments in the modeling of different types of stable and metastable phases, in the coupling of thermodynamics and kinetics, and in the scope of application software are needed. Thermochemical calculations have very great potential for providing a sound and inexpensive basis for materials and process development in many areas of technology. SGTE is making a major effort to provide comprehensive, high-quality, self-consistent, computerized thermodynamic databases both for pure substance and for mixtures of all types and is playing a leading role in establishing methods for data evaluation and modeling of solution phases.

3.51 Program for Material and Process Development

Computer-Assisted Materials and Process Development (CAMPADA) is a consortium that has created a general computer-based tool for materials and process development. The software includes databases for thermodynamic data of metallic phases, oxides, slag phases, and aqueous solutions. Also included are computation programs for equilibrium calculations and for simulation of kinetic processes which can be linked together into more complex metallurgical and chemical processes.

The center of the consortium is the Department of Materials Science at the Royal Institute of Technology, Stockholm. The consortium has been able to establish a large number of alloy systems for the iron and steel industry and other metallurgical industries. Moreover, alloy systems that include semiconductors, geological systems, and aqueous solutions have been assessed.

A database structure and a database model for aqueous solutions have been implemented in the Thermo-Calc system. The first part of a more user-friendly interface for this system has been developed and the structure of a computer program for the simulation of metallurgical processes has been devised.

CAMPADA is composing a toolbox that can simulate and unify complex processes. The work will lead to an extension of databases for equilibria in aqueous solutions that can allow description of hydrometallurgical processes and corrosion in aqueous solutions.

Another area to be treated concerns diffusion and diffusion-controlled transformations. A package of subroutines will include models for interphase-controlled reactions. Different tools can then be used to simulate the transformation and structural changes that occur during heating of steel, metal alloys, composites, and ceramic materials. Simulation of processes for the production of steel and other metals will be possible by linking subprocesses. It will also be possible to simulate the effect of new, untried methods and processes and the reactions of materials in various environments.

3.52 Physical Properties and Pollution from Biogenic Sources

This work originated when those connected with the National Research Council (NRC) advised that part of the reason the EPA could not always have a very clear record on ozone concentration in the troposphere (calculations and monitoring) was neglect of biogenic compounds. Biogenic compounds are those arising from trees, etc. and are capable of reacting with ozone and NOx compounds. However, the people at NRC never saw the work done with THERM and Envirochemkin but only the results from the "official" modeling programs done by consultants to the Research Triangle and the EPA regions across the country. As a matter of fact, the biogenic compounds, such as isoprene, pinene, limonene, etc., had their properties created by THERM and when used in Envirochemkin did show significant results on tropospheric ozone concentration. It is important to include such work, because the creation of the thermodynamic properties by the method shown here is no different than the methods used for more well-known species and will continue to be important in observing the modeling and control of global warming and other phenomena that could become much more deleterious to humankind.

3.53 Variations on the Game of Life: Self-Interest and Resource Optimization

The standard 2D game of life has been extended by including the prisoner-like strategies. Such modifications and observation of population values have led to surprising and counterintuitive results. Adjusting the levels of selfish nodes, through a parameter which introduces a random percentage of selfish births, has led to five conclusions.

1. For certain classes of resource allocation problems, efficient utilization of available resources dictates a mix of strategies; i.e., selfish and selfless.
2. The optimal amount of selfishness, although small, is not zero.
3. In extreme cases of resources availability, some degree of selfish behavior appears to be crucial to the survival of the whole.
4. There are benefits from the inclusion of some selfishness in terms of stability, as measured by average longevity, as well as in terms of overall population levels.
5. Combining selfishness into an otherwise totally cooperative population not only increases productivity, but it also actually decreases the overall mortality rate.

3.54 Affinity Fingerprints

The lead discovery technology called TRAP has a biological descriptor defined as the pattern of affinities of a particular molecule for a set of proteins. The binding affinity determined constitutes a molecular descriptor. If an invariable set of proteins is used for all the compounds in a library, the collection of affinities for each compound provides a uniform molecular descriptor that permits establishing similarities and differences among chemicals. In this way, the affinity patterns constitute an identifier of the molecule. With an appropriate panel, essentially all organic compounds show unique patterns of binding, justifying calling these patterns affinity fingerprints.

The affinity fingerprints can be utilized in the same manner as other more traditional descriptors based on physiochemical properties. Maximally diverse sets of compounds can be established by selecting compounds with significant differences in their affinity fingerprints. Such maximally diverse sets can be used in a screen process. Alternatively, compounds with similar affinity fingerprints can be expected to display similar biological properties. In this way, if only a maximally diverse subset is initially screened, and a hit is found in this process, other compounds in the library with similar affinity fingerprints can be retrieved, maximizing the chances of finding additional compounds similar to the initial lead.

3.55 Compound Properties Using Data from Sufficiently Similar Chemicals

The prediction of compound properties from chemical structure can be improved by correcting for an expected structure-specific calculation error, which is derived from structurally similar compounds selected from a data set with validated experimental values. The identification of proper reference compounds by a computer program requires a suitable algorithm. Two approaches were tested for various group contribution methods to calculate the water solubility and octanol/water partition coefficient of organic chemicals. One algorithm is based on a hierarchical classification scheme, and the other evaluates structural similarity through comparison of atom-centered fragments. The latter is superior.

There are many estimation methods for compound properties that are helpful to screen the fate and ecotoxicological potential of chemicals in the environment. Most are based on group contribution schemes, and comparative evaluations have been performed for a number of properties. Recently, quantum chemical continuum-solvation models offer a promising alternative to calculate various partition coefficients of compounds from chemical structure.

Prediction of compound properties from chemical structure can be improved through correction by substance-specific method errors estimated from the average error or bias with structurally similar reference compounds. This reference method makes use of the physicochemical knowledge that is implicitly enclosed in (sufficiently) large and reliable databases. The suitability of this approach depends on the precision of the original method as compared with both the structural resolution power of the algorithm to characterize molecular similarity and the experimental accuracy of the property of interest. For calculation models with a relatively high overall precision, application of the reference method is not recommended.

The concept of atom-centered fragments (ACFs) has proved useful to characterize molecular similarity for the identification of suitable reference compounds in the context of group contribution schemes. For future applications with more-elaborated calculation methods, extensions of the ACF concept like inclusion of greater fragments or consideration of more specific structural details are easily possible.

3.56 Multisimplex

Multisimplex can optimize almost any technical system in a quick and easy way and can optimize up to 15 control and response variables simultaneously. Its main variables include continuous multivariate online optimization, handling an unlimited number of control variables, handling an unlimited number of response variables and constraints, multiple optimization sessions, fuzzy set membership functions, etc. It is a Windows-based software for experimental design and optimization. Only one property or measure seldom defines the production process or the quality of a manufactured product, so that in optimization more than one response variable must be considered simultaneously. Multisimplex uses the approach of fuzzy set theory, with membership functions, to form a realistic description of the optimization objectives. Different response variables, with separate scales and optimization objectives, can then be combined into a joint measure called the aggregated value of membership.

3.57 The Pricing Structure of Industry Changes in Response to the Increased Restrictions on the Use of Toxic Chemicals

Technology assessment concerns answers to the following questions:

1. What are the ultimate performance characteristics of the industry in terms of feedstock, energy, and investment capital use?

2. Is it possible to anticipate the effects on the industry of changing product lines and raw material sources?
3. What are the characteristics of new processes and products that lead to their eventual adoption by the economy?
4. How is the pricing structure of the industry affected by the forces of the marketplace?

One thing that comes out of this study is that the price of a compound, chemical, or product in industry can be considered as a physical property and is every bit as important and in many cases can be even more important. Another thing to consider is that the personal agenda or preferences of the CEO or the people at the top is most important to the company even when faced with the most powerful regulating agency of the land.

In considering answers to the questions above, two models are proposed for industry: the economic model and the integrated model. The first provides a reasonable representation of the total industry with the careful use of constraints and bounds, and it can be a useful tool to project future industrial development for a variety of economic and technological scenarios. The model is a useful tool for long-range planning studies, such as testing relative attractiveness of new technologies and their impact on processes and feedstocks; testing impacts of changes in feed supply, product demand, and process economics on process selections, and also studying the impact of economic growth, market variations, and price fluctuations on the chemicals industry.

The economic model of the industry uses estimates of the market prices of intermediate chemicals and, as a result, must be considered as valid only for perturbations away from a stable industry. The integrated model represents an ideal situation in which no chemical value is assigned to intermediate chemicals, and they become valuable as they become integrated into the chains of processes leading to valuable final products. The fact that both these models largely encompass the same projected industry indicates that the current industry in the U.S. has been well integrated through the forces of the marketplace.

TSCA and other such acts have given the EPA great authority to regulate the manufacture, processing, use and disposal of chemicals that are perceived as dangerous to the health of humankind and the environment. Benzene-based chemicals have been scrutinized, and decisions made on such chemicals have serious effects on the growth and developments of certain sectors of the chemical industry.

Restrictions on the use of toxic chemicals will influence the choice of process routes to commercial products or the development of new uses for existing products or new products to displace more hazardous ones. Since chemical prices combine consumer demand, availability, and function in the economy, such restrictions would sufficiently disrupt the supply and demand balances in the economy and therefore create new pricing structures.

The role of benzene in the U.S. petrochemical industry in the U.S. economy is variable, using the applicable economic environment and assuming that the industry has total freedom, regardless of the installed capacities, to select the optimal technologies in order to minimize its total production cost. However, the price of benzene has increased. Thus, different economic environments allow one to observe the behavior of the industry with respect to increased restrictions on the production of benzene.

A systems analysis of the U.S. petrochemical industry gives insight into complex interactions between the supply and demand for intermediate chemicals used captively by the industry. Changes in technology can greatly alter pricing structures. Programs such as Envirochemkin and Process Synthesis are a major step forward in giving swift insight into the price changes and can provide further information upon which to base decision making.

3.58 Structure–Reactivity Maps as a Tool for Reaction Site Descriptors

A new family of reaction site descriptors, the n-level reaction site representations, containing both electronic and steric information, has been shown as a three-level system. The structure map with its associated "colorings" can display complex data. It has been used to facilitate the analysis of documentation metabolic pathway information. It tells, given a data set composed of compounds, which reaction will take place in a majority of cases. An atlas of structure–reactivity maps constructed from a metabolic pathway database would provide a useful tool for researchers interested in predicting drug metabolism. By examining these structure maps, it may be possible to determine which structural modifications are necessary to influence the metabolism of a new compound.

A structure map bears many analogies with an intercity map. In both cases, the underlying graph is called a proximity graph. In a structure map, the vertices of the proximity graph are molecular structures rather than cities, and the edges denote pair-wise molecular similarities, distances, or dissimilarities. The many molecular similarity indexes that might be used to generate the underlying pair-wise distance or proximity matrix graph comes in a variety of forms, such as nearest neighbor graphs, minimum spanning trees, relative neighborhood graphs, sphere of influence graphs, etc. Here a similarity measure based on common substructures and a proximity graph closely related to the neighborhood proximity graph was used. The structure map seems to be an efficient way to visualize the increasing amount of information contained in chemical activity databases. The structure map of a set of compounds in conjunction with a given activity of these compounds constitute a "map" of the chemical structure–activity space. The collection of these structure–activity maps for an entire database constitutes an "atlas" for the chemist interested in chemical structure–activity relationships. The task of predicting properties of new compounds is facilitated with a suitable set of structure–activity maps. The activity may be predicted either directly from the structure–activity map, using local averaging methods such as nearest neighbor predictors, or by using QSAR methods employing descriptors suggested by visual inspection of the structure–activity map.

Just as the region surrounding a city in an intercity distance map might be colored to indicate altitude at that city, we can color a structure map to indicate the value of some property associated with the structures. In general, there are many properties that might be associated with the structures, giving rise to many different colorings of a structure map. We can combine these different colorings using one of the many methods available for representing multivariate data.

3.59 The Cost of Natural Resource Damage

NRDA, the Natural Resources Damage Estimate, is a part of Superfund and of late has been used to achieve a quick and efficient restoration of publicly owned natural resources. Earlier NRDAs were conducted in an adversarial atmosphere, focused more on estimating the value of the lost resources. Some potentially responsible parties label the NRDA program a sleeping giant that could rival the Superfund remediation program in terms of cost and litigation.

The NRDA for the Exxon Valdez oil spill cost Exxon close to $1 billion in addition to a cleanup bill of nearly $2 billion. However, NRDAs at hazardous waste sites have been much more modest. The majority are settled as part of a Superfund cleanup agreement negotiated with the EPA.

The aim of NRDAs is to restore damaged natural resources so that they can continue to be enjoyed by the public. Thus, agencies must assess the extent of the injury and link it to a toxic release, devise a restoration plan, and determine the full value of what the public has lost. Three of the principal activities are assessing injury, planning for restoration, and determining damages.

Determining injury assessments has been difficult. Common tasks are the determination of sampling strategies, the collection of samples, and conducting toxicity tests. Baselines need to be established to measure resource injuries and the significance of individual organisms vs. populations and communities of organisms.

Damage assessments evaluate two environmental problems: restoration of the damaged natural resources (primary restoration) and the compensatory damages. Compensatory damages are aimed at compensating the public for the loss of a resource from the time the damage occurred until restoration.

Contingent valuation is an economic technique used to provide monetary value for goods, services, and public programs for which market data do not exist. The technique determines the value of goods and services based on the results of carefully designed opinion surveys. The contingent valuation studies can produce estimates reliable enough to be the starting point of a judicial process of damage assessment.

3.60 The Value of the World's Ecosystem Services

The services of ecological systems and the natural capital stocks that produce them are critical to the functioning of Earth's life-support system. They contribute to human welfare, and thus represent part of the total economic value of the planet. For the entire biosphere the value (mostly outside the market) is estimated to be in the range of U.S. $16 to 54 trillion ($10^{12}$) per year, with an average of U.S. $33 trillion per year. This is uncertain and considered to be a minimum estimate. Global gross national product total is around U.S. $18 trillion per year.

Traditionally, ecological concerns have been considered to be a financial burden on economic development; however, the ecosystem functions do not compete with economic values. Instead, they are intimately linked with and are positive influences on the global economy.

This study involved a wide range of studies on the value of a broad range of ecosystem services. They included food production, raw materials, recreation, water supply, as well as regulation of climate and atmospheric gases, water cycling, erosion control, soil formation, nutrient cycling, and the purification of wastes. Estimates included 17 categories of services for the range of environments on Earth, including both marine and terrestrial environments. Coastal environments, including estuaries, coastal wetlands, beds of sea grass and algae, coral reefs, and continental shelfs are of disproportionately high value. They are particularly valuable in regulating the cycling of nutrients, which control the productivity of plants on land and in the sea. The majority of the value from ecosystem services is currently outside the market system.

The economic estimates ignore the fact that many ecosystem services are literally irreplaceable. One practical use of the estimates, the authors write, is to help modify systems of national accounting to reflect better the value of ecosystem services and natural capital. Another use is for weighing the ecosystem services lost against the benefits of a particular project or policy. Making good decisions on questions ranging from whether to drain a local wetland, whether to curb global fossil fuel consumption in order to limit climate change, all depend on adequately valuing ecosystem services.

3.61 Summary Report — Symposium on Computational Thermochemistry

This meeting encompassed everything from empirical group methods to high-level *ab initio* calculations. Topic areas included

- Advances in group contribution techniques
- New software to predict thermochemistry
- New compound techniques
- Validation of predictive methods
- Applications of techniques to practical problems

The topic areas for the panel included, among others, the following areas:

- Thermochemical data needs in industry, academia, the military, etc.
- Bottlenecks in generating data
- Experimental verification needs
- Suggestions for techniques development (e.g., user-friendliness)
- Condensed phase data (e.g., vaporization heats, sublimation heats)
- Common, consensus test data set for model verification

A summary of the panel discussion follows:

1. A need exists to develop and make available a critically evaluated database of thermochemical data for technique developers to use to test the accuracy of their models.
2. Efforts need to be focused on the accurate prediction of other properties of engineering importance and extension of current techniques to a wider variety of molecular types.
3. These techniques need to be disseminated to the engineering community.
4. There needs to be an effort to make private but nonproprietary (industrial) data public.

Appendix

Included Computer Programs

SYNPROPS

A computer disk is attached to this book, which contains the computer program called SYNPROPS. This is a spreadsheet program attached to an optimizer program. The optimizer program is part of the spreadsheet program supplied with the Quattro Pro (version 5.7) program. There are two versions of the SYNPROPS program. One is called the linear model and one is called the hierarchical model. The difference is the chemical groups that are considered, which are different for each case. Please study Tables 1 and 2. They are the versions of the programs that do not include the formulas. After you have done so, you may load the programs into a spreadsheet. Please load them into the same cells that are on the spreadsheets on your disk.

This can be determined from a set of tables for the linear model and another set of tables for the hierarchical model. These tables are the formula printouts of the tables, and the formulas are those in the cells which you do not see in the printouts that you see as the initial tables. What you do not see in the initial tables are the values of the formulas embedded in those cells. The values of the parameters B, C, D, E, and F are, respectively, the sum of the entries in the cells of the columns above the parameters, obtained from the numbers in the B column multiplied by the constant values in the column to the left of the column in which the parameter value is given, etc. The formulas on the O and Q columns for the properties look like QSAR equations. They were obtained by Cramer by careful statistical work. Some of the equations or properties were added using the same Tools menu as is used for the optimizer program, but under the Regression heading.

To run the program hit Optimizer under the Tools menu with your mouse and bring forth Options. Set Max. Time (~100), Max. Iterations (~100), Precision (1E-06), and Tolerance (0.05). Estimates can be set as Tangent or Quadratic. Derivatives can be set as Forward or Central. Search can be set as Newton or Conjugate. You can show iteration results. Assume linear or automatic scaling. You can Save Model, Load Model, or generate a Report. In the optimizer itself you can search for the maximum or the minimum or a target value (which is set to 0 when your goal is a maximum or a minimum). In the constraints box you can fix a set of cells to be greater than or equal to or smaller than some set number. You can also require that all your answers be integers (although this will restrict some of your intermediate calculation results appearing elsewhere on your spreadsheet). In order to register your constraints, you choose add, change, or delete for your list of constraints, and another option box appears which facilitates the registration of constraints. It is important when starting your calculation to pick a proper estimate for the number of groups (in the B column). Also, the author usually uses larger numbers for the maximum time and iterations shown above. A run on a 486 should not often be more than 5 min. When you are done, you will have a set of numbers for certain groups which you hope to put together to form a molecule. Knowledge of chemistry helps. If you cannot do so, you may have to start again with a different set of constraints, etc.

There are many controlling constraints that can be added simply in each computer run. As examples, you can set the molecular result to possess a molecular weight within a specified range and you can fix the answer to possess a fixed number of carbon or hydrogen or oxygen or nitrogen atoms, etc. You can set the number of groups to be below some number or above some number or equal to some number. You can set a large number of properties to be above some number or below some number or equal to some number, all simultaneously. You can set a maximum property or a minimum property or a target property at the same time as all the other constraints. You can set the parameters (B, C, D, E, and F) at preset values. There are many such devices that can help you home in on the answer you seek.

Matrix Mechanics

The formation of matrices and then operations of matrices are easily carried out on personal computers possessing the present Lotus or Quattro Pro programs. Such programs allow the execution of matrix multiplication and inversion. Some of the uses of such manipulations of data and the correlation of properties of molecules with their structure was discussed in Section 2.15. We wish to extend such work now to cases where

1. The use of matrix mechanics and SYNPROPS is used to help the determination of molecular structure that will possess the physical or toxic properties desired.
2. The use of matrix mechanics is used to ascertain the toxicity of a mixture of molecular species.
3. Matrix mechanics is used to ascertain the quantities of specific gases above a pool of a liquid mixture. Obviously, Henry's law constants are also used here.
4. The use of matrix mechanics and additional mathematics is used to ascertain the location of a source for hazardous material detected at some distance away in the field.

Case 1

The basic matrix operation involved is shown below. Let **S** be a matrix denoting a structure or structures if a set of molecules. Let **P** be a matrix that denotes the properties of a molecule or a set of molecules. Let **A** be a matrix that converts the structure matrix to the property matrix. Then we have the matrix equation:

$$\mathbf{SA} = \mathbf{P}$$

Now if we multiply the equation above by $\mathbf{S}^{-1}$, we obtain

$$\mathbf{S}^{-1}\mathbf{SA} = \mathbf{S}^{-1}\mathbf{P}, \text{ or } \mathbf{A} = \mathbf{S}^{-1}\mathbf{P}$$

likewise

$$\mathbf{SAA}^{-1} = \mathbf{PA}^{-1}, \text{ or } \mathbf{S} = \mathbf{PA}^{-1}$$

when the inverse of **A** multiplies the equation on the right hand side.

It is very important to know that if you take the inverse of a matrix that matrix must be square or have as many rows as it has columns in all of the matrix algebra given below.

In case 1 we have

$$\begin{matrix} 1 & B1 & C1 & D1 & E1 & F1 \\ 1 & B2 & C2 & D2 & E2 & F2 \\ 1 & B3 & C3 & D3 & E3 & F3 \\ 1 & B4 & C4 & D4 & E4 & F4 \\ 1 & B5 & C5 & D5 & E5 & F5 \\ 1 & B6 & C6 & D6 & E6 & F6 \end{matrix} \quad \begin{matrix} a1 & a2 & a3 & a4 & a5 & a6 \\ b1 & b2 & b3 & b4 & b5 & b6 \\ c1 & c2 & c3 & c4 & c5 & c6 \\ d1 & d2 & d3 & d4 & d5 & d6 \\ e1 & e2 & e3 & e4 & e5 & e6 \\ f1 & f2 & f3 & f4 & f5 & f6 \end{matrix} = \begin{matrix} P11 & P12 & P13 & P14 & P15 & P16 \\ P21 & P22 & P23 & P24 & P25 & P26 \\ P31 & P32 & P33 & P34 & P35 & P36 \\ P41 & P42 & P43 & P44 & P45 & P46 \\ P51 & P52 & P53 & P54 & P55 & P56 \\ P61 & P62 & P63 & P64 & P65 & P66 \end{matrix}$$

The first matrix on the left of the equal sign is the structure matrix for six molecules (1 to 6). The lead number is unity so that when it is multiplied by the second matrix on the left the lead number in the resultant equations will be a1 or a2 or a3 or a4 or a5 or a6, the constant term in the equation. The second matrix on the left is the conversion matrix for converting the structure matrix to the property matrix, which is the matrix for six properties for each of the six molecules defined by the parameters in the structure matrix. This means that there are 36 properties in all. Thus, the item P34 in the property matrix means the fourth property of the third molecule, etc.

Now the necessary items needed to do the matrix calculation are the inverse and multiply parts found in the Tools menu in the Quattro Pro spreadsheet menu or higher version than 5. One defines the matrix by the lead cell address and the end cell address with two dots in between and also the destination lead cell in the spreadsheet where the inverse or product will go. Now the third equation above indicates that you know the structure to property equations or matrix **A** and the properties you wish for the property matrix **P**. Then you find the inverse of **A**, which is $\mathbf{A}^{-1}$, as indicated above, and multiply that times the **P** matrix to obtain the **S** matrix (using the multiply function found in Quattro Pro).

Now the **S** matrix will contain the five parameters for six molecules which are then inserted in the SYNPROPS program to find the group constitutions for the sought six molecules.

Case 2

The risk concentrations for given chemicals are given in updated tables from Roy Smith, PhD., toxicologist and formerly of the EPA. These numbers k_i are divided into the measured concentrations of a particular chemical on a site to obtain a measure of its toxicity to humans. Thus if compound x has a concentration of 20 μg/m³ and the value of k_i for compound x is 40 μg/m³, the measure is 0.5. Now suppose y and z are also present with y at 100 μg/m³ and z at 200 μg/m³, respectively, and with k_i at 500 and 80 ug/m³, resp. Then we put this in matrix form as

20 0 0 1/40 0 0 0.5 0 0

0 100 0 0 1/500 0 = 0 0.2 0

0 0 200 0 0 1/80 0 0 2.5 2.7

where the number 2.7 is the sum of the diagonal elements of the product matrix (its trace). Since it is greater than unity, the site is not healthy.

Case 3

Another case may occur when, during the treatment of a hazardous waste site, a pump-and-treat method is used. Suppose there are three significant hazardous species and concentrations are taken every 5 h to see when the sum of each of the species is below its risk concentration and also the sum of the risk concentrations of the three species is below unity.

Then

X_1 X_2 X_3

Time = 0 h 5 10 15 1/5 1/5 1/5 2.1 2.1 2.1

5 h 3 7 10 1/20 1/20 1/20 1.35 1.35 1.35

10 h 2 4 5 1/25 1/25 1/25 0.8 0.8 0.8

This shows that after 10 h of pumping the sum of the risk concentrations is below unity and also that the risk concentration of each species is below unity.

Case 4

Another case occurs when there is more than one type of toxicity standard for each set of species that can be acceptable. Then we may have the following criteria:

	VCL	PCE	Benzene
Human health	x	x	x
MCL	x	x	x
Human inhalation risk	x	x	x
Ecol Inhalation risk	x	x	x
NAITCH	x	x	x
Unit risk	x	x	x
Air risk	x	x	x
RAL	x	x	x

Now if we have 25 μg/l of vinyl chloride, 50 μg/l of perchloroethylene, and 25 μg/l of benzene and we know all the toxicity values for these compounds for the sets shown above, we can form the **P** matrix and the **Q** matrix as shown below.

	VCL	PCE	Benzene	VCL	PCE	Benzene
Conc liq	1.00E+00	0.00E+00	1.00E+00	2.50E+01	0.00E+00	0.00E+00
Human health	5.00E-07	1.25E-06	8,33E-07	0.00E+00	5.00E+01	0.00E+00
MCL	5.00E-01	2.50E-01	2.50E-01	0.00E+00	0.00E+00	5.00E+01
C gas	1.47E-01	1.06E-02	4.06E-09	Conc. matrix μg/l		
Inhal R human	1.23E-05	1.00E-08	4.06E-09			
Inhal R eco	4.66E-10	1.00E-09	1.05E-11			
NAITCH PA	2.19E-02	1.32E-06	5.65E-11			
Unit risk	6.16E-06	1.01E-08	3.37E-14			
Air risk	6.37E+00	1.01E-02	3.39E-08			
RAL	3.19E+00	5.07E-03	1.69E-08			

Multiplying the **P** matrix and the **Q** matrix together yields the **Cg** matrix shown below which allows one to read the gaseous concentration and eight different health (or toxicity) for vinyl chloride, PCE, and benzene from a particular concentration of each in the liquid phase. This is shown below.

Mol Wt	6.25E+01	1.66E+02	7.80E+01
Cliq μg/l	2.50E+01	5.00E+01	5.00E+01
Human health μg/l	1.25E-05	6.25E-05	4.17E-05
MCL	1.25E+01	1.25E+01	1.25E+01
Cgas μg/m^3	3.67E+00	5.32E-01	2.03E-07
Human inhal. Risk	3.08E-04	5.03E-07	2.03E-07
Ecol. Inhal. Risk	1.16E-08	5.00E-08	5.27E-10
NAITCH PA	5.47E-01	6.62E-05	2.82E-09
Unit risk	1.54E-04	5.05E-07	1.69E-12
Air risk	1.59E+02	5.07E-01	1.69E-06
RAL	7.97E+01	2.53E-01	8.47E-07

Cg Matrix

There is also the case where the hazardous materials are in the liquid state and we wish to measure or calculate them in the gaseous state. The very approximate method used is to use the formula:

Cg = Constant x Henry's Constant x CL

or the constant in the gas phase is equal to a constant times the Henry's law constant times the concentration in the liquid phase. The risk concentration can then be calculated as Cg/ki the concentration in the gas phase divided by the risk concentration for that species. This is done in Table 44 where each species is considered at three concentrations in the liquid phase; 5000, 500, and 50 μg/l and the results are sorted from highest to lowest Cg/ki. This method, although approximate, has been valuable for screening the most important hazardous species found on Superfund sites.

Case 5

In this case we consider a method for obtaining the location of a source of pollution in the field and the concentration of each pollutant at the source.

This involves stochastic processes and, within this framework, Markov chains. Thus, it is supposed that the probability of outcome on any given experiment depends only on the outcome of the immediately preceding experiment. P_{ij} is the probability of outcome aj, given that outcome ai occurred in the preceding experiment. The numbers pij are called transition probabilities. If we begin in some particular state, then we have enough information and can calculate probabilities of statements relating to the overall sequence of experiments. In this Markov chain process, we can display a square array or matrix as

$$\mathbf{p} = \begin{matrix} p11 & p12 & p13 \\ p21 & p22 & p23 \\ p31 & p32 & p33 \end{matrix}$$

This is a transition matrix. We assign the numbers 1, 2, 3 to represent the pollutants we are measuring in the field at two points X1 and X2, where X2 – X1 is the distance from X1 to X2. Then P11 represents the probability that the concentration changes from X1 to X2, so if C1 is the concentration at X1, then the concentration at X2 will be p11.C1, etc. Now this transition matrix represents just one link (i) in the chain from receptor to source, so we have identified the matrix as P(i) and each element in it as pij(i). However, these matrices are all considered identical. Later it is possible to allow for reflecting barriers, absorbing barriers, etc., where one can differentiate them.

Now in going from source to receptor, we have

$$P(1)P(2)P(3) \cdots S = R$$

where P(i) is the transition matrix at each increment (I) from source to receptor, S is the matrix for concentrations at the source, and R is the matrix for concentrations at the receptor. Then

$$\ldots P(3)^{-1}P(2)^{-1}P(1)^{-1}P(1)P(2)P(3) \cdots S = \ldots P(3)^{-1}P(2)^{-1}P(1)^{-1}R$$

Now everything on the left-hand side of the above equation drops out except S. So if you multiply your field concentration matrix **R** by the product of the inverses of the transition matrices, then you have the concentrations at the source. Our new versions of LotusTM or Quatro Pro version 5.7 allows us to multiply and take inverses of matrices. This derivation also gives the identical analytic equations as used in the ISC model currently used by the EPA Air Division with its various modifications.

The method of taking measurements in the field is known as triangulation. Moving from one point to another we define an equilateral triangle and take samples at each point for analytical determination. We repeat this at another position trying to get two of the apices of the triangle to have equal and lower concentrations than the third apex. Then a line bisecting the base is dropped from each triangle. These lines indicate the direction of the source, and when they are extended they indicate the source location as the point where they cross.

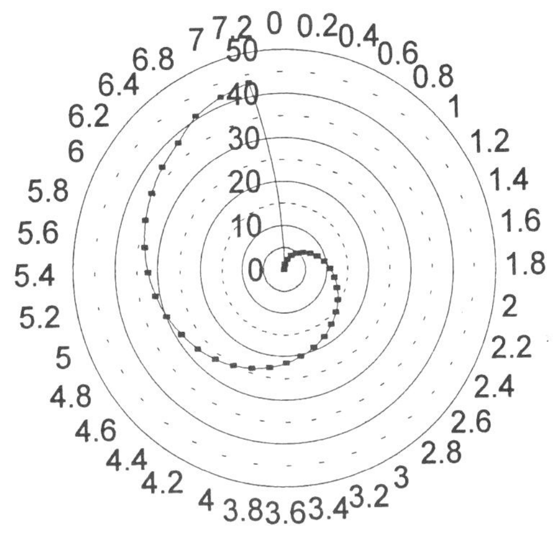

FIGURE 1 Logarithmic Spiral.

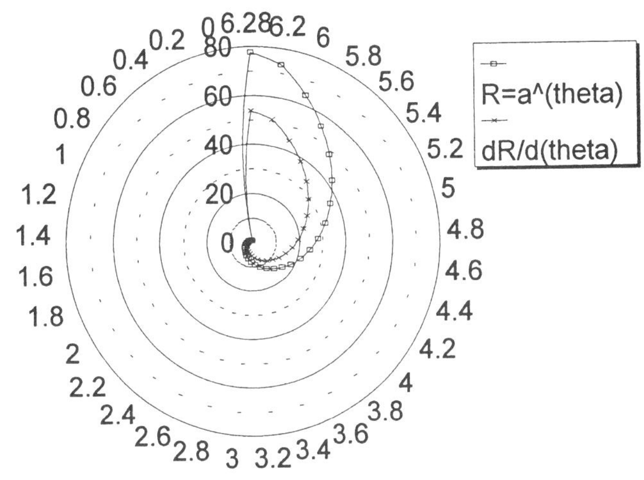

FIGURE 2 Logarithmic spiral.

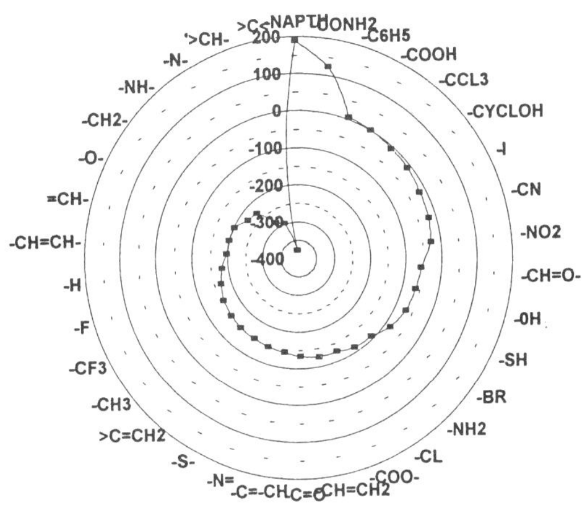

FIGURE 3 Boiling point.

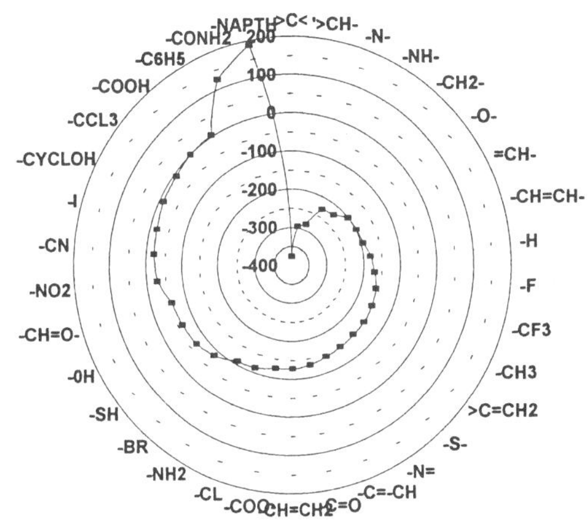

FIGURE 4 Boiling point (mirror image).

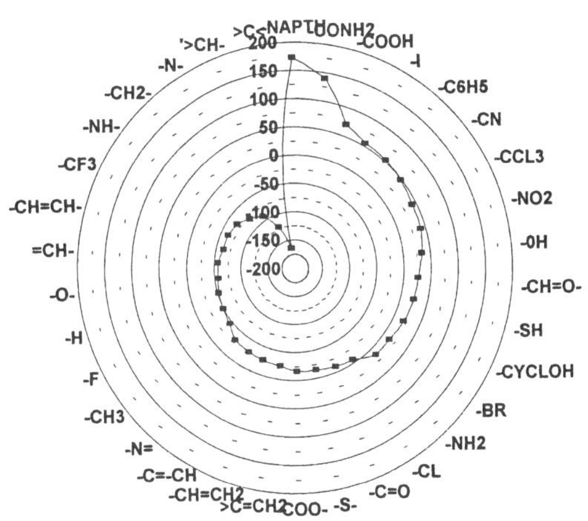

FIGURE 5 Melting point.

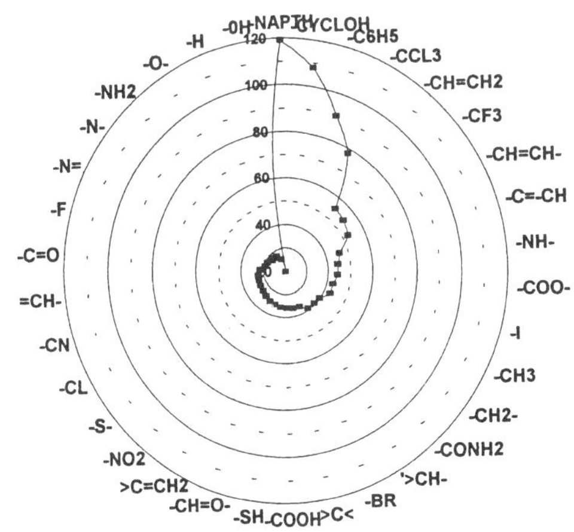

FIGURE 6 Molar volume.

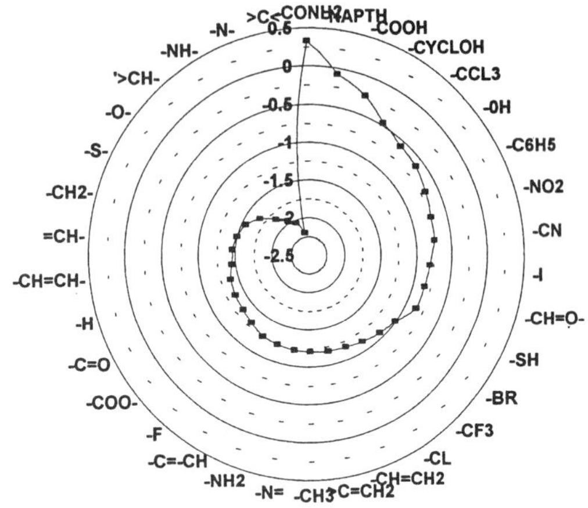

FIGURE 7 Experimental volume.

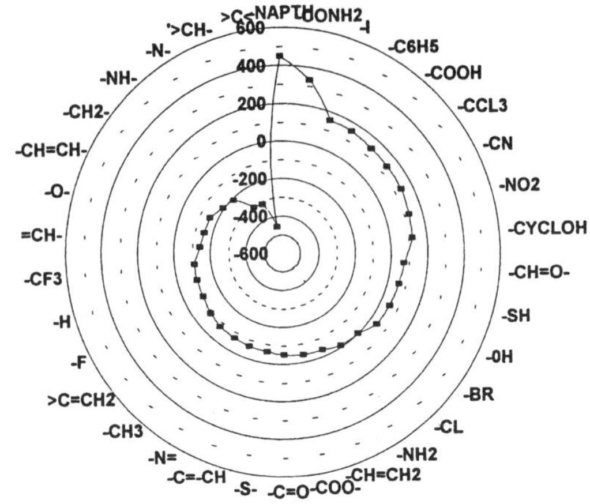

FIGURE 8 Critical temperature.

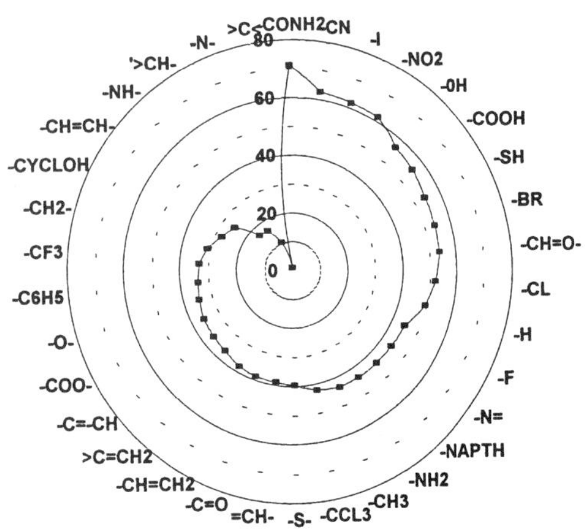

FIGURE 9 Critical pressure.

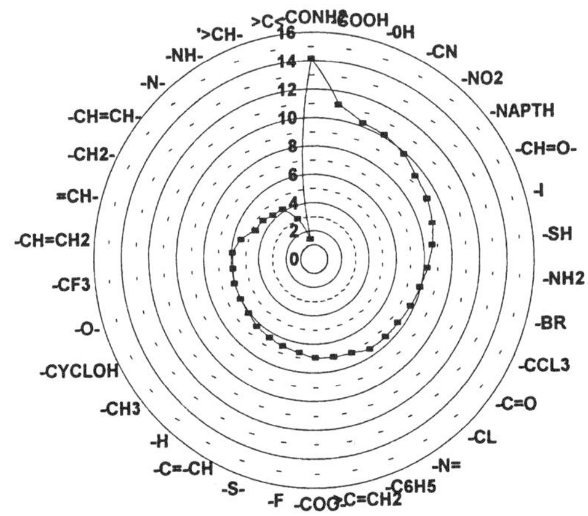

FIGURE 10 Solubility parameter.

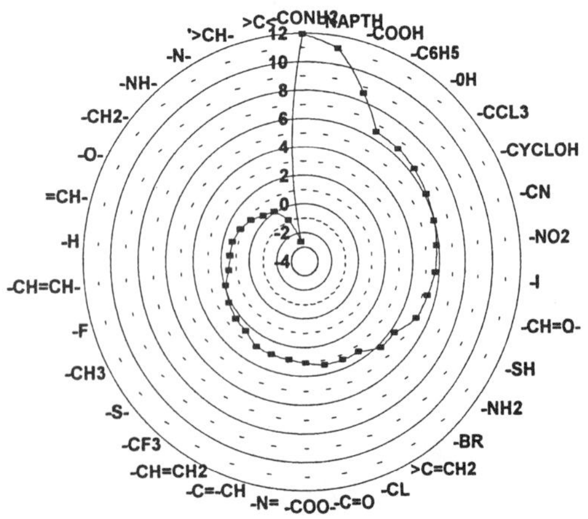

FIGURE 11 Heat of vaporization.

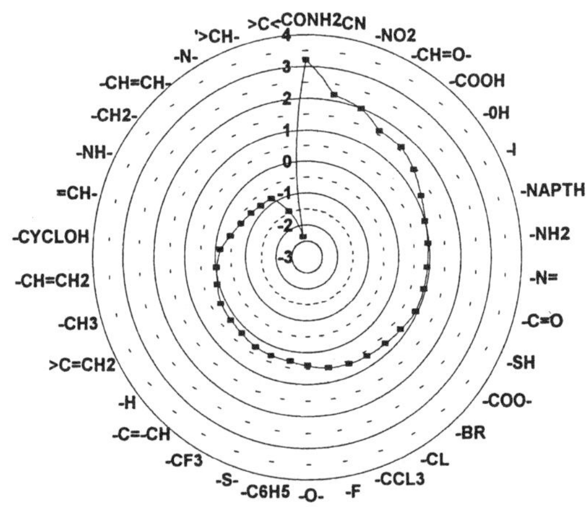

FIGURE 12 Dipole moments.

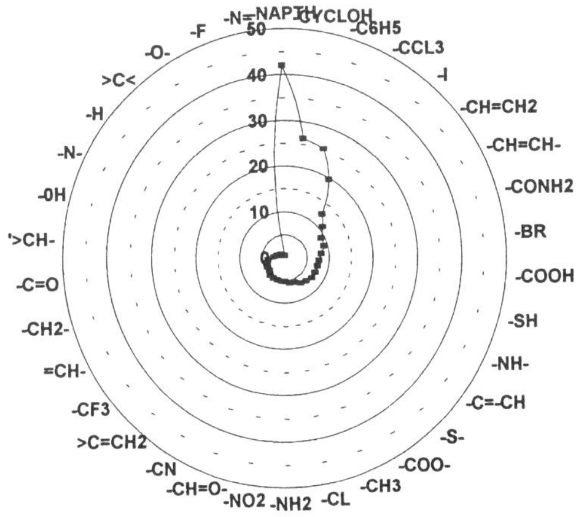

FIGURE 13 Molar refractivity.

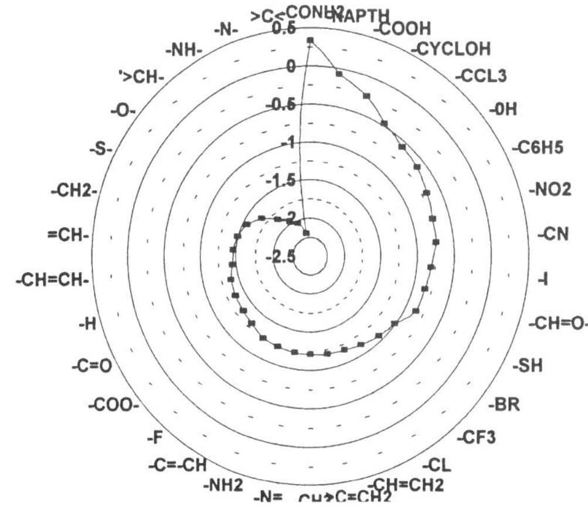

FIGURE 14 Log viscosity.

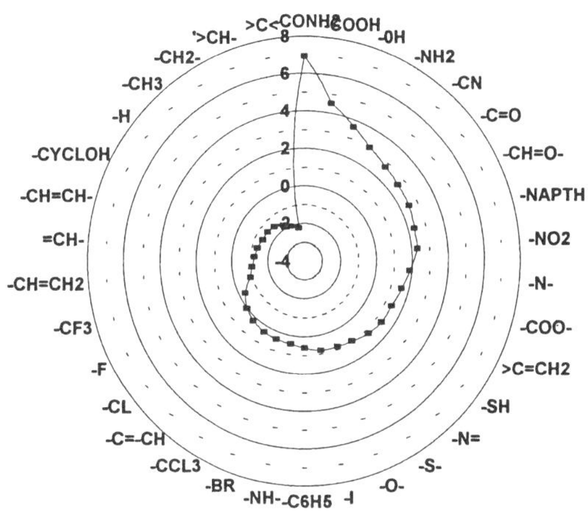

FIGURE 15 Log activity coefficient.

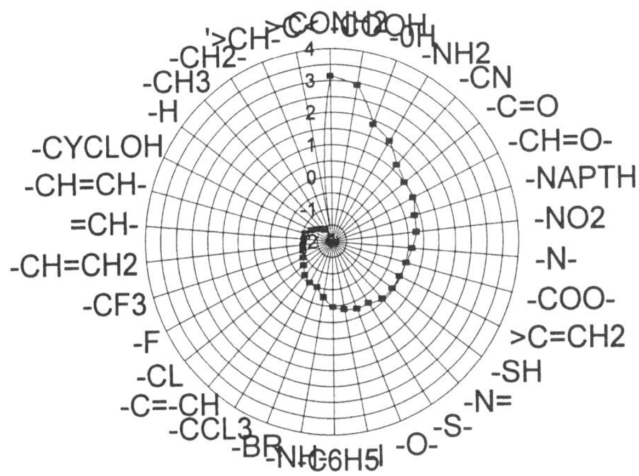

FIGURE 16 Log partition coefficient.

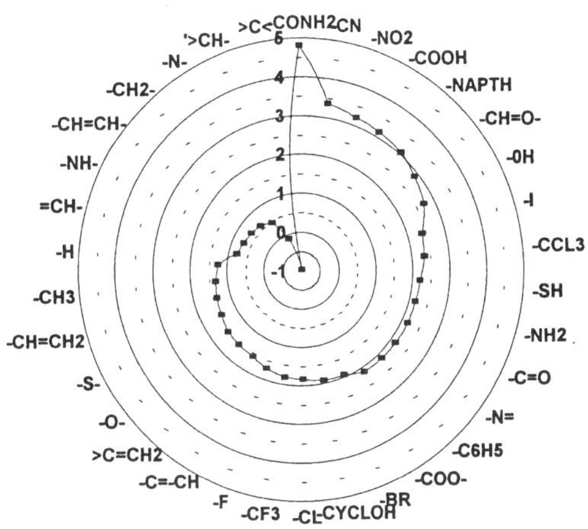

FIGURE 17 Thermal conductivity.

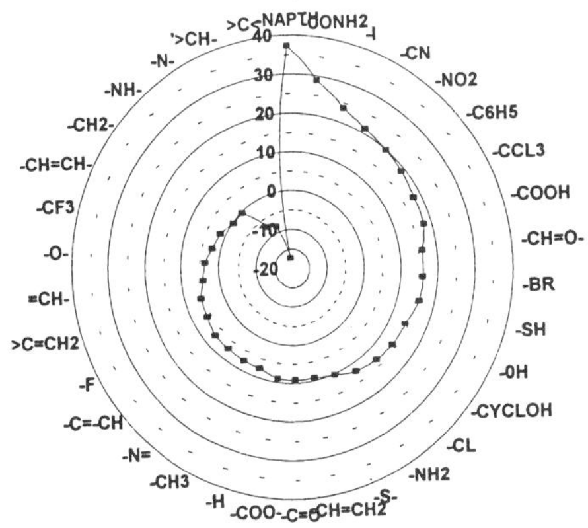

FIGURE 18 Surface tension.

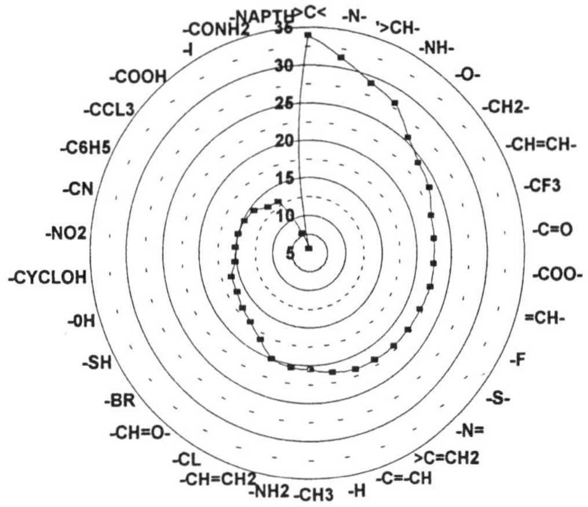

FIGURE 19 Isothermal compressibility.

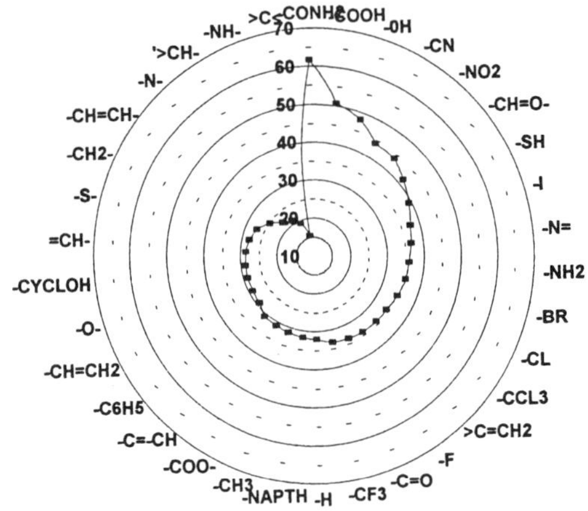

FIGURE 20 E_T.

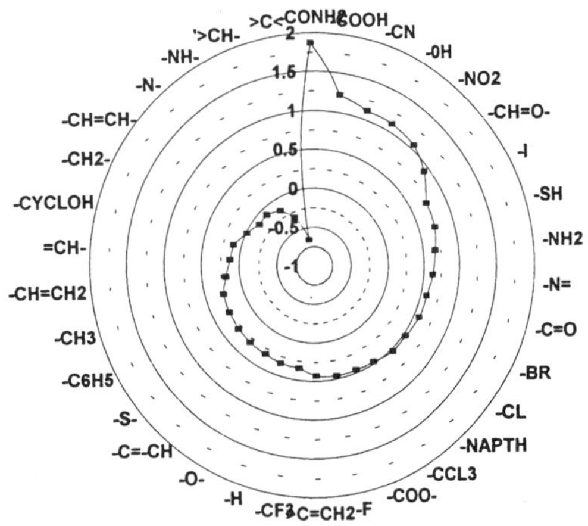

FIGURE 21 Log dielectric constant.

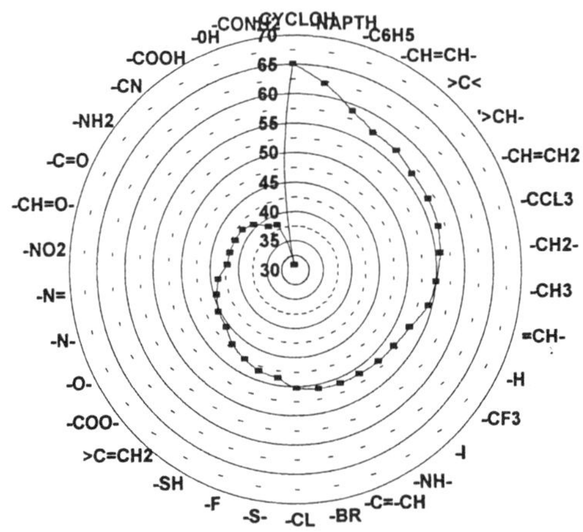

FIGURE 22 Chi (magnetic susceptibility).

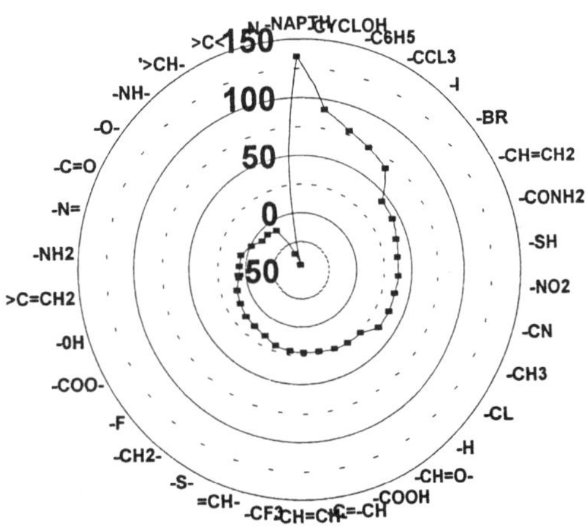

FIGURE 23 Molecular weight.

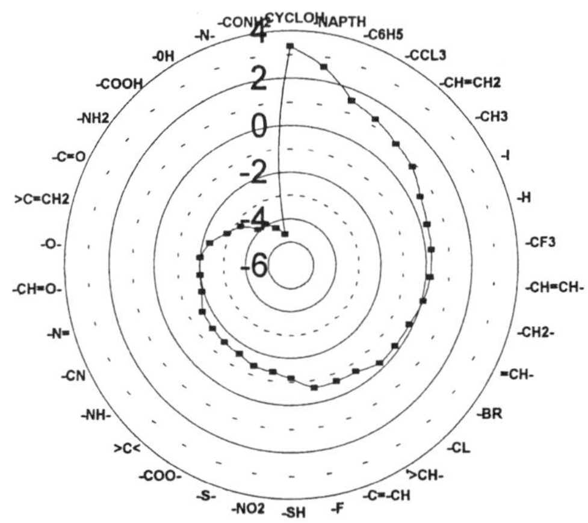

FIGURE 24 Solubility in CCL_4.

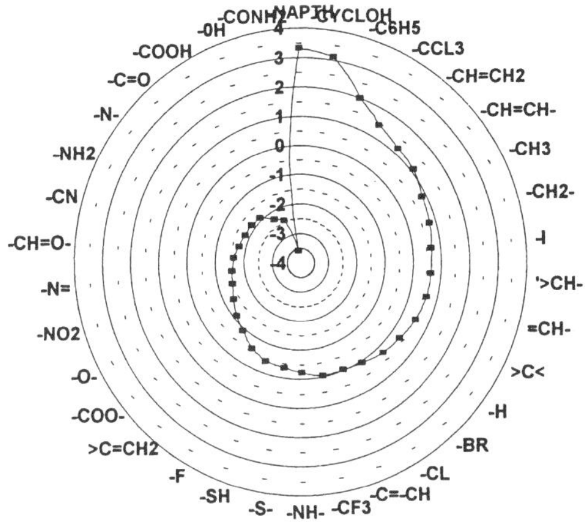

FIGURE 25 Solubility in oil.

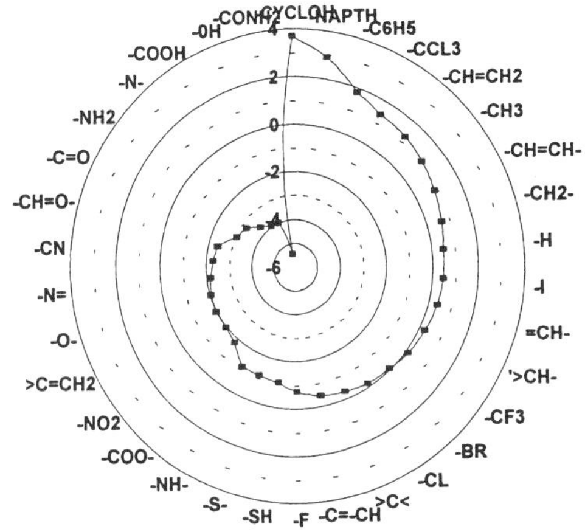

FIGURE 26 Solubility in cyclohexane.

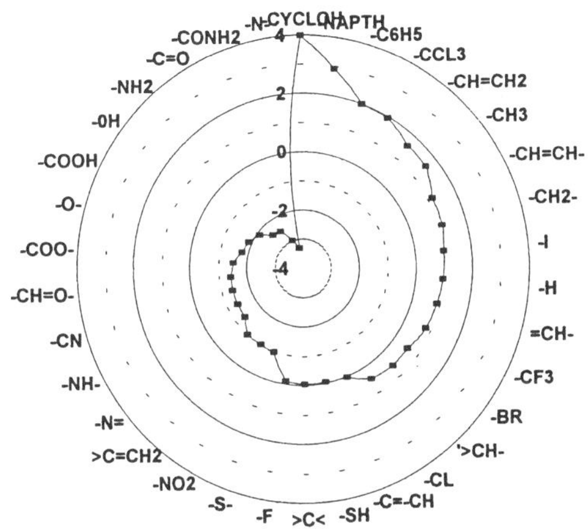

FIGURE 27 Solubility in ethyl alcohol.

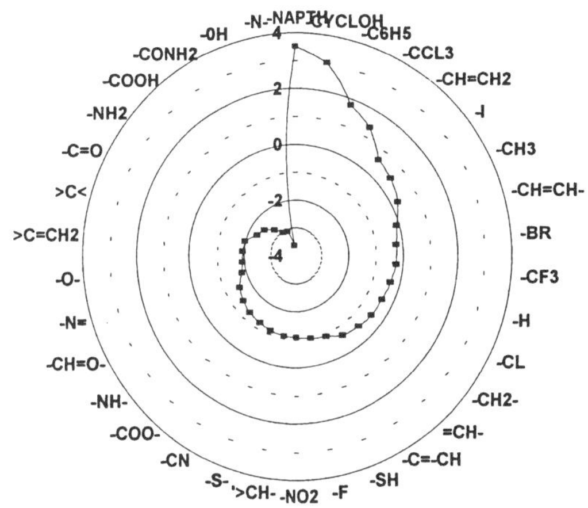

FIGURE 28 Solubility in chloroform.

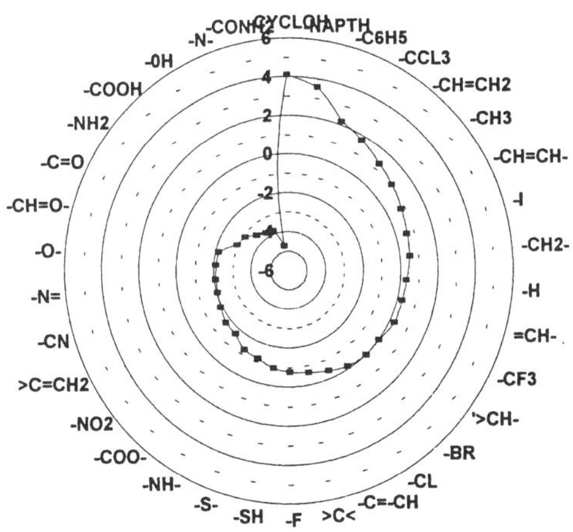

FIGURE 29 Solubility in benzene.

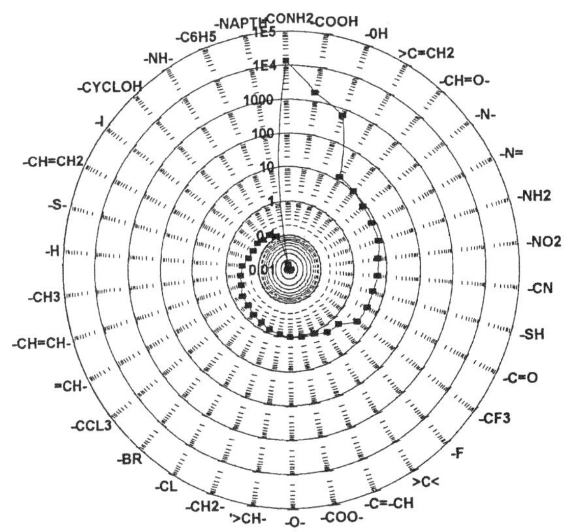

FIGURE 30 Risk-based concentrations: ambient air.

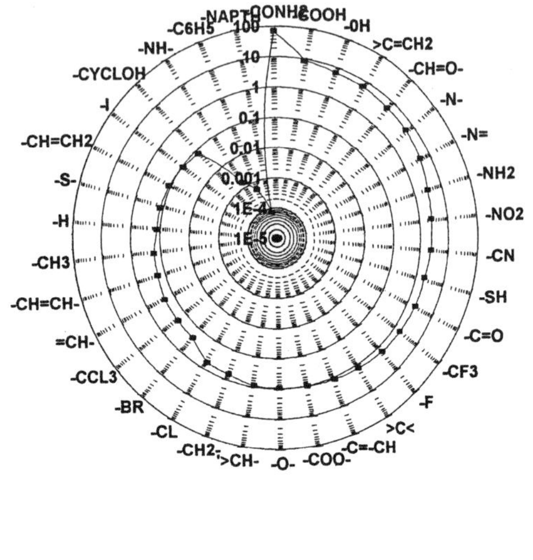

FIGURE 31 Risk-based concentrations: tap water.

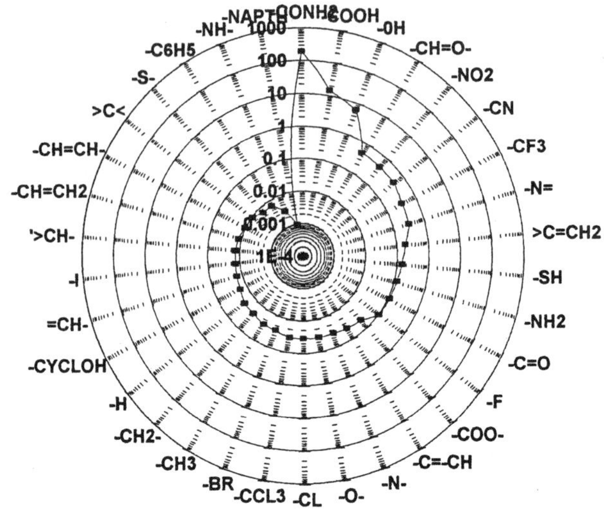

FIGURE 32 Risk-based concentrations: MCL.

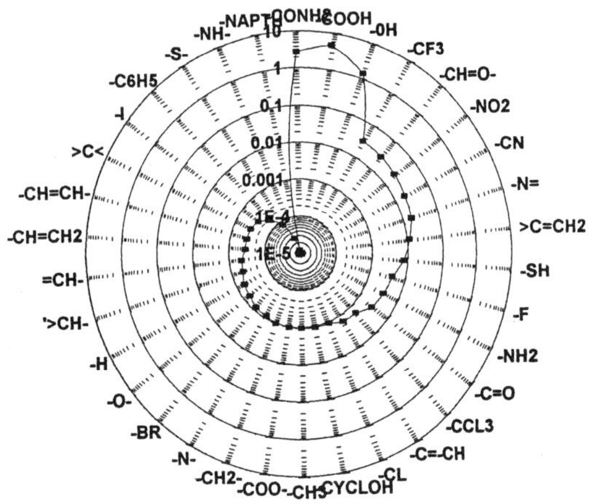

FIGURE 33 Risk-based concentrations: commercial soil.

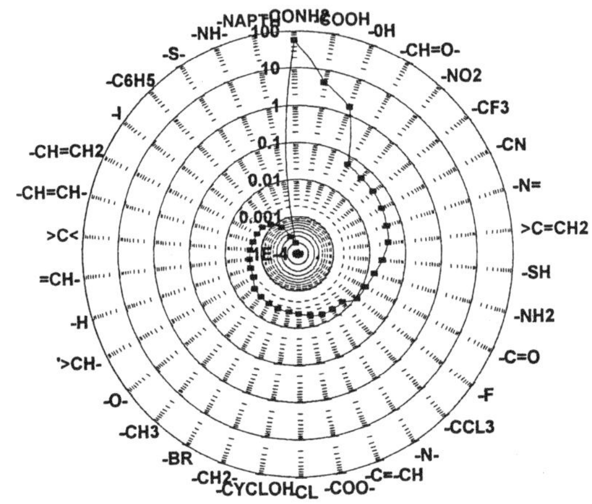

FIGURE 34 Risk-based concentrations: residential soil.

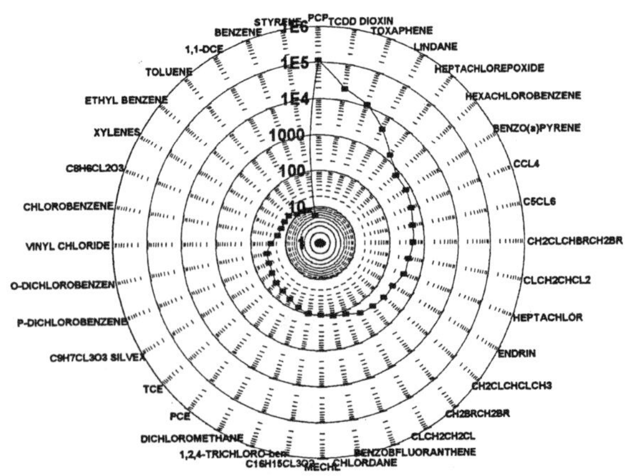

FIGURE 35 Predicted risk-based concentrations for 36 hazardous compounds in tap water.

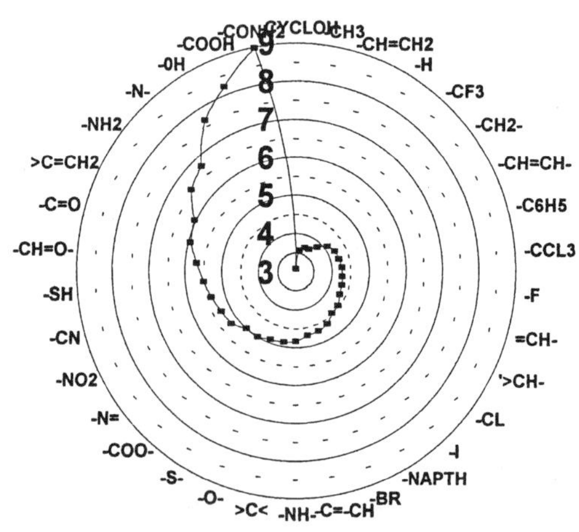

FIGURE 36 Drug activity 1.

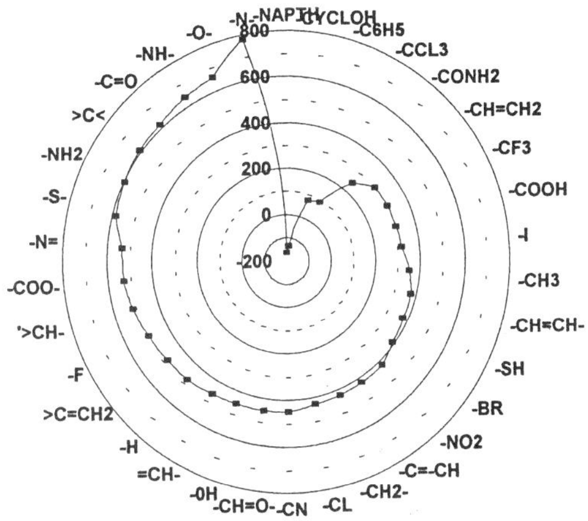

FIGURE 37 Drug activity 2.

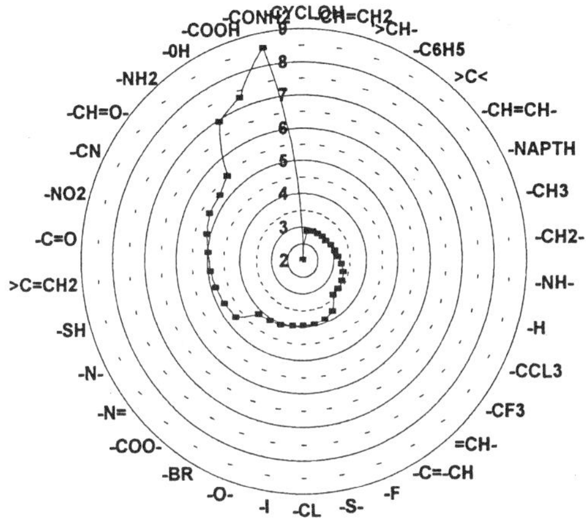

FIGURE 38 Drug activity 3.

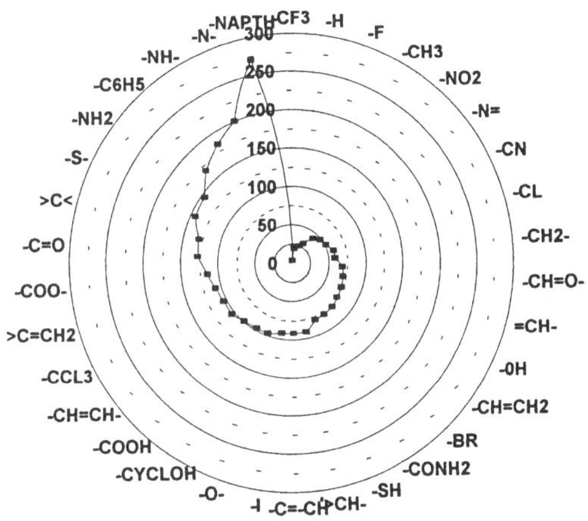

FIGURE 39 Drug activity 4.

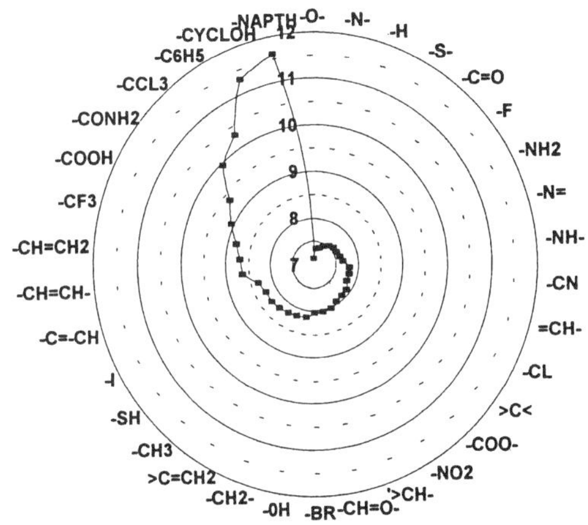

FIGURE 40 Drug activity 5.

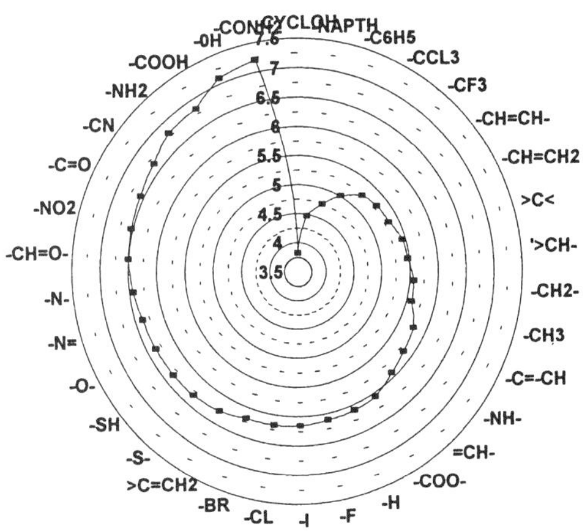

FIGURE 41 Drug activity 6.

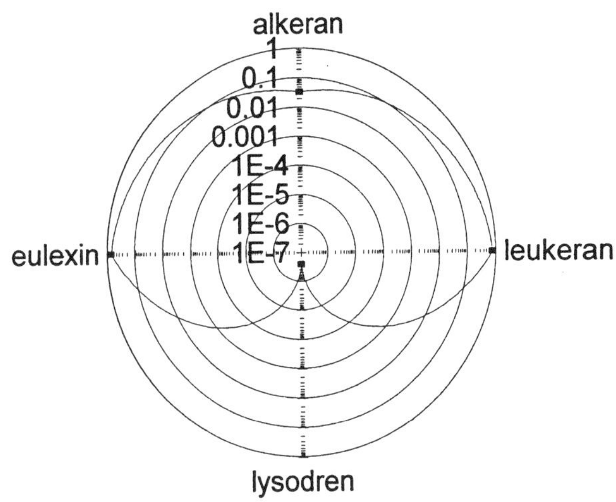

FIGURE 42 Risk-based concentrations (tap water): antineoplastic drugs.

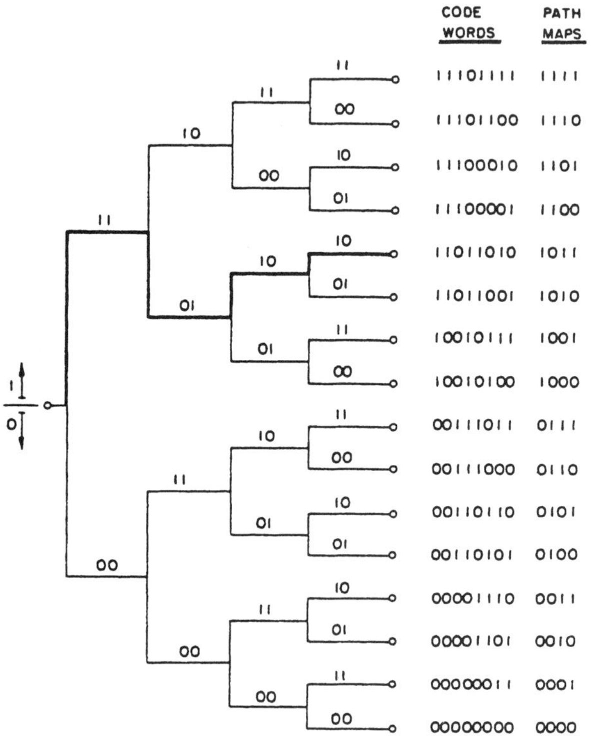

FIGURE 43 A binary tree code.

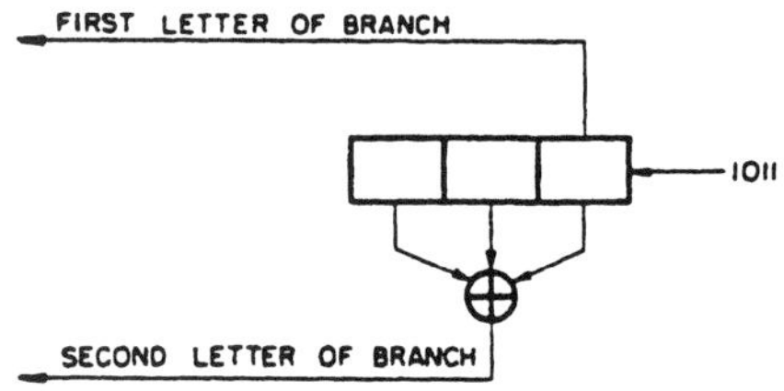

FIGURE 44 Code word–generating circuit for the tree code of Figure 43.

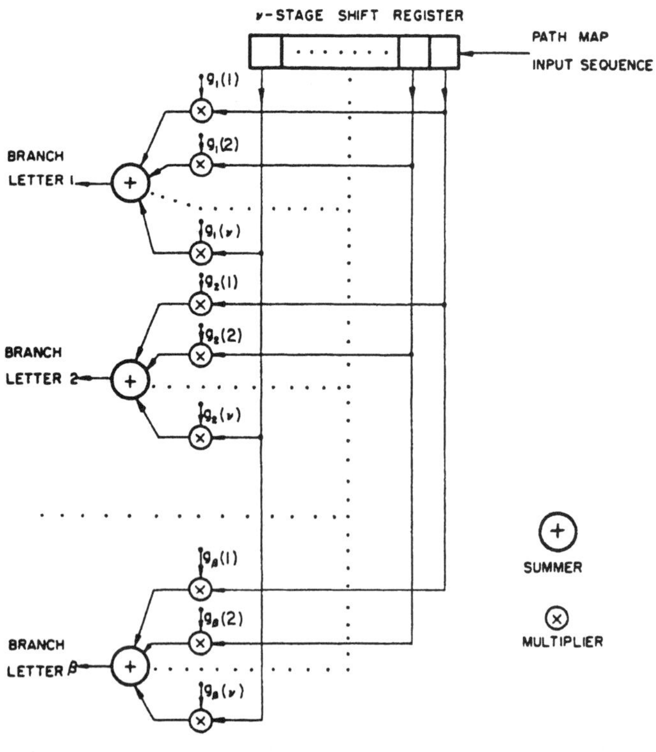

FIGURE 45 Code word–generating circuit (source decoder) for a general tree code.

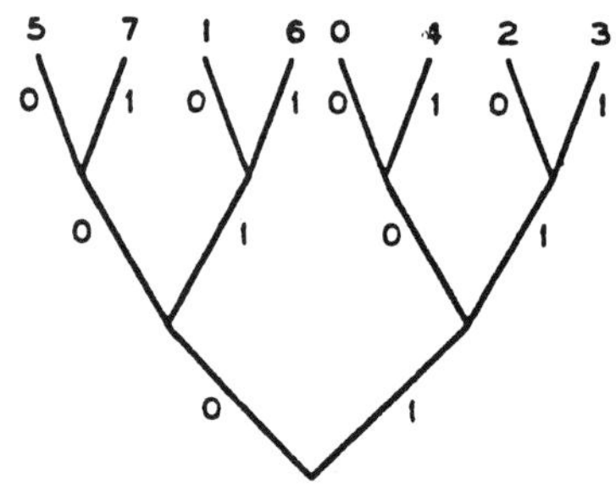

FIGURE 46 Tree of choice.

A	AI	AJ	AK	AL	AM	AN	AO	AP	AQ	AR
1	conh2/cn/l/no2/oh/cooh/sh/br/cho/napth/c6h5/ccl3/cyclo/ch=ch2/cf3/ch=ch/c=-c/nh/nh2/ch3/h/ch2/c=o/>c</>ch-/>c=ch2/-n-/-n=/-s-/=c-/coo/-o-/									
2	11111:11110:00000:00000:00000:00000:00	Pc								
3	11110:10011:11000:00000:00000:00000:00	Tb								
4		molar V								
5	11111:11011:00000:00000:00000:00000:00	sol par								
6	11011:10001:11100:00000:00000:00000:00	exp V								
7	11111:10011:00000:00010:00000:00000:00	dipole m								
8	11011:10001:11100:00000:00000:00000:00	Hv								
9	10100:00101:11111:00000:00000:00000:00	mol refrac								
10	11111:10001:11000:00000:00000:00000:00	Tm								
11	11011:10001:11100:00000:00000:00000:00	logvisc								
12	11011:10011:00000:00010:00100:00000:00	logactiv								
13	11111:10011:01000:00000:00000:00000:00	thermcond								
14	00100:00001:11111:0001:10000:00000:00	solccl4								
15	conh2/cn/l/no2/oh/cooh/sh/br/cho/napth/c6h5/ccl3/cyclo/ch=ch2/cf3/ch=ch/c=-c/nh/nh2/ch3/h/ch2/c=o/>c</>ch-/>c=ch2/-n-/-n=/-s-/=c-/coo/-o-/									
16	00100:00001:11110:10001:01000:00000:00	soloil								
17	00000:00001:11110:10001:11000:00000:00	solcyclo								
18	00100:00001:11110:100001:01000:00000:0(	soletoh								
19	00100:00101:11110:10001:00000:00000:00	solchcl3								
20	00000:00001:11110:10001:00011:00000:00	chi								
21	10011:10010:00000:00010:00000:11100:00	air								
22	10011:10010:00000:00010:00000:11100:00	tap								
23	11011:10010:00000:00010:00000:10100:00	mcl								
24	11011:10010:00010:00000:00000:10100:00	com								
25	11011:10010:00010:00000:00000:10100:00	res								
26		drug1								
27		drug2								
28		drug3								
29	00100:00001:11110:10001:01000:00000:00	solbenz								
30		drug4								
31		drug5								
32		drug6								
33	11110:10001:11100:00000:00000:00000:00	Tc								
34	conh2/cn/l/no2/oh/cooh/sh/br/cho/napth/c6h5/ccl3/cyclo/ch=ch2/cf3/ch=ch/c=-c/nh/nh2/ch3/h/ch2/c=o/>c</>ch-/>c=ch2/-n-/-n=/-s-/=c-/coo/-o-/									
35	00000:00000:00001:10100:01110:01000:11	ISO COMI	0	0	0	0	0	0	0	0
36	11111:10110:00000:00000:00000:00100:00	Et								
37	10100:01101:11110:00000:00000:00000:00	mol. wt.								
38	10100:00101:11111:00000:00000:00000:00	mag. susc.								
39	11111:11010:00000:00010:00000:00000:00	log(di con)								
40	11110:10011:11000:00000:00000:00000:00	surf ten								

FIGURE 47 Codes for 15 groups that make each property highest.

A	AI	AJ	AK	AL	AM	AN	AO	AP	AQ	AR	AS
1	conh2/cn/l/no2/oh/cooh/sh/br/cho/napth/c6h5/ccl3/cyclo/ch=ch2/cf3/ch=ch/c=-c/nh/nh2/ch3/h/ch2/c=o/>c</>ch-/>c=ch2/-n-/-n=/-s-/=c-/coo/-o-/										
2	12345:67890:00000:00000:00000:00000:00	max		min	Pc						
3	28790:40000:13560:00000:00000:00000:00	Tb									
4	00000:00001:34256:78900:00000:00000:00	molar V									
5	14853:29076:00000:00000:00000:00000:00	sol par									
6	190186:30002:75400:00000:00000:00000:00	exp V									
7	12736:50048:00000:00090:00000:00000:00	dipole m									
8	18095:30002:46700:00000:00000:00000:00	Hv									
9	80500:00901:34260:70000:00000:00000:00	mol refrac									
10	26489:30001:57000:00000:00000:00000:00	Tm									
11	19086:30002:71540:00000:00000:00000:00	logvisc									
12	15093:20078:00000:00040:00100:00000:00	logactiv									
13	12876:39045:08000:00000:00000:00000:00	thermcond									
14	00700:00002:34159:00006:80000:00000:00	solccl4									
15	conh2/cn/l/no2/oh/cooh/sh/br/cho/napth/c6h5/ccl3/cyclo/ch=ch2/cf3/ch=ch/c=-c/nh/nh2/ch3/h/ch2/c=o/>c</>ch-/>c=ch2/-n-/-n=/-s-/=c-/coo/-o-/										
16	00900:00001:34250:60007:08000:00000:00	soloil									
17	00000:00002:34150:70006:98000:00000:00	solcyclo									
18	00900:00002:34150:70006:08000:00000:00	soletoh									
19	00600:00901:34250:80007:00000:00000:00	solchcl3									
20	00000:00002:38170:40000:09056:00000:00	chi									
21	10093:20050:00000:00080:00000:46700:00	air									
22	10093:20050:00000:00080:00000:46700:00	tap									
23	16053:20040:00007:00000:00000:90800:00	mcl									
24	17063:20050:00004:00000:00000:90800:00	com									
25	17053:20040:0000600000:00000:90800:00	res									
26		drug1									
27		drug2									
28		drug3									
29	00800:00002:34150:70006:09000:00000:00	solbenz									
30		drug4									
31		drug5									
32		drug6									
33	27380:50001:46900:00000:00000:00000:00	Tc									
34	conh2/cn/l/no2/oh/cooh/sh/br/cho/napth/c6h5/ccl3/cyclo/ch=ch2/cf3/ch=ch/c=-c/nh/nh2/ch3/h/ch2/c=o/>c</>ch-/>c=ch2/-n-/-n=/-s-/=c-/coo/-										-o-
35	00000:00000:000080:70040:06913:02000:05	ISO COM	0	0	0	0	0	0	0	0	19989
36	14853:27060:00000:00000:00000:00900:00	Et									
37	80500:09601:34270:00000:00000:00000:00	mol. wt.									
38	70500:00801:34260:100000:00000:00000:00	mag. susc.									
39	13754:28060:00000:00090:00000:00000:00	log(di con)									
40	24350:80091:67000:00000:00000:00000:00	surf ten									

FIGURE 48 Codes for 15 groups that make each property highest.

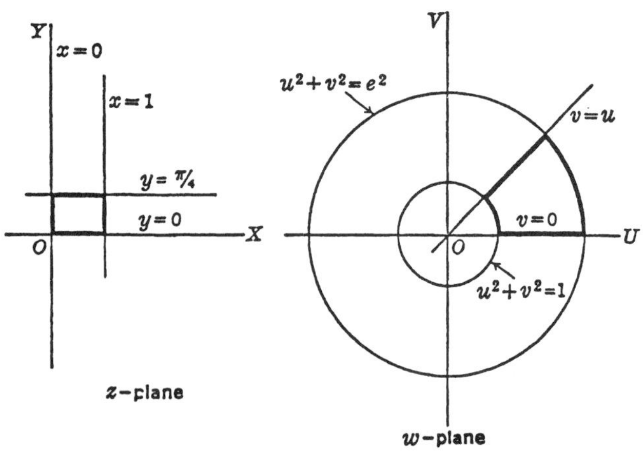

FIGURE 49 Mapping of a rectangle into the sector of a ring.

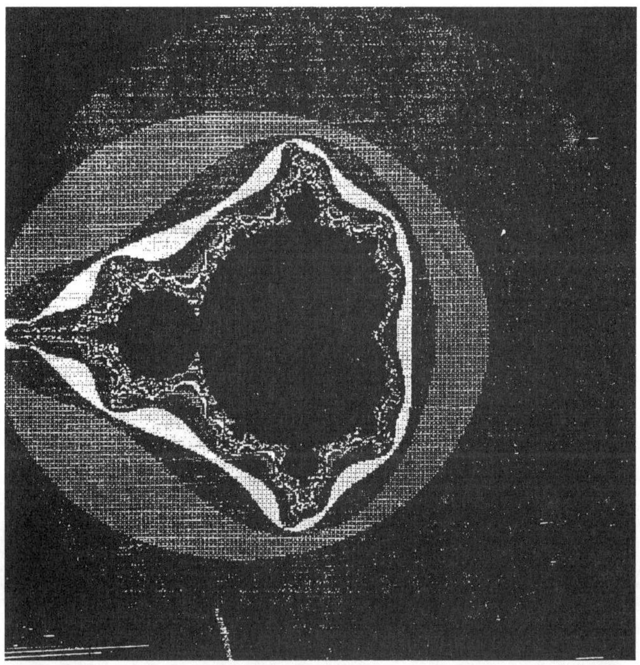

FIGURE 50 The mandelbrot set.

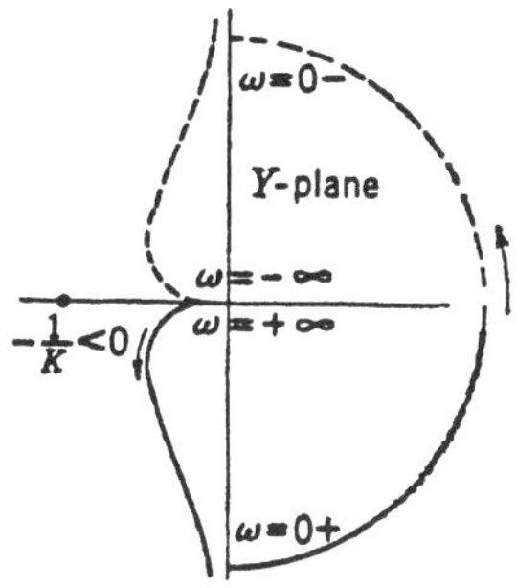

FIGURE 51 A servo system that is stable according to the nyquist criterion.

1. Hypothetical Structure, Drug 1& 3	$NH_2C_6H(CL)(X)(Y)(Z)$
2. 6-Deoxytetracycline, Drug 2	$C_{18}H_4(X)(Y)(R)(OH)_4(=O)_2(CONH_2)(N(CH_3)_2)$
3. See 1.	See 1.
4. Piperidinesulfamylsemicarbazides Drug 4	$(R2)(R1)C_6H_9NSO_2NHCONHNR3$
5. N,N-Dimethyl-2-Bromophenethylamines Drug 5	$(Z)(Y)C_6H_3CH(BR)CH_2N(CH_3)_2.HBR$
6. N-(phenoxyethyl) cyclopropylamines Drug 6	$(X)C_6H_4OCH_2CH_2NC_3H_5$
7. Mephalan (Alkeran)	$(CLCH_2CH_2)_2NC_6H_4CH_2CH(NH_2)COOH$
8. Chlorambucil (Leukeran)	$(CLCH_2CH_2)_2NC_6H_4CH_2CH_2CH_2COOH$
9. Mitotane (Lysodren)	$CLC_6H_4CH(CHCL_2)C_6H_4CL$
10. Flutamide (Eulexin)	$(CH_3)_2CHCONHC_6H_3CF_3(NO_2)$

FIGURE 52 Drug structures.

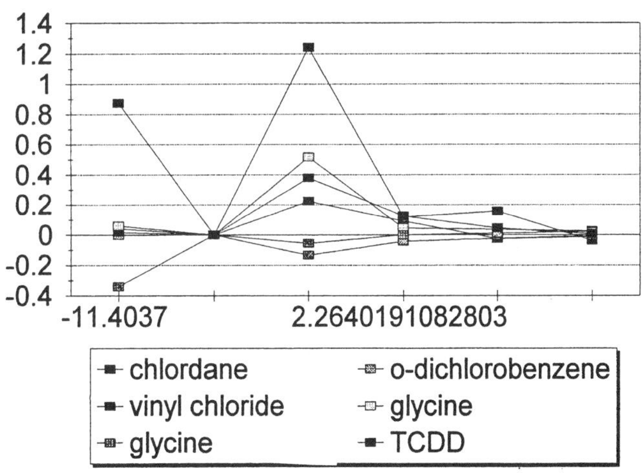

FIGURE 53 QSARs — air toxicity.

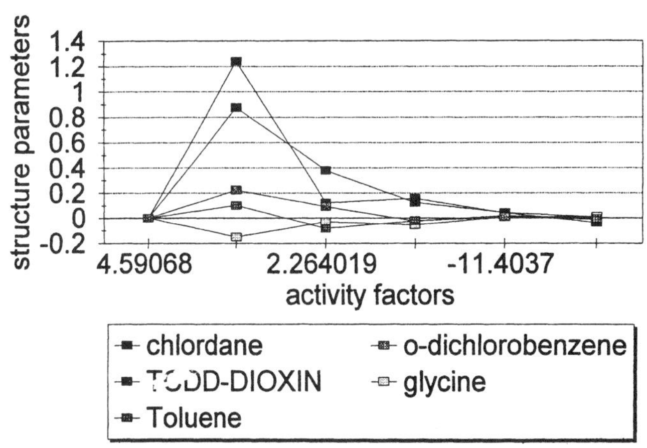

FIGURE 54 QSARs — tap water toxicity.

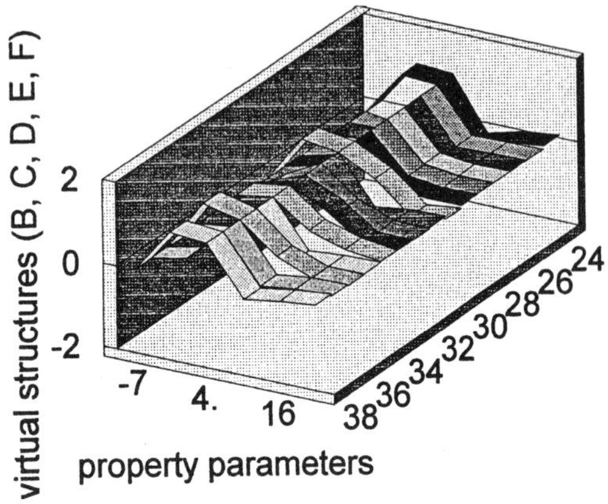

FIGURE 55 Structure vs. properties — last 15 compounds — tap risk concentration.

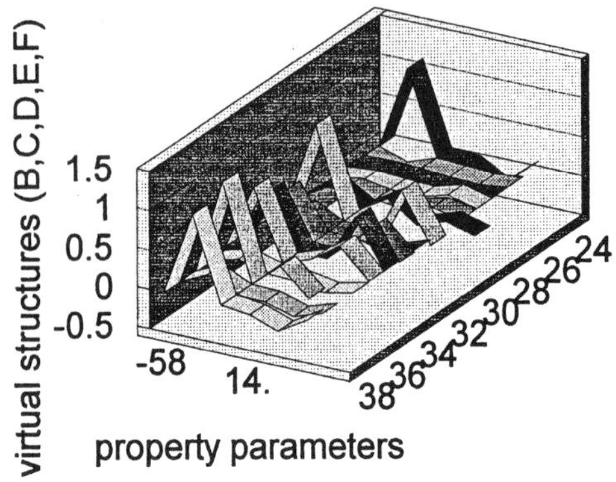

FIGURE 56 Structure vs. properties — last 15 compounds — air risk concentration.

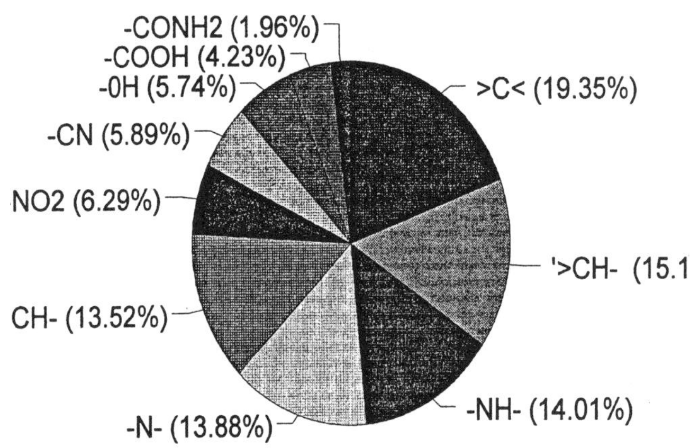

FIGURE 57 Solubility parameter.

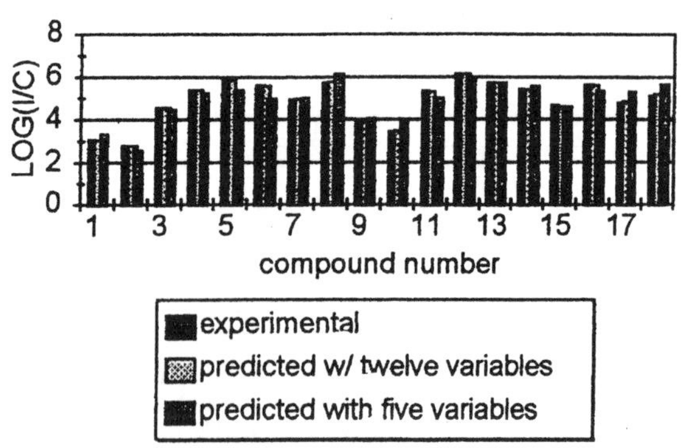

FIGURE 58 Drug design — predicted vs. experimental.

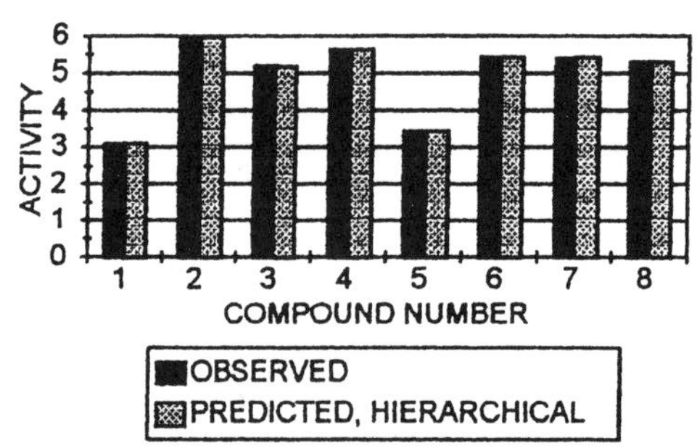

FIGURE 59 Drug design for 6-deoxytetracycline derivatives.

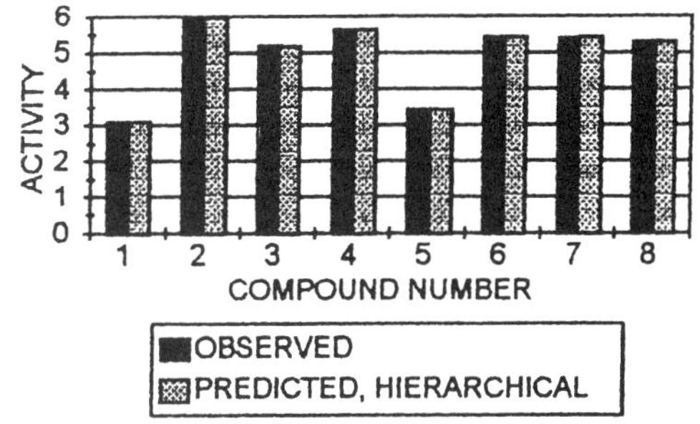

FIGURE 60 Biological activity — substituent at X, Y, or Z, predicted vs. observed.

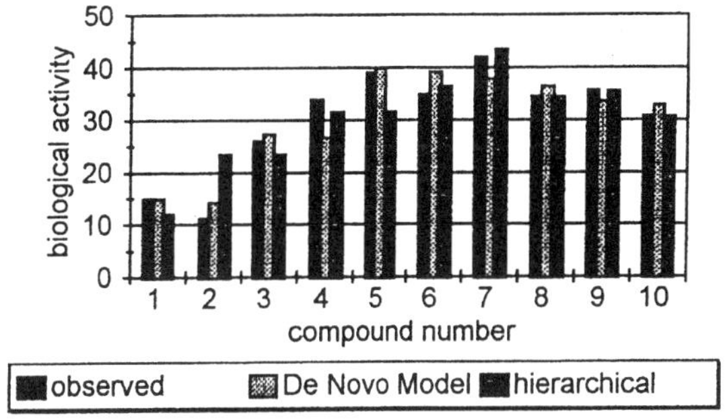

FIGURE 61 Hypoglycemic activities for piperidine-sulfamylsemicarbazides.

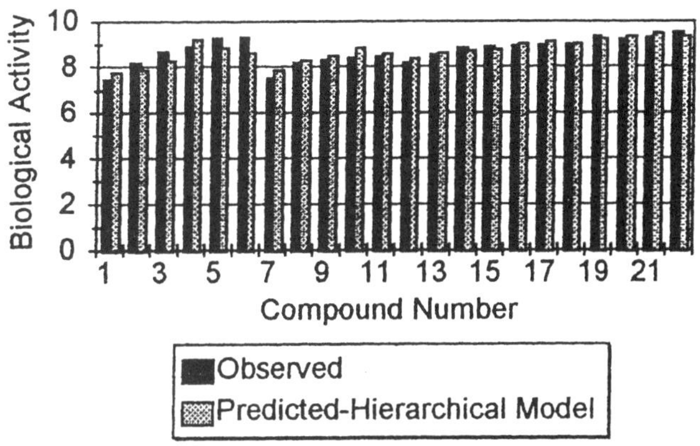

FIGURE 62 Blocking adrenaline with *N,N*-dimethyl-2-bromophenethylamines.

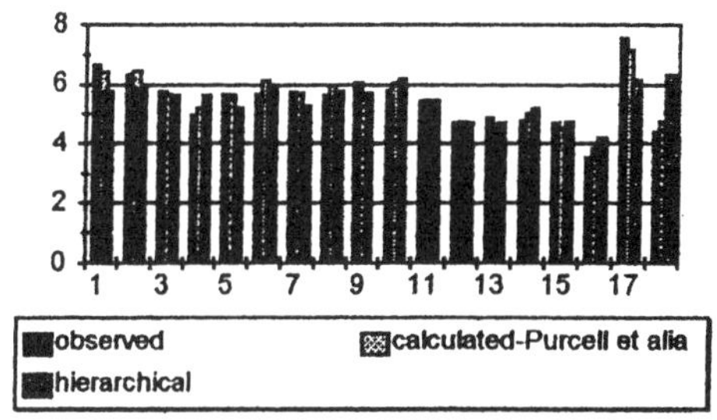

FIGURE 63 Cyclopropylamine inhibitors of monamine oxidase.

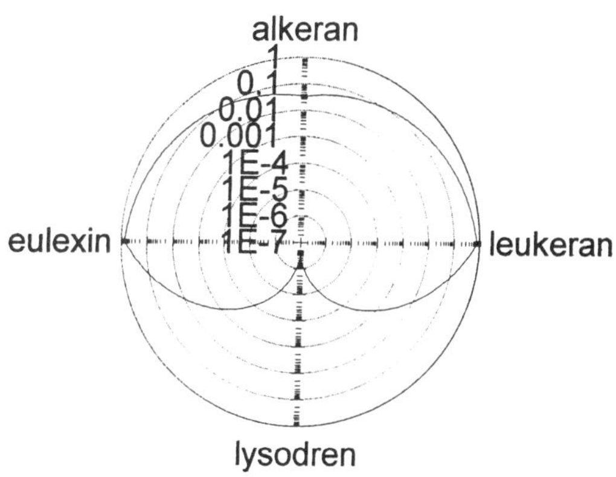

FIGURE 64 Tap water toxicity for four cancer drugs.

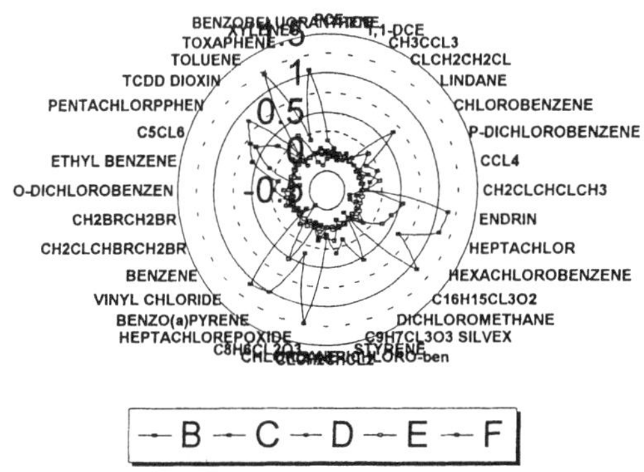

FIGURE 65 Values of B, C, D, E, and F for 36 hazardous compounds.

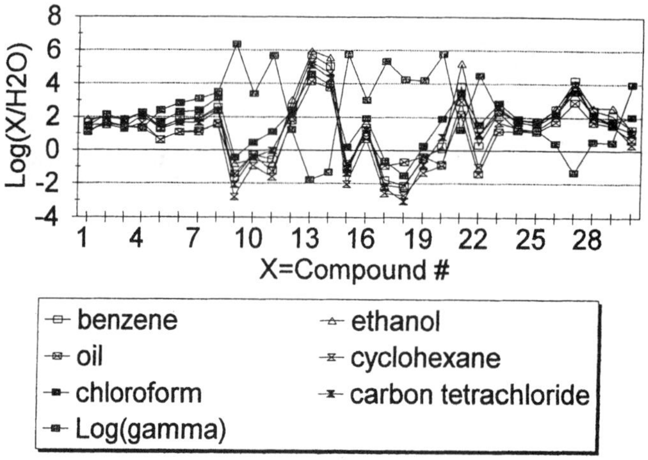

FIGURE 66 Log (X/H_2O) vs. compound.

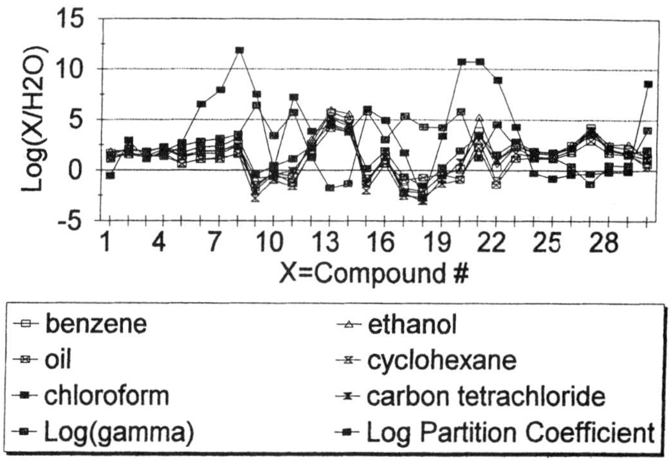

FIGURE 67 Log (X/H_2O) vs. compound.

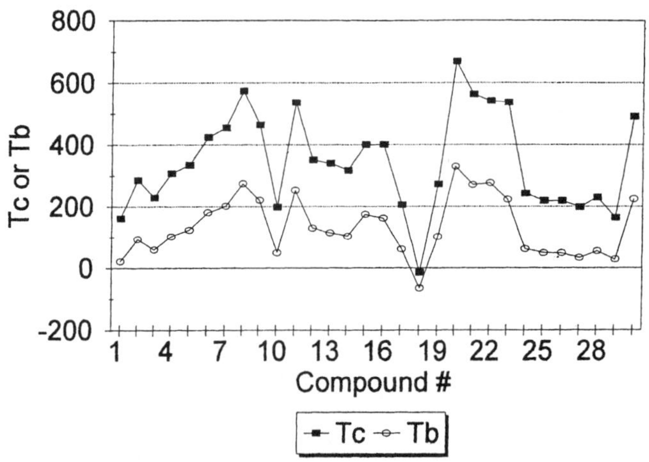

FIGURE 68 T_c and T_b vs. compound number.

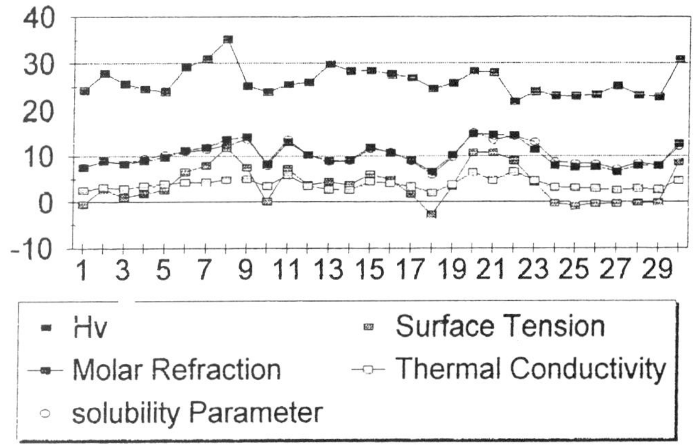

FIGURE 69 Physical properties vs. compound number.

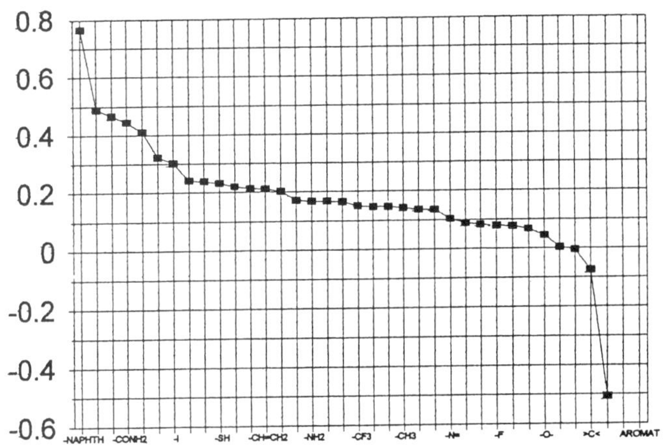

FIGURE 70 B parameter vs. group.

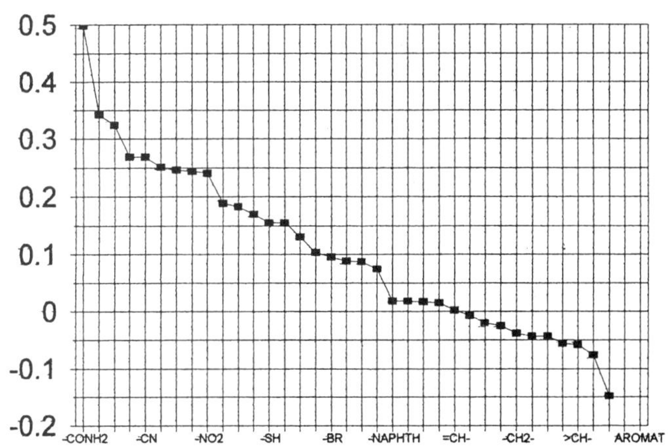

FIGURE 71 C parameter vs. group.

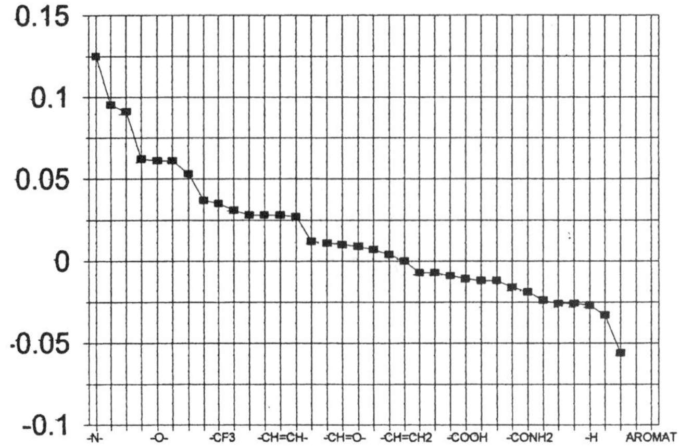

FIGURE 72 D parameter vs. group.

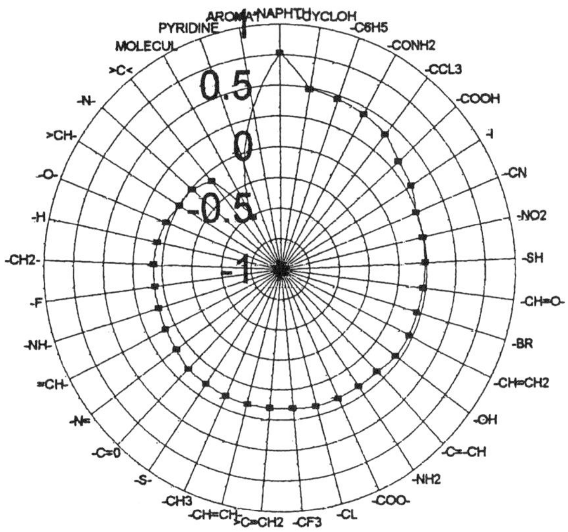

FIGURE 73 B parameter vs. group.

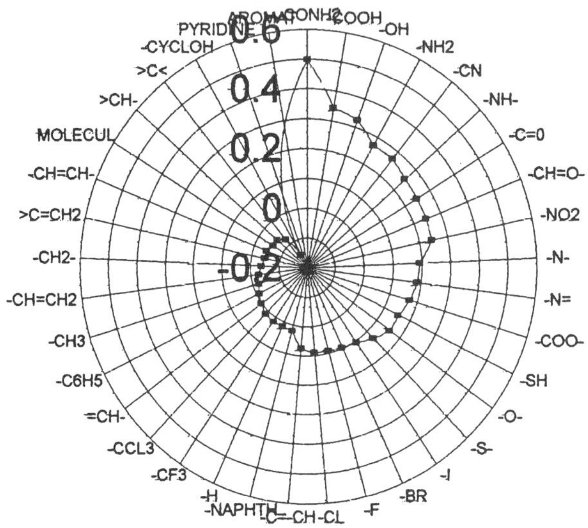

FIGURE 74 C parameter vs. group.

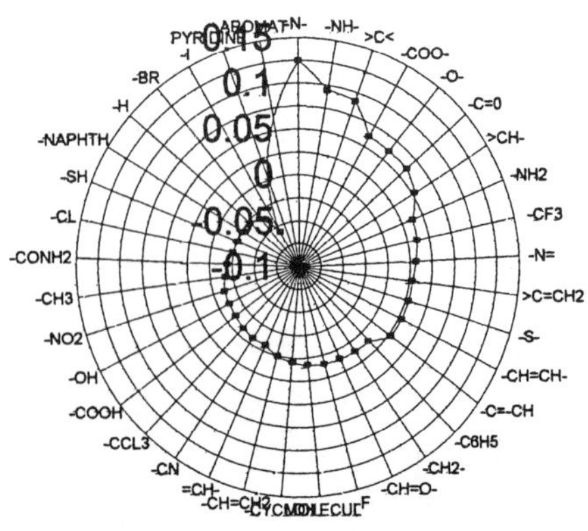

FIGURE 75 D parameter vs. group.

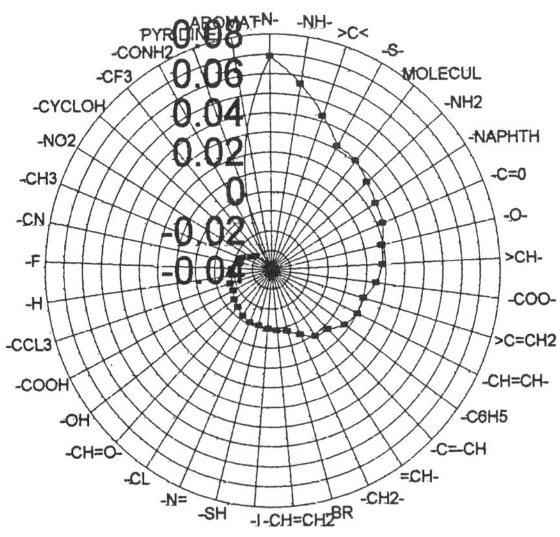

FIGURE 76 E parameter vs. group.

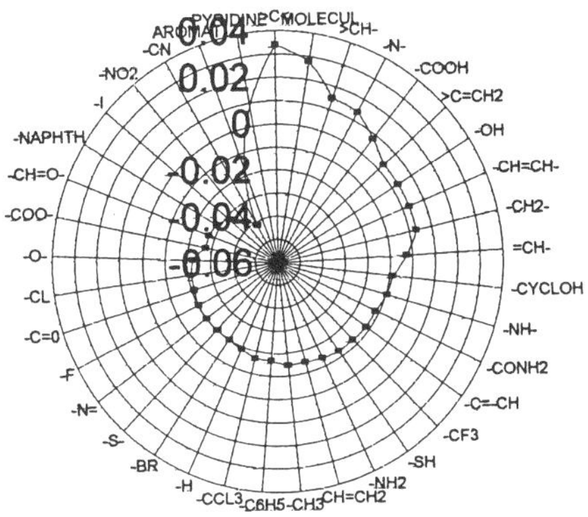

FIGURE 77 F parameter vs. group.

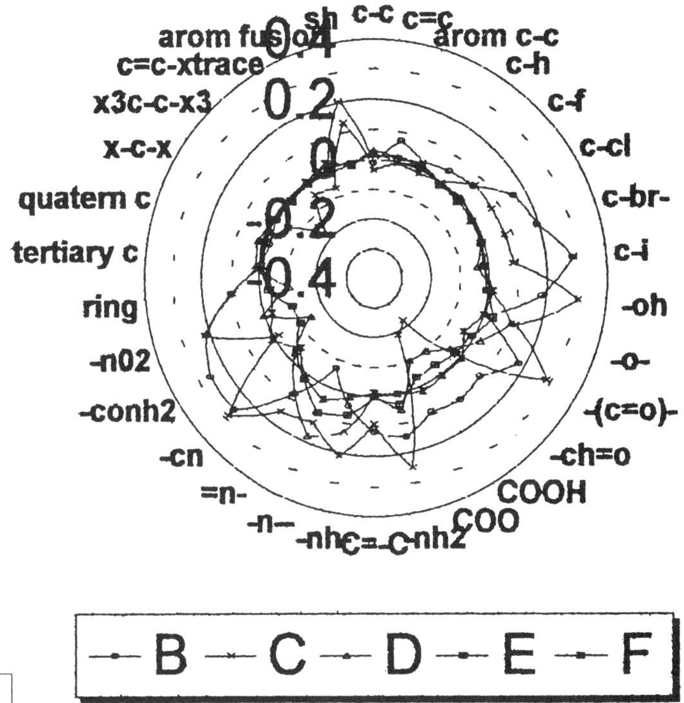

FIGURE 78 Parameter values B, C, D, E, F, shape; and size vs. group contributions.

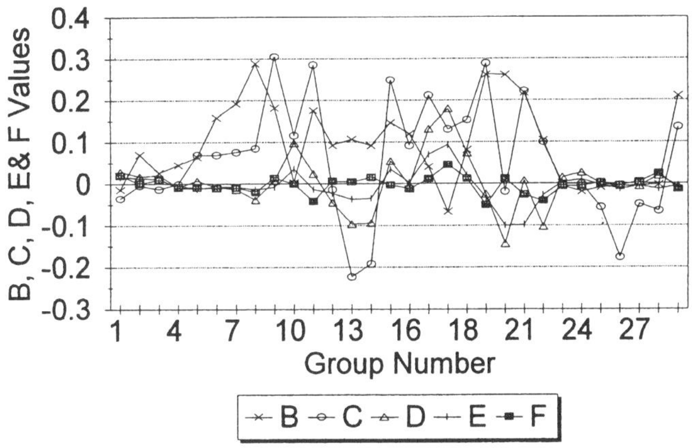

FIGURE 79 B, C, D, E, and F parameters for chemical groups.

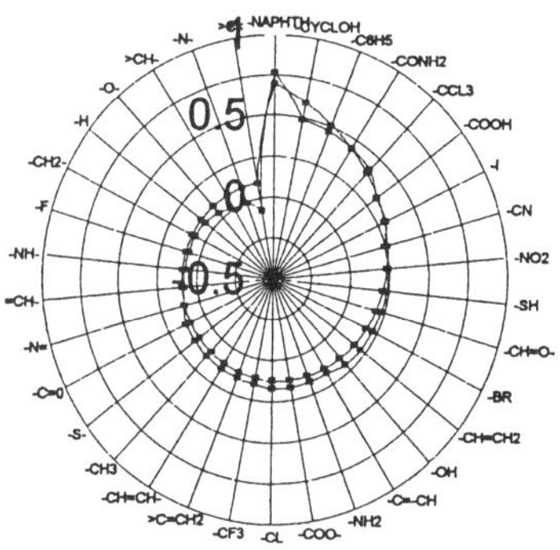

FIGURE 80 B parameters vs. group and fit.

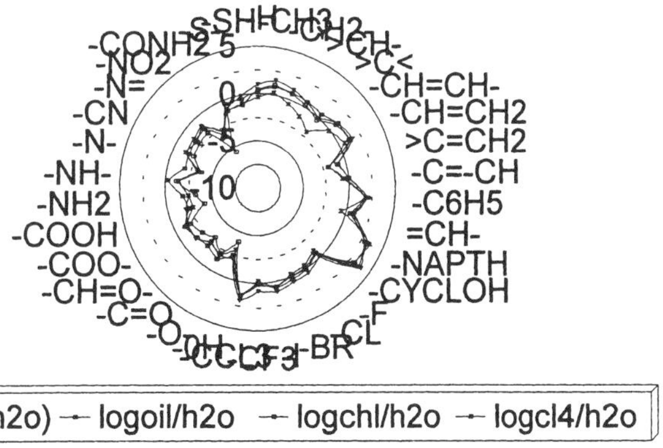

FIGURE 81 Group solubility in solvents.

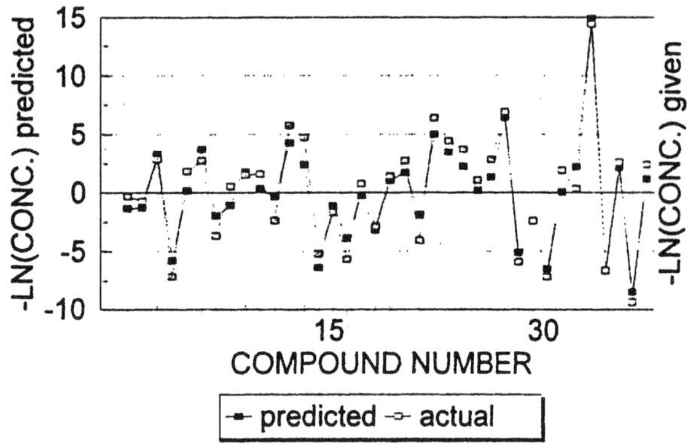

FIGURE 82 Tap water risk-based concentration, cited and predicted.

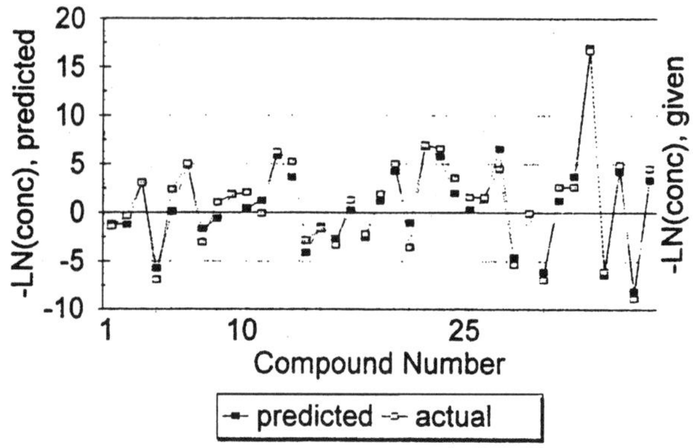

FIGURE 83 Ambient air risk-based concentration, cited and predicted.

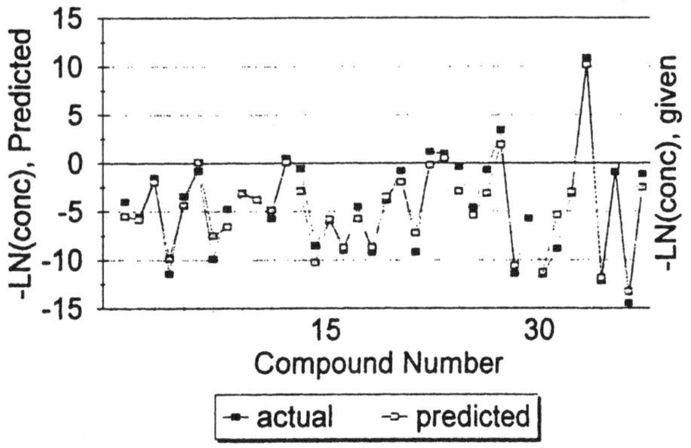

FIGURE 84 Commercial soil risk-based concentration, cited and predicted.

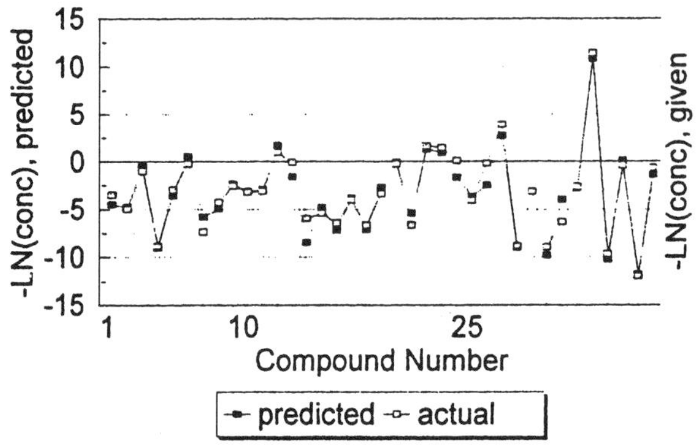

FIGURE 85 Residential soil risk-based concentration, cited and predicted.

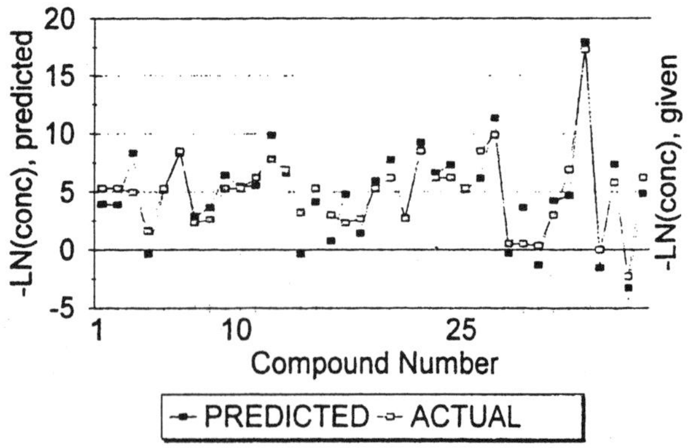

FIGURE 86 MCL Risk-based concentration, cited and predicted.

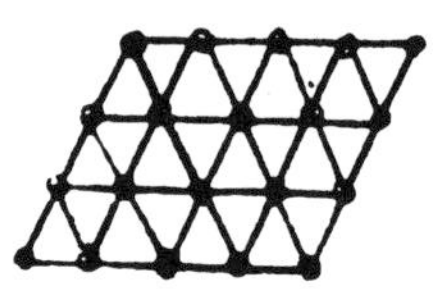

FIGURE 87 The triangular lattice.

FIGURE 88 Essential overlap figures.

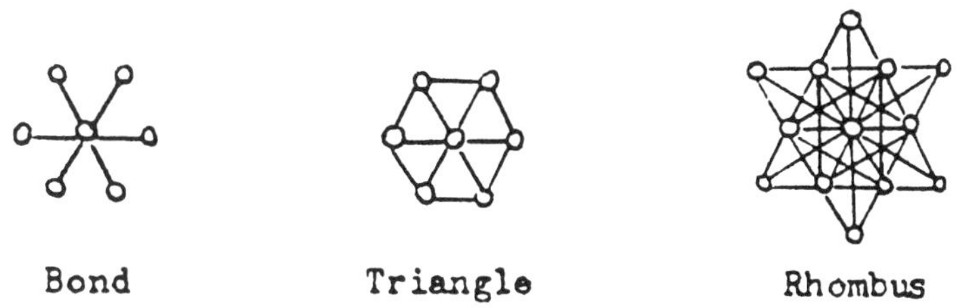

FIGURE 89 Effect of considering larger basic figures.

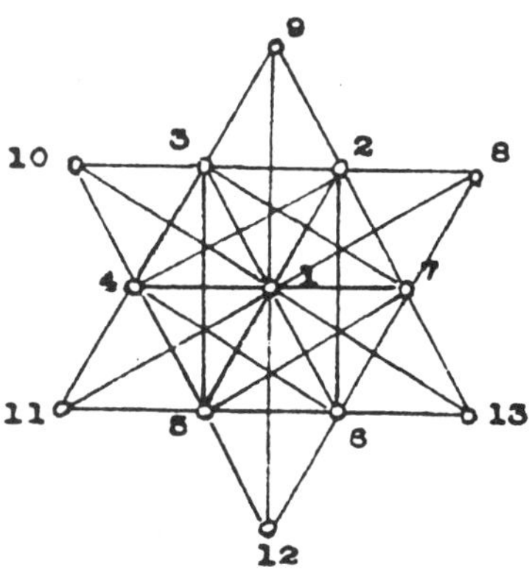

FIGURE 90 The rhombus approximation.

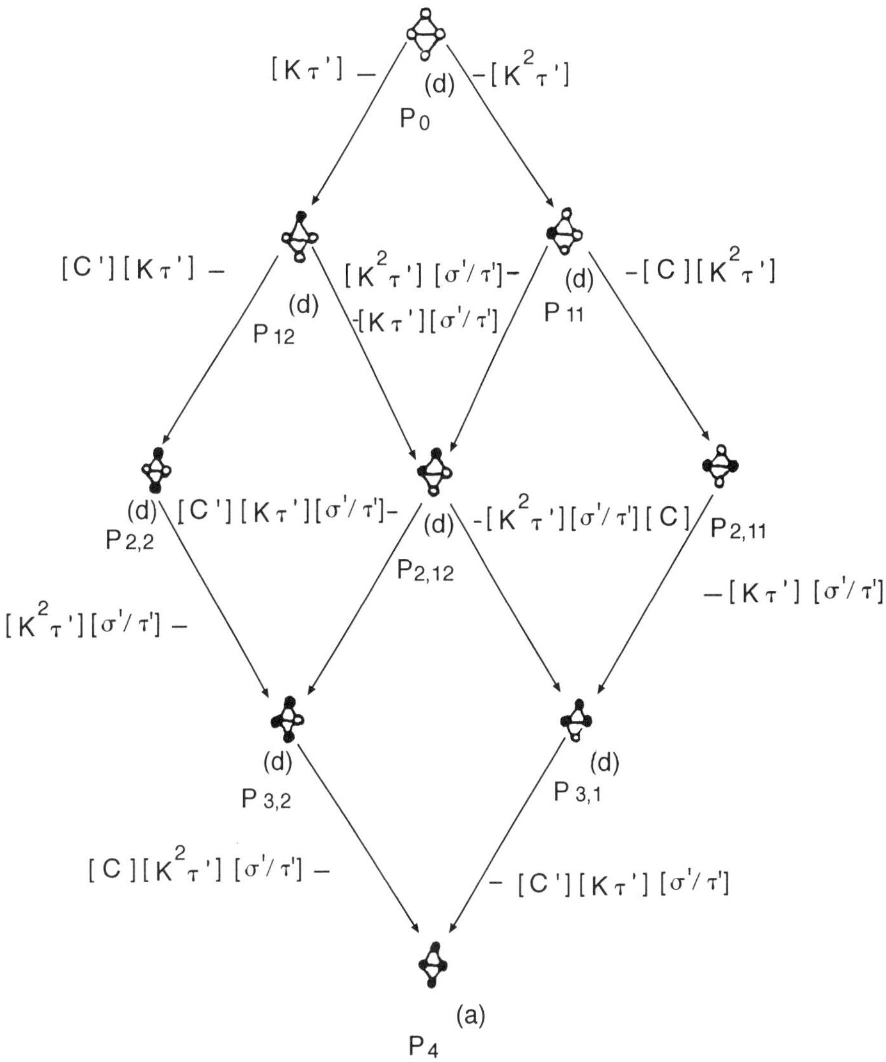

FIGURE 91 The successive filling of rhombus sites.

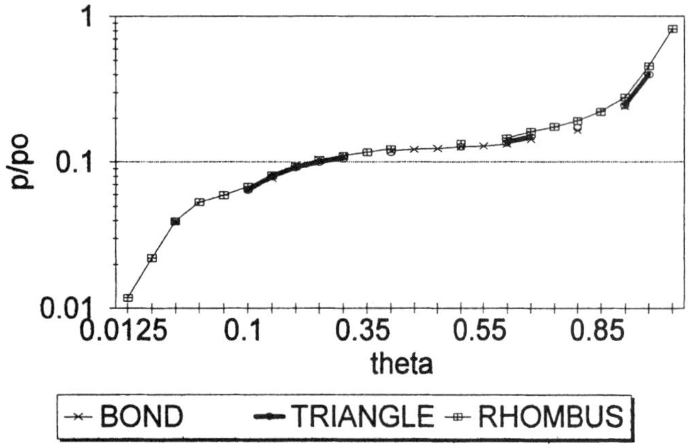

FIGURE 92 Order and Complexity.

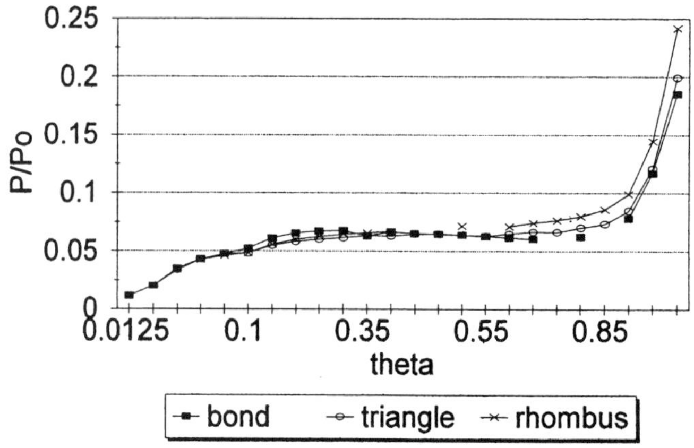

FIGURE 93 Order–disorder, c = 2.5, bond, triangle, rhombus.

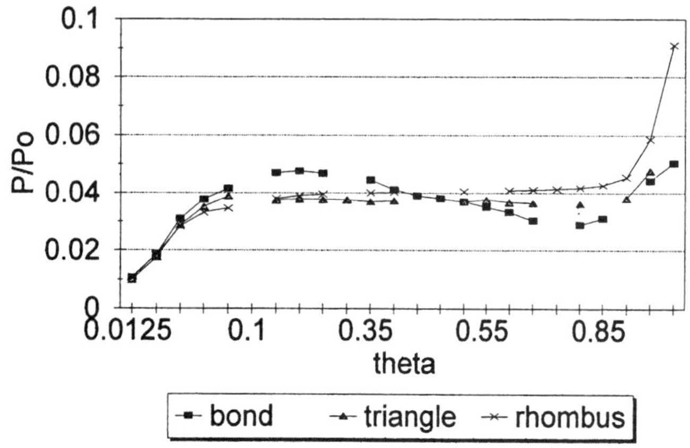

FIGURE 94 Order–disorder, c = 3, Z = 6, bond, triangle, rhombus.

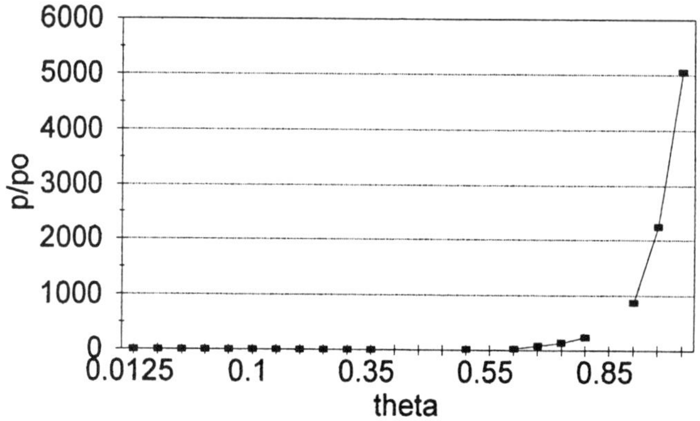

FIGURE 95 p/p^0 for rhombus, repulsive interaction, c = 0.36, c = 1, Z = 6.

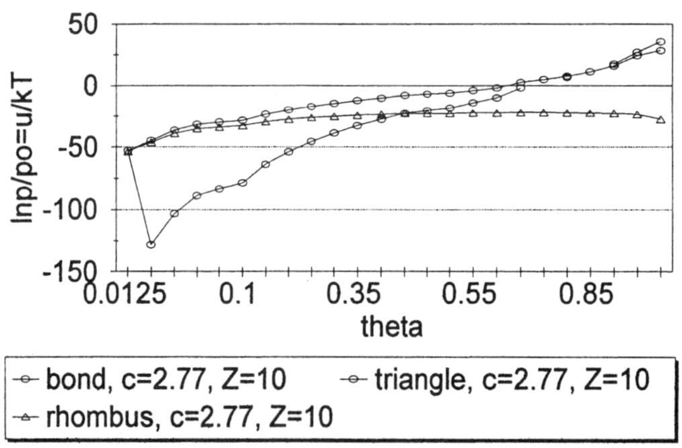

FIGURE 96 μ/kT vs. occupancy for the bond, triangle, and rhombus.

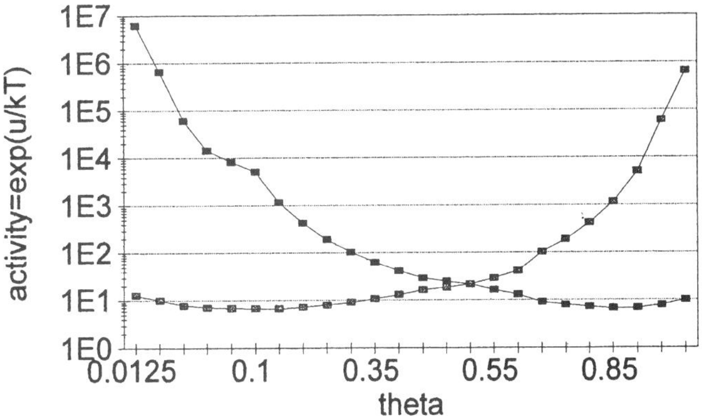

FIGURE 97 Activity vs. θ, two components, c = 2.77, Z = 12.

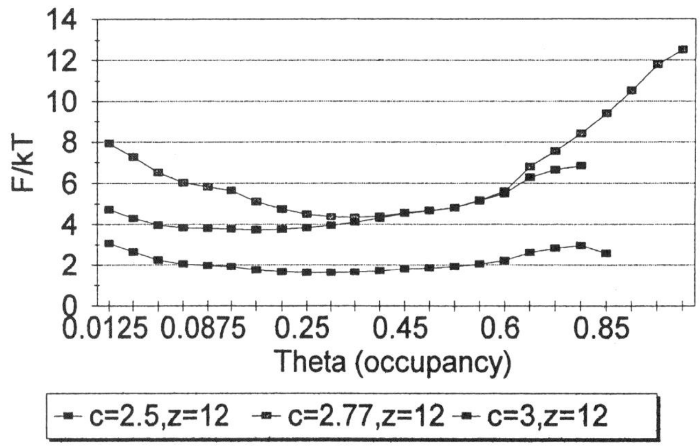

FIGURE 98 F/kT: Bond figure, c = 2.5, 2.77 and 3 and Z = 12.

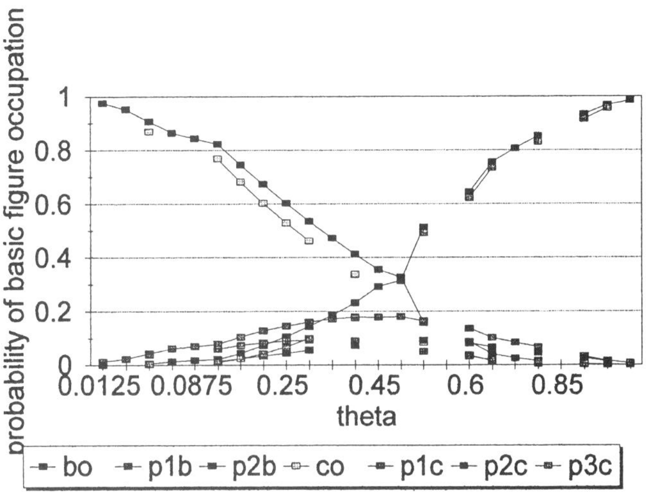

FIGURE 99 Probability Vs. θ, bond and triangle, c = 2.77, Z = 12.

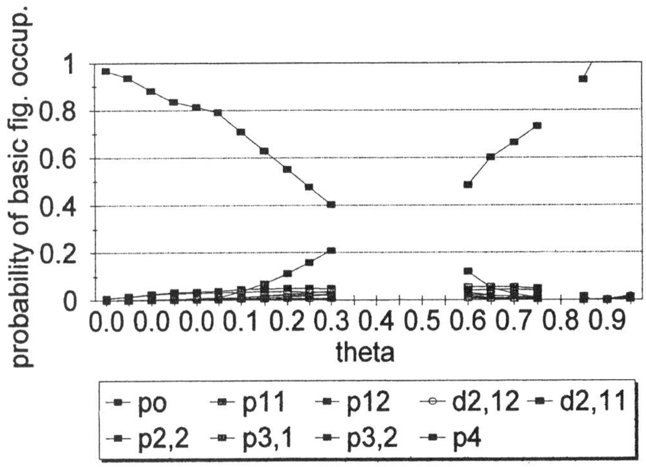

FIGURE 100 Probability vs. θ, for the rhombus, c = 3. C = 1.2, Z = 6.

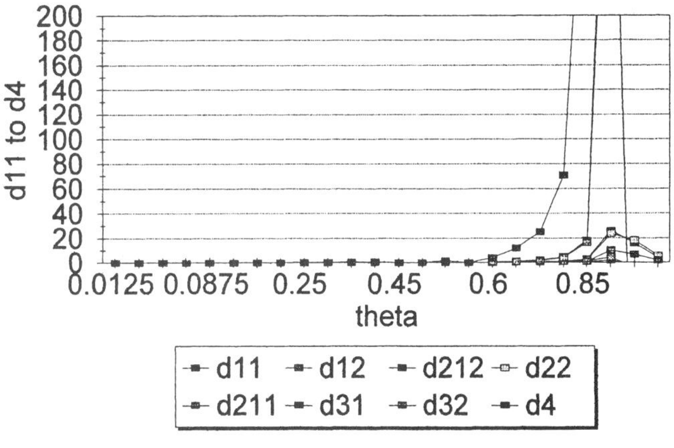

FIGURE 101 d vs. θ for the rhombus, c = 3, c = 1.2, Z = 6.

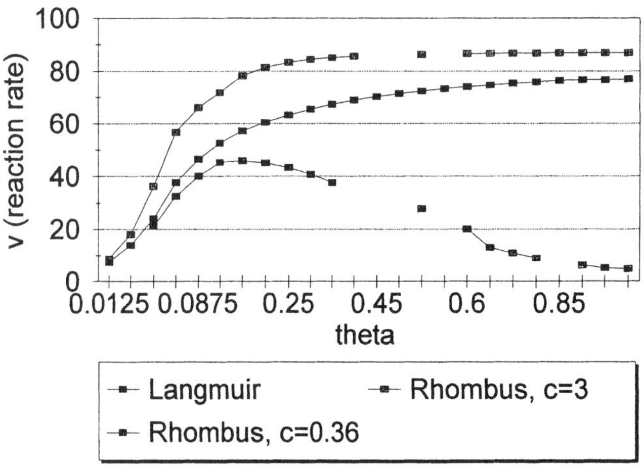

FIGURE 102 d for the rhombus, repulsive interactions, c = 0.36, c = 0.6.

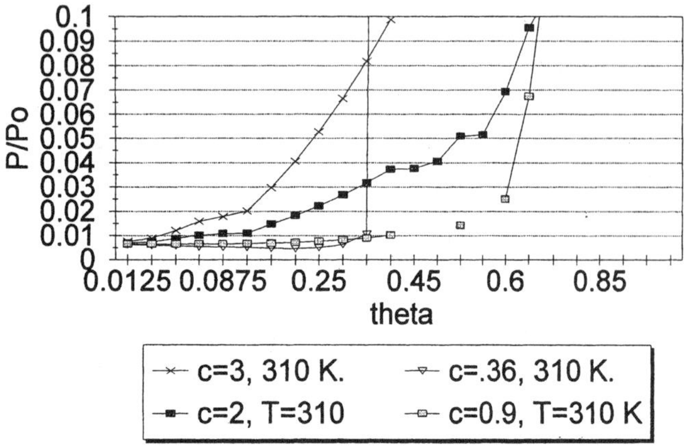

FIGURE 103 Metastasis/repulsion, c = 0.36, Z = 6, T = 1 and 310 K.

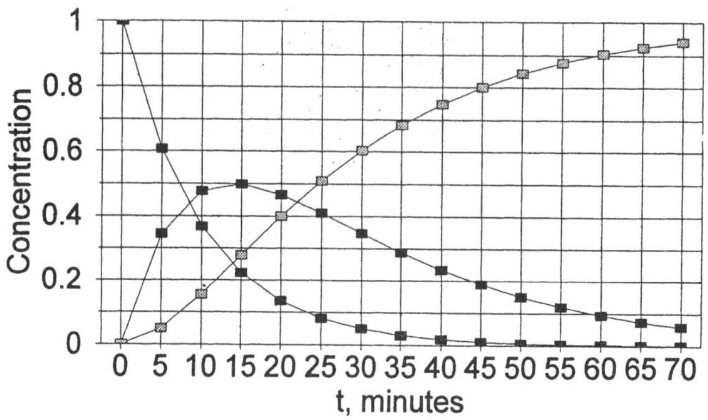

FIGURE 104 Concentration–time curves.

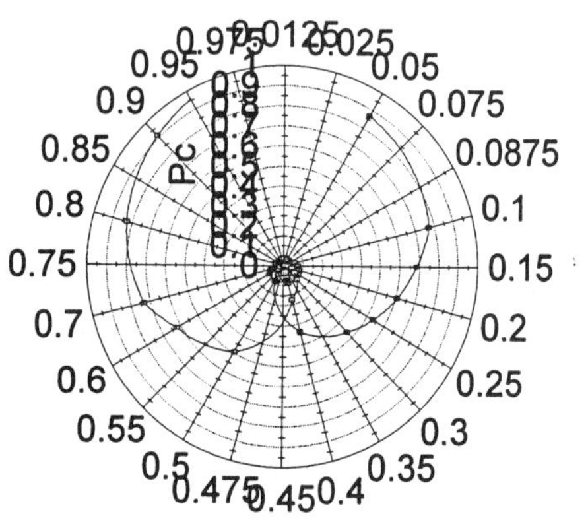

FIGURE 105 Triangle approximation, p vs. θ.

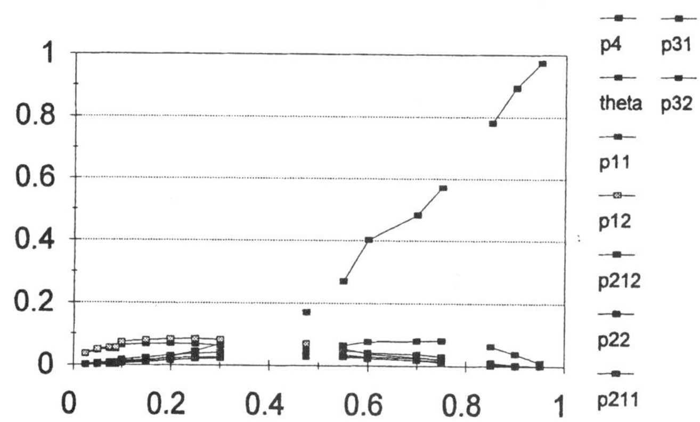

FIGURE 106 Rhombus probabilities, c = 0.36, c = 0.82151.

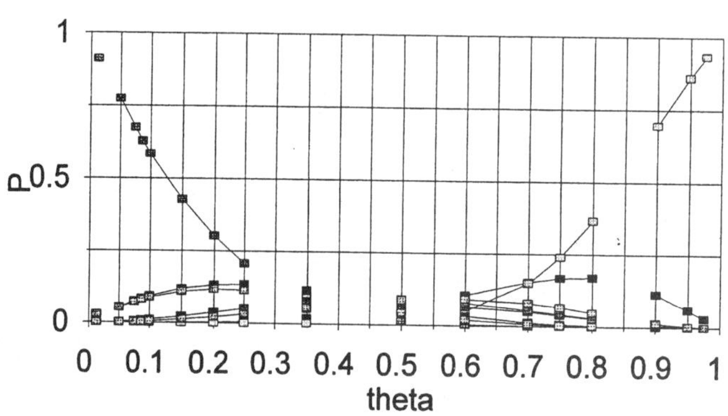

FIGURE 107 Rhombus probabilities, p vs. θ, c = 0.36, c = 0.82151.

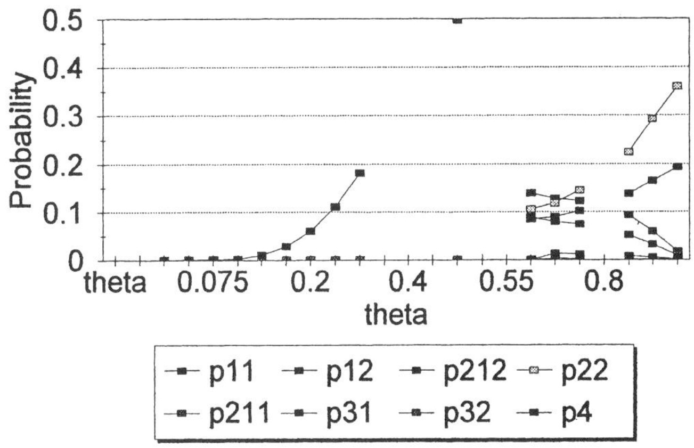

FIGURE 108 Rhombus probabilities, c = 0.9, c = 0.97993.

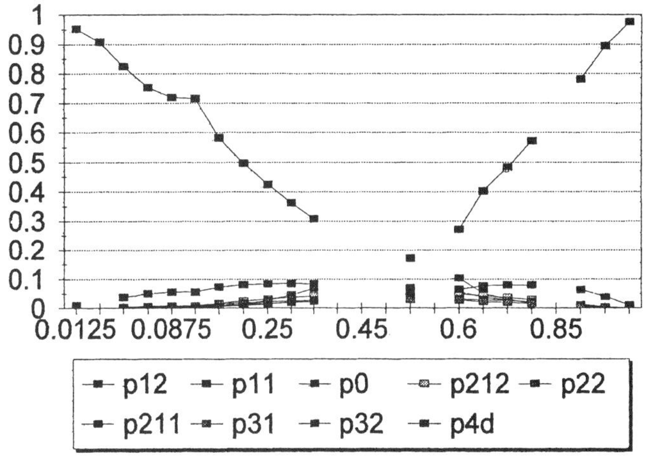

FIGURE 109 Rhombus probabilities, c = 2, c = 1.02605.

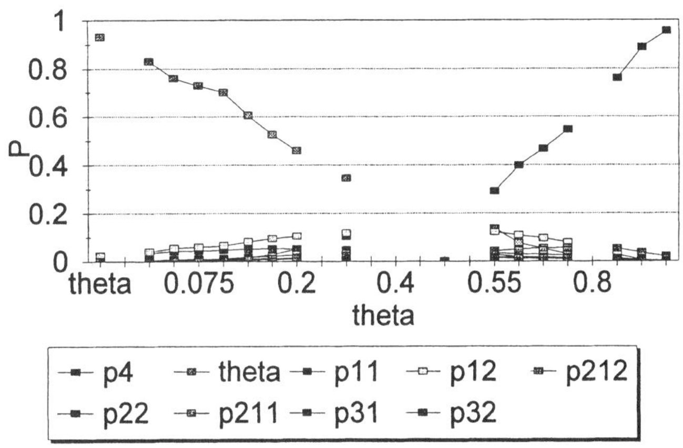

FIGURE 110 Rhombus probabilities, p vs. θ, c = 2.5, c = 1.03452.

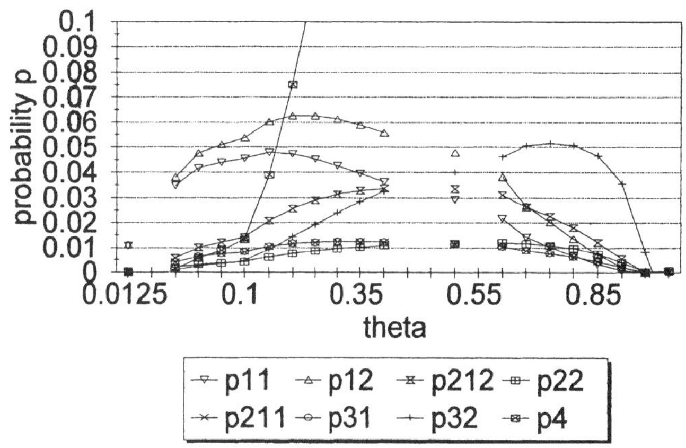

FIGURE 111 Probabilities for rhombus, c = 2.77, c = 1.0385.

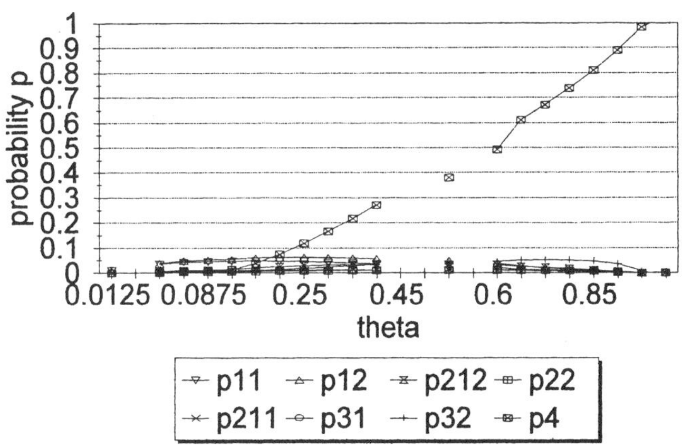

FIGURE 112 Probabilities for rhombus, c = 2.77, c = 1.0385.

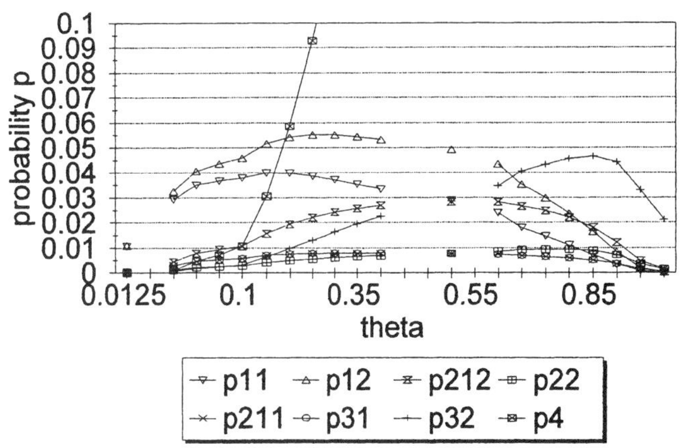

FIGURE 113 Probabilities for rhombus, c = 3, c = 1.04177.

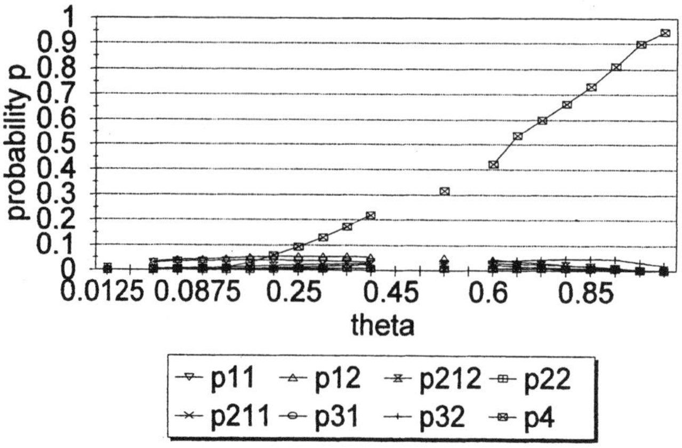

FIGURE 114 Probabilities for rhombus, c = 3, c = 1.04177.

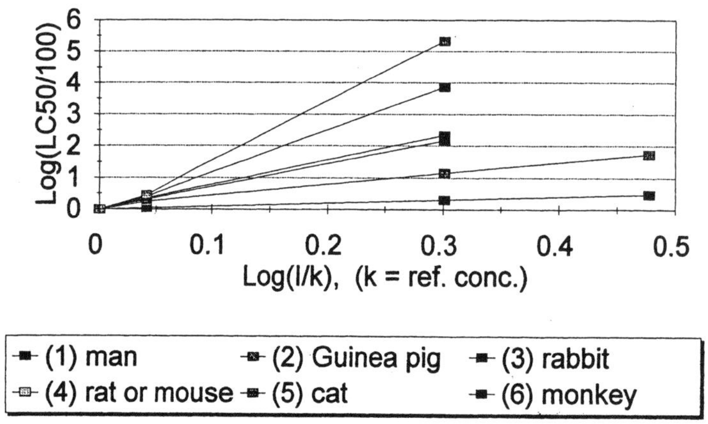

FIGURE 115 $LC_{50/100}$.

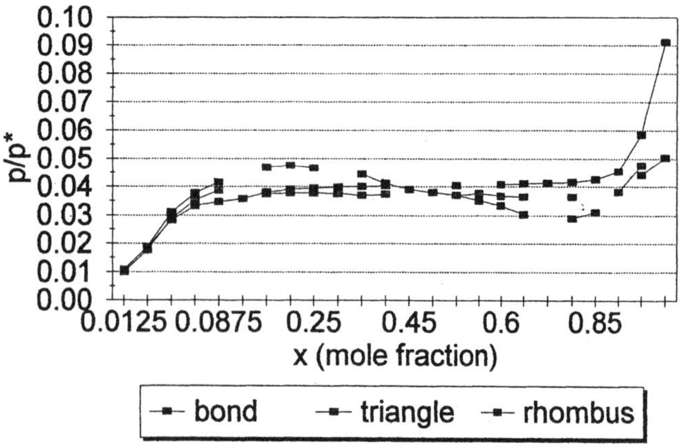

FIGURE 116 Mole fraction vs. p_a/p_a^* and p_b/p_b^*, c = 2.77, Z = 10.

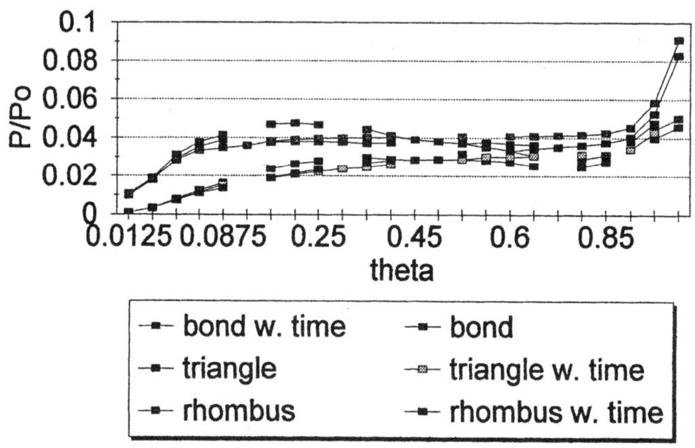

FIGURE 117 Effect of time on results.

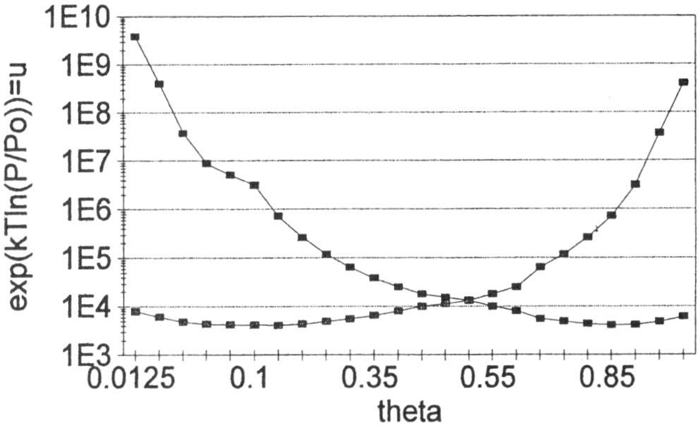

FIGURE 118 Chemical potential vs.θ, c = 2.77, Z = 12.

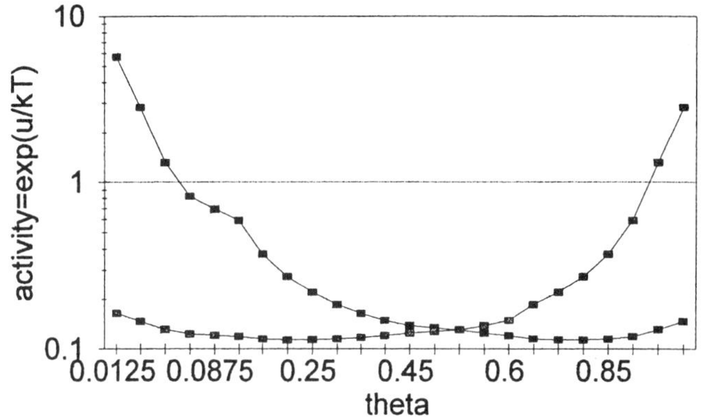

FIGURE 119 Activity vs. θ, c = 2,77, Z = 12.

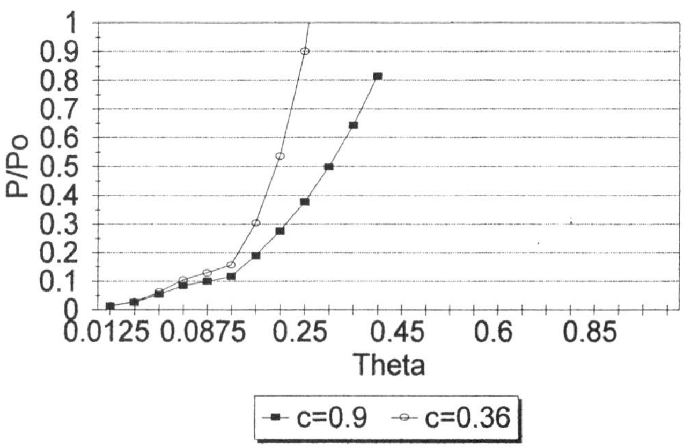

FIGURE 120 Rhombus repulsion, c = 0.9 and 0.36.

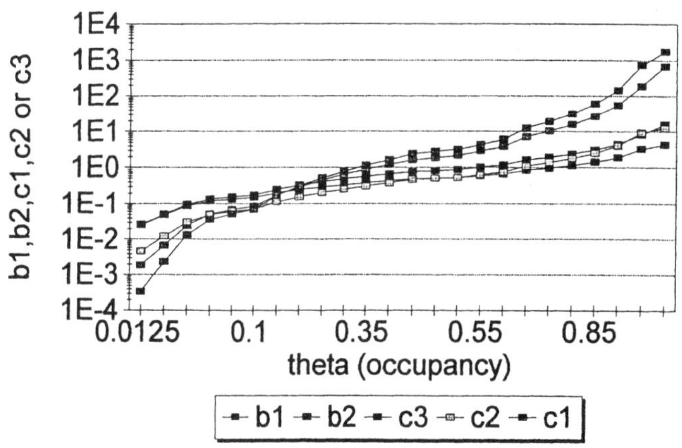

FIGURE 121 Cancer model, c = 2.77, Z = 12, bond and triangle.

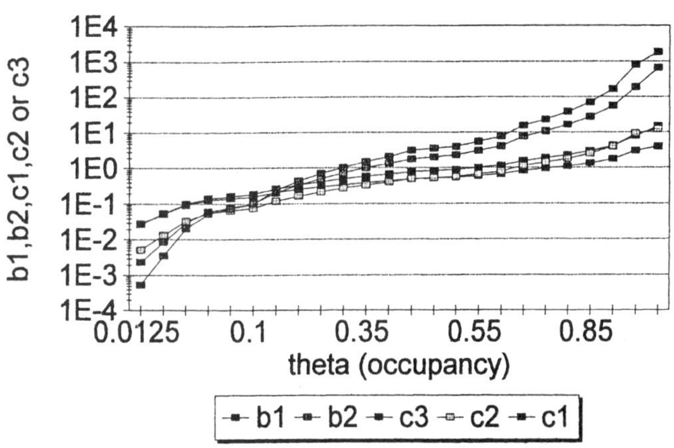

FIGURE 122 Cancer model, c = 3, Z = 12, bond and triangle.

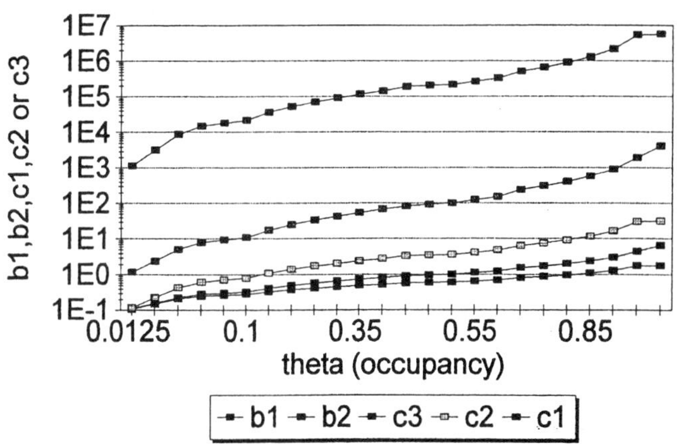

FIGURE 123 Cancer model, c = 100, Z = 50.

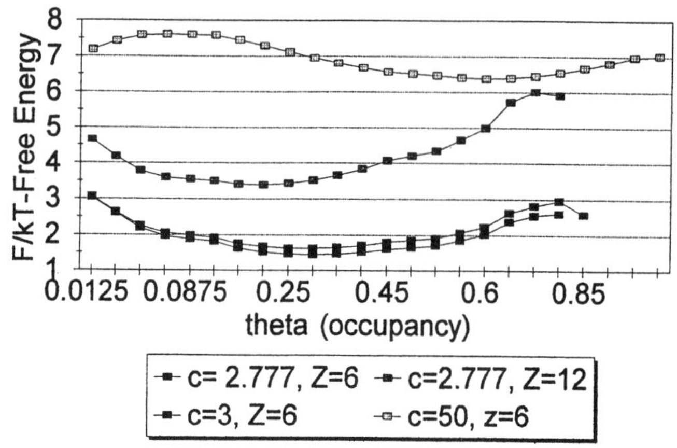

FIGURE 124 Order–disorder/cancer cells.

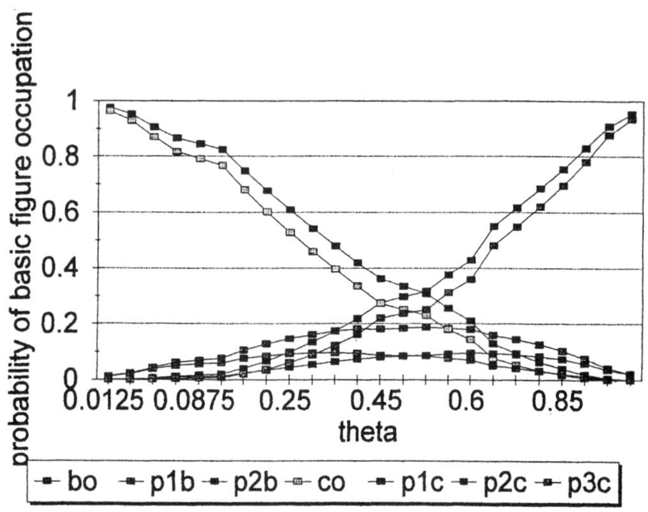

FIGURE 125 Probabilities vs. θ, bond and triangle, c = 2.77, Z = 12.

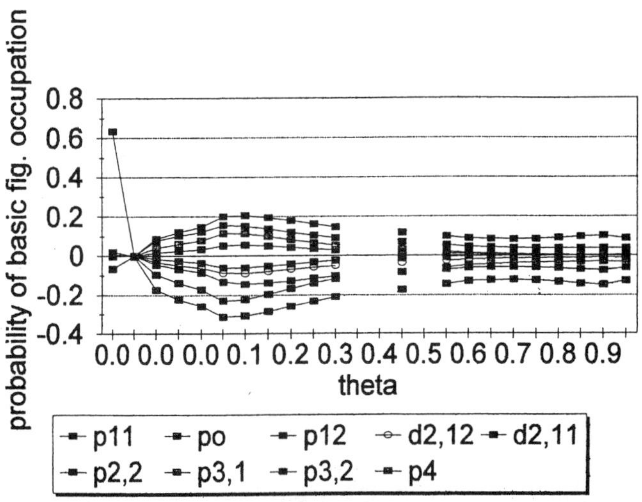

FIGURE 126 Probabilities vs. θ, K_r = 3, c = 3, c = 1.042.

TABLE 1 Linear Model

E	A	B	C	D	E	F	G	H	I	J	K	L	M	N	O	P	Q
73	TABLE I - GROUP PARAMETERS / B, C, D, E, F					LINEAR	MODEL							B	0.021788	A^(1/2)	1.019675
74	-H	1	0.066	0.066	0.018	0.018	-0.027	-0.027	-0.019	-0.019	-0.019	-0.019		-H	1	mol wt	1.0079
75	-CH3	0	0.142	0	-0.02	0	-0.016	0	-0.023	0	-0.015	0		-CH3	0	E rho g/cc	1.035896
76	-CH2-	0	0.076	0	-0.038	0	0.011	0	-0.004	0	0.003	0		-CH2-	0	LOG(E)	0.335615
77	>CH-	0	0.003	0	-0.058	0	0.053	0	0.018	0	0.015	0		>CH-	0	CHI	52.4805
78	>C<	0	-0.075	0	-0.076	0	0.091	0	0.043	0	0.034	0		>C<	0	-LOGKb	-0.56798
79	-CH=CH-	0	0.147	0	-0.043	0	0.028	0	0.01	0	0.003	0		-CH=CH-	0	DIPOLE N	0.20565
80	-CH=CH2	0	0.212	0	-0.025	0	0	0	-0.009	0	-0.015	0		-CH=CH2	0	SUM INT	1
81	>C=CH2	0	0.147	0	-0.043	0	0.028	0	0.01	0	0.003	0		>C=CH2	0	Tb-Tm	-128.725
82	-C=-CH	0	0.171	0	0.074	0	0.027	0	0.002	0	-0.012	0		-C=-CH	0	MOL. WT	26.4226
83	-C6H5	0	0.467	0	-0.007	0	0.012	0	0.007	0	-0.017	0		-C6H5	0	COUNT	0
84	=CH-	0	0.088	0	0.002	0	-0.007	0	0.001	0	-0.003	0	#GRP/	=CH-	0	logben/h2	0.3314
85	AROMAT	0		0		0		0		0		0	CPD	AROM	0	Tm	-50.8455
86	-NAPHTH	0	0.766	0	0.018	0	-0.026	0	0.024	0	-0.028	0		-NAPTH	0	B rho g/cc	3.666394
87	-CYCLOH	0	0.489	0	-0.148	0	0.004	0	-0.029	0	-0.009	0		-CYCLOH	0	B	-0.4405
88	-F	0	0.078	0	0.088	0	0.009	0	-0.019	0	-0.02	0		-F	0	C	-0.038
89	-CL	0	0.165	0	0.087	0	-0.024	0	-0.012	0	-0.021	0		-CL	0	D	-0.02
90	-BR	0	0.213	0	0.095	0	-0.033	0	-0.008	0	-0.02	0		-BR	0	E	0.012
91	-I	0	0.302	0	0.103	0	-0.056	0	-0.01	0	-0.031	0		-I	0	F	0.009
92	-CF3	0	0.15	0	0.017	0	0.035	0	-0.037	0	-0.013	0	.	CF3	0	Tb	-179.57
93	-CCL3	0	0.41	0	0.015	0	-0.009	0	-0.017	0	-0.017	0		-CCL3	0	SOL PAR	6.28669
94	-OH	0	0.202	0	0.324	0	-0.012	0	-0.015	0	0.003	0		-OH	0	Tc	-102.365
95	-O-	0	0.044	0	0.155	0	0.061	0	0.019	0	-0.022	0		-O-	0	**Pc**	49.46995
96	-C=0	0	0.135	0	0.246	0	0.061	0	0.023	0	-0.021	0		-C=O	0	log AC C(	-1.43319
97	-CH=O-	0	0.219	0	0.244	0	0.01	0	-0.014	0	-0.027	0		-CH=O	0	log par co	0.444145
98	-COO-	0	0.167	0	0.17	0	0.062	0	0.015	0	-0.027	0		-COO-	0	Hv	1.39767
99	-COOH	0	0.323	0	0.342	0	-0.011	0	-0.017	0	0.008	0		-COOH	0	Surf Ten	-3.9E+13
100	-NH2	0	0.167	0	0.269	0	0.037	0	0.027	0	-0.014	0		-NH2	0	log visc	-1.38556
101	-NH-	0	0.082	0	0.251	0	0.095	0	0.056	0	-0.01	0		-NH-	0	log(et/h2o	0.98335
102	-N-	0	-0.006	0	0.189	0	0.125	0	0.069	0	0.014	0		-N-	0	logoil/h2o	0.34285
103	-CN	0	0.241	0	0.269	0	-0.007	0	-0.023	0	-0.041	0		-CN	0	logcyh/h2(	0.47705
104	-N=	0	0.102	0	0.183	0	0.031	0	-0.011	0	-0.02	0		-N=	0	logchl/h2o	-0.343
105	PYRIDINE	0		0		0		0		0		0		PYRIDINE	0	logcl4/h2o	0.27635
106	-NO2	0	0.238	0	0.241	0	-0.012	0	-0.027	0	-0.037	0		-NO2	0	vol.Bondi	7.2067
107	-CONH2	0	0.444	0	0.499	0	-0.019	0	-0.039	0	-0.012	0		-CONH2	0	v exp	25.507
108	-S-	0	0.136	0	0.13	0	0.028	0	0.032	0	-0.02	0		-S-	0	C	0
109	-SH	0	0.231	0	0.155	0	-0.026	0	-0.011	0	-0.013	0		-SH	0	H	1
110	MOLECUI	1	-0.5065	-0.5065	-0.056	-0.056	0.007	0.007	0.031	0.031	0.028	0.028		MOLECUI	1	O	0
111		0	B =	-0.4405	C =	-0.038	D =	-0.02	E =	0.012	F =	0.009		S	0	N	0
112		0												CL	0	F	0
113														BR	0	I	*0*
114														loggamma	-1.43439	D1	-0.01003
115														m.refract	1.272815	LC50mg/L	1.064864
116														thermcond	1.262905	isothcom	21.1437
117														COST $/g	ERR	w	-1.68893
118														MCL	45.83906	Zc	0.426114
119														TAP	3.133201	AIR	6.073325
120														COM	1583.394	RES	424.8742
121														drug1	3.694379	drug2	479.5124
122														drug4	19.56298	drug3	3.339547
123														drug6	6.089324	drug5	7.383173

TABLE 2 Hierarchical Model

B	A	B	C	D	E	F	G	H	I	J	K	L	M	N	O	P	Q
72														B/A^.5	0.011302		
73														B	-39.8587	A^(1/2)	-3526.72
74							HIERARCHICAL MODEL							rho (B-E)	-152.734	mol wt	596.8299
75	coefficient	0												coefficiem	0	E rho g/cc	141.7423
76	intercept	1	-50.647	-50.647	-5.605	-5.605	0.663	0.663	3.12	3.12	2.8	2.8		intercept	1	LOG(E)	-764.543
77	mol wt	151.0887	0.2251	34.01007	0.0294	4.442009	-0.0669	-10.1078	-0.0051	-0.77055	-0.014	-2.11524		mol. wt.	151.0887	CHI	4347.026
78	O atoms	6.046346	0.82	4.958003	-0.32	-1.93483	-0.31	-1.87437	-0.39	-2.35807	-0.28	-1.69298		# o	6.046346	-LOGKb	-60913.6
79	N atoms	0	1.17	0	-1.31	0	0.55	0	-0.53	0	-0.3	0		# n	0	DIPOLE M	-6789.97
80	S atoms	0	2.31	0	8.1	0	8.67	0	4.73	0	0.58	0		# s	0	SUM INT	12
81	c-c bonds	0	-4.09	0	-3.91	0	3.69	0	1.67	0	2.11	0		c-c	0	Tb-Tm	-129431
82	arom. c-c	0	-0.08	0	-1.71	0	2.77	0	1.27	0	1.08	0		arom. c-c	0	MOL. WT	-128597
83	c=c bonds	0	4.36	0	-0.85	0	2.47	0	1.32	0	0.16	0		c=c	0	COUNT	12.56689
84	c=-bonds	0	9.26	0	8.8	0	0.34	0	0.34	0	-1.14	0	#GRP/	c=-c	0	logben/h2	-6002.83
85	c-h bonds	0	4.28	0	-0.15	0	-0.83	0	-1.05	0	-0.78	0	CPD	c-h	0	Tm	-135174
86	c-f bonds	0	1.48	0	6.24	0	2.2	0	-1.17	0	-0.66	0		c-f	0	B rho g/cc	-10.9916
87	c-cl bonds	0	6.44	0	5.7	0	1.84	0	-0.59	0	-0.57	0		c-cl	0	B	-0.13914
88	c-br bonds	0	1.3	0	5.23	0	3.89	0	-0.37	0	0.15	0		c-br	0	C	0.61
89	c-i bonds	3.512414	-0.38	-1.33472	4.64	16.2976	4.78	16.78934	-0.84	-2.95043	-0.24	-0.84298		c-I	3.512414	D	0.3292
90	-oh grps	0	13.52	0	30.3	0	2.05	0	-0.35	0	1.84	0		oh	0	E	0.045
91	-o- grps	3.482321	-4.14	-14.4168	11.39	39.66364	11.2	39.002	3.89	13.54623	0.44	1.532221		o	3.482321	F	333
92	-(c=o)- gr	0.122784	10.43	1.280635	27.99	3.436718	4.59	0.563578	0.88	0.10805	-3.56	-0.43711		c=o	0.122784	Tb	-264605
93	-ch=o grp	2.44124	1.85	4.516295	-2.04	-4.98013	-2.37	-5.78574	-1.8	-4.39423	1.23	3.002726		ch=o	2.44124	SOL PAR	-2553.99
94	cooh	0	-1.11	0	-22.99	0	-6.02	0	-2.71	0	1.61	0		cooh	0	Tc	-602688
95	-coo- grps	0	-2.43	0	-19.86	0	-5.91	0	-2.53	0	2.67	0		coo	0	**Pc**	-76590.7
96	-sh grps	0	11.28	0	4.44	0	-7.22	0	-5.15	0	-0.41	0		sh	0	log AC CC	2471.08
97	-NH2	0	9.84	0	25.76	0	6.05	0	4.02	0	0.17	0		nh2	0	log par co	1689.49
98	-nh- grps	0	-0.46	0	22.02	0	13.6	0	7.68	0	1.62	0		nh	0	Hv	4837.254
99	-n- - grps	0	-11.06	0	13.95	0	18.39	0	9.9	0	5.08	0		-n--	0	Surf Ten	-5.70454
100	=n- grps	0	3.82	0	16.12	0	7.76	0	2.55	0	1.72	0		=n-	0	log visc	1684.505
101	-cn grps	0	19.11	0	29.38	0	-1.38	0	-2.72	0	-4.5	0		cn	0	log(et/h2o	324.1547
102	-conh2 gr	0	14.22	0	-1.59	0	-11.7	0	-8.91	0	2.34	0		conh2	0	logoil/h2o	-1006.41
103	-no2 grps	0	8.55	0	22.73	0	3.82	0	-8.35	0	-1.18	0		no2	0	logcyh/h2	-6337.1
104	cycles(ring	1	10.45	10.45	9.96	9.96	-10.34	-10.34	-2.85	-2.85	-4.2	-4.2		cycles(ring	1	logchl/h2o	-8663.59
105	tertiary C	1	-0.74	-0.74	-0.12	-0.12	1.44	1.44	0.3	0.3	-0.67	-0.67		tert C	1	logcl4/h2o	-10997
106	quatern c	1	-1.99	-1.99	-0.16	-0.16	2.57	2.57	0.95	0.95	-0.61	-0.61		quat C	1	vol.Bondi	11699.49
107	x-c-x trace	0	-0.99	0	-5.7	0	-0.11	0	-0.44	0	0.19	0		x-c-x	0	v exp	-907.256
108	X3-C-C-X3	0	-0.77	0	-17.62	0	-0.63	0	-1.42	0	-0.51	0		x3-c-x3	0	C	4.320231
109	c=c-x trac	0	-0.47	0	-4.96	0	-0.74	0	-0.58	0	0.31	0		c=c-x	0	H	2.44124
110	arom fusio	2E-06	-0.31	-6.2E-07	-6.57	-1.3E-05	1.81	3.6E-06	-1.4	-2.8E-06	2.41	4.8E-06		arom. fusi	2E-06	O	6.046346
111		1	B=	-0.13914	C=	0.61	D=	0.3292	E=	0.04701	F=	-0.03233		S	0	N	0
112														CL	0	F	0
113														BR	0	I	*3.512414*
114														loggamma	2471.076	D1	8.3E-06
115														m.refract	-3995.09	D12	
116														thermcond	-4822.93	isothcom	7353.613
117														COST $/g	0.00201	w	ERR
118														MCL	0	Zc	ERR
119														TAP	0	AIR	0
120														COM	0	RES	0
121														drug1	13475.72	drug2	-685477
122														drug4	98165.33	drug3	8984.079
123														drug6	-135.681	drug5	6560.408

TABLE 3 Values for B, C, D, E, and F for 36 Compounds

C	A	B	C	D	E	F	G
1	NAME		B	C	D	E	F
2	PCE		0.12942	-0.10019	-0.03781	-0.01863	-0.00022
3	TCE		0.04489	-0.06222	-0.03297	-0.01174	-0.00249
4	1,1-DCE		-0.04955	-0.08125	-0.0923	-0.00882	-0.00287
5	CH3CCL3		0.04482	-0.06043	-0.01872	-0.01988	-0.00438
6	CLCH2CH2CL		-0.02461	0.04194	-0.01907	-0.01085	-0.00735
7	LINDANE		0.650492	0.227454	-0.00934	-0.01033	-0.00912
8	CHLOROBENZENE		0.124999	0.023542	-0.02897	0.01476	-0.00966
9	P-DICHLOROBENZE		0.224135	0.092169	-0.02532	0.017603	-0.01238
10	CCL4		0.097386	0.217174	-0.02268	-0.00025	-0.01634
11	CH2CLCHCLCH3		0.051662	0.003968	-0.00816	-0.01596	-0.00382
12	ENDRIN		1.144564	0.538338	-0.1744	0.003374	-0.07533
13	HEPTACHLOR		1.094874	0.435606	-0.15102	-0.00224	-0.05697
14	HEXACHLOROBENZ		0.620679	0.366677	-0.01069	0.028976	-0.02327
15	C16H15CL3O2		1.087393	0.121172	0.156889	0.040672	-0.03669
16	DICHLOROMETHAN		-0.12069	-0.03408	-0.03219	-0.01473	-0.00709
17	C9H7CL3O3 SILVEX		0.5402	-0.02161	0.010827	0.017955	0.003468
18	STYRENE		0.172774	-0.08803	-0.00505	0.019488	-0.00348
19	1,2,4-TRICHLORO-		0.323271	0.160796	-0.02166	0.020446	-0.0151
20	CLCH2CHCL2		0.074524	0.110571	-0.01542	-0.0081	-0.01008
21	CHLORDANE		1.240548	0.120289	0.158875	0.040825	-0.03627
22	C8H6CL2O3		0.36479	-0.05226	-0.00375	0.020127	0.002655
23	HEPTACHLOREPOX		1.013189	0.52041	-0.04063	0.035445	-0.03811
24	BENZO(a)PYRENE		1.062891	-0.24077	-0.02347	-0.01537	0.068776
25	VINYL CHLORIDE		-0.13409	-0.04328	-0.02438	-0.00192	-0.00515
26	BENZENE		0.025863	-0.04509	-0.03263	0.011994	-0.00694
27	CH2CLCHBRCH2BR		0.02346	0.087566	-0.01896	-0.01295	-0.0054
28	CH2BRCH2BR		0.072707	0.058681	-0.03755	-0.01108	-0.0054
29	O-DICHLOROBENZ		0.224135	0.092169	-0.02532	0.017603	-0.01238
30	ETHYL BENZENE		0.178412	-0.12104	-0.0108	0.001886	0.000137
31	C5CL6		0.562942	0.331445	-0.00876	0.029889	-0.01989
32	PENTACHLORPPHE		0.685816	0.604054	0.00065	0.028417	0.000613
33	***TCDD DIOXIN***		0.875294	0.378511	0.125929	0.044079	0.008024
34	TOLUENE		0.102138	-0.08306	-0.02171	0.006901	-0.0034
35	TOXAPHENE		1.211823	0.306311	-0.01911	-0.0189	-0.02493
36	XYLENES		0.178412	-0.12104	-0.0108	0.001886	0.000137
37	BENZOBFLUORANT		1.065991	-0.17507	-0.04157	-0.00137	0.044676

TABLE 4 Optimizer Template

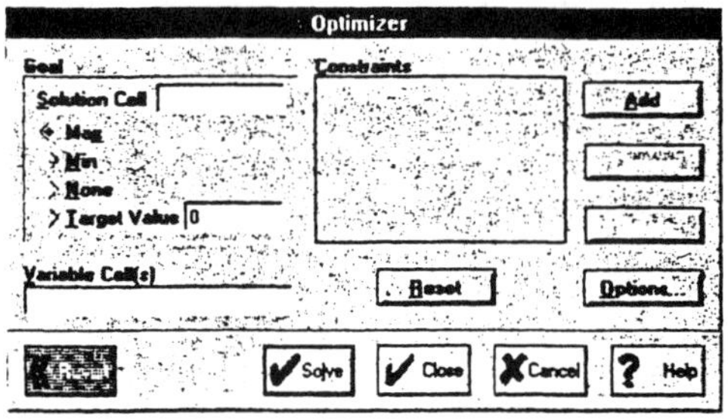

TABLE 5 Tap Water and Air Risk for 36 Compounds

C	AF	AG	AA	AB
1	NAME		TAP	AIR
2	PCE		0.019267	3.906958
3	TCE		0.02052	3.641055
4	1,1-DCE		0.096603	15.11521
5	CH3CCL3		0.009844	2.161108
6	CLCH2CH2CL		0.005495	0.839702
7	LINDANE		0.000243	0.025296
8	CHLOROBENZENE		0.054369	7.079197
9	P-DICHLOROBENZE		0.025889	2.97164
10	CCL4		0.001602	0.173443
11	CH2CLCHCLCH3		0.004219	0.708738
12	ENDRIN		0.004032	1.428365
13	HEPTACHLOR		0.002805	0.741185
14	HEXACHLOROBENZ		0.001331	0.092243
15	C16H15CL3O2		0.009971	4.086027
16	DICHLOROMETHAN		0.016032	3.129434
17	C9H7CL3O3 SILVEX		0.023588	2.430721
18	STYRENE		0.169453	25.26631
19	1,2,4-TRICHLORO-I		0.012327	1.247303
20	CLCH2CHCL2		0.002599	0.350615
21	CHLORDANE		0.008684	3.650515
22	C8H6CL2O3		0.064101	6.824843
23	HEPTACHLOREPOX		0.000708	0.050402
24	BENZO(a)PYRENE		0.001355	0.031382
25	VINYL CHLORIDE		0.036227	5.796446
26	BENZENE		0.114859	16.9519
27	CH2CLCHBRCH2BR		0.002131	0.266965
28	CH2BRCH2BR		0.004835	0.63097
29	O-DICHLOROBENZE		0.02589	2.971593
30	ETHYL BENZENE		0.068196	12.14341
31	C5CL6		0.001941	0.129135
32	PENTACHLORPPHE		8.6E-06	0.000103
33	*TCDD DIC*	*IOXIN*	4.7E-05	0.001042
34	TOLUENE		0.088236	14.31039
35	TOXAPHENE		8.6E-05	0.016522
36	XYLENES		0.068196	12.14341
37	BENZOBFLUORANT		0.00794	0.319483

TABLE 6 Freon Compound, HFC-338mccq, $CF_3CF_2CF_2CH_2F$

E	A	B	C	D	E	F	G	H	I	J	K	L	M	N	O	P	Q
72														B/A^.5	0.033932		
73	HFC-338mccq CF3CF2CF2CH2F					LINEAR	MODEL							B	0.133458	A^(1/2)	3.933055
74	-H	0	0.066	0	0.018	0	-0.027	0	-0.019	0	-0.019	0		-H	0	mol wt	202.047
75	-CH3	0	0.142	0	-0.02	0	-0.016	0	-0.023	0	-0.015	0		-CH3	0	E rho g/cc	0.091531
76	-CH2-	1	0.076	0.076	-0.038	-0.038	0.011	0.011	-0.004	-0.004	0.003	0.003		-CH2-	1	LOG(E)	0.899965
77	>CH-	0	0.003	0	-0.058	0	0.053	0	0.018	0	0.015	0		>CH-	0	CHI	38.2469
78	>C<	2	-0.075	-0.15	-0.076	-0.152	0.091	0.182	0.043	0.086	0.034	0.068		>C<	2	-LOGKb	18.49398
79	-CH=CH-	0	0.147	0	-0.043	0	0.028	0	0.01	0	0.003	0		-CH=CH-	0	DIPOLE M	2.43235
80	-CH=CH2	0	0.212	0	-0.025	0	0	0	-0.009	0	-0.015	0		-CH=CH2	0	SUM INT	9
81	>C=CH2	0	0.147	0	-0.043	0	0.028	0	0.01	0	0.003	0		>C=CH2	0	Tb-Tm	21.4714
82	-C=-CH	0	0.171	0	0.074	0	0.027	0	0.002	0	-0.012	0		-C=-CH	0	MOL. WT.	11.6889
83	-C6H5	0	0.467	0	-0.007	0	0.012	0	0.007	0	-0.017	0		-C6H5	0	COUNT	5
84	=CH-	0	0.088	0	0.002	0	-0.007	0	0.001	0	-0.003	0	#GRP/	=CH-	0	logben/h2	-2.3714
85	AROMAT	0		0		0		0		0		0	CPD	AROM	0	Tm	-3.1055
86	-NAPHTH	0	0.766	0	0.018	0	-0.026	0	0.024	0	-0.028	0	log1/s25C	-NAPTH	0	B rho g/cc	0.198469
87	-CYCLOH	0	0.489	0	-0.148	0	0.004	0	-0.029	0	-0.009	0	-3.09795	-CYCLOH	0	B	-0.0405
88	-F	5	0.078	0.39	0.088	0.44	0.009	0.045	-0.019	-0.095	-0.02	-0.1	logBCF	-F	5	C	0.211
89	-CL	0	0.165	0	0.087	0	-0.024	0	-0.012	0	-0.021	0	-1.47013	-CL	0	D	0.245
90	-BR	0	0.213	0	0.095	0	-0.033	0	-0.008	0	-0.02	0	logKoc	-BR	0	E	-0.019
91	-I	0	0.302	0	0.103	0	-0.056	0	-0.01	0	-0.031	0	0.489325	-I	0	F	-0.014
92	-CF3	1	0.15	0.15	0.017	0.017	0.035	0.035	-0.037	-0.037	-0.013	-0.013		CF3	1	Tb	18.3659
93	-CCL3	0	0.41	0	0.015	0	-0.009	0	-0.017	0	-0.017	0		-CCL3	0	SOL PAR	7.98959
94	-OH	0	0.202	0	0.324	0	-0.012	0	-0.015	0	0.003	0		-OH	0	Tc	89.3099
95	-O-	0	0.044	0	0.155	0	0.061	0	0.019	0	-0.022	0		-O-	0	Pc	3.09975
96	-C=0	0	0.135	0	0.246	0	0.061	0	0.023	0	-0.021	0		-C=O	0	log AC CC	4.479865
97	-CH=O-	0	0.219	0	0.244	0	0.01	0	-0.014	0	-0.027	0		-CH=O	0	log par co	-1.63175
98	-COO-	0	0.167	0	0.17	0	0.062	0	0.015	0	-0.027	0		-COO-	0	Hv	7.684555
99	-COOH	0	0.323	0	0.342	0	-0.011	0	-0.017	0	0.008	0		-COOH	0	Surf Ten	-1.6605
100	-NH2	0	0.167	0	0.269	0	0.037	0	0.027	0	-0.014	0		-NH2	0	log visc	-0.2102
101	-NH-	0	0.082	0	0.251	0	0.095	0	0.056	0	-0.01	0		-NH-	0	log(et/h2o	-2.80765
102	-N-	0	-0.006	0	0.189	0	0.125	0	0.069	0	0.014	0		-N-	0	logoil/h2o	-3.08745
103	-CN	0	0.241	0	0.269	0	-0.007	0	-0.023	0	-0.041	0		-CN	0	logcyh/h2	-3.43575
104	-N=	0	0.102	0	0.183	0	0.031	0	-0.011	0	-0.02	0		-N=	0	logchl/h2o	-0.7773
105	PYRIDINE	0		0		0		0		0		0		PYRIDINE	0	logcl4/h2o	-1.55165
106	-NO2	0	0.238	0	0.241	0	-0.012	0	-0.027	0	-0.037	0		-NO2	0	vol.Bondi	58.8952
107	-CONH2	0	0.444	0	0.499	0	-0.019	0	-0.039	0	-0.012	0		-CONH2	0	v exp	127.704
108	-S-	0	0.136	0	0.13	0	0.028	0	0.032	0	-0.02	0		-S-	0	C	4
109	-SH	0	0.231	0	0.155	0	-0.026	0	-0.011	0	-0.013	0		-SH	0	H	2
110	MOLECUI	1	-0.5065	-0.5065	-0.056	-0.056	0.007	0.007	0.031	0.031	0.028	0.028		MOLECUI	1	O	0
111	9	9	B =	-0.0405	C =	0.211	D =	0.28	E =	-0.019	F =	-0.014		S	0	N	0
112		0							9	Pvp mm	Kirchhoff	Pvpr298	3365.072	CL	0	F	8
113			B	C	D	E	F		0.717794	Pvp mm	riedel-P-M	10-1500n	0.997927	BR	0	I	0
114	-H	0	-0.4405	-0.038	-0.02	0.012	0.009			Pvp mm	riedel-P-M	>1500mm	0.998087	loggamma	4.481765	D1	-0.06613
115	-CH3	0	-0.3645	-0.076	-0.009	0.008	0.013			T	500	G1	0.237158	m.refract	15.85591	LC50mg/L	762.0577
116	ch2	0	-0.4305	-0.094	0.018	0.027	0.031					G2	0.217158	thermcon	4.958605	isothcom	23.8315
117	'>CH-	0	-0.5035	-0.114	0.06	0.049	0.043					h	-0.02197	COST $/g	ERR	w	0.054512
118	>C<	0	-0.5815	-0.132	0.098	0.074	0.062					g1	-2.0575	MCL	0.00519	Zc	0.286639
119	-CH=CH-	0	-0.3595	-0.099	0.035	0.041	0.031					g2	-2.07102	TAP	0.001242	AIR	0.001001
120	-CH=CH2	0	-0.2945	-0.081	0.007	0.022	0.013							COM	0.012543	RES	0.035112
121	>C=CH2	0	-0.3595	0.099	0.035	0.041	0.031							drug1	5.191229	drug2	84.7716
122	-C=-CH	0	-0.3355	0.018	0.034	0.033	0.016							drug4	48.85788	drug3	4.550646
123	-C6H5	0	-0.0395	-0.063	0.019	0.038	0.011							drug6	3.905667	drug5	11.74573
124	=CH-	0	-0.4185	-0.054	0	0.032	0.025							C-E	0.23	B-D	-0.2855
125	-NAPTH	0	0.2595	-0.038	-0.019	0.055	0							B-C	-0.2515	sum grps	9

TABLE 7 Freon Compound, R32, CH_2F_2

E	A	B	C	D	E	F	G	H	I	J	K	L	M	N	O	P	Q
72														B/A^.5	0.025507		
73	R32 CH2F2					LINEAR	MODEL							B	0.058768	A^(1/2)	2.303975
74	-H	0	0.066	0	0.018	0	-0.027	0	-0.019	0	-0.019	0		-H	0	mol wt	52.02361
75	-CH3	0	0.142	0	-0.02	0	-0.016	0	-0.023	0	-0.015	0		-CH3	0	E rho g/cc	0.644371
76	-CH2-	1	0.076	0.076	-0.038	-0.038	0.011	0.011	-0.004	-0.004	0.003	0.003		-CH2-	1	LOG(E)	0.790615
77	>CH-	0	0.003	0	-0.058	0	0.053	0	0.018	0	0.015	0		>CH-	0	CHI	46.0346
78	>C<	0	-0.075	0	-0.076	0	0.091	0	0.043	0	0.034	0		>C<	0	-LOGKb	11.42374
79	-CH=CH-	0	0.147	0	-0.043	0	0.028	0	0.01	0	0.003	0		-CH=CH-	0	DIPOLE M	1.56785
80	-CH=CH2	0	0.212	0	-0.025	0	0	0	-0.009	0	-0.015	0		-CH=CH2	0	SUM INT	3
81	>C=CH2	0	0.147	0	-0.043	0	0.028	0	0.01	0	0.003	0		>C=CH2	0	Tb-Tm	-60.8252
82	-C=-CH	0	0.171	0	0.074	0	0.027	0	0.002	0	-0.012	0		-C=-CH	0	MOL. WT	34.9642
83	-C6H5	0	0.467	0	-0.007	0	0.012	0	0.007	0	-0.017	0		-C6H5	0	COUNT	1
84	=CH-	0	0.088	0	0.002	0	-0.007	0	0.001	0	-0.003	0	#GRP/	=CH-	0	logben/h2	-0.5342
85	AROMAT	0		0		0		0		0		0	CPD	AROM	0	Tm	-3.8925
86	-NAPHTH	0	0.766	0	0.018	0	-0.026	0	0.024	0	-0.028	0	log1/s25C	-NAPTH	0	B rho g/cc	1.528737
87	-CYCLOH	0	0.489	0	-0.148	0	0.004	0	-0.029	0	-0.009	0	-1.45843	-CYCLOH	0	B	-0.2745
88	-F	2	0.078	0.156	0.088	0.176	0.009	0.018	-0.019	-0.038	-0.02	-0.04	logBCF	-F	2	C	0.082
89	-CL	0	0.165	0	0.087	0	-0.024	0	-0.012	0	-0.021	0	-0.43906	-CL	0	D	0.036
90	-BR	0	0.213	0	0.095	0	-0.033	0	-0.008	0	-0.02	0	logKoc	-BR	0	E	-0.011
91	-I	0	0.302	0	0.103	0	-0.056	0	-0.01	0	-0.031	0	1.227354	-I	0	F	-0.009
92	-CF3	0	0.15	0	0.017	0	0.035	0	-0.037	0	-0.013	0	.	CF3	0	Tb	-64.7177
93	-CCL3	0	0.41	0	0.015	0	-0.009	0	-0.017	0	-0.017	0		-CCL3	0	SOL PAR	8.20051
94	-OH	0	0.202	0	0.324	0	-0.012	0	-0.015	0	0.003	0		-OH	0	Tc	48.9254
95	-O-	0	0.044	0	0.155	0	0.061	0	0.019	0	-0.022	0		-O-	0	Pc	46.74035
96	-C=0	0	0.135	0	0.246	0	0.061	0	0.023	0	-0.021	0		-C=O	0	log AC CO	0.932295
97	-CH=O-	0	0.219	0	0.244	0	0.01	0	-0.014	0	-0.027	0		-CH=O	0	log par co	-0.27508
98	-COO-	0	0.167	0	0.17	0	0.062	0	0.015	0	-0.027	0		-COO-	0	Hv	4.60967
99	-COOH	0	0.323	0	0.342	0	-0.011	0	-0.017	0	0.008	0		-COOH	0	Surf Ten	-11.2545
100	-NH2	0	0.167	0	0.269	0	0.037	0	0.027	0	-0.014	0		-NH2	0	log visc	-0.78674
101	-NH-	0	0.082	0	0.251	0	0.095	0	0.056	0	-0.01	0		-NH-	0	log(et/h2o	-0.17165
102	-N-	0	-0.006	0	0.189	0	0.125	0	0.069	0	0.014	0		-N-	0	logoil/h2o	-0.94215
103	-CN	0	0.241	0	0.269	0	-0.007	0	-0.023	0	-0.041	0		-CN	0	logcyh/h2o	-0.85375
104	-N=	0	0.102	0	0.183	0	0.031	0	-0.011	0	-0.02	0		-N=	0	logchl/h2o	-0.2894
105	PYRIDINE	0		0		0		0		0		0		PYRIDINE	0	logcl4/h2o	-0.15105
106	-NO2	0	0.238	0	0.241	0	-0.012	0	-0.027	0	-0.037	0		-NO2	0	vol.Bondi	22.8713
107	-CONH2	0	0.444	0	0.499	0	-0.019	0	-0.039	0	-0.012	0		-CONH2	0	v exp	54.261
108	-S-	0	0.136	0	0.13	0	0.028	0	0.032	0	-0.02	0		-S-	0	C	1
109	-SH	0	0.231	0	0.155	0	-0.026	0	-0.011	0	-0.013	0		-SH	0	H	2
110	MOLECUI	1	-0.5065	-0.5065	-0.056	-0.056	0.007	0.007	0.031	0.031	0.028	0.028		MOLECUI	1	O	0
111	3	3	B =	-0.2745	C =	0.082	D =	0.036	E =	-0.011	F =	-0.009		S	0	N	0
112		0							3	Pvp mm	Kirchhoff	Pvpr298	3136.981	CL	0	F	2
113			B	C	D	E	F		0.717794	Pvp mm	riedel-P-M	10-1500n	1,00834	BR	0	I	0
114	-H	0	-0.4405	-0.038	-0.02	0.012	0.009			Pvp mm	riedel-P-M	>1500mm	1.008055	loggamma	0.933395	D1	-0.01767
115	-CH3	0	-0.3645	-0.076	-0.009	0.008	0.013			T	500	G1	0.3555	m.refract	5.992275	LC50mg/L	48.44092
116	ch2	0	-0.4305	-0.094	0.018	0.027	0.031					G2	0.3355	thermcon	3.014645	isothcom	18.6727
117	'>CH-	0	-0.5035	-0.114	0.06	0.049	0.043					h	0.239558	COST $/g	ERR	w	-0.40751
118	>C<	0	-0.5815	-0.132	0.098	0.074	0.062					g1	0.184724	MCL	1.612128	Zc	0.323601
119	-CH=CH-	0	-0.3595	-0.099	0.035	0.041	0.031					g2	0.19217	TAP	0.248395	AIR	0.395249
120	-CH=CH2	0	-0.2945	-0.081	0.007	0.022	0.013							COM	23.09488	RES	15.2078
121	>C=CH2	0	-0.3595	0.099	0.035	0.041	0.031							drug1	4.156151	drug2	320.2231
122	-C=-CH	0	-0.3355	0.018	0.034	0.033	0.016							drug4	8.081216	drug3	4.044674
123	-C8H5	0	-0.0395	-0.063	0.019	0.038	0.011							drug8	5.634648	drug5	8.658213
124	=CH-	0	-0.4185	-0.054	0	0.032	0.025							C-E	0.093	B-D	-0.3105
125	-NAPTH	0	0.2595	-0.038	-0.019	0.055	0							B-C	-0.3565	sum grps	3

TABLE 8 Freon Compound, HFC-245fa, $CF_3CH_2CHF_2$

E	A	B	C	D	E	F	G	H	I	J	K	L	M	N	O	P	Q
72														B/A^.5	0.030553		
73	hfc-245fa CF3CH2CHF2					LINEAR	MODEL							B	0.101683	A^(1/2)	3.328105
74	-H	0	0.066	0	0.018	0	-0.027	0	-0.019	0	-0.019	0		-H	0	mol wt	134.0487
75	-CH3	0	0.142	0	-0.02	0	-0.016	0	-0.023	0	-0.015	0		-CH3	0	E rho g/cc	0.500178
76	-CH2-	1	0.076	0.076	-0.038	-0.038	0.011	0.011	-0.004	-0.004	0.003	0.003		-CH2-	1	LOG(E)	0.759115
77	>CH-	1	0.003	0.003	-0.058	-0.058	0.053	0.053	0.018	0.018	0.015	0.015		>CH-	1	CHI	48.2602
78	>C<	0	-0.075	0	-0.076	0	0.091	0	0.043	0	0.034	0		>C<	0	-LOGKb	12.63346
79	-CH=CH-	0	0.147	0	-0.043	0	0.028	0	0.01	0	0.003	0		-CH=CH-	0	DIPOLE M	1.66205
80	-CH=CH2	0	0.212	0	-0.025	0	0	0	-0.009	0	-0.015	0		-CH=CH2	0	SUM INT	5
81	>C=CH2	0	0.147	0	-0.043	0	0.028	0	0.01	0	0.003	0		>C=CH2	0	Tb-Tm	-11.1442
82	-C=-CH	0	0.171	0	0.074	0	0.027	0	0.002	0	-0.012	0		-C=-CH	0	MOL. WT	47.2443
83	-C6H5	0	0.467	0	-0.007	0	0.012	0	0.007	0	-0.017	0		-C6H5	0	COUNT	1.5
84	=CH-	0	0.088	0	0.002	0	-0.007	0	0.001	0	-0.003	0	#GRP/	=CH-	0	logben/h2	0.3262
85	AROMAT	0		0		0		0		0		0	CPD	AROM	0	Tm	2.5565
86	-NAPHTH	0	0.766	0	0.018	0	-0.026	0	0.024	0	-0.028	0	log1/s25C	-NAPTH	0	B rho g/cc	1.090146
87	-CYCLOH	0	0.489	0	-0.148	0	0.004	0	-0.029	0	-0.009	0	-0.68775	-CYCLOH	0	B	-0.1215
88	-F	2	0.078	0.156	0.088	0.176	0.009	0.018	-0.019	-0.038	-0.02	-0.04	logBCF	-F	2	C	0.041
89	-CL	0	0.165	0	0.087	0	-0.024	0	-0.012	0	-0.021	0	0.005049	-CL	0	D	0.089
90	-BR	0	0.213	0	0.095	0	-0.033	0	-0.008	0	-0.02	0	logKoc	-BR	0	E	-0.03
91	-I	0	0.302	0	0.103	0	-0.056	0	-0.01	0	-0.031	0	1.545246	-I	0	F	-0.007
92	-CF3	1	0.15	0.15	0.017	0.017	0.035	0.035	-0.037	-0.037	-0.013	-0.013		CF3	1	Tb	-8.5877
93	-CCL3	0	0.41	0	0.015	0	-0.009	0	-0.017	0	-0.017	0		-CCL3	0	SOL PAR	7.93777
94	-OH	0	0.202	0	0.324	0	-0.012	0	-0.015	0	0.003	0		-OH	0	Tc	101.2705
95	-O-	0	0.044	0	0.155	0	0.061	0	0.019	0	-0.022	0		-O-	0	Pc	30.97985
96	-C=0	0	0.135	0	0.246	0	0.061	0	0.023	0	-0.021	0		-C=O	0	log AC CO	1.217115
97	-CH=O-	0	0.219	0	0.244	0	0.01	0	-0.014	0	-0.027	0		-CH=O	0	log par co	0.309275
98	-COO-	0	0.167	0	0.17	0	0.062	0	0.015	0	-0.027	0		-COO-	0	Hv	6.342705
99	-COOH	0	0.323	0	0.342	0	-0.011	0	-0.017	0	0.008	0		-COOH	0	Surf Ten	-4.9815
100	-NH2	0	0.167	0	0.269	0	0.037	0	0.027	0	-0.014	0		-NH2	0	log visc	-0.35088
101	-NH-	0	0.082	0	0.251	0	0.095	0	0.056	0	-0.01	0		-NH-	0	log(et/h2o	0.41885
102	-N-	0	-0.006	0	0.189	0	0.125	0	0.069	0	0.014	0		-N-	0	logoil/h2o	-0.61995
103	-CN	0	0.241	0	0.269	0	-0.007	0	-0.023	0	-0.041	0		-CN	0	logcyh/h2(	-0.28445
104	-N=	0	0.102	0	0.183	0	0.031	0	-0.011	0	-0.02	0		-N=	0	logchl/h2o	0.6671
105	PYRIDINE	0		0		0		0		0		0		PYRIDINE	0	logcl4/h2o	0.71825
106	-NO2	0	0.238	0	0.241	0	-0.012	0	-0.027	0	-0.037	0		-NO2	0	vol.Bondi	43.3376
107	-CONH2	0	0.444	0	0.499	0	-0.019	0	-0.039	0	-0.012	0		-CONH2	0	v exp	94.455
108	-S-	0	0.136	0	0.13	0	0.028	0	0.032	0	-0.02	0		-S-	0	C	3
109	-SH	0	0.231	0	0.155	0	-0.026	0	-0.011	0	-0.013	0		-SH	0	H	3
110	MOLECUL	1	-0.5065	-0.5065	-0.056	-0.056	0.007	0.007	0.031	0.031	0.028	0.028		MOLECUL	1	O	0
111	5	5	B =	-0.1215	C =	0.041	D =	0.124	E =	-0.03	F =	-0.007		S	0	N	0
112		0							5	Pvp mm	Kirchhoff	Pvpr298	16816.29	CL	0	F	5
113			B	C	D	E	F		0.717794	Pvp mm	riedel-P-M	10-1500n	0.999789	BR	0	I	0
114	-H	0	-0.4405	-0.038	-0.02	0.012	0.009			Pvp mm	riedel-P-M	>1500mm	0.999811	loggamma	1.220115	D1	-0.03961
115	-CH3	0	-0.3645	-0.076	-0.009	0.008	0.013			T	500	G1	0.261632	m.refract	13.57992	LC50mg/l	139.0596
116	ch2	0	-0.4305	-0.094	0.018	0.027	0.031					G2	0.241632	thermcond	3.651315	isothcom	18.5455
117	>CH-	0	-0.5035	-0.114	0.06	0.049	0.043					h	0.032114	COST $/g	0.21	w	-0.04995
118	>C<	0	-0.5815	-0.132	0.098	0.074	0.062					g1	-0.6734	MCL	0.246849	Zc	0.294996
119	-CH=CH-	0	-0.3595	-0.099	0.035	0.041	0.031					g2	-0.66477	TAP	0.060913	AIR	0.072817
120	-CH=CH2	0	-0.2945	-0.081	0.007	0.022	0.013							COM	1.341192	RES	1.706598
121	>C=CH2	0	-0.3595	0.099	0.035	0.041	0.031							drug1	3.911777	drug2	59.54923
122	-C=-CH	0	-0.3355	0.018	0.034	0.033	0.016							drug4	4.275751	drug3	3.656492
123	-C6H5	0	-0.0395	-0.063	0.019	0.038	0.011							drug6	4.475855	drug5	10.53676
124	=CH-	0	-0.4185	-0.054	0	0.032	0.025							C-E	0.071	B-D	-0.2105
125	-NAPTH	0	0.2595	-0.038	-0.019	0.055	0							B-C	-0.1625	sum grps	5

TABLE 9 Freon Compound, R125, CF_3CHF_2

													B/A^.5	0.028651		
R125 CF3CHF2					LINEAR	MODEL							B	0.081073	A^(1/2)	2.829615
-H	0	0.066	0	0.018	0	-0.027	0	-0.019	0	-0.019	0		-H	0	mol wt	120.0219
-CH3	0	0.142	0	-0.02	0	-0.016	0	-0.023	0	-0.015	0		-CH3	0	E rho g/cc	0.475841
-CH2-	0	0.076	0	-0.038	0	0.011	0	-0.004	0	0.003	0		-CH2-	0	LOG(E)	0.822765
>CH-	1	0.003	0.003	-0.058	-0.058	0.053	0.053	0.018	0.018	0.015	0.015		>CH-	1	CHI	45.7351
>C<	0	-0.075	0	-0.076	0	0.091	0	0.043	0	0.034	0		>C<	0	-LOGKb	13.35729
-CH=CH-	0	0.147	0	-0.043	0	0.028	0	0.01	0	0.003	0		-CH=CH-	0	DIPOLE M	1.77575
-CH=CH2	0	0.212	0	-0.025	0	0	0	-0.009	0	-0.015	0		-CH=CH2	0	SUM INT	4
>C=CH2	0	0.147	0	-0.043	0	0.028	0	0.01	0	0.003	0		>C=CH2	0	Tb-Tm	-30.124
-C=-CH	0	0.171	0	0.074	0	0.027	0	0.002	0	-0.012	0		-C=-CH	0	MOL. WT.	36.7663
-C6H5	0	0.467	0	-0.007	0	0.012	0	0.007	0	-0.017	0		-C6H5	0	COUNT	1.5
=CH-	0	0.088	0	0.002	0	-0.007	0	0.001	0	-0.003	0	#GRP/	=CH-	0	logben/h2	-0.3898
AROMAT	0		0		0		0		0		0	CPD	AROM	0	Tm	-5.0445
-NAPHTH	0	0.766	0	0.018	0	-0.026	0	0.024	0	-0.028	0	log1/s25C	-NAPTH	0	B rho g/cc	1.087857
-CYCLOH	0	0.489	0	-0.148	0	0.004	0	-0.029	0	-0.009	0	-1.39349	-CYCLOH	0	B	-0.1975
-F	2	0.078	0.156	0.088	0.176	0.009	0.018	-0.019	-0.038	-0.02	-0.04	logBCF	-F	2	C	0.079
-CL	0	0.165	0	0.087	0	-0.024	0	-0.012	0	-0.021	0	-0.39156	-CL	0	D	0.078
-BR	0	0.213	0	0.095	0	-0.033	0	-0.008	0	-0.02	0	logKoc	-BR	0	E	-0.026
-I	0	0.302	0	0.103	0	-0.056	0	-0.01	0	-0.031	0	1.261359	-I	0	F	-0.01
-CF3	1	0.15	0.15	0.017	0.017	0.035	0.035	-0.037	-0.037	-0.013	-0.013		CF3	1	Tb	-35.1685
-CCL3	0	0.41	0	0.015	0	-0.009	0	-0.017	0	-0.017	0		-CCL3	0	SOL PAR	8.13735
-OH	0	0.202	0	0.324	0	-0.012	0	-0.015	0	0.003	0		-OH	0	Tc	71.1129
-O-	0	0.044	0	0.155	0	0.061	0	0.019	0	-0.022	0		-O-	0	**Pc**	36.81405
-C=0	0	0.135	0	0.246	0	0.061	0	0.023	0	-0.021	0		-C=O	0	log AC CC	1.313385
-CH=O-	0	0.219	0	0.244	0	0.01	0	-0.014	0	-0.027	0		-CH=O	0	log par co	-0.21257
-COO-	0	0.167	0	0.17	0	0.062	0	0.015	0	-0.027	0		-COO-	0	Hv	5.568225
-COOH	0	0.323	0	0.342	0	-0.011	0	-0.017	0	0.008	0		-COOH	0	Surf Ten	-8.0975
-NH2	0	0.167	0	0.269	0	0.037	0	0.027	0	-0.014	0		-NH2	0	log visc	-0.52568
-NH-	0	0.082	0	0.251	0	0.095	0	0.056	0	-0.01	0		-NH-	0	log(et/h2o	-0.17715
-N-	0	-0.006	0	0.189	0	0.125	0	0.069	0	0.014	0		-N-	0	logoil/h2o	-1.13405
-CN	0	0.241	0	0.269	0	-0.007	0	-0.023	0	-0.041	0		-CN	0	logcyh/h2c	-0.90925
-N=	0	0.102	0	0.183	0	0.031	0	-0.011	0	-0.02	0		-N=	0	logchl/h2o	0.0344
PYRIDINE	0		0		0		0		0		0		PYRIDINE	0	logcl4/h2o	0.10815
-NO2	0	0.238	0	0.241	0	-0.012	0	-0.027	0	-0.037	0		-NO2	0	vol.Bondi	33.797
-CONH2	0	0.444	0	0.499	0	-0.019	0	-0.039	0	-0.012	0		-CONH2	0	v exp	77.266
-S-	0	0.136	0	0.13	0	0.028	0	0.032	0	-0.02	0		-S-	0	C	2
-SH	0	0.231	0	0.155	0	-0.026	0	-0.011	0	-0.013	0		-SH	0	H	1
MOLECUL	1	-0.5065	-0.5065	-0.056	-0.056	0.007	0.007	0.031	0.031	0.028	0.028		MOLECUL	1	O	0
4	4	B =	-0.1975	C =	0.079	D =	0.113	E =	-0.026	F =	-0.01		S	0	N	0
	0							4	Pvp mm	Kirchhoff	Pvpr298	6961.661	CL	0	F	5
		B	C	D	E	F		0.717794	Pvp mm	riedel-P-M	10-1500n	1.000793	BR	0	I	0
-H	0	-0.4405	-0.038	-0.02	0.012	0.009			Pvp mm	riedel-P-M	>1500mm	1.000795	loggamma	1.315985	D1	-0.02644
-CH3	0	-0.3645	-0.076	-0.009	0.008	0.013			T	500	G1	0.305207	m.refract	9.064385	LC50mg/L	143.012
ch2	0	-0.4305	-0.094	0.018	0.027	0.031					G2	0.285207	thermcond	3.547475	isothcom	19.1971
'>CH-	0	-0.5035	-0.114	0.06	0.049	0.043					h	0.128412	COST $/g	0.18	w	-0.22208
>C<	0	-0.5815	-0.132	0.098	0.074	0.062					g1	-0.03793	MCL	0.288156	Zc	0.308767
-CH=CH-	0	-0.3595	-0.099	0.035	0.041	0.031					g2	-0.02472	TAP	0.064154	AIR	0.089323
-CH=CH2	0	-0.2945	-0.081	0.007	0.022	0.013							COM	2.0279	RES	2.357789
>C=CH2	0	-0.3595	0.099	0.035	0.041	0.031							drug1	4.068509	drug2	184.0354
-C=-CH	0	-0.3355	0.018	0.034	0.033	0.016							drug4	-5.28664	drug3	3.949993
-C6H5	0	-0.0395	-0.063	0.019	0.038	0.011							drug6	4.979829	drug5	9.716284
=CH-	0	-0.4185	-0.054	0	0.032	0.025							C-E	0.105	B-D	-0.2755
-NAPTH	0	0.2595	-0.038	-0.019	0.055	0							B-C	-0.2765	sum grps	4

TABLE 10 Freon Compound, R134a, CF_3CH_2F

E	A	B	C	D	E	F	G	H	I	J	K	L	M	N	O	P	Q
72														B/A^.5	0.028651		
73	R134a CF3CH2F					LINEAR	MODEL							B	0.081073	A^(1/2)	2.829615
74	-H	0	0.066	0	0.018	0	-0.027	0	-0.019	0	-0.019	0		-H	0	mol wt	120.0219
75	-CH3	0	0.142	0	-0.02	0	-0.016	0	-0.023	0	-0.015	0		-CH3	0	E rho g/cc	0.475841
76	-CH2-	0	0.076	0	-0.038	0	0.011	0	-0.004	0	0.003	0		-CH2-	0	LOG(E)	0.822765
77	>CH-	1	0.003	0.003	-0.058	-0.058	0.053	0.053	0.018	0.018	0.015	0.015		>CH-	1	CHI	45.7351
78	>C<	0	-0.075	0	-0.076	0	0.091	0	0.043	0	0.034	0		>C<	0	-LOGKb	13.35729
79	-CH=CH-	0	0.147	0	-0.043	0	0.028	0	0.01	0	0.003	0		-CH=CH-	0	DIPOLE M	1.77575
80	-CH=CH2	0	0.212	0	-0.025	0	0	0	-0.009	0	-0.015	0		-CH=CH2	0	SUM INT	4
81	>C=CH2	0	0.147	0	-0.043	0	0.028	0	0.01	0	0.003	0		>C=CH2	0	Tb-Tm	-30.124
82	-C=-CH	0	0.171	0	0.074	0	0.027	0	0.002	0	-0.012	0		-C=-CH	0	MOL. WT.	36.7663
83	-C6H5	0	0.467	0	-0.007	0	0.012	0	0.007	0	-0.017	0		-C6H5	0	COUNT	1.5
84	=CH-	0	0.088	0	0.002	0	-0.007	0	0.001	0	-0.003	0	#GRP/	=CH-	0	logben/h2	-0.3898
85	AROMAT	0		0		0		0		0		0	CPD	AROM	0	Tm	-5.0445
86	-NAPHTH	0	0.766	0	0.018	0	-0.026	0	0.024	0	-0.028	0	log1/s25C	-NAPTH	0	B rho g/cc	1.087857
87	-CYCLOH	0	0.489	0	-0.148	0	0.004	0	-0.029	0	-0.009	0	-1.39349	-CYCLOH	0	B	-0.1975
88	-F	2	0.078	0.156	0.088	0.176	0.009	0.018	-0.019	-0.038	-0.02	-0.04	logBCF	-F	2	C	0.079
89	-CL	0	0.165	0	0.087	0	-0.024	0	-0.012	0	-0.021	0	-0.39156	-CL	0	D	0.078
90	-BR	0	0.213	0	0.095	0	-0.033	0	-0.008	0	-0.02	0	logKoc	-BR	0	E	-0.026
91	-I	0	0.302	0	0.103	0	-0.056	0	-0.01	0	-0.031	0	1.261359	-I	0	F	-0.01
92	-CF3	1	0.15	0.15	0.017	0.017	0.035	0.035	-0.037	-0.037	-0.013	-0.013		CF3	1	Tb	-35.1685
93	-CCL3	0	0.41	0	0.015	0	-0.009	0	-0.017	0	-0.017	0		-CCL3	0	SOL PAR	8.13735
94	-OH	0	0.202	0	0.324	0	-0.012	0	-0.015	0	0.003	0		-OH	0	Tc	71.1129
95	-O-	0	0.044	0	0.155	0	0.061	0	0.019	0	-0.022	0		-O-	0	Pc	36.81405
96	-C=0	0	0.135	0	0.246	0	0.061	0	0.023	0	-0.021	0		-C=O	0	log AC CO	1.313385
97	-CH=O-	0	0.219	0	0.244	0	0.01	0	-0.014	0	-0.027	0		-CH=O	0	log par co	-0.21257
98	-COO-	0	0.167	0	0.17	0	0.062	0	0.015	0	-0.027	0		-COO-	0	Hv	5.568225
99	-COOH	0	0.323	0	0.342	0	-0.011	0	-0.017	0	0.008	0		-COOH	0	Surf Ten	-8.0975
100	-NH2	0	0.167	0	0.269	0	0.037	0	0.027	0	-0.014	0		-NH2	0	log visc	-0.52568
101	-NH-	0	0.082	0	0.251	0	0.095	0	0.056	0	-0.01	0		-NH-	0	log(et/h2o	-0.17715
102	-N-	0	-0.006	0	0.189	0	0.125	0	0.069	0	0.014	0		-N-	0	logoil/h2o	-1.13405
103	-CN	0	0.241	0	0.269	0	-0.007	0	-0.023	0	-0.041	0		-CN	0	logcyh/h2o	-0.90925
104	-N=	0	0.102	0	0.183	0	0.031	0	-0.011	0	-0.02	0		-N=	0	logchl/h2o	0.0344
105	PYRIDINE	0		0		0		0		0		0		PYRIDINE	0	logcl4/h2o	0.10815
106	-NO2	0	0.238	0	0.241	0	-0.012	0	-0.027	0	-0.037	0		-NO2	0	vol.Bondi	33.797
107	-CONH2	0	0.444	0	0.499	0	-0.019	0	-0.039	0	-0.012	0		-CONH2	0	v exp	77.266
108	-S-	0	0.136	0	0.13	0	0.028	0	0.032	0	-0.02	0		-S-	0	C	2
109	-SH	0	0.231	0	0.155	0	-0.026	0	-0.011	0	-0.013	0		-SH	0	H	1
110	MOLECUL	1	-0.5065	-0.5065	-0.056	-0.056	0.007	0.007	0.031	0.031	0.028	0.028		MOLECUL	1	O	0
111	4	4	B =	-0.1975	C =	0.079	D =	0.113	E =	-0.026	F =	-0.01		S	0	N	0
112		0							4	Pvp mm	Kirchhoff	Pvpr298	6961.661	CL	0	F	5
113			B	C	D	E	F		0.717794	Pvp mm	riedel-P-M	10-1500m	1.000793	BR	0	I	0
114	-H	0	-0.4405	-0.038	-0.02	0.012	0.009			Pvp mm	riedel-P-M	>1500mm	1.000795	loggamma	1.315985	D1	-0.02644
115	-CH3	0	-0.3645	-0.076	-0.009	0.008	0.013			T	500	G1	0.305207	m.refract	9.064385	LC50mg/L	143.012
116	ch2	0	-0.4305	-0.094	0.018	0.027	0.031					G2	0.285207	thermcond	3.547475	isothcom	19.1971
117	'>CH-	0	-0.5035	-0.114	0.06	0.049	0.043					h	0.128412	COST $/g	0.18	w	-0.22208
118	>C<	0	-0.5815	-0.132	0.098	0.074	0.062					g1	-0.03793	MCL	0.288156	Zc	0.308767
119	-CH=CH-	0	-0.3595	-0.099	0.035	0.041	0.031					g2	-0.02472	TAP	0.064154	AIR	0.089323
120	-CH=CH2	0	-0.2945	-0.081	0.007	0.022	0.013							COM	2.0279	RES	2.357789
121	>C=CH2	0	-0.3595	0.099	0.035	0.041	0.031							drug1	4.068509	drug2	184.0354
122	-C=-CH	0	-0.3355	0.018	0.034	0.033	0.016							drug4	-5.28664	drug3	3.949993
123	-C6H5	0	-0.0395	-0.063	0.019	0.038	0.011							drug6	4.979829	drug5	9.716284
124	=CH-	0	-0.4185	-0.054	0	0.032	0.025							C-E	0.105	B-D	-0.2755
125	-NAPTH	0	0.2595	-0.038	-0.019	0.055	0							B-C	-0.2765	sum grps	4

TABLE 11 Hypothetical Example

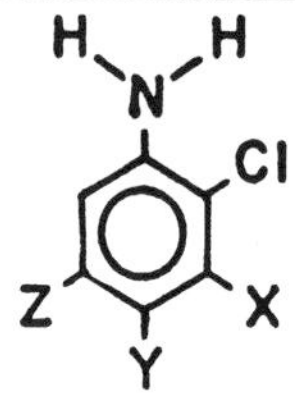

I

Compound	Substituents at Position: X^a H	X^a F	X^a Cl	X^a Br	Y H	Y NO_2	Y CN	Y $COCH_3$	Z^a H	Z^a CH_3	Z^a C_2H_5	Z^a OCH_3	Log 1/C Observed	Log 1/C Calculated
1	X				X				X				3.095	3.120
2		X				X			X				6.000	6.002
3			X				X				X		5.201	5.203
4				X				X				X	5.631	5.593
5	X				X							X	3.425	3.387
6		X				X					X		5.420	5.451
7			X				X			X			5.401	5.377
8				X				X	X				5.390	5.327
P-1		X				X						X		6.269
P-2		X					X					X		6.156
P-3			X			X				X				5.487

[a] *Refers to substituent positions in 1.*

TABLE 12 Tetracycline with Additions at the X, Y, and R Positions

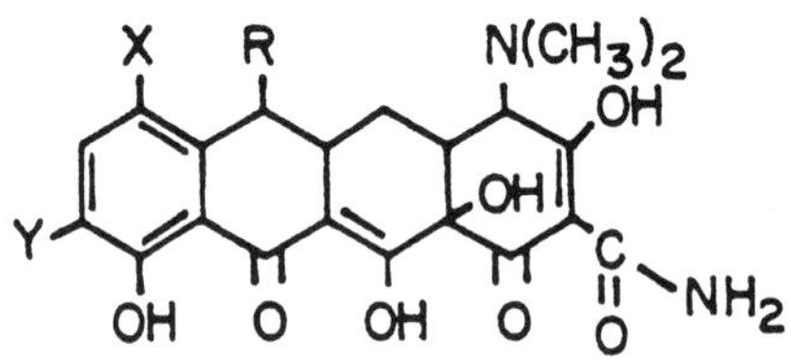

Compound Number	Substituents
I	R = H, X = Y = NO_2
II	R = H, X = Cl, Y = NO_2
III	R = H, X = Br, Y = NO_2
IV	R = H, X = Cl, Y = NH_2
V	R = H, X = Br, Y = NH_2
VI	R = H, X = NO_2, Y = NH_2
VII	R = CH_3, X = NO_2, Y = NH_2
VIII	R = CH_3, X = NO_2, Y = CH_3CONH
IX	R = CH_3, X = Br, Y = NH_2
X	R = CH_3, X = Br, Y = CH_3CONH

TABLE 13 Biological Activities of Compounds of Table 12

	Substituent Positions								Biological Activity	
	R		X			Y				
Cpd.	H	CH_3	NO_2	Cl	Br	NO_2	NH_2	CH_3CONH	Observed	Calculated
I	1		1			1			60	–8
II	1			1		1			21	102
III	1				1	1			15	2
IV	1			1			1		525	444
V	1				1		1		320	343
VI	1		1				1		275	333
VII		1	1				1		160	145
VIII		1	1					1	15	40
IX		1			1		1		140	155
X		1			1			1	75	50
P-I	1				1			1		2.37
P-II	1			1				1		

TABLE 14 Another Hypothetical System

	X				Y				Z				log 1/C		
Cpd.	H	F	Cl	Br	H	NO_2	CN	$COCH_3$	H	CH_3	C_2H_5	OCH_3	obs	calc	Δ log 1/C
1	1				1				1				3.095	3.086	0.009
2	1				1					1			2.763	2.759	0.004
3	1					1					1		4.556	4.556	–0.010
4	1					1						1	5.405	5.407	–0.002
5		1					1		1				5.921	5.921	0.0
6		1					1			1			5.625	5.593	0.032
7		1						1			1		4.931	4.912	0.019
8		1						1				1	5.701	5.752	–0.051
9			1		1				1				3.852	3.826	0.026
10			1		1					1			3.460	3.498	–0.038
11			1			1					1		5.332	5.306	0.026
12			1			1						1	6.132	6.146	–0.014
13				1			1		1				5.701	5.735	–0.035
14				1			1			1			5.409	5.407	0.002
15				1				1			1		4.690	4.725	–0.035
16				1				1				1	5.631	5.565	0.066
17	1						1					1	4.793	4.793	0.0
18			1			1			1				5.121	5.121	0.0
P-1		1					1					1		6.945	
P-2		1				1						1		7.559	
P-3				1			1					1		6.759	
P-4				1		1						1		7.373	

TABLE 15 Derivatives of Piperidinesulfamylsemicarbazide

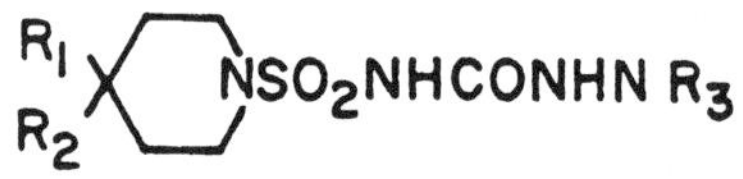

Compound	Substituents at R_1, R_2: H	CH_3	C_2H_5	$(CH_2)_2$	Substituents at R_3: $(CH_2)_5$	$(CH_2)_6$	Biological Activity
I	2				1		14.8
II	2					1	11.1
III	1	1			1		26.1
IV	1	1				1	33.9
V		2			1		39.1
VI		2				1	34.9
VII		1	1			1	42.0
VIII			2			1	34.4
IX				2	1		35.6
X				2		1	30.8

Final Matrix of the Observed Hypoglycemic Activities

Compound	Substituents at R_1, R_2: H	CH_3	$(CH_2)_2$[a]	Substituent at R_3: $(CH_2)_6$	Biological Activity[b]
I	2			−1.5	14.8
II	2			1	11.1
III	1	1		−1.5	26.1
IV	1	1		1	33.9
V		2		−1.5	39.1
VI		2		1	34.9
VII	−2	−1.33	−1.33	1	42.0
VIII	−4	−4.33	−2.66	1	34.4
IX			2	−1.5	35.6
X			2	1	30.8

TABLE 16 Derivatives of N,N-Dimethyl-2-Bromophenethylamines

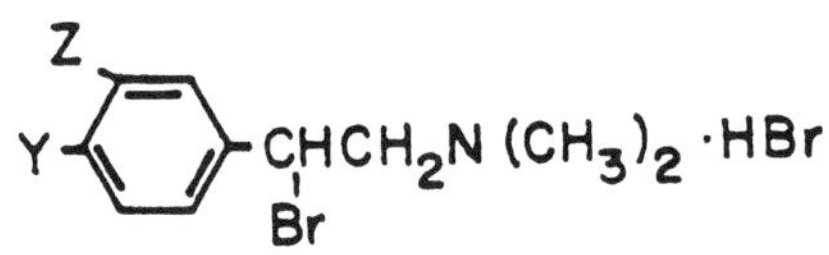

Compound	Substituent at X	Y	Log 1/C Observed	Calculated	Δ log 1/C[c]
9	H	H	7.46	7.89	−0.43
10	F	H	8.16	7.98	0.18
11	Cl	H	8.68	8.38	0.30
12	Br	H	8.89	8.77	0.12
13	I	H	9.25	8.98	0.27
14	CH_3	H	9.30	8.79	0.51
15	H	F	7.52	7.51	0.01
16	H	Cl	8.16	8.23	−0.07
17	H	Br	8.30	8.42	−0.12
18	H	I	8.40	8.74	−0.34
19	H	CH_3	8.46	8.62	−0.16
20	F	Cl	8.19	8.32	−0.13
21	F	Br	8.57	8.51	0.06
22	F	CH_3	8.82	8.71	0.11
23	Cl	Cl	8.89	8.72	0.17
24	Cl	Br	8.92	8.91	0.01
25	Cl	CH_3	8.96	9.11	−0.15
26	Br	Cl	9.00	9.11	−0.11
27	Br	Br	9.35	9.30	0.05
28	Br	CH_3	9.22	9.49	−0.27
29	CH_3	CH_3	9.30	9.53	−0.23
30	CH_3	Br	9.52	9.32	0.20
P-4	I	I		9.83	
P-5	I	CH_3		9.72	

TABLE 17 Inhibitors of Monoamine Oxidase

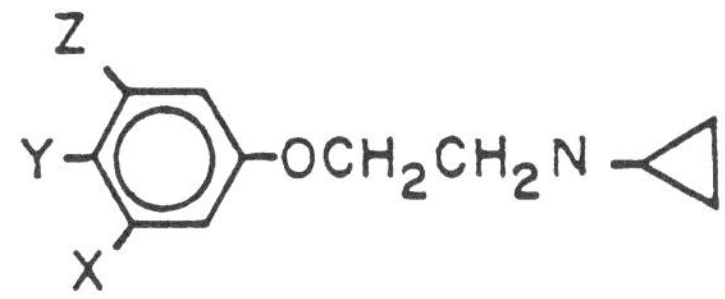

Compound	Substituent at X	Y	Z	Log 1/C Observed	Calculated	Δ log 1/C[c]
31	H	Br	H	6.64	6.40	0.24
32	Cl	Cl	H	6.30	6.46	–0.16
33	NO_2	H	H	5.76	5.68	0.08
34	CF_3	H	H	4.98	5.22	–0.24
35	H	CH_3	H	5.69	5.65	0.04
36	Cl	H	Cl	5.68	6.10	–0.42
37	Cl	CH_3	H	5.75	5.72	0.03
38	Br	H	H	5.64	5.95	–0.31
39	CH_3	Cl	H	6.06	5.47	0.59
40	Cl	H	H	5.82	6.04	–0.22
41	H	OCH_3	H	5.46	5.47	–0.01
42	CH_3	CH_3	H	4.71	4.73	–0.02
43	CH_3	H	CH_3	4.85	4.12	0.73
44	CH_3	H	H	4.78	5.05	–0.27
45	CH_3	Cl	CH_3	4.70	4.55	0.15
46	CH_3	CH_3	CH_3	3.54	3.81	–0.27
47	H	$N{=}NC_6H_5$	H	7.56	7.16	0.40
48	H	NH_2	H	4.40	4.74	–0.34

TABLE 18 Reaction File

```
ELEMENTS
C H
END
SPECIES
C2H6 CH3 CH4 C2H5 C2H4 H2 H
END
REACTIONS
C2H6=2CH3                1.2E12     0.0     59218.
C2H6+CH3=C2H5+CH4        8.01E8     0.0     11943.
C2H5=C2H4+H              3E11       0.0     34974.
H+C2H6=C2H5+H2           1.3E11     0.0      8982.
H+H=H2                   6.99E13    0.0         0.
END
```

TABLE 19 Input File

```
0
1162 .4507
0.0d-6 20000000d-6 1000000d-6
C2H6
1.0 011
C2H5
0.0 011
C2H4
0.0 011
CH4
0.0 011
CH3
0.0 011
H2
0.0 011
H
0.0 011
END
```

TABLE 20 Spec.out

The running time in DOS on a 486 PC is less than 2 secs. The output (Spec.out) is

t[usec]	T[K]	P[atm}	C2H6	CH3	CH4	C2H5	C2H4
			H2	H			
0	1900.0	.451	1.000E+00	0.000E+00	0.000E+00	0.000E+00	0.000E+00
			0.000E+00	0.000E+00			
30000.0	1900.0	.451	1.932E-01	1.086E-01	1.613E-01	1.290E-05	3.086E-01
			2.277E-01	5.656E-04			
60000.0	1900.0	.451	7.380E-02	6.712E-02	1.773E-01	1.802E-05	3.850E-01
			2.961E-01	6.408E-04			
90000.0	1900.0	.451	3.421E-01	4.579E-02	1.798E-01	2.027E-05	4.149E-01
			3.247E-01	6.703E-04			

from which we can plot 2 and 3 dimensional graphs.

TABLE 21 Intp.out

```
 OUTPUT FROM INTERPRETER


        ELEMENTS CONSIDERED

            ELEMENTS            ATOMIC WEIGHT

          1.  C                12.0112
          2.  H                1.00797


        SPECIES CONSIDERED

            SPECIES       PHASE  CHARGE  MOLECULAR WEIGHT  LOW TEMP  HIGH TEMP

          1.  C2H6          G       0       30.0701         300.00    4000.00
          2.  CH3           G       0       15.0351         300.00    5000.00
          3.  CH4           G       0       16.0430         300.00    5000.00
          4.  C2H5          G       0       29.0622         300.00    5000.00
          5.  C2H4          G       0       28.0542         300.00    5000.00
          6.  H2            G       0       2.01594         300.00    5000.00
          7.  H             G       0       1.00797         300.00    5000.00


        REACTIONS CONSIDERED

1.  C2H6=2CH3
2.  C2H6+CH3=C2H5+CH4
3.  C2H5=C2H4+H
4.  H+C2H6=C2H5+H2
5.  H+H=H2

        NO ERRORS WERE ENCOUNTERED ON INPUT CARDS.
        ALL INPUT WAS PROCESSED, AND THE LINKING FILE CREATED
```

TABLE 22 NASA File

```
  File  Edit  Search  Options                                                Help
                                      SANDIA.DAT
C2H4                 121286C   2H    4          G  0300.00   5000.00  1000.00
 3.52841878E+00 1.14851845E-02-4.41838529E-06 7.84460053E-10-5.26684849E-14
 4.42828857E+03 2.23038912E+00-8.61487985E-01 2.79616285E-02-3.38867721E-05
 2.78515220E-08-9.73787891E-12 5.57304590E+03 2.42114868E+01
C2H5                  12387C   2H    5          G  0300.00   5000.00  1000.00
 7.19048023E+00 6.48407731E-03-6.42806469E-07-2.34787939E-10 3.88087728E-14
 1.06745488E+04-1.47808924E+01 2.69070172E+00 8.71913321E-03 4.41983866E-06
 9.33870314E-10-3.92777347E-12 1.28704043E+04 1.21381950E+01
C2H6                 121686C   2H    6          G  0300.00   4000.00  1000.00
 4.82593823E+00 1.38404286E-02-4.55725876E-06 6.72496725E-10-3.59816140E-14
-1.27177930E+04-5.23950672E+00 1.46253872E+00 1.54946670E-02 5.78050731E-06
-1.25783188E-08 4.58626713E-12-1.12391758E+04 1.44322949E+01
C2N                  121286C   2N    1          G  0300.00   5000.00  1000.00
 6.15156126E+00 1.51164969E-03-6.62936202E-07 1.28614841E-10-9.16082988E-15
 6.48431797E+04-8.17784977E+00 3.49854446E+00 8.55443254E-03-6.28869702E-06
 8.63847771E-10 4.91599600E-13 6.55661094E+04 5.54837418E+00
C2N2                 121286C   2N    2          G  0300.00   5000.00  1000.00
 6.54800272E+00 3.98470741E-03-1.63421635E-06 3.03859660E-10-2.11106902E-14
 3.49071641E+04-9.73579025E+00 4.26545906E+00 1.19225690E-02-1.34201418E-05
 9.19229670E-09-2.77894179E-12 3.54788750E+04 1.71321237E+00
C2O                  121286C   2O    1          G  0300.00   5000.00  1000.00
```

TABLE 23 Rates.out File

```
 time (sec) =    0.0000E+00   T [K] =  1162.0   P (atm) =  .4507
R#    For rate     Rev rate    log Net dif/sum Rank Reaction
 1    4.122E-05    1.152E-46    -4.38    1.000   1   C2H6=2CH3
 2    2.147E-29    2.824E-55   -28.67    1.000   4   C2H6+CH3=C2H5+CH4
 3    7.918E-26    6.615E-49   -25.10    1.000   2   C2H5=C2H4+H
 4    1.256E-26    7.418E-54   -25.90    1.000   3   H+C2H6=C2H5+H2
 5    6.990E-47    6.543E-36   -35.18   -1.000   5   2H=H2

 time (sec) =    1.0000E+00   T [K] =  1162.0   P (atm) =  .4507
R#    For rate     Rev rate    log Net dif/sum Rank Reaction
 1    3.898E-05    3.897E-05    -8.23     .000   4   C2H6=2CH3
 2    1.181E-08    3.819E-14    -7.93    1.000   3   C2H6+CH3=C2H5+CH4
 3    8.814E-07    7.605E-07    -6.92     .074   1   C2H5=C2H4+H
 4    1.091E-07    9.830E-12    -6.96    1.000   2   H+C2H6=C2H5+H2
 5    5.901E-09    7.789E-13    -8.23    1.000   5   2H=H2

 time (sec) =    2.0000E+00   T [K] =  1162.0   P (atm) =  .4507
R#    For rate     Rev rate    log Net dif/sum Rank Reaction
 1    3.699E-05    3.699E-05    -8.26     .000   4   C2H6=2CH3
 2    1.092E-08    1.227E-13    -7.96    1.000   3   C2H6+CH3=C2H5+CH4
 3    1.492E-06    1.381E-06    -6.96     .038   1   C2H5=C2H4+H
 4    9.961E-08    3.140E-11    -7.00     .999   2   H+C2H6=C2H5+H2
 5    5.457E-09    1.470E-12    -8.26     .999   5   2H=H2

 time (sec) =    3.0000E+00   T [K] =  1162.0   P (atm) =  .4507
R#    For rate     Rev rate    log Net dif/sum Rank Reaction
 1    3.521E-05    3.521E-05    -8.29     .000   4   C2H6=2CH3
 2    1.014E-08    2.343E-13    -7.99    1.000   3   C2H6+CH3=C2H5+CH4
 3    1.993E-06    1.891E-06    -6.99     .026   1   C2H5=C2H4+H
 4    9.139E-08    5.960E-11    -7.04     .999   2   H+C2H6=C2H5+H2
 5    5.069E-09    2.089E-12    -8.30     .999   5   2H=H2

 time (sec) =    4.0000E+00   T [K] =  1162.0   P (atm) =  .4507
R#    For rate     Rev rate    log Net dif/sum Rank Reaction
 1    3.361E-05    3.361E-05    -8.32     .000   4   C2H6=2CH3
 2    9.457E-09    3.605E-13    -8.02    1.000   3   C2H6+CH3=C2H5+CH4
 3    2.408E-06    2.314E-06    -7.03     .020   1   C2H5=C2H4+H
 4    8.425E-08    9.120E-11    -7.07     .998   2   H+C2H6=C2H5+H2
 5    4.728E-09    2.646E-12    -8.33     .999   5   2H=H2

 time (sec) =    5.0000E+00   T [K] =  1162.0   P (atm) =  .4507
R#    For rate     Rev rate    log Net dif/sum Rank Reaction
 1    3.216E-05    3.216E-05    -8.35     .000   4   C2H6=2CH3
 2    8.851E-09    4.934E-13    -8.05    1.000   3   C2H6+CH3=C2H5+CH4
 3    2.753E-06    2.666E-06    -7.06     .016   1   C2H5=C2H4+H
 4    7.799E-08    1.242E-10    -7.11     .997   2   H+C2H6=C2H5+H2
 5    4.426E-09    3.150E-12    -8.35     .999   5   2H=H2
```

TABLE 24 Sense.out File

```
 time (sec) =    0.0000E+00   T [K] =  1162.0   P (atm) =  .4507
R#  Reaction                 C2H6       C2H5       C2H4       CH4        CH3
  1 C2H6=2CH3             -4.727E-06  0.000E+00  0.000E+00  0.000E+00  9.454E-06
  2 C2H6+CH3=C2H5+CH4      0.000E+00  0.000E+00  0.000E+00  0.000E+00  0.000E+00
  3 C2H5=C2H4+H            0.000E+00  0.000E+00  0.000E+00  0.000E+00  0.000E+00
  4 H+C2H6=C2H5+H2         0.000E+00  0.000E+00  0.000E+00  0.000E+00  0.000E+00
  5 2H=H2                  0.000E+00  0.000E+00  0.000E+00  0.000E+00  0.000E+00

 time (sec) =    1.0000E+00   T [K] =  1162.0   P (atm) =  .4507
R#  Reaction                 C2H6       C2H5       C2H4       CH4        CH3
  1 C2H6=2CH3             -6.777E-10  0.000E+00  0.000E+00  0.000E+00  1.355E-09
  2 C2H6+CH3=C2H5+CH4     -2.600E-15  2.600E-15  0.000E+00  2.600E-15 -2.600E-15
  3 C2H5=C2H4+H            0.000E+00 -1.527E-12  1.527E-12  0.000E+00  0.000E+00
  4 H+C2H6=C2H5+H2        -4.107E-17  4.107E-17  0.000E+00  0.000E+00  0.000E+00
  5 2H=H2                  0.000E+00  0.000E+00  0.000E+00  0.000E+00  0.000E+00

 time (sec) =    2.0000E+00   T [K] =  1162.0   P (atm) =  .4507
R#  Reaction                 C2H6       C2H5       C2H4       CH4        CH3
  1 C2H6=2CH3             -6.267E-10  0.000E+00  0.000E+00  0.000E+00  1.253E-09
  2 C2H6+CH3=C2H5+CH4     -2.404E-15  2.404E-15  0.000E+00  2.404E-15 -2.404E-15
  3 C2H5=C2H4+H            0.000E+00 -1.395E-12  1.395E-12  0.000E+00  0.000E+00
  4 H+C2H6=C2H5+H2        -3.747E-17  3.747E-17  0.000E+00  0.000E+00  0.000E+00
  5 2H=H2                  0.000E+00  0.000E+00  0.000E+00  0.000E+00  0.000E+00

 time (sec) =    3.0000E+00   T [K] =  1162.0   P (atm) =  .4507
R#  Reaction                 C2H6       C2H5       C2H4       CH4        CH3
  1 C2H6=2CH3             -5.815E-10  0.000E+00  0.000E+00  0.000E+00  1.163E-09
  2 C2H6+CH3=C2H5+CH4     -2.233E-15  2.233E-15  0.000E+00  2.233E-15 -2.233E-15
  3 C2H5=C2H4+H            0.000E+00 -1.281E-12  1.281E-12  0.000E+00  0.000E+00
  4 H+C2H6=C2H5+H2        -3.437E-17  3.437E-17  0.000E+00  0.000E+00  0.000E+00
  5 2H=H2                  0.000E+00  0.000E+00  0.000E+00  0.000E+00  0.000E+00

 time (sec) =    4.0000E+00   T [K] =  1162.0   P (atm) =  .4507
R#  Reaction                 C2H6       C2H5       C2H4       CH4        CH3
  1 C2H6=2CH3             -5.427E-10  0.000E+00  0.000E+00  0.000E+00  1.085E-09
  2 C2H6+CH3=C2H5+CH4     -2.082E-15  2.082E-15  0.000E+00  2.082E-15 -2.082E-15
  3 C2H5=C2H4+H            0.000E+00 -1.182E-12  1.182E-12  0.000E+00  0.000E+00
  4 H+C2H6=C2H5+H2        -3.167E-17  3.167E-17  0.000E+00  0.000E+00  0.000E+00
  5 2H=H2                  0.000E+00  0.000E+00  0.000E+00  0.000E+00  0.000E+00

 time (sec) =    5.0000E+00   T [K] =  1162.0   P (atm) =  .4507
R#  Reaction                 C2H6       C2H5       C2H4       CH4        CH3
  1 C2H6=2CH3             -5.077E-10  0.000E+00  0.000E+00  0.000E+00  1.015E-09
  2 C2H6+CH3=C2H5+CH4     -1.949E-15  1.949E-15  0.000E+00  1.949E-15 -1.949E-15
  3 C2H5=C2H4+H            0.000E+00 -1.095E-12  1.095E-12  0.000E+00  0.000E+00
  4 H+C2H6=C2H5+H2        -2.930E-17  2.930E-17  0.000E+00  0.000E+00  0.000E+00
  5 2H=H2                  0.000E+00  0.000E+00  0.000E+00  0.000E+00  0.000E+00
```

TABLE 25 Estimated Rate Constants for Reactions

E	A	B	C	D	E	F	G	H
154	formula						cc/mole-sec	
155	2 atoms						1E+16	1E+15
156	atom&linear molecule, linear complex						1E+12	1E+11
157	atom&linear molecule, linear complex						3.5E+09	3.5E+08
158	atom&linear molecule, nonlinear complex						1E+14	1E+13
159	atom&linear molecule, nonlinear complex						6E+12	6E+11
160	atom&nonlinear molecule, nonlinear complex						1E+12	1E+11
161	atom&nonlinear molecule, nonlinear complex						3.5E+09	3.5E+08
162	two linear molecules, linear complex						1E+08	1E+07
163	two linear molecules, linear complex						1160	116
164	two linear molecules, nonlinear complex						1E+10	1E+09
165	two linear molecules, nonlinear complex						2010000	201000
166	one linear&one non linear molecule, nonlinear complex						1E+08	1E+07
167	one linear&one nonlinear molecule, nonlinear complex						1160	116
168	two nonlinear molecules. nonlinear complex						1040000	104000
169	two nonlinear molecules, nonlinear complex						0.11	0.0111

TABLE 26 Thermodynamic Data at 298 K

THERMODYNAMIC DATA AT 298 K

The GRI-Mech thermochemistry is based upon standard databases, including the NASA-Lewis and Technion archives. This was supplemented by a critical review of the standard enthalpies of formation for radical species.

Clicking on the species symbols in the following summary table will bring up a full table of the molar heat capacity, entropy and standard enthalpy of formation values for that species at 100 K intervals from 200 to 3000 K, as computed from the polynomial coefficients , and citations for the source of the data.

Species Cp
(cal/mol K) S
(cal/mol K) delta-Hf
(kcal/mol) Ar 4.97 36.98 0.00 C 4.98 37.79 171.29 C2H 10.04 50.99 135.33 C2H2 10.52 48.02 54.54 C2H3 10.21 55.90 71.64 C2H4 10.25 52.42 12.55 C2H5 12.07 59.06 28.36 C2H6 12.55 54.79 -20.04 CH 6.97 43.75 142.77 CH2 8.37 46.47 93.77 CH2(S) 8.00 45.23 102.75 CH2CO 12.37 57.81 -11.40 CH2O 8.46 52.29 -25.95 CH2OH 11.56 58.30 -3.50 CH3 9.18 46.37 35.11 CH3O 9.05 54.58 3.89 CH3OH 10.53 57.32 -48.03 CH4 8.53 44.54 -17.83 CN 6.97 48.43 104.85 CO 6.97 47.24 -26.42 CO2 8.88 51.10 -94.05 H 4.97 27.42 52.10 H2 6.89 31.23 0.00 H2CN 9.14 53.59 59.11 H2O 8.03 45.13 -57.80 H2O2 10.13 56.05 -33.48 HCCO 11.57 58.96 42.40 HCCOH 13.93 56.25 18.70 HCN 8.57 48.23 32.27 HCNN 11.73 59.77 110.45 HCNO 11.18 58.00 40.88 HCO 8.27 53.62 10.04 HNCO 11.11 57.53 -28.22 HNO 8.10 52.80 25.40 HO2 8.34 54.76 3.00 HOCN 11.06 57.86 -2.82 N 4.97 36.64 112.97 N2 6.96 45.76 0.00 N2O 9.23 52.58 19.50 NCO 9.58 55.54 31.50 NH 6.98 43.31 85.30 NH2 8.09 46.60 45.90 NH3 8.52 46.07 -10.97 NNH 8.28 53.65 59.64 NO 7.14 50.37 21.81 NO2 8.88 57.40 8.17 O 5.24 38.49 59.55 O2 7.02 49.03 0.00 OH 7.14 43.92 9.40

[the thermo data file | GRI-Mech main menu]

TABLE 27A C–H–O Reactions

Number Reaction Rate Coefficient A T^n exp(-E/RT) Served as Optimization Variable in GRI-Mech Releases A (mol,cm3,s) n (T in K) E (cal/mol) 1 O + O + M -> O2 + M 1.20E+17 -1.0 0.0   2 O + H + M -> OH + M 5.00E+17 -1.0 0.0   3 O + H2 -> H + OH 5.00E+04 2.67 6290   4 O + HO2 -> OH + O2 2.00E+13 0.0 0.0   5 O + H2O2 -> OH + HO2 9.63E+06 2.0 4000   6 O + CH -> H + CO 5.70E+13 0.0 0.0   7 O + CH2 -> H + HCO 8.00E+13 0.0 0.0   8 O + CH2(S) -> H2 + CO 1.50E+13 0.0 0.0   9 O + CH2(S) -> H + HCO 1.50E+13 0.0 0.0   10 O + CH3 -> H + CH2O 8.43E+13 0.0 0.0 1.1; 1.2; 2.1 11 O + CH4 -> OH + CH3 1.02E+09 1.5 8600 1.2; 2.1 12 O + CO + M -> CO2 + M 6.02E+14 0.0 3000   13 O + HCO -> OH + CO 3.00E+13 0.0 0.0   14 O + HCO -> H + CO2 3.00E+13 0.0 0.0   15 O + CH2O -> OH + HCO 3.90E+13 0.0 3540   16 O + CH2OH -> OH + CH2O 1.00E+13 0.0 0.0   17 O + CH3O -> OH + CH2O 1.00E+13 0.0 0.0   18 O + CH3OH -> OH + CH2OH 3.88E+05 2.5 3100   19 O + CH3OH -> OH + CH3O 1.30E+05 2.5 5000   20 O + C2H -> CH + CO 5.00E+13 0.0 0.0   21 O + C2H2 -> H + HCCO 1.02E+07 2.0 1900   22 O + C2H2 -> OH + C2H 4.60E+19 -1.41 28950   23 O + C2H2 -> CO + CH2 1.02E+07 2.0 1900   24 O + C2H3 -> H + CH2CO 3.00E+13 0.0 0.0   25 O + C2H4 -> CH3 + HCO 1.92E+07 1.83 220   26 O + C2H5 -> CH3 + CH2O 1.32E+14 0.0 0.0   27 O + C2H6 -> OH + C2H5 8.98E+07 1.92 5690   28 O + HCCO -> H + CO + CO 1.00E+14 0.0 0.0   29 O + CH2CO -> OH + HCCO 1.00E+13 0.0 8000   30 O + CH2CO -> CH2 + CO2 1.75E+12 0.0 1350   31 O2 + CO -> O + CO2 2.50E+12 0.0 47800   32 O2 + CH2O -> HO2 + HCO 1.00E+14 0.0 40000 1.2; 2.1 33 H + O2 + M -> HO2 + M 2.80E+18 -0.86 0.0 2.1 34 H + O2 + O2 -> HO2 + O2 3.00E+20 -1.72 0.0   35 H + O2 + H2O -> HO2 + H2O 9.38E+18 -0.76 0.0 1.1; 1.2; 2.1 36 H + O2 + N2 -> HO2 + N2 3.75E+20 -1.72 0.0 2.1 37 H + O2 + AR -> HO2 + AR 7.00E+17 -0.8 0.0 1.2; 2.1 38 H + O2 -> O + OH 8.30E+13 0.0 14413 1.1; 1.2; 2.1 39 H + H + M -> H2 + M 1.00E+18 -1.0 0.0   40 H + H + H2 -> H2 + H2 9.00E+16 -0.6 0.0   41 H + H + H2O -> H2 + H2O 6.00E+19 -1.25 0.0   42 H + H + CO2 -> H2 + CO2 5.50E+20 -2.0 0.0   43 H + OH + M -> H2O + M 2.20E+22 -2.0 0.0 1.1; 1.2; 2.1 44 H + HO2 -> O + H2O 3.97E+12 0.0 671   45 H + HO2 -> O2 + H2 2.80E+13 0.0 1068   46 H + HO2 -> OH + OH 1.34E+14 0.0 635   47 H + H2O2 -> HO2 + H2 1.21E+07 2.0 5200   48 H + H2O2 -> OH + H2O 1.00E+13 0.0 3600   49 H + CH -> C + H2 1.10E+14 0.0 0.0 1.2; 2.1 50 H + CH2 (+M) -> CH3 (+M) pressure dependent 1.2; 2.1 51 H + CH2(S) -> CH + H2 3.00E+13 0.0 0.0   52 H + CH3 (+M) -> CH4 (+M) pressure dependent 1.1; 1.2; 2.1 53 H + CH4 -> CH3 + H2 6.60E+08 1.62 10840 1.1; 1.2; 2.1 54 H + HCO (+M) -> CH2O (+M) pressure dependent   55 H + HCO -> H2 + CO 7.34E+13 0.0 0.0 1.1; 1.2; 2.1 56 H + CH2O (+M) -> CH2OH (+M) pressure dependent   57 H + CH2O (+M) -> CH3O (+M) pressure dependent   58 H + CH2O -> HCO + H2 2.30E+10 1.05 3275   59 H + CH2OH (+M) -> CH3OH (+M) pressure dependent   60 H + CH2OH -> H2 + CH2O 2.00E+13 0.0 0.0   61 H + CH2OH -> OH + CH3 1.20E+13 0.0 0.0   62 H + CH2OH -> CH2(S) + H2O 6.00E+12 0.0 0.0   63 H + CH3O (+M) -> CH3OH (+M) pressure dependent   64 H + CH2OH -> CH2(S) + H2O 3.40E+06 1.6 0.0   65 H + CH3O -> H2 + CH2O 2.00E+13 0.0 0.0   66 H + CH3O -> OH + CH3 3.20E+13 0.0 0.0   67 H + CH3O -> CH2(S) + H2O 1.60E+13 0.0 0.0   68 H + CH3OH -> CH2OH + H2 1.70E+07 2.1 4870   69 H + CH3OH -> CH3O + H2 4.20E+06 2.1 4870   70 H + C2H (+M) -> C2H2 (+M) pressure dependent   71 H + C2H2 (+M) -> C2H3 (+M) pressure dependent   72 H + C2H3 (+M) -> C2H4 (+M) pressure dependent   73 H + C2H3 -> H2 + C2H2 3.00E+13 0.0 0.0   74 H +

TABLE 27B C–H–O Reactions

0.0   78 H + C2H6 -> C2H5 + H2 1.15E+08 1.9 7530 1.2; 2.1 79 H + HCCO -> CH2(S) +CO 1.00E+14 0.0 0.0   80 H + CH2CO -> HCCO + H2 5.00E+13 0.0 8000   81 H + CH2CO -> CH3 + CO 1.13E+13 0.0 3428   82 H + HCCOH -> H + CH2CO 1.00E+13 0.0 0.0   83 H2 + CO (+M) -> CH2O (+M) pressure dependent   84 OH + H2 -> H + H2O 2.16E+08 1.51 3430   85 OH + OH (+M) -> H2O2 (+M) pressure dependent   86 OH + OH -> O + H2O 3.57E+04 2.4 -2110   87 OH + HO2 -> O2 + H2O 2.90E+13 0.0 -500 1.2; 2.1 88 OH + H2O2 -> HO2 + H2O 1.75E+12 0.0 320   89 OH + H2O2 -> HO2 + H2O 5.80E+14 0.0 9560   90 OH + C -> H + CO 5.00E+13 0.0 0.0   91 OH + CH -> H + HCO 3.00E+13 0.0 0.0   92 OH + CH2 -> H + CH2O 2.00E+13 0.0 0.0   93 OH + CH2 -> CH + H2O 1.13E+07 2.0 3000   94 OH + CH2(S) -> H + CH2O 3.00E+13 0.0 0.0   95 OH + CH3 (+M) -> CH3OH (+M) pressure dependent   96 OH + CH3 -> CH2 + H2O 5.60E+07 1.6 5420   97 OH + CH3 -> CH2(S) + H2O 2.50E+13 0.0 0.0 1.1; 1.2; 2.1 98 OH + CH4 -> CH3 + H2O 1.00E+08 1.6 3120 1.1; 1.2; 2.1 99 OH + CO -> H + CO2 4.76E+07 1.23 70 1.1; 1.2; 2.1 100 OH + HCO -> H2O + CO 5.00E+13 0.0 0.0   101 OH + CH2O -> HCO + H2O 3.43E+09 1.18 -447   102 OH + CH2OH -> H2O + CH2O 5.00E+12 0.0 0.0   103 OH + CH3O -> H2O + CH2O 5.00E+12 0.0 0.0   104 OH + CH3OH -> CH2OH + H2O 1.44E+06 2.0 -840   105 OH + CH3OH -> CH3O + H2O 6.30E+06 2.0 1500   106 OH + C2H -> H + HCCO 2.00E+13 0.0 0.0   107 OH + C2H2 -> H + CH2CO 2.18E-04 4.5 -1000   108 OH + C2H2 -> H + HCCOH 5.04E+05 2.3 13500   109 OH + C2H2 -> C2H + H2O 3.37E+07 2.0 14000   110 OH + C2H2 -> CH3 + CO 4.83E-04 4.0 -2000   111 OH + C2H3 -> H2O + C2H2 5.00E+12 0.0 0.0   112 OH + C2H4 -> C2H3 + H2O 3.60E+06 2.0 2500   113 OH + C2H6 -> C2H5 + H2O 3.54E+06 2.12 870   114 OH + CH2CO -> HCCO + H2O 7.50E+12 0.0 2000   115 HO2 + HO2 -> O2 + H2O2 1.30E+11 0.0 -1630   116 HO2 + HO2 -> O2 + H2O2 4.20E+14 0.0 12000   117 HO2 + CH2 -> OH + CH2O 2.00E+13 0.0 0.0   118 HO2 + CH3 -> O2 + CH4 1.00E+12 0.0 0.0 1.1; 1.2; 2.1 119 HO2 + CH3 -> OH + CH3O 2.00E+13 0.0 0.0 1.1; 1.2; 2.1 120 HO2 + CO -> OH + CO2 1.50E+14 0.0 23600   121 HO2 + CH2O -> HCO + H2O2 1.00E+12 0.0 8000   122 C + O2 -> O + CO 5.80E+13 0.0 576   123 C + CH2 -> H + C2H 5.00E+13 0.0 0.0   124 C + CH3 -> H + C2H2 5.00E+13 0.0 0.0   125 CH + O2 -> O + HCO 3.30E+13 0.0 0.0 2.1 126 CH + H2 -> H + CH2 1.11E+08 1.79 1670 1.2; 2.1 127 CH + H2O -> H + CH2O 1.71E+13 0.0 -755 2.1 128 CH + CH2 -> H + C2H2 4.00E+13 0.0 0.0   129 CH + CH3 -> H + C2H3 3.00E+13 0.0 0.0   130 CH + CH4 -> H + C2H4 6.00E+13 0.0 0.0   131 CH + CO (+M) -> HCCO (+M) pressure dependent   132 CH + CO2 -> HCO + CO 3.40E+12 0.0 690   133 CH + CH2O -> H + CH2CO 9.46E+13 0.0 -515   134 CH + HCCO -> CO + C2H2 5.00E+13 0.0 0.0   135 CH2 + O2 -> OH + HCO 1.32E+13 0.0 1500 1.1; 1.2; 2.1 136 CH2 + H2 -> H + CH3 5.00E+05 2.0 7230   137 CH2 + CH2 -> H2 + C2H2 3.20E+13 0.0 0.0   138 CH2 + CH3 -> H + C2H4 4.00E+13 0.0 0.0   139 CH2 + CH4 -> CH3 + CH3 2.46E+06 2.0 8270   140 CH2 + CO (+M) -> CH2CO (+M) pressure dependent   141 CH2 + HCCO -> C2H3 + CO 3.00E+13 0.0 0.0   142 CH2(S) + N2 -> CH2 + N2 1.50E+13 0.0 600   143 CH2(S) + AR -> CH2 + AR 9.00E+12 0.0 600   144 CH2(S) + O2 -> H + OH + CO 2.80E+13 0.0 0.0   145 CH2(S) + O2 -> CO + H2O 1.20E+13 0.0 0.0   146 CH2(S) + H2 -> CH3 + H 7.00E+13 0.0 0.0 1.1; 1.2; 2.1 147 CH2(S) + H2O (+M) -> CH3OH (+M) pressure dependent   148 CH2(S) + H2O -> CH2 + H2O 3.00E+13 0.0 0.0   149 CH2(S) + CH3 -> H + C2H4 1.20E+13 0.0 -570   150 CH2(S) + CH4 -> CH3 + CH3 1.60E+13 0.0 -570   151 CH2(S) + CO -> CH2 + CO 9.00E+12 0.0 0.0   152 CH2(S) + CO2 -> CH2 + CO2 7.00E+12 0.0 0.0   153 CH2(S) + CO2 -> CO + CH2O 1.40E+13 0.0 0.0   154 CH2(S) + C2H6 -> CH3 + C2H5 4.00E+13 0.0 -550   155 CH3 + O2 -> O + CH3O 2.68E+13 0.0 28800 1.1; 1.2; 2.1 156 CH3 + O2 -> OH + CH2O 3.60E+10 0.0 8940 1.1; 1.2; 2.1 157 CH3 + H2O2 -> HO2 + CH4 2.45E+04 2.47 5180   158 CH3 + CH3 (+M) -> C2H6 (+M) pressure dependent 1.1; 1.2; 2.1 159 CH3 + CH3 -> H + C2H5 5.00E+12 0.1 10600 1.1; 1.2; 2.1 160 CH3 + HCO -> CH4 + CO 2.65E+13 0.0 0.0 1.1; 1.2; 2.1 161 CH3 + CH2O -> HCO + CH4 3.32E+03 2.81 5860   162 CH3 + CH3OH -> CH2OH + CH4 3.00E+07 1.5 9940   163 CH3 + CH3OH -> CH3O + CH4 1.00E+07 1.5 9940   164 CH3 + C2H4 -> C2H3 + CH4 2.27E+05 2.0 9200   164 CH3 + C2H4 -> C2H3 + CH4 2.27E+05 2.0 9200   165 CH3 + C2H6 -> C2H5 + CH4 6.14E+06 1.74 10450   166 HCO + H2O -> H + CO + H2O 2.24E+18 -1.0 17000 1.1; 1.2; 2.1 167 HCO + M -> H + CO + M 1.87E+17

C2H3 + O2 -> HCO + CH2O 3.98E+12 0.0 -240   174 C2H4 (+M) -> H2 + C2H2 (+M) pressure dependent   175 C2H5 + O2 -> HO2 + C2H4 8.40E+11 0.0 3875   176 HCCO + O2 -> OH + CO + CO 1.60E+12 0.0 854   177 HCCO + HCCO -> CO + CO + C2H2 1.00E+13 0.0 0.0

[the mechanism file in Chemkin format | GRI-Mech main menu]

TABLE 28 N-Containing Reactions

Number Reaction Rate Coefficient
A T^n exp(-E/RT) Served as
Optimization
Variable in
GRI-Mech
Releases A
(mol,cm3,s) n
(T in K) E
(cal/mol) 178 N + NO -> N2 +O 3.50E+13 0.0 330 2.1 179
N + O2 -> NO +O 2.65E+12 0.0 6400   180
N + OH -> NO +H 7.33E+13 0.0 1120 2.1 181
N2O + O -> N2 +O2 1.40E+12 0.0 10810   182
N2O + O -> NO + NO 2.90E+13 0.0 23150   183
N2O + H -> N2 +OH 4.40E+14 0.0 18880 2.1 184
N2O + OH -> N2 +HO2 2.00E+12 0.0 21060   185
N2O (+M) -> N2 +O (+M) pressure dependent   186
HO2 + NO -> NO2 +OH 2.11E+12 0.0 -480   187
NO + O + M -> NO2 +M 1.06E+20 -1.41 0.0   188
NO2 + O -> NO +O2 3.90E+12 0.0 -240   189
NO2 + H -> NO +OH 1.32E+14 0.0 360   190
NH + O -> NO +H 5.00E+13 0.0 0.0 2.1 191
NH + H -> N +H2 3.20E+13 0.0 330   192
NH + OH -> HNO +H 2.00E+13 0.0 0.0   193
NH + OH -> N +H2O 2.00E+09 1.2 0.0   194
NH + O2 -> HNO +O 4.61E+05 2.0 6500   195
NH + O2 -> NO +OH 1.28E+06 1.5 100   196
NH + N -> N2 +H 1.50E+13 0.0 0.0   197
NH + H2O -> HNO +H2 2.00E+13 0.0 13850   198
NH + NO -> N2 +OH 2.16E+13 -0.23 0.0   199
NH + NO -> N2O +H 4.16E+14 -0.45 0.0 2.1 200
NH2 + O -> OH +NH 7.00E+12 0.0 0.0   201
NH2 + O -> H +HNO 4.60E+13 0.0 0.0   202
NH2 + H -> NH +H2 4.00E+13 0.0 3650   203
NH2 + OH -> NH +H2O 9.00E+07 1.5 -460   204
NNH -> N2 +H 3.30E+08 0.0 0.0   205
NNH + M -> N2 +H +M 1.30E+14 -0.11 4980   206
NNH + O2 -> HO2 +N2 5.00E+12 0.0 0.0   207
NNH + O -> OH +N2 2.50E+13 0.0 0.0   208
NNH + O -> NH +NO 7.00E+13 0.0 0.0   209
NNH + H -> H2 +N2 5.00E+13 0.0 0.0   210
NNH + OH -> H2O +N2 2.00E+13 0.0 0.0   211
NNH + CH3 -> CH4 +N2 2.50E+13 0.0 0.0   212
H + NO + M -> HNO +M 8.95E+19 -1.32 740   213
HNO + O -> NO +OH 2.50E+13 0.0 0.0   214
HNO + H -> H2 +NO 4.50E+11 0.72 660   215
HNO + OH -> NO +H2O 1.30E+07 1.9 -950   216
HNO + O2 -> HO2 +NO 1.00E+13 0.0 13000   217

TABLE 29A GRI-MECH Versi on 2.11

```
! GRI-Mech Version 2.11 11/3/95  1st release 9/6/95  CHEMKIN-II format
! corrected sign of E for NH3 + H, NH3 + OH
! See README21 file at anonymous FTP site unix.sri.com, directory gri;
! WorldWideWeb home page through http://www.gri.org , under 'Basic
! Research', for additional information, contacts, and disclaimer
ELEMENTS
O  H  C  N  AR
END
SPECIES
H2      H       O       O2      OH      H2O     HO2     H2O2
C       CH      CH2     CH2(S)  CH3     CH4     CO      CO2
HCO     CH2O    CH2OH   CH3O    CH3OH   C2H     C2H2    C2H3
C2H4    C2H5    C2H6    HCCO    CH2CO   HCCOH   N       NH
NH2     NH3     NNH     NO      NO2     N2O     HNO     CN
HCN     H2CN    HCNN    HCNO    HOCN    HNCO    NCO     N2
AR
END
!THERMO
! Insert GRI-Mech thermodynamics here or use in default file
!END
REACTIONS
2O+M<=>O2+M                              1.200E+17   -1.000        .00
H2/ 2.40/ H2O/15.40/ CH4/ 2.00/ CO/ 1.75/ CO2/ 3.60/ C2H6/ 3.00/ AR/  .83/
O+H+M<=>OH+M                             5.000E+17   -1.000        .00
H2/2.00/ H2O/6.00/ CH4/2.00/ CO/1.50/ CO2/2.00/ C2H6/3.00/ AR/ .70/
O+H2<=>H+OH                              5.000E+04    2.670    6290.00
O+HO2<=>OH+O2                            2.000E+13     .000        .00
O+H2O2<=>OH+HO2                          9.630E+06    2.000    4000.00
O+CH<=>H+CO                              5.700E+13     .000        .00
O+CH2<=>H+HCO                            8.000E+13     .000        .00
O+CH2(S)<=>H2+CO                         1.500E+13     .000        .00
O+CH2(S)<=>H+HCO                         1.500E+13     .000        .00
O+CH3<=>H+CH2O                           8.430E+13     .000        .00
O+CH4<=>OH+CH3                           1.020E+09    1.500    8600.00
O+CO+M<=>CO2+M                           6.020E+14     .000    3000.00
H2/2.00/ O2/6.00/ H2O/6.00/ CH4/2.00/ CO/1.50/ CO2/3.50/ C2H6/3.00/ AR/ .50/
O+HCO<=>OH+CO                            3.000E+13     .000        .00
O+HCO<=>H+CO2                            3.000E+13     .000        .00
O+CH2O<=>OH+HCO                          3.900E+13     .000    3540.00
O+CH2OH<=>OH+CH2O                        1.000E+13     .000        .00
O+CH3O<=>OH+CH2O                         1.000E+13     .000        .00
O+CH3OH<=>OH+CH2OH                       3.880E+05    2.500    3100.00
O+CH3OH<=>OH+CH3O                        1.300E+05    2.500    5000.00
O+C2H<=>CH+CO                            5.000E+13     .000        .00
O+C2H2<=>H+HCCO                          1.020E+07    2.000    1900.00
O+C2H2<=>OH+C2H                          4.600E+19   -1.410   28950.00
O+C2H2<=>CO+CH2                          1.020E+07    2.000    1900.00
O+C2H3<=>H+CH2CO                         3.000E+13     .000        .00
O+C2H4<=>CH3+HCO                         1.920E+07    1.830     220.00
O+C2H5<=>CH3+CH2O                        1.320E+14     .000        .00
O+C2H6<=>OH+C2H5                         8.980E+07    1.920    5690.00
O+HCCO<=>H+2CO                           1.000E+14     .000        .00
O+CH2CO<=>OH+HCCO                        1.000E+13     .000    8000.00
O+CH2CO<=>CH2+CO2                        1.750E+12     .000    1350.00
O2+CO<=>O+CO2                            2.500E+12     .000   47800.00
O2+CH2O<=>HO2+HCO                        1.000E+14     .000   40000.00
H+O2+M<=>HO2+M                           2.800E+18    -.860        .00
O2/ .00/ H2O/ .00/ CO/ .75/ CO2/1.50/ C2H6/1.50/ N2/ .00/ AR/ .00/
H+2O2<=>HO2+O2                           3.000E+20   -1.720        .00
```

TABLE 29B GRI-MECH Versi on 2.11

```
H+O2+AR<=>HO2+AR                           7.000E+17     -.800        .00
H+O2<=>O+OH                                8.300E+13      .000   14413.00
2H+M<=>H2+M                                1.000E+18    -1.000        .00
H2/ .00/ H2O/ .00/ CH4/2.00/ CO2/ .00/ C2H6/3.00/ AR/ .63/
2H+H2<=>2H2                                9.000E+16     -.600        .00
2H+H2O<=>H2+H2O                            6.000E+19    -1.250        .00
2H+CO2<=>H2+CO2                            5.500E+20    -2.000        .00
H+OH+M<=>H2O+M                             2.200E+22    -2.000        .00
H2/ .73/ H2O/3.65/ CH4/2.00/ C2H6/3.00/ AR/ .38/
H+HO2<=>O+H2O                              3.970E+12      .000     671.00
H+HO2<=>O2+H2                              2.800E+13      .000    1068.00
H+HO2<=>2OH                                1.340E+14      .000     635.00
H+H2O2<=>HO2+H2                            1.210E+07     2.000    5200.00
H+H2O2<=>OH+H2O                            1.000E+13      .000    3600.00
H+CH<=>C+H2                                1.100E+14      .000        .00
H+CH2(+M)<=>CH3(+M)                        2.500E+16     -.800        .00
     LOW  /  3.200E+27    -3.140    1230.00/
     TROE/   .6800    78.00  1995.00  5590.00 /
H2/2.00/ H2O/6.00/ CH4/2.00/ CO/1.50/ CO2/2.00/ C2H6/3.00/ AR/ .70/
H+CH2(S)<=>CH+H2                           3.000E+13      .000        .00
H+CH3(+M)<=>CH4(+M)                        1.270E+16     -.630     383.00
     LOW  /  2.477E+33    -4.760    2440.00/
     TROE/   .7830    74.00  2941.00  6964.00 /
H2/2.00/ H2O/6.00/ CH4/2.00/ CO/1.50/ CO2/2.00/ C2H6/3.00/ AR/ .70/
H+CH4<=>CH3+H2                             6.600E+08     1.620   10840.00
H+HCO(+M)<=>CH2O(+M)                       1.090E+12      .480    -260.00
     LOW  /  1.350E+24    -2.570    1425.00/
     TROE/   .7824   271.00  2755.00  6570.00 /
H2/2.00/ H2O/6.00/ CH4/2.00/ CO/1.50/ CO2/2.00/ C2H6/3.00/ AR/ .70/
H+HCO<=>H2+CO                              7.340E+13      .000        .00
H+CH2O(+M)<=>CH2OH(+M)                     5.400E+11      .454    3600.00
     LOW  /  1.270E+32    -4.820    6530.00/
     TROE/   .7187   103.00  1291.00  4160.00 /
H2/2.00/ H2O/6.00/ CH4/2.00/ CO/1.50/ CO2/2.00/ C2H6/3.00/
H+CH2O(+M)<=>CH3O(+M)                      5.400E+11      .454    2600.00
     LOW  /  2.200E+30    -4.800    5560.00/
     TROE/   .7580    94.00  1555.00  4200.00 /
H2/2.00/ H2O/6.00/ CH4/2.00/ CO/1.50/ CO2/2.00/ C2H6/3.00/
H+CH2O<=>HCO+H2                            2.300E+10     1.050    3275.00
H+CH2OH(+M)<=>CH3OH(+M)                    1.800E+13      .000        .00
     LOW  /  3.000E+31    -4.800    3300.00/
     TROE/   .7679   338.00  1812.00  5081.00 /
H2/2.00/ H2O/6.00/ CH4/2.00/ CO/1.50/ CO2/2.00/ C2H6/3.00/
H+CH2OH<=>H2+CH2O                          2.000E+13      .000        .00
H+CH2OH<=>OH+CH3                           1.200E+13      .000        .00
H+CH2OH<=>CH2(S)+H2O                       6.000E+12      .000        .00
H+CH3O(+M)<=>CH3OH(+M)                     5.000E+13      .000        .00
     LOW  /  8.600E+28    -4.000    3025.00/
     TROE/   .8902   144.00  2838.00 45569.00 /
H2/2.00/ H2O/6.00/ CH4/2.00/ CO/1.50/ CO2/2.00/ C2H6/3.00/
H+CH3O<=>H+CH2OH                           3.400E+06     1.600        .00
H+CH3O<=>H2+CH2O                           2.000E+13      .000        .00
H+CH3O<=>OH+CH3                            3.200E+13      .000        .00
H+CH3O<=>CH2(S)+H2O                        1.600E+13      .000        .00
H+CH3OH<=>CH2OH+H2                         1.700E+07     2.100    4870.00
H+CH3OH<=>CH3O+H2                          4.200E+06     2.100    4870.00
H+C2H(+M)<=>C2H2(+M)                       1.000E+17    -1.000        .00
     LOW  /  3.750E+33    -4.800    1900.00/
     TROE/   .6464   132.00  1315.00  5566.00 /
H2/2.00/ H2O/6.00/ CH4/2.00/ CO/1.50/ CO2/2.00/ C2H6/3.00/ AR/ .70/
```

TABLE 29C GRI-MECH Versi on 2.11

```
     TROE/   .7507   98.50  1302.00  4167.00 /
H2/2.00/ H2O/6.00/ CH4/2.00/ CO/1.50/ CO2/2.00/ C2H6/3.00/ AR/ .70/
H+C2H3(+M)<=>C2H4(+M)                    6.080E+12     .270     280.00
     LOW  /  1.400E+30   -3.860   3320.00/
     TROE/   .7820  207.50  2663.00  6095.00 /
H2/2.00/ H2O/6.00/ CH4/2.00/ CO/1.50/ CO2/2.00/ C2H6/3.00/ AR/ .70/
H+C2H3<=>H2+C2H2                         3.000E+13     .000        .00
H+C2H4(+M)<=>C2H5(+M)                    1.080E+12     .454    1820.00
     LOW  /  1.200E+42   -7.620   6970.00/
     TROE/   .9753  210.00   984.00  4374.00 /
H2/2.00/ H2O/6.00/ CH4/2.00/ CO/1.50/ CO2/2.00/ C2H6/3.00/ AR/ .70/
H+C2H4<=>C2H3+H2                         1.325E+06    2.530   12240.00
H+C2H5(+M)<=>C2H6(+M)                    5.210E+17    -.990    1580.00
     LOW  /  1.990E+41   -7.080   6685.00/
     TROE/   .8422  125.00  2219.00  6882.00 /
H2/2.00/ H2O/6.00/ CH4/2.00/ CO/1.50/ CO2/2.00/ C2H6/3.00/ AR/ .70/
H+C2H5<=>H2+C2H4                         2.000E+12     .000        .00
H+C2H6<=>C2H5+H2                         1.150E+08    1.900    7530.00
H+HCCO<=>CH2(S)+CO                       1.000E+14     .000        .00
H+CH2CO<=>HCCO+H2                        5.000E+13     .000    8000.00
H+CH2CO<=>CH3+CO                         1.130E+13     .000    3428.00
H+HCCOH<=>H+CH2CO                        1.000E+13     .000        .00
H2+CO(+M)<=>CH2O(+M)                     4.300E+07    1.500   79600.00
     LOW  /  5.070E+27   -3.420  84350.00/
     TROE/   .9320  197.00  1540.00 10300.00 /
H2/2.00/ H2O/6.00/ CH4/2.00/ CO/1.50/ CO2/2.00/ C2H6/3.00/ AR/ .70/
OH+H2<=>H+H2O                            2.160E+08    1.510    3430.00
2OH(+M)<=>H2O2(+M)                       7.400E+13    -.370        .00
     LOW  /  2.300E+18    -.900  -1700.00/
     TROE/   .7346   94.00  1756.00  5182.00 /
H2/2.00/ H2O/6.00/ CH4/2.00/ CO/1.50/ CO2/2.00/ C2H6/3.00/ AR/ .70/
2OH<=>O+H2O                              3.570E+04    2.400   -2110.00
OH+HO2<=>O2+H2O                          2.900E+13     .000    -500.00
OH+H2O2<=>HO2+H2O                        1.750E+12     .000     320.00
 DUPLICATE
OH+H2O2<=>HO2+H2O                        5.800E+14     .000    9560.00
 DUPLICATE
OH+C<=>H+CO                              5.000E+13     .000        .00
OH+CH<=>H+HCO                            3.000E+13     .000        .00
OH+CH2<=>H+CH2O                          2.000E+13     .000        .00
OH+CH2<=>CH+H2O                          1.130E+07    2.000    3000.00
OH+CH2(S)<=>H+CH2O                       3.000E+13     .000        .00
OH+CH3(+M)<=>CH3OH(+M)                   6.300E+13     .000        .00
     LOW  /  2.700E+38   -6.300   3100.00/
     TROE/   .2105   83.50  5398.00  8370.00 /
H2/2.00/ H2O/6.00/ CH4/2.00/ CO/1.50/ CO2/2.00/ C2H6/3.00/
OH+CH3<=>CH2+H2O                         5.600E+07    1.600    5420.00
OH+CH3<=>CH2(S)+H2O                      2.501E+13     .000        .00
OH+CH4<=>CH3+H2O                         1.000E+08    1.600    3120.00
OH+CO<=>H+CO2                            4.760E+07    1.228      70.00
OH+HCO<=>H2O+CO                          5.000E+13     .000        .00
OH+CH2O<=>HCO+H2O                        3.430E+09    1.180    -447.00
OH+CH2OH<=>H2O+CH2O                      5.000E+12     .000        .00
OH+CH3O<=>H2O+CH2O                       5.000E+12     .000        .00
OH+CH3OH<=>CH2OH+H2O                     1.440E+06    2.000    -840.00
OH+CH3OH<=>CH3O+H2O                      6.300E+06    2.000    1500.00
OH+C2H<=>H+HCCO                          2.000E+13     .000        .00
OH+C2H2<=>H+CH2CO                        2.180E-04    4.500   -1000.00
OH+C2H2<=>H+HCCOH                        5.040E+05    2.300   13500.00
OH+C2H2<=>C2H+H2O                        3.370E+07    2.000   14000.00
```

TABLE 29D GRI-MECH Versi on 2.11

```
OH+C2H4<=>C2H3+H2O                         3.600E+06    2.000     2500.00
OH+C2H6<=>C2H5+H2O                         3.540E+06    2.120      870.00
OH+CH2CO<=>HCCO+H2O                        7.500E+12     .000     2000.00
2HO2<=>O2+H2O2                             1.300E+11     .000    -1630.00
 DUPLICATE
2HO2<=>O2+H2O2                             4.200E+14     .000    12000.00
 DUPLICATE
HO2+CH2<=>OH+CH2O                          2.000E+13     .000         .00
HO2+CH3<=>O2+CH4                           1.000E+12     .000         .00
HO2+CH3<=>OH+CH3O                          2.000E+13     .000         .00
HO2+CO<=>OH+CO2                            1.500E+14     .000    23600.00
HO2+CH2O<=>HCO+H2O2                        1.000E+12     .000     8000.00
C+O2<=>O+CO                                5.800E+13     .000      576.00
C+CH2<=>H+C2H                              5.000E+13     .000         .00
C+CH3<=>H+C2H2                             5.000E+13     .000         .00
CH+O2<=>O+HCO                              3.300E+13     .000         .00
CH+H2<=>H+CH2                              1.107E+08    1.790     1670.00
CH+H2O<=>H+CH2O                            1.713E+13     .000     -755.00
CH+CH2<=>H+C2H2                            4.000E+13     .000         .00
CH+CH3<=>H+C2H3                            3.000E+13     .000         .00
CH+CH4<=>H+C2H4                            6.000E+13     .000         .00
CH+CO(+M)<=>HCCO(+M)                       5.000E+13     .000         .00
     LOW  /  2.690E+28   -3.740   1936.00/
     TROE/   .5757  237.00  1652.00  5069.00 /
H2/2.00/ H2O/6.00/ CH4/2.00/ CO/1.50/ CO2/2.00/ C2H6/3.00/ AR/ .70/
CH+CO2<=>HCO+CO                            3.400E+12     .000      690.00
CH+CH2O<=>H+CH2CO                          9.460E+13     .000     -515.00
CH+HCCO<=>CO+C2H2                          5.000E+13     .000         .00
CH2+O2<=>OH+HCO                            1.320E+13     .000     1500.00
CH2+H2<=>H+CH3                             5.000E+05    2.000     7230.00
2CH2<=>H2+C2H2                             3.200E+13     .000         .00
CH2+CH3<=>H+C2H4                           4.000E+13     .000         .00
CH2+CH4<=>2CH3                             2.460E+06    2.000     8270.00
CH2+CO(+M)<=>CH2CO(+M)                     8.100E+11     .500     4510.00
     LOW  /  2.690E+33   -5.110   7095.00/
     TROE/   .5907  275.00  1226.00  5185.00 /
H2/2.00/ H2O/6.00/ CH4/2.00/ CO/1.50/ CO2/2.00/ C2H6/3.00/ AR/ .70/
CH2+HCCO<=>C2H3+CO                         3.000E+13     .000         .00
CH2(S)+N2<=>CH2+N2                         1.500E+13     .000      600.00
CH2(S)+AR<=>CH2+AR                         9.000E+12     .000      600.00
CH2(S)+O2<=>H+OH+CO                        2.800E+13     .000         .00
CH2(S)+O2<=>CO+H2O                         1.200E+13     .000         .00
CH2(S)+H2<=>CH3+H                          7.000E+13     .000         .00
CH2(S)+H2O(+M)<=>CH3OH(+M)                 2.000E+13     .000         .00
     LOW  /  2.700E+38   -6.300   3100.00/
     TROE/   .1507  134.00  2383.00  7265.00 /
H2/2.00/ H2O/6.00/ CH4/2.00/ CO/1.50/ CO2/2.00/ C2H6/3.00/
CH2(S)+H2O<=>CH2+H2O                       3.000E+13     .000         .00
CH2(S)+CH3<=>H+C2H4                        1.200E+13     .000     -570.00
CH2(S)+CH4<=>2CH3                          1.600E+13     .000     -570.00
CH2(S)+CO<=>CH2+CO                         9.000E+12     .000         .00
CH2(S)+CO2<=>CH2+CO2                       7.000E+12     .000         .00
CH2(S)+CO2<=>CO+CH2O                       1.400E+13     .000         .00
CH2(S)+C2H6<=>CH3+C2H5                     4.000E+13     .000     -550.00
CH3+O2<=>O+CH3O                            2.675E+13     .000    28800.00
CH3+O2<=>OH+CH2O                           3.600E+10     .000     8940.00
CH3+H2O2<=>HO2+CH4                         2.450E+04    2.470     5180.00
2CH3(+M)<=>C2H6(+M)                        2.120E+16    -.970      620.00
     LOW  /  1.770E+50   -9.670   6220.00/
     TROE/   .5325  151.00  1038.00  4970.00 /
```

TABLE 29E GRI-MECH Versi on 2.11

CH3+HCO<=>CH4+CO	2.648E+13	.000	.00
CH3+CH2O<=>HCO+CH4	3.320E+03	2.810	5860.00
CH3+CH3OH<=>CH2OH+CH4	3.000E+07	1.500	9940.00
CH3+CH3OH<=>CH3O+CH4	1.000E+07	1.500	9940.00
CH3+C2H4<=>C2H3+CH4	2.270E+05	2.000	9200.00
CH3+C2H6<=>C2H5+CH4	6.140E+06	1.740	10450.00
HCO+H2O<=>H+CO+H2O	2.244E+18	-1.000	17000.00
HCO+M<=>H+CO+M	1.870E+17	-1.000	17000.00
H2/2.00/ H2O/ .00/ CH4/2.00/ CO/1.50/ CO2/2.00/ C2H6/3.00/			
HCO+O2<=>HO2+CO	7.600E+12	.000	400.00
CH2OH+O2<=>HO2+CH2O	1.800E+13	.000	900.00
CH3O+O2<=>HO2+CH2O	4.280E-13	7.600	-3530.00
C2H+O2<=>HCO+CO	5.000E+13	.000	1500.00
C2H+H2<=>H+C2H2	4.070E+05	2.400	200.00
C2H3+O2<=>HCO+CH2O	3.980E+12	.000	-240.00
C2H4(+M)<=>H2+C2H2(+M)	8.000E+12	.440	88770.00
LOW / 7.000E+50 -9.310 99860.00/			
TROE/ .7345 180.00 1035.00 5417.00 /			
H2/2.00/ H2O/6.00/ CH4/2.00/ CO/1.50/ CO2/2.00/ C2H6/3.00/ AR/ .70/			
C2H5+O2<=>HO2+C2H4	8.400E+11	.000	3875.00
HCCO+O2<=>OH+2CO	1.600E+12	.000	854.00
2HCCO<=>2CO+C2H2	1.000E+13	.000	.00
N+NO<=>N2+O	3.500E+13	.000	330.00
N+O2<=>NO+O	2.650E+12	.000	6400.00
N+OH<=>NO+H	7.333E+13	.000	1120.00
N2O+O<=>N2+O2	1.400E+12	.000	10810.00
N2O+O<=>2NO	2.900E+13	.000	23150.00
N2O+H<=>N2+OH	4.400E+14	.000	18880.00
N2O+OH<=>N2+HO2	2.000E+12	.000	21060.00
N2O(+M)<=>N2+O(+M)	1.300E+11	.000	59620.00
LOW / 6.200E+14 .000 56100.00/			
H2/2.00/ H2O/6.00/ CH4/2.00/ CO/1.50/ CO2/2.00/ C2H6/3.00/ AR/ .70/			
HO2+NO<=>NO2+OH	2.110E+12	.000	-480.00
NO+O+M<=>NO2+M	1.060E+20	-1.410	.00
H2/2.00/ H2O/6.00/ CH4/2.00/ CO/1.50/ CO2/2.00/ C2H6/3.00/ AR/ .70/			
NO2+O<=>NO+O2	3.900E+12	.000	-240.00
NO2+H<=>NO+OH	1.320E+14	.000	360.00
NH+O<=>NO+H	5.000E+13	.000	.00
NH+H<=>N+H2	3.200E+13	.000	330.00
NH+OH<=>HNO+H	2.000E+13	.000	.00
NH+OH<=>N+H2O	2.000E+09	1.200	.00
NH+O2<=>HNO+O	4.610E+05	2.000	6500.00
NH+O2<=>NO+OH	1.280E+06	1.500	100.00
NH+N<=>N2+H	1.500E+13	.000	.00
NH+H2O<=>HNO+H2	2.000E+13	.000	13850.00
NH+NO<=>N2+OH	2.160E+13	-.230	.00
NH+NO<=>N2O+H	4.160E+14	-.450	.00
NH2+O<=>OH+NH	7.000E+12	.000	.00
NH2+O<=>H+HNO	4.600E+13	.000	.00
NH2+H<=>NH+H2	4.000E+13	.000	3650.00
NH2+OH<=>NH+H2O	9.000E+07	1.500	-460.00
NNH<=>N2+H	3.300E+08	.000	.00
NNH+M<=>N2+H+M	1.300E+14	-.110	4980.00
H2/2.00/ H2O/6.00/ CH4/2.00/ CO/1.50/ CO2/2.00/ C2H6/3.00/ AR/ .70/			
NNH+O2<=>HO2+N2	5.000E+12	.000	.00
NNH+O<=>OH+N2	2.500E+13	.000	.00
NNH+O<=>NH+NO	7.000E+13	.000	.00
NNH+H<=>H2+N2	5.000E+13	.000	.00
NNH+OH<=>H2O+N2	2.000E+13	.000	.00
NNH+CH3<=>CH4+N2	2.500E+13	.000	.00

TABLE 29F GRI-MECH Versi on 2.11

```
HNO+O<=>NO+OH                              2.500E+13     .000         .00
HNO+H<=>H2+NO                              4.500E+11     .720      660.00
HNO+OH<=>NO+H2O                            1.300E+07    1.900     -950.00
HNO+O2<=>HO2+NO                            1.000E+13     .000    13000.00
CN+O<=>CO+N                                7.700E+13     .000         .00
CN+OH<=>NCO+H                              4.000E+13     .000         .00
CN+H2O<=>HCN+OH                            8.000E+12     .000     7460.00
CN+O2<=>NCO+O                              6.140E+12     .000     -440.00
CN+H2<=>HCN+H                              2.100E+13     .000     4710.00
NCO+O<=>NO+CO                              2.350E+13     .000         .00
NCO+H<=>NH+CO                              5.400E+13     .000         .00
NCO+OH<=>NO+H+CO                           2.500E+12     .000         .00
NCO+N<=>N2+CO                              2.000E+13     .000         .00
NCO+O2<=>NO+CO2                            2.000E+12     .000    20000.00
NCO+M<=>N+CO+M                             8.800E+16    -.500    48000.00
H2/2.00/ H2O/6.00/ CH4/2.00/ CO/1.50/ CO2/2.00/ C2H6/3.00/ AR/ .70/
NCO+NO<=>N2O+CO                            2.850E+17   -1.520      740.00
NCO+NO<=>N2+CO2                            5.700E+18   -2.000      800.00
HCN+M<=>H+CN+M                             1.040E+29   -3.300   126600.00
H2/2.00/ H2O/6.00/ CH4/2.00/ CO/1.50/ CO2/2.00/ C2H6/3.00/ AR/ .70/
HCN+O<=>NCO+H                              1.107E+04    2.640     4980.00
HCN+O<=>NH+CO                              2.767E+03    2.640     4980.00
HCN+O<=>CN+OH                              2.134E+09    1.580    26600.00
HCN+OH<=>HOCN+H                            1.100E+06    2.030    13370.00
HCN+OH<=>HNCO+H                            4.400E+03    2.260     6400.00
HCN+OH<=>NH2+CO                            1.600E+02    2.560     9000.00
H+HCN+M<=>H2CN+M                           1.400E+26   -3.400     1900.00
H2/2.00/ H2O/6.00/ CH4/2.00/ CO/1.50/ CO2/2.00/ C2H6/3.00/ AR/ .70/
H2CN+N<=>N2+CH2                            6.000E+13     .000      400.00
C+N2<=>CN+N                                6.300E+13     .000    46020.00
CH+N2<=>HCN+N                              2.857E+08    1.100    20400.00
CH+N2(+M)<=>HCNN(+M)                       3.100E+12     .150         .00
     LOW  /  1.300E+25    -3.160      740.00/
     TROE/    .6670  235.00  2117.00  4536.00 /
H2/2.00/ H2O/6.00/ CH4/2.00/ CO/1.50/ CO2/2.00/ C2H6/3.00/ AR/ .70/
CH2+N2<=>HCN+NH                            1.000E+13     .000    74000.00
CH2(S)+N2<=>NH+HCN                         1.000E+11     .000    65000.00
C+NO<=>CN+O                                1.900E+13     .000         .00
C+NO<=>CO+N                                2.900E+13     .000         .00
CH+NO<=>HCN+O                              5.000E+13     .000         .00
CH+NO<=>H+NCO                              2.000E+13     .000         .00
CH+NO<=>N+HCO                              3.000E+13     .000         .00
CH2+NO<=>H+HNCO                            3.100E+17   -1.380     1270.00
CH2+NO<=>OH+HCN                            2.900E+14    -.690      760.00
CH2+NO<=>H+HCNO                            3.800E+13    -.360      580.00
CH2(S)+NO<=>H+HNCO                         3.100E+17   -1.380     1270.00
CH2(S)+NO<=>OH+HCN                         2.900E+14    -.690      760.00
CH2(S)+NO<=>H+HCNO                         3.800E+13    -.360      580.00
CH3+NO<=>HCN+H2O                           9.600E+13     .000    28800.00
CH3+NO<=>H2CN+OH                           1.000E+12     .000    21750.00
HCNN+O<=>CO+H+N2                           2.200E+13     .000         .00
HCNN+O<=>HCN+NO                            2.000E+12     .000         .00
HCNN+O2<=>O+HCO+N2                         1.200E+13     .000         .00
HCNN+OH<=>H+HCO+N2                         1.200E+13     .000         .00
HCNN+H<=>CH2+N2                            1.000E+14     .000         .00
HNCO+O<=>NH+CO2                            9.800E+07    1.410     8500.00
HNCO+O<=>HNO+CO                            1.500E+08    1.570    44000.00
HNCO+O<=>NCO+OH                            2.200E+06    2.110    11400.00
HNCO+H<=>NH2+CO                            2.250E+07    1.700     3800.00
HNCO+H<=>H2+NCO                            1.050E+05    2.500    13300.00
```

TABLE 29G GRI-MECH Versi on 2.11

```
HNCO+M<=>NH+CO+M                                  1.180E+16     .000    84720.00
H2/2.00/ H2O/6.00/ CH4/2.00/ CO/1.50/ CO2/2.00/ C2H6/3.00/ AR/ .70/
HCNO+H<=>H+HNCO                                   2.100E+15    -.690     2850.00
HCNO+H<=>OH+HCN                                   2.700E+11     .180     2120.00
HCNO+H<=>NH2+CO                                   1.700E+14    -.750     2890.00
HOCN+H<=>H+HNCO                                   2.000E+07    2.000     2000.00
HCCO+NO<=>HCNO+CO                                 2.350E+13     .000         .00
CH3+N<=>H2CN+H                                    6.100E+14    -.310      290.00
CH3+N<=>HCN+H2                                    3.700E+12     .150      -90.00
NH3+H<=>NH2+H2                                    5.400E+05    2.400     9915.00
NH3+OH<=>NH2+H2O                                  5.000E+07    1.600      955.00
NH3+O<=>NH2+OH                                    9.400E+06    1.940     6460.00
END
```

TABLE 30A MCM Partial List

```
* ***********************************************************************;
* INORGANIC CHEMISTRY ;
* ***********************************************************************;
*;
* Ox/NOx CHEMISTRY ;
*;
% J<1>                          : O3 = O1D          ;
% J<2>                          : O3 = O            ;
% 6.00D-34*O2*O2*((TEMP/300)@-2.8)    : O = O3          ;
% 5.60D-34*O2*N2*((TEMP/300)@-2.8)    : O = O3          ;
% 8.00D-12*EXP(-2060/TEMP)      : O + O3 =          ;
% KMT01                         : O + NO = NO2      ;
% 6.50D-12*EXP(120/TEMP)        : O + NO2 = NO      ;
% KMT02                         : O + NO2 = NO3     ;
% 3.20D-11*O2*EXP(70/TEMP)            : O1D = O         ;
% 1.80D-11*N2*EXP(110/TEMP)           : O1D = O         ;
% 1.80D-12*EXP(-1370/TEMP)            : NO + O3 = NO2   ;
% 1.20D-13*EXP(-2450/TEMP)            : NO2 + O3 = NO3      ;
% 3.30D-39*EXP(530/TEMP)*O2     : NO + NO = NO2 + NO2       ;
% 1.67D-06*(H0/H)               : NO2 = HONO        ;
% 1.80D-11*EXP(110/TEMP)              : NO + NO3 = NO2 + NO2
     ;
% 4.50D-14*EXP(-1260/TEMP)            : NO2 + NO3 = NO + NO2
     ;
% KMT03 % KMT04                 : NO2 + NO3 = N2O5  ;
% 4.00D-04                      : N2O5 = NA + NA    ;
*;
% J<4>                          : NO2 = NO + O      ;
% J<5>                          : NO3 = NO          ;
% J<6>                          : NO3 = NO2 + O     ;
*;
* HOx FORMATION, INTERCONVERSION AND REMOVAL ;
*;
% 2.20D-10*H2O                  : O1D = OH + OH     ;
% 1.90D-12*EXP(-1000/TEMP)            : OH + O3 = HO2   ;
% 7.70D-12*EXP(-2100/TEMP)            : OH + H2 = HO2   ;
% 1.50D-13*KMT05                : OH + CO = HO2     ;
% 2.90D-12*EXP(-160/TEMP)       : OH + H2O2 = HO2   ;
% 1.40D-14*EXP(-600/TEMP)       : HO2 + O3 = OH     ;
% 4.80D-11*EXP(250/TEMP)              : OH + HO2 =          ;
% 2.20D-13*KMT06*EXP(600/TEMP)        : HO2 + HO2 = H2O2    ;
```

TABLE 30B MCM Partial List

```
% 1.90D-33*M*KMT06*EXP(980/TEMP)     : HO2 + HO2 = H2O2             ;
% KMT07                   : OH + NO = HONO            ;
% KMT08                   : OH + NO2 = HNO3           ;
% 2.30D-11                : OH + NO3 = HO2 + NO2           ;
% 3.70D-12*EXP(240/TEMP)             : HO2 + NO = OH + NO2
      ;
% KMT09 % KMT10                 : HO2 + NO2 = HO2NO2            ;
% 3.50D-12                : HO2 + NO3 = OH + NO2           ;
% 1.80D-11*EXP(-390/TEMP)       : OH + HONO = NO2              ;
% KMT11                   : OH + HNO3 = NO3           ;
% 1.50D-12*EXP(360/TEMP)        : OH + HO2NO2 = NO2            ;
% 6.00D-06              : HNO3 = NA             ;
*;
% J<3>                    : H2O2 = OH + OH            ;
% J<7>                    : HONO = OH + NO            ;
% J<8>                    : HNO3 = OH + NO2           ;
*;
* SOx CHEMISTRY ;
*;
% 4.00D-32*EXP(-1000/TEMP)*M         : O + SO2 = SO3           ;
% KMT12                         : OH + SO2 = HSO3         ;
% 1.30D-12*EXP(-330/TEMP)*O2         : HSO3 = HO2 + SO3            ;
% 1.20D-15*H2O                  : SO3 = SA             ;
*;
* ********************************************************************** ;
* ALKANES ;
* ********************************************************************** ;
*;
* METHANE ;
*;
% 7.44D-18*TEMP@2*EXP(-1361/TEMP)  : OH + CH4 = CH3O2            ;
% KRO2NO*0.999              : CH3O2 + NO = CH3O + NO2      ;
% KRO2NO*0.001              : CH3O2 + NO = CH3NO3          ;
% 7.20D-14*EXP(-1080/TEMP)*O2     : CH3O = HCHO + HO2         ;
% KMT13 % KMT14               : CH3O2 + NO2 = CH3O2NO2       ;
% KRO2NO3                   : CH3O2 + NO3 = CH3O + NO2     ;
% 4.10D-13*EXP(790/TEMP)         : CH3O2 + HO2 = CH3OOH       ;
% 1.82D-13*EXP(416/TEMP)*0.33*RO2  : CH3O2 = CH3O              ;
% 1.82D-13*EXP(416/TEMP)*0.335*RO2 : CH3O2 = HCHO              ;
% 1.82D-13*EXP(416/TEMP)*0.335*RO2 : CH3O2 = CH3OH             ;
% 1.00D-14*EXP(1060/TEMP)        : OH + CH3NO3 = HCHO + NO2      ;
```

TABLE 30C MCM Partial List

```
% J<51>                        : CH3NO3 = CH3O + NO2          ;
% 1.90D-12*EXP(190/TEMP)       : OH + CH3OOH = CH3O2          ;
% 1.00D-12*EXP(190/TEMP)       : OH + CH3OOH = HCHO + OH      ;
% J<41>                        : CH3OOH = CH3O + OH           ;
*;
* ETHANE ;
*;
% 1.51D-17*TEMP@2*EXP(-492/TEMP)  : OH + C2H6 = C2H5O2        ;
% KRO2NO*0.991                 : C2H5O2 + NO = C2H5O + NO2    ;
% KRO2NO*0.009                 : C2H5O2 + NO = C2H5NO3        ;
% 6.00D-14*EXP(-550/TEMP)*O2   : C2H5O = CH3CHO + HO2         ;
% KRO2NO3                      : C2H5O2 + NO3 = C2H5O + NO2   ;
% 7.50D-13*EXP(700/TEMP)       : C2H5O2 + HO2 = C2H5OOH       ;
% 3.10D-13*0.6*RO2             : C2H5O2 = C2H5O               ;
% 3.10D-13*0.2*RO2             : C2H5O2 = CH3CHO              ;
% 3.10D-13*0.2*RO2             : C2H5O2 = C2H5OH              ;
% 4.40D-14*EXP(720/TEMP)       : OH + C2H5NO3 = CH3CHO + NO2  ;
% J<52>                        : C2H5NO3 = C2H5O + NO2        ;
% 1.90D-12*EXP(190/TEMP)       : OH + C2H5OOH = C2H5O2        ;
% 1.00D-11                     : OH + C2H5OOH = CH3CHO + OH   ;
% J<41>                        : C2H5OOH = C2H5O + OH         ;
*;
* PROPANE ;
*;
% 1.50D-17*TEMP@2*EXP(-44/TEMP)*0.307 : OH + C3H8 = NC3H7O2   ;
% 1.50D-17*TEMP@2*EXP(-44/TEMP)*0.693 : OH + C3H8 = IC3H7O2   ;
*;
% KRO2NO*0.71*0.98             : NC3H7O2 + NO = NC3H7O + NO2  ;
% KRO2NO*0.71*0.02             : NC3H7O2 + NO = NC3H7NO3      ;
% 3.70D-14*EXP(-460/TEMP)*O2   : NC3H7O = C2H5CHO + HO2       ;
% KRO2NO3                      : NC3H7O2 + NO3 = NC3H7O + NO2 ;
% KRO2HO2*0.64                 : NC3H7O2 + HO2 = NC3H7OOH     ;
% 6.00D-13*0.6*RO2             : NC3H7O2 = NC3H7O             ;
% 6.00D-13*0.2*RO2             : NC3H7O2 = C2H5CHO            ;
% 6.00D-13*0.2*RO2             : NC3H7O2 = NPROPOL            ;
% 7.30D-13                     : OH + NC3H7NO3 = C2H5CHO + NO2 ;
% J<53>                        : NC3H7NO3 = NC3H7O + NO2      ;
% 1.90D-12*EXP(190/TEMP)       : OH + NC3H7OOH = NC3H7O2      ;
% 1.53D-11                     : OH + NC3H7OOH = C2H5CHO + OH ;
% J<41>                        : NC3H7OOH = NC3H7O + OH       ;
*;
```

TABLE 30D MCM Partial List

```
% J<53>                        : HO1C4NO3 = HO1C4O + NO2                 ;
% 1.90D-12*EXP(190/TEMP)       : OH + NC4H9OOH = NC4H9O2                 ;
% 1.67D-11                     : OH + NC4H9OOH = C3H7CHO + OH            ;
% J<41>                        : NC4H9OOH = NC4H9O + OH                  ;
% 1.90D-12*EXP(190/TEMP)       : OH + HO1C4OOH = HO1C4O2                 ;
% 2.06D-11                     : OH + HO1C4OOH = HOC3H6CHO + OH          ;
% J<41>                        : HO1C4OOH = HO1C4O + OH                  ;
% 1.02D-11                     : OH + HOC4H8OH = HOC3H6CHO + HO2         ;
*;
% J<15>                        : HOC3H6CHO = HO1C3O2 + HO2 + CO          ;
% 3.04D-11                     : OH + HOC3H6CHO = HOC3H6CO3              ;
% KNO3AL                       : NO3+HOC3H6CHO = HOC3H6CO3+HNO3          ;
% KRO2NO*2.7                   : HOC3H6CO3 + NO = HO1C3O2 + NO2          ;
% KRO2NO*0.71*0.981            : HO1C3O2 + NO = HO1C3O + NO2             ;

% KRO2NO*0.71*0.019            : HO1C3O2 + NO = HO1C3NO3                 ;
% 3.70D-14*EXP(-460/TEMP)*O2   : HO1C3O = HOC2H4CHO + HO2                ;
% KFPAN % KBPAN                : HOC3H6CO3 + NO2 = C4PAN1                ;
% KRO2NO3                      : HOC3H6CO3 + NO3 = HO1C3O2 + NO2         ;
% KAPHO2*0.71                  : HOC3H6CO3+HO2 = HOC3H6CO3H              ;

% KAPHO2*0.29                  : HOC3H6CO3+HO2 = HOC3H6CO2H + O3         ;
% 5.00D-12*0.7*RO2             : HOC3H6CO3 = HO1C3O2                     ;
% 5.00D-12*0.3*RO2             : HOC3H6CO3 = HOC3H6CO2H                  ;
% KRO2NO3                      : HO1C3O2 + NO3 = HO1C3O + NO2            ;
% KRO2HO2*0.64                 : HO1C3O2 + HO2 = HO1C3OOH                ;
% 6.00D-13*0.6*RO2             : HO1C3O2 = HO1C3O                        ;
% 6.00D-13*0.2*RO2             : HO1C3O2 = HOC2H4CHO                     ;
% 6.00D-13*0.2*RO2             : HO1C3O2 = HOC3H6OH                      ;
% 9.60D-12                     : OH + C4PAN1 = HOC3H6CO3 + NO2           ;
% 4.23D-12                     : OH + HO1C3NO3 = HOC2H4CHO + NO2         ;
% J<53>                        : HO1C3NO3 = HO1C3O + NO2                 ;
% 1.32D-11                     : OH + HOC3H6CO3H = HOC3H6CO3             ;
% J<41>                        : HOC3H6CO3H = HO1C3O2 + OH               ;
% 1.04D-11                     : OH + HOC3H6CO2H = HO1C3O2               ;
% 1.90D-12*EXP(190/TEMP)       : OH + HO1C3OOH = HO1C3O2                 ;
% 1.92D-11                     : OH + HO1C3OOH = HOC2H4CHO + OH          ;
% J<41>                        : HO1C3OOH = HO1C3O + OH                  ;
% 9.10D-12                     : OH + HOC3H6OH = HOC2H4CHO + HO2         ;
*;
% J<15>                        : HOC2H4CHO = HOCH2CH2O2+HO2+CO           ;
```

TABLE 30E MCM Partial List

```
% KRO2NO*0.71*0.958              : IC3H7O2 + NO = IC3H7O + NO2       ;
% KRO2NO*0.71*0.042              : IC3H7O2 + NO = IC3H7NO3           ;
% 1.50D-14*EXP(-200/TEMP)*O2     : IC3H7O = CH3COCH3 + HO2           ;
% KRO2NO3                        : IC3H7O2 + NO3 = IC3H7O + NO2      ;
% KRO2HO2*0.64                   : IC3H7O2 + HO2 = IC3H7OOH          ;
% 4.00D-14*0.6*RO2               : IC3H7O2 = IC3H7O                  ;
% 4.00D-14*0.2*RO2               : IC3H7O2 = CH3COCH3                ;
% 4.00D-14*0.2*RO2               : IC3H7O2 = IPROPOL                 ;
% 4.90D-13                       : OH + IC3H7NO3 = CH3COCH3 + NO2    ;
% J<54>                          : IC3H7NO3 = IC3H7O + NO2           ;
% 1.90D-12*EXP(190/TEMP)         : OH + IC3H7OOH = IC3H7O2           ;
% 2.42D-11                       : OH + IC3H7OOH = CH3COCH3 + OH     ;
% J<41>                          : IC3H7OOH = IC3H7O + OH            ;
*;
* BUTANE (N-BUTANE) ;
*;
% 1.51D-17*TEMP@2*EXP(190/TEMP)*0.147 : OH + NC4H10 = NC4H9O2
        ;
% 1.51D-17*TEMP@2*EXP(190/TEMP)*0.853 : OH + NC4H10 = SC4H9O2
        ;
*;
% KRO2NO*0.60*0.967              : NC4H9O2 + NO = NC4H9O + NO2       ;
% KRO2NO*0.60*0.033              : NC4H9O2 + NO = NC4H9NO3           ;
% 3.70D-14*EXP(-460/TEMP)*O2     : NC4H9O = C3H7CHO + HO2            ;
% 1.30D+11*EXP(-4127/TEMP)       : NC4H9O = HO1C4O2                  ;
% KRO2NO*0.60*0.987              : HO1C4O2 + NO = HO1C4O + NO2       ;
% KRO2NO*0.60*0.013              : HO1C4O2 + NO = HO1C4NO3           ;
% 8.4D+10*EXP(-3523/TEMP)        : HO1C4O = HOC3H6CHO + HO2          ;
% KRO2NO3                        : NC4H9O2 + NO3 = NC4H9O + NO2      ;
% KRO2HO2*0.74                   : NC4H9O2 + HO2 = NC4H9OOH          ;
% 1.30D-12*0.6*RO2               : NC4H9O2 = NC4H9O                  ;
% 1.30D-12*0.2*RO2               : NC4H9O2 = C3H7CHO                 ;
% 1.30D-12*0.2*RO2               : NC4H9O2 = NBUTOL                  ;
% KRO2NO3                        : HO1C4O2 + NO3 = HO1C4O + NO2      ;
% KRO2HO2*0.74                   : HO1C4O2 + HO2 = HO1C4OOH          ;
% 1.30D-12*0.6*RO2               : HO1C4O2 = HO1C4O                  ;
% 1.30D-12*0.2*RO2               : HO1C4O2 = HOC3H6CHO               ;
% 1.30D-12*0.2*RO2               : HO1C4O2 = HOC4H8OH                ;
% 1.78D-12                       : OH + NC4H9NO3 = C3H7CHO + NO2     ;
% J<53>                          : NC4H9NO3 = NC4H9O + NO2           ;
% 5.62D-12                       : OH + HO1C4NO3 = HOC3H6CHO + NO2   ;
```

/

TABLE 31A Ideal Gas Thermodynamic Data for Combustion and Air-Pollution Use by Alexander Burcat

```
        Ideal Gas Thermodynamic Data for Combustion and Air-Pollution Use.
                        Database Author: Alexander Burcat
                   This Database was last updated on March 1997
                            Discard Previous Versions
```

```
This database is provided free of charge for non commercial use, provided
proper quotation will be included in the pertinent publications.
IT IS STRICTLY FORBIDDEN TO INCLUDE THIS DATABASE AS IS OR PARTS OF IT IN
ANY COMMERCIAL DATABASE, SOFTWARE, FIRMWARE OR HARDWARE AND ANY OTHER TYPE OF
COMMERCIAL USE WITHOUT WRITTEN PERMISSION FROM THE AUTHOR.

The latest print quotation to be made for this database is:
Alexander Burcat and Bonnie McBride
"1995 Ideal Gas Thermodynamic Data for Combustion and Air-Pollution Use"
Technion Aerospace Engineering (TAE) Report # 732 January 1995.
or

Alexander Burcat's Ideal Gas Thermochemical Database
http://ftp.technion.ac.il/pub/supported/aetdd/thermodynamics

The following are changes introduced to the database since the last print:

CORRECTED SPECIES Norbornane,C8H7N,C9H8,C4H5N,C5H5N,C6H5NO,C2H6N,CClF,C3H6Ox2,
C4H9x2,C5H7Cl2,C2Br2F4,NCN,NO,NO+,NH3,NF3,NF2,NO2-,NOCl,NO2Cl,NO2F,N2F2,O3,OD,
OF,CF3Br,HNO,HNO2,HNO3,HD,HDO,HBr,HCl,HF,HOF,H3O+,H2S,H2SO4,H2F2,H3F3,H4F4,
H5F5,H6F6,H7F7,CCl2F,H,CHCLF2,C4H7,HSO3,C8H6,C8H9,C8H10,C16H10,FO,C6F6,C4H8O2,
C5H11x2,C4H4O,C4H6O,C4H8O,C4H4S

ADDED SPECIES 3xC7H8,C7H12,C7H14,C4H9N,NO3-,NHF,NHF2,NH2F,NO3F,N2F4,C5H8,
iC5H10x3,iC6H11x4,iC6H12x4,iC6H13x4,O2+,O2-,O-,OH+,OH-,OD-,2C6H13,HDO2,HOCl,
HPO,C9H10,H+,H-,2xC3H3Cl,C4H7O,C4H8O,C4H9Ox4,C3H5x3,C4H7x2,C3H7O,C5H9x3,C4H10O
x3,H2NCOx3,C4H4N2x2,CH3N2,CH4N,CH5N2,C5H6N2,C4H10FO2P,C9H7,C3H7Ix2,C3F6,C3H4N2,
C5H8N4O12,C2H6O2,C2H5O2Cl,C2H4O2Cl2,F2O,FO2,F2O2,BrO,BrO2x2,BrO3,Br2Ox2,IO,IO2x2,
IO3,I2Ox2,C6H5OO,C5H11,C6H6,C4H6S,C4H8S,C4H8S2x2

ORIGIN OF CALCULATIONS: A-Species calculated by M.B. Colket    C-CHEMKIN database
J-JANAF tables. L-NASA Lewis Research Center P- Thermodynamic Research Center
(Formerly American Petroleum Institute)    R-Russian Tables (TSIV)
T- Technion-Israel Inst. Technology

AL(cr) REF ELEMEN CODA89AL 1.   0.   0.   0.S   200.000   933.610     26.98154 1
 0.00000000E+00 0.00000000E+00 0.00000000E+00 0.00000000E+00 0.00000000E+00    2
 0.00000000E+00 0.00000000E+00 1.01040191E+00 1.20769743E-02-2.62083556E-05    3
 2.64282413E-08-9.01916513E-12-6.54454196E+02-5.00471254E+00 0.00000000E+00    4

AL(L) REF ELEMENT CODA89AL 1.   0.   0.   0.L   933.610  6000.000     26.98154 1
 3.81862551E+00 0.00000000E+00 0.00000000E+00 0.00000000E+00 0.00000000E+00    2
-9.49651808E+01-1.75229704E+01 3.81862551E+00 0.00000000E+00 0.00000000E+00    3
 0.00000000E+00 0.00000000E+00-9.49651808E+01-1.75229704E+01 0.00000000E+00    4

Al  HF298=10.711+/-0.21 KJ   REF=JANAF
AL               J 6/83AL 1.   0.   0.   0.G   200.000  6000.000  B  26.98154 1
```

TABLE 31B Ideal Gas Thermodynamic Data for Combustion and Air-Pollution Use by Alexander Burcat

```
EXTRAPOLATED AND ESTIMATED BY M.B.COLKET (UNPUBLISHED). HF298=345.1+/-6.0 KJ
N-C4H5           N-A 8/83C   4H   5    0    0G   300.       3000.       E  53.0835  1
  0.7255918E 01  0.1714131E-01 -0.7435045E-05  0.1641052E-08 -0.1536078E-12    2
  0.3822741E 05 -0.1242283E 02 -0.2032928E 01  0.5363053E-01 -0.6338269E-04    3
  0.4064144E-07 -0.1039802E-10  0.4021965E 05  0.3278783E 02  0.4150575E+05    4

I-C4H5  CH2=CHC=CH2 RADICAL  CALCULATED USING BENSON'S GROUP ADDITIVITY METHOD.
EXTRAPOLATED AND ESTIMATED BY M.B.COLKET (UNPUBLISHED). HF298=330.8 KJ
I-C4H5           I-A 8/83C   4H   5    0    0G   300.       3000.       E  53.0835  1
  0.7255918E 01  0.1714131E-01 -0.7435045E-05  0.1641052E-08 -0.1536078E-12    2
  0.3633986E 05 -0.1342058E 02  0.1025140E 02 -0.2562767E-01  0.1071509E-03    3
 -0.1109019E-06  0.3755005E-10  0.3712855E 05 -0.2057453E 02  0.3978587E+05    4

T-C4H5  CH2CH=C=CH2 RADICAL  CALCULATED USING BENSON'S GROUP ADDITIVITY METHOD.
EXTRAPOLATED AND ESTIMATED BY M.B.COLKET (UNPUBLISHED). HF298=302.2 KJ
T-C4H5           T-A 8/83C   4H   5    0    0G   300.       3000.       E  53.0835  1
  0.4073421E 01  0.2490737E-01 -0.1322954E-04  0.3415906E-08 -0.3487446E-12    2
  0.3397542E 05  0.4325293E 01 -0.1950923E 01  0.5328144E-01 -0.6530677E-04    3
  0.4699422E-07 -0.1420488E-10  0.3504846E 05  0.3254246E 02  0.3634610E+05    4

C4H5N PYRROLE (CY) (AZOLE, IMIDOLE)  SIGMA=2  IAIBIC=1586.5  NU=3527,3148,3140,
3125,3116,1530,1467,1422,1382,1287,1144,1134,1074,1048,1016,881,869,865,826,721,
710,618,601,474  HF298=108.18+/-0.81 KJ REF= Das et.al JPCRD 22,(1993),659  Max
Lst Sq Error Cp @ 200 **2.1%**
C4H5N PYRROLE      T 8/95C   4H   5N   1    0G   200.000   6000.000   B  67.09044 1
 0.97727000D+01 0.16111112D-01-0.57690298D-05 0.92973976D-09-0.55604978D-13    2
 0.82631988D+04-0.30359085D+02 0.38558756D+00 0.18943518D-01 0.52673009D-04    3
-0.91458921D-07 0.40445724D-10 0.11750328D+05 0.23102254D+02 0.13010989D+05    4

C4H6  2-BUTAYN (DIMETHYLACETYLENE) STATWT=1  SIGMA=6  IA=1.0605  IB=IC=25.0029
IR=.530  V(3)=500. (ONE ROTATION ONLY)  NU=371(2),725,1029(2),1050(2),1380(2),
1126,1448(2),1468(2),2270,2916,2966(2),2976(3),213(2)  REF=YOST OSBORNE &GARNER
MAX LST SQ ERROR CP # 1300 K  0.87 % .HF298=34.97 KCAL REF= Stull, Westrum &
Sinke
C4H6               T12/82C   4H   6    0    0G   300.000   5000.000   B  54.091   1
 0.80465832E+01 0.16485251E-01-0.55222272E-05 0.81235929E-09-0.42950784E-13    2
 0.13701305E+05-0.18017740E+02 0.31971083E+01 0.20255916E-01 0.65101922E-05    3
-0.16584423E-07 0.64002822E-11 0.15715203E+05 0.99088230E+01 0.17597979E+05    4

1,3-C4H6  1,3-BUTADIENE  CALCULATED USING BENSON'S GROUP ADDITIVITY METHOD.
EXTRAPOLATED AND ESTIMATED BY M.B.COLKET (UNPUBLISHED). HF298=26.5 KCAL
REF= Weisman & Benson 1989
1,3-C4H6           A 8/83C   4H   6    0    0G   300.       3000.       E  54.0914  1
  0.7130943E 01  0.2064008E-01 -0.9323906E-05  0.2147307E-08 -0.2061361E-12    2
  0.9612654E 04 -0.1417222E 02 -0.2530900E 01  0.5387969E-01 -0.5173933E-04    3
  0.2510720E-07 -0.4324726E-11  0.1187691E 05  0.3391388E 02  0.1333562E+05    4

1,2-C4H6  1,2-BUTADIENE  CALCULATED USING BENSON'S GROUP ADDITIVITY METHOD.
EXTRAPOLATED AND ESTIMATED BY M.B.COLKET (UNPUBLISHED). HF298=162.2 KJ  REF=
Stull Westrum & Sinke 1969.
1,2-C4H6           A 8/83C   4H   6    0    0G   300.       3000.       E  54.0914  1
  0.1781557E 02 -0.4257502E-02  0.1051185E-04 -0.4473844E-08  0.5848138E-12    2
  0.1267342E 05 -0.6982662E 02  0.1023467E 01  0.3495919E-01 -0.2200905E-04    3
  0.6942272E-08 -0.7879187E-12  0.1811799E 05  0.1975066E 02  0.1950807E+05    4
```

TABLE 32 Matrix of Properties for Groups

1	A	B	C	D	E	F	G	H	I	J	K	L	M	N	O	P	Q	R
271		1	2	3	4	5	6	7	8	9	10	11	12	13	14	15	16	17
272					TABLE II													
273		c-c	c=c	arom c-c	c-h	c-f	c-cl	c-br-	c-i	-oh	-o-	-(c=o)-	-ch=o	COOH	COO	-nh2	C≡C	-nh-
274	mol wt	12.011	12.011	12.011	1.0079	22.00115	41.4585	79.904	129.9073	17.0073	15.9994	28.0104	29.0183	45.0177	44.0098	16.0225	12.011	15.0146
275	E rho g/cc	0.770876	0.922208	0.861909	1.044688	0.976679	1.076944	1.096062	1.2308	0.934364	0.563807	1.05755	1.166174	1.348291	1.319753	0.806346	1.078636	0.424951
276	LOG(E)	0.443986	0.658445	0.586257	0.840311	1.016923	1.083306	1.125191	1.312812	1.62385	0.663259	1.690747	0.938897	0.578767	0.612922	1.184604	1.059116	0.655936
277	CHI	56.74281	55.9138	55.90297	54.74477	50.95373	52.45278	52.63093	53.21137	40.58349	48.51207	40.27918	56.22255	67.36511	65.77323	43.86395	51.22721	45.15278
278	-LOGKb	3.184525	9.877409	7.130923	13.61889	17.04318	18.71576	19.27559	24.22923	23.99519	10.46887	33.47262	13.15477	7.20745	6.068194	19.50588	18.25117	8.95659
279	DIPOLE M	0.530933	1.261783	0.965003	1.672549	2.085383	2.240947	2.294646	2.813482	2.930329	1.435	3.986274	1.584592	0.805283	0.696814	2.443673	2.212933	1.317844
280	SUM INT	1	1	1	1	1	1	1	1	1	1	1	1	1	1	1	1	1
281	Tb-Tm	-29.9583	-0.28007	-14.1315	8.474846	18.38823	37.1615	44.4312	74.02068	41.57655	-19.0278	65.96483	18.6812	20.71445	12.8291	16.85098	23.39769	-30.6715
282	MOL. WT	74.43891	95.05693	84.80987	99.63565	93.77089	112.4903	117.7819	141.3493	84.40717	56.64702	102.0556	111.2252	141.9228	132.648	78.99083	104.7441	43.07013
283	COUNT	0.5	1	2	-0.5	0.5	0.5	0.5	0.5	0.5	1	1	0.5	0.5	1	0.5	1	1
284	logben/h2	1.695789	2.087869	1.822429	2.220977	1.314522	1.853779	1.919842	2.568517	-1.68803	-0.27861	-0.5613	2.5932	5.727194	5.045897	-1.10436	1.391469	-1.79053
285	Tm	50.99679	93.99579	73.99569	93.88489	105.9016	143.2545	156.5707	199.3873	179.0963	68.77127	186.0085	111.7716	93.84132	90.77313	157.1746	138.5635	91.54221
286	B rho g/cc	1.424904	1.656836	1.569027	1.915616	1.776719	1.871334	1.87518	2.050918	1.494474	1.043644	1.816518	2.061809	2.454068	2.374953	1.332039	1.863104	0.738287
287	B	-0.01386	0.070637	0.026237	0.045069	0.064325	0.157723	0.192864	0.28862	0.181683	0.002815	0.175551	0.09202	0.106635	0.091166	0.146167	0.119637	0.040898
288	C	-0.03557	-0.00497	-0.01357	-0.0012	0.068868	0.069189	0.075792	0.084593	0.3048	0.115404	0.284935	-0.01507	-0.22306	-0.19206	0.249211	0.091531	0.211514
289	D	0.028865	0.016665	0.019665	-0.00897	0.007281	-0.00934	-0.01456	-0.03911	0.006022	0.098196	0.024061	-0.04621	-0.09652	-0.09474	0.055281	-0.00464	0.131455
290	E	0.016099	0.012599	0.012099	-0.01055	-0.0128	-0.00797	-0.0077	-0.0149	-0.00825	0.0342	-0.0141	-0.02335	-0.03715	-0.0353	0.034099	0.002799	0.070749
291	F	0.019418	-8.2E-05	0.009118	-0.00794	-0.00968	-0.0115	-0.00969	-0.02059	0.013219	-0.00084	-0.04232	0.005437	0.004198	0.014939	-0.00354	-0.01308	0.011098
292	Tb	21.03847	93.71571	59.86419	102.3597	124.2898	180.416	201.0019	273.408	220.6728	49.74352	251.9733	130.4528	114.5558	103.6022	174.0256	161.9612	60.8707
293	SOL PAR	7.677077	8.866727	8.421509	9.522555	10.26952	10.96197	11.28602	12.43745	13.52078	8.47188	13.48095	10.22057	8.722228	8.883435	11.50663	10.83618	8.772861
294	Tc	162.7164	285.1565	228.6011	306.318	333.4262	423.8675	454.7357	574.5046	464.198	198.168	536.8621	348.9081	338.6178	316.126	399.5599	399.9567	205.921
295	Pc	32.09804	40.84591	37.85833	50.45764	50.8236	54.38974	55.34758	64.00271	58.05685	27.30696	65.67316	55.9699	57.38951	56.24577	44.91046	54.58701	19.79236
296	log AC CC	1.040892	1.675969	1.408132	1.29215	2.365114	2.80899	3.075079	3.468696	6.35351	3.380111	5.66295	1.228438	-1.78726	-1.34804	5.755165	2.996437	5.32707
297	log par co	1.754022	1.799909	1.733488	1.792084	1.224155	1.647205	1.764019	2.133792	-0.03235	0.152774	-0.287	2.355479	4.288782	4.038747	-0.11736	1.298699	-0.58692
298	Hv	7.490724	8.906187	8.27664	8.950174	9.783611	11.22606	11.87878	13.52432	14.08705	8.109118	12.92853	10.17573	8.95851	9.143559	11.84001	10.67395	9.035392
299	Surf Ten	-0.56839	2.896107	1.075707	1.84782	2.637308	6.466646	7.90742	11.83342	7.449021	0.115401	7.197608	3.772828	4.372029	3.737808	5.992833	4.905107	1.676812
300	log visc	-0.36792	-0.18204	-0.25343	-0.06784	0.061581	0.257167	0.363085	0.634063	0.725411	-0.45029	0.434814	0.235639	0.162015	0.20103	0.094559	0.092732	-0.50418
301	log(et/h2o	1.863257	1.933567	1.865447	2.236377	1.386084	1.813179	1.907263	2.413758	-0.82304	-0.58872	-0.81028	3.024556	5.962877	5.54934	-1.16705	1.307907	-2.12485
302	logoil/h2o	1.462733	1.554363	1.430583	1.364544	0.57351	1.085161	1.177833	1.55365	-1.4269	-0.26142	-1.28461	1.755847	4.143956	3.750252	-0.79516	0.822713	-0.95452
303	logcyh/h2	1.21785	1.49556	1.28154	1.632445	0.603118	1.060206	1.080307	1.61937	-2.79542	-0.9519	-1.62811	1.957851	5.311107	4.611242	-2.03514	0.65238	-2.5581
304	logchl/h2o	1.430783	2.133483	1.733223	2.249721	1.700012	2.281673	2.372581	3.149067	-0.45177	0.43515	1.087233	2.389024	4.566784	3.916323	0.15808	1.904513	-0.65932
305	logcl4/h2o	1.104656	1.710646	1.359926	2.072337	1.270722	1.667644	1.658738	2.358478	-2.06945	-0.43067	-0.04018	2.173683	5.131704	4.320988	-1.37499	1.232706	-2.23608
306	vol.Bondi	52.24135	57.37255	54.05251	52.01233	52.77757	60.11236	62.81097	68.92005	56.47953	54.27812	56.18198	53.94545	57.83165	55.85288	59.30068	56.2202	58.33788
307	molvolexp	96.56404	103.0753	98.39774	95.37355	96.00992	104.4532	107.4591	114.8435	90.33649	100.4723	96.50188	95.37613	105.2613	100.5097	97.96144	97.10794	101.3532
308	C	1	1	1	0	0.25	0.5	0	0.25	0	0	1	1	1	1	0	1	0
309	H	0	0	0	1	0	0	0	0	1	0	0	1	1	0	2	0	1
310	O	0	0	0	0	0	0	0	0	1	1	1	1	2	2	0	0	0
311	N	0	0	0	0	0	0	0	0	0	0	0	0	0	0	1	0	1
312	F	0	0	0	0	1	0	0	0	0	0	0	0	0	0	0	0	0
313	I	0	0	0	0	0	0	0	1	0	0	0	0	0	0	0	0	0
314	D1	-0.03778	-0.07636	-0.05485	-0.2049	0.22572	0.05405	0.038283	0.021922	0.019162	-0.03087	0.031968	0.058989	0.085795	0.069144	0.146999	0.149894	-0.02757
315	D12																	
316	isothcom	14.56436	11.15396	12.61998	9.786768	9.468121	6.409687	5.28656	1.294226	4.74905	16.30826	4.854978	6.883966	5.723468	6.09708	9.08648	7.383675	16.98557
317	w	0.063641	0.209797	0.152047	0.215085	0.254699	0.317603	0.339503	0.389159	0.388353	0.143632	0.379054	0.255925	0.219114	0.208922	0.330691	0.291651	0.17985
318	Zc	0.285909	0.274216	0.278836	0.273793	0.270624	0.265592	0.26384	0.259867	0.259932	0.279509	0.260676	0.270526	0.273471	0.274286	0.264545	0.267668	0.276612
319	AIR	0.429925	1.610378	0.730323	1.542002	0.351504	0.57717	0.430837	0.925835	0.000479	0.215396	0.132667	0.383754	17.43731	3.612272	0.038136	0.836395	0.045561
320	RES	43.82368	106.4487	57.73098	53.68093	13.55889	24.41758	19.78934	30.04912	0.073091	28.35416	4.227332	14.34311	272.3464	83.25222	7.445183	44.935	22.087
321	drug1	4.648172	4.311632	4.515396	3.963299	4.522954	4.648883	4.843039	4.663765	7.790254	5.214485	5.381901	4.447386	2.542756	3.229118	6.655875	4.784981	6.706092
322	drug4	131.3643	138.6708	129.782	81.97639	82.90482	114.1847	123.6276	127.8199	131.7043	170.8797	101.4075	69.63231	39.03401	43.12285	205.964	127.8082	261.825
323	drug6	4.584142	4.588508	4.635058	4.734052	4.877768	4.793347	4.773676	4.713379	5.735594	4.841552	5.496924	4.750141	4.016615	4.201741	5.449069	5.037096	5.170799
324	rho (B-E)	0.654028	0.734628	0.707119	0.870928	0.800039	0.79439	0.779118	0.820118	0.56011	0.479837	0.758968	0.895634	1.105777	1.055201	0.525693	0.784468	0.313337
325	coefficien	0	0	0	0	0	0	0	0	0	0	0	0	0	0	0	0	0
326	intercept	0	0	0	0	0	0	0	0	0	0	0	0	0	0	0	0	0
327	mol. wt.	12.011	12.011	12.011	1.0079	22.00115	41.4585	79.904	129.9068	17.0073	15.9994	28.0104	29.0183	45.0177	44.0098	16.0225	12.011	15.0146
328	# o	0	0	0	0	0	0	0	0	1	1	1	1	2	2	0	0	0
329	# n	0	0	0	0	0	0	0	0	0	0	0	0	0	0	1	0	1
330	# s	0	0	0	0	0	0	0	0	0	0	0	0	0	0	0	0	0
331	c-c	1	0	0	0	0	0	0	0	0	0	0	0	0	0	0	0	0
332	arom. c-c	0	0	1	0	0	0	0	0	0	0	0	0	0	0	0	0	0
333	c=c	0	1	0	0	0	0	0	0	0	0	0	0	0	0	0	0	0
334	c≡c	0	0	0	0	0	0	0	0	0	0	0	0	0	0	0	1	0
335	c-h	0	0	0	1	0	0	0	0	0	0	0	0	0	0	0	0	0
336	c-f	0	0	0	0	1	0	0	0	0	0	0	0	0	0	0	0	0
337	c-cl	0	0	0	0	0	1	0	0	0	0	0	0	0	0	0	0	0
338	c-br	0	0	0	0	0	0	1	0	0	0	0	0	0	0	0	0	0
339	c-I	0	0	0	0	0	0	0	1	0	0	0	0	0	0	0	0	0
340	oh	0	0	0	0	0	0	0	0	1	0	0	0	0	0	0	0	0
341	o	0	0	0	0	0	0	0	0	0	1	0	0	0	0	0	0	0
342	c=o	0	0	0	0	0	0	0	0	0	0	1	0	0	0	0	0	0
343	ch=o	0	0	0	0	0	0	0	0	0	0	0	1	0	0	0	0	0
344	cooh	0	0	0	0	0	0	0	0	0	0	0	0	1	0	0	0	0
345	coo	0	0	0	0	0	0	0	0	0	0	0	0	0	1	0	0	0
346	sh	0	0	0	0	0	0	0	0	0	0	0	0	0	0	0	0	0
347	nh2	0	0	0	0	0	0	0	0	0	0	0	0	0	0	1	0	0
348	nh	0	0	0	0	0	0	0	0	0	0	0	0	0	0	0	0	1
349	-n--	0	0	0	0	0	0	0	0	0	0	0	0	0	0	0	0	0
350	=n-	0	0	0	0	0	0	0	0	0	0	0	0	0	0	0	0	0
351	cn	0	0	0	0	0	0	0	0	0	0	0	0	0	0	0	0	0
352	conh2	0	0	0	0	0	0	0	0	0	0	0	0	0	0	0	0	0
353	no2	0	0	0	0	0	0	0	0	0	0	0	0	0	0	0	0	0
354	cycles(rin	0	0	0	0	0	0	0	0	0	0	0	0	0	0	0	0	0
355	tert C	0	0	0	0	0	0	0	0	0	0	0	0	0	0	0	0	0
356	quat C	0	0	0	0	0	0	0	0	0	0	0	0	0	0	0	0	0
357	x-c-x	0	0	0	0	0	0	0	0	0	0	0	0	0	0	0	0	0
358	x3-c-x3	0	0	0	0	0	0	0	0	0	0	0	0	0	0	0	0	0
359	c=c-x	0	0	0	0	0	0	0	0	0	0	0	0	0	0	0	0	0
360	arom. fusi	0	0	0	0	0	0	0	0	0	0	0	0	0	0	0	0	0
361	S	0	0	0	0	0	0	0	0	0	0	0	0	0	0	0	0	0
362	CL	0	0	0	0	0	1	0	0	0	0	0	0	0	0	0	0	0
363	BR	0	0	0	0	0	0	1	0	0	0	0	0	0	0	0	0	0
364	loggamma	1.039282	1.674709	1.406922	1.293205	2.366394	2.809788	3.075849	3.470186	6.354335	3.376691	5.66436	1.230773	-1.78354	-1.34451	5.751755	2.996157	5.319995
365	m.refract	24.20837	27.90766	25.60517	24.52532	23.93287	29.22132	30.91682	35.21717	25.14838	23.87867	25.50721	25.92008	29.83105	28.3769	28.41878	27.6641	26.87328
366	thermcon	2.532252	3.242837	2.942001	3.543918	4.009717	4.236935	4.337763	4.884368	5.051047	3.462347	5.862533	3.528077	2.733216	2.646942	4.47508	4.106527	3.433524
367	COST $/g	0.000833	0.000833	0.000833	0.009922	0.001477	0.000603	0.000501	0.000404	0.001176	0.000625	0.000714	0.001034	0.000889	0.000682	0.001872	0.000833	0.001332
368	MCL	0.01078	0.019502	0.012213	0.008657	0.00227	0.004386	0.003901	0.005452	2.7E-05	0.004636	0.000449	0.003476	0.060968	0.023537	0.001501	0.007186	0.004596

1	A	B	C	D	E	F	G	H	I	J	K	L	M	N	O	P	Q	R
369	TAP	0.561817	2.135003	0.941921	1.633037	0.389566	0.748394	0.598521	1.350623	0.000778	0.312559	0.155402	0.428924	17.15002	3.718476	0.067266	1.057422	0.093334
370	COM	131.0224	395.6865	181.8877	134.7881	25.16012	58.63032	47.06098	78.66333	0.055684	91.47317	6.708379	25.48711	774.7182	188.7817	21.78315	135.4037	104.0392
371	drug2	-15.779	-59.3603	-35.733	-68.5847	-59.0981	-141.613	-178.757	-279.029	-97.728	118.8647	-7.51009	-180.662	-329.967	-313.795	47.9936	-56.2983	193.8081
372	drug3	3.56135	3.434071	3.575146	3.493934	4.10953	4.121146	4.263955	4.222147	7.102065	4.168846	5.374595	3.947531	2.105729	2.683402	5.63247	4.206068	5.105538
373	drug5	10.93643	11.01456	10.93496	10.73377	10.83465	11.25692	11.47698	11.79481	11.49552	10.67174	10.62689	11.20008	11.50578	11.52918	10.95202	10.81734	10.78824

TABLE 33 Illustrative Optimizer Run, Linear Model

	A	B	C	D	E	F	G	H	I	J	K	L	M	N	O	P	Q
73						LINEAR	MODEL							B	0.132063	A^(1/2)	4.717035
74	-H	0	0.066	0	0.018	0	-0.027	0	-0.019	0	-0.019	0		-H	0	mol wt	153.823
75	-CH3	0	0.142	0	-0.02	0	-0.016	0	-0.023	0	-0.015	0		-CH3	0	E rho g/cc	1.124358
76	-CH2-	0	0.076	0	-0.038	0	0.011	0	-0.004	0	0.003	0		-CH2-	0	LOG(E)	0.932935
77	>CH-	0	0.003	0	-0.058	0	0.053	0	0.018	0	0.015	0		>CH-	0	CHI	53.2076
78	>C<	0	-0.075	0	-0.076	0	0.091	0	0.043	0	0.034	0		>C<	0	-LOGKb	15.29356
79	-CH=CH-	0	0.147	0	-0.043	0	0.028	0	0.01	0	0.003	0		-CH=CH-	0	DIPOLE M	1.87225
80	-CH=CH2	0	0.212	0	-0.025	0	0	0	-0.009	0	-0.015	0		-CH=CH2	0	SUM INT	2
81	>C=CH2	0	0.147	0	-0.043	0	0.028	0	0.01	0	0.003	0		>C=CH2	0	Tb-Tm	8.2144
82	-C=-CH	0	0.171	0	0.074	0	0.027	0	0.002	0	-0.012	0		-C=-CH	0	MOL. WT.	102.4088
83	-C6H5	0	0.467	0	-0.007	0	0.012	0	0.007	0	-0.017	0		-C6H5	0	COUNT	1
84	=CH-	0	0.088	0	0.002	0	-0.007	0	0.001	0	-0.003	0	#GRP/	=CH-	0	logben/h2	1.7862
85	AROMAT	0		0		0		0		0		0	CPD	AROM	0	Tm	118.6005
86	-NAPHTH	0	0.766	0	0.018	0	-0.026	0	0.024	0	-0.028	0		-NAPTH	0	B rho g/cc	1.971216
87	-CYCLOH	0	0.489	0	-0.148	0	0.004	0	-0.029	0	-0.009	0		-CYCLOH	0	B	0.0685
88	-F	0	0.078	0	0.088	0	0.009	0	-0.019	0	-0.02	0		-F	0	C	0.046
89	-CL	1	0.165	0.165	0.087	0.087	-0.024	-0.024	-0.012	-0.012	-0.021	-0.021		-CL	1	D	-0.017
90	-BR	0	0.213	0	0.095	0	-0.033	0	-0.008	0	-0.02	0		-BR	0	E	0.002
91	-I	0	0.302	0	0.103	0	-0.056	0	-0.01	0	-0.031	0		-I	0	F	-0.01
92	-CF3	0	0.15	0	0.017	0	0.035	0	-0.037	0	-0.013	0		CF3	0	Tb	126.8149
93	-CCL3	1	0.41	0.41	0.015	0.015	-0.009	-0.009	-0.017	-0.017	-0.017	-0.017		-CCL3	1	SOL PAR	10.19037
94	-OH	0	0.202	0	0.324	0	-0.012	0	-0.015	0	0.003	0		-OH	0	Tc	350.878
95	-O-	0	0.044	0	0.155	0	0.061	0	0.019	0	-0.022	0		-O-	0	Pc	54.72785
96	-C=0	0	0.135	0	0.246	0	0.061	0	0.023	0	-0.021	0		-C=O	0	log AC CC	2.095345
97	-CH=O-	0	0.219	0	0.244	0	0.01	0	-0.014	0	-0.027	0		-CH=O	0	log par co	1.547725
98	-COO-	0	0.167	0	0.17	0	0.062	0	0.015	0	-0.027	0		-COO-	0	Hv	9.6279
99	-COOH	0	0.323	0	0.342	0	-0.011	0	-0.017	0	0.008	0		-COOH	0	Surf Ten	2.8085
100	-NH2	0	0.167	0	0.269	0	0.037	0	0.027	0	-0.014	0		-NH2	0	log visc	-0.05572
101	-NH-	0	0.082	0	0.251	0	0.095	0	0.056	0	-0.01	0		-NH-	0	log(et/h2o	1.77065
102	-N-	0	-0.006	0	0.189	0	0.125	0	0.069	0	0.014	0		-N-	0	logoil/h2o	1.18205
103	-CN	0	0.241	0	0.269	0	-0.007	0	-0.023	0	-0.041	0		-CN	0	logcyh/h2	1.17095
104	-N=	0	0.102	0	0.183	0	0.031	0	-0.011	0	-0.02	0		-N=	0	logchl/h2	2.0157
105	PYRIDINE	0		0		0		0		0		0		PYRIDINE	0	logcl4/h2	1.58195
106	-NO2	0	0.238	0	0.241	0	-0.012	0	-0.027	0	-0.037	0		-NO2	0	vol.Bondi	51.9521
107	-CONH2	0	0.444	0	0.499	0	-0.019	0	-0.039	0	-0.012	0		-CONH2	0	v exp	91.082
108	-S-	0	0.136	0	0.13	0	0.028	0	0.032	0	-0.02	0		-S-	0	C	1
109	-SH	0	0.231	0	0.155	0	-0.026	0	-0.011	0	-0.013	0		-SH	0	H	0
110	MOLECU	1	-0.5065	-0.5065	-0.056	-0.056	0.007	0.007	0.031	0.031	0.028	0.028		MOLECU	1	O	0
111		0	B =	0.0685	C =	0.046	D =	-0.026	E =	0.002	F =	-0.01		S	0	N	0
112		0								Pvp mm	Kirchhoff	Pvpr298	81739.32	CL	4	F	0
113			B	C	D	E	F		0.717794	Pvp mm	riedel-P-M	10-1500n	-1.43408	BR	0	I	0
114	-H	0	-0.4405	-0.038	-0.02	0.012	0.009			Pvp mm	riedel-P-M	>1500mm	-1.40421	loggamm	2.095145	D1	-0.24946
115	-CH3	0	-0.3645	-0.076	-0.009	0.008	0.013			T	500	G1	-0.42178	m.refract	25.78612	LC50mg/l	99.85523
116	ch2	0	-0.4305	-0.094	0.018	0.027	0.031					G2	-0.44178	thermcon	3.678415	isothcom	8.3779
117	>CH-	0	-0.5035	-0.114	0.06	0.049	0.043					h	-1.47819	COST $/g	ERR	w	0.421622
118	>C<	0	-0.5815	-0.132	0.098	0.074	0.062					g1	5.255786	MCL	13.77378	Zc	0.25727
119	-CH=CH-	0	-0.3595	-0.099	0.035	0.041	0.031					g2	4.866707	TAP	2.017863	AIR	1.766189
120	-CH=CH2	0	-0.2945	-0.081	0.007	0.022	0.013							COM	295.4122	RES	88.37437
121	>C=CH2	0	-0.3595	0.099	0.035	0.041	0.031							drug1	4.446984	drug2	-29.2305
122	-C=-CH	0	-0.3355	0.018	0.034	0.033	0.016							drug4	114.0132	drug3	3.868534
123	-C6H5	0	-0.0395	-0.063	0.019	0.038	0.011							drug6	5.046468	drug5	10.53103
124	=CH-	0	-0.4185	-0.054	0	0.032	0.025							C-E	0.044	B-D	0.0855
125	-NAPTH	0	0.2595	-0.038	-0.019	0.055	0							B-C	0.0225	sum grps	2

TABLE 34 Properties for Aqueous Solutions

A	P	Q	R	S	T	U	V	W	X	Y	Z	AA	AB	AC	AD	AE
50		0	0.001	0.05	0.01	0.05	0.1	0.2	0.3	0.4	0.5	0.6	0.7	0.8	0.9	1
51		WATER														SOLUT X
52	LOG(E)	1.8938	1.893724	1.889993	1.893039	1.889993	1.886185	1.878571	1.870956	1.863342	1.855727	1.848113	1.840498	1.832884	1.825269	1.817655
53	CHI	-12.97	-12.9076	-9.84962	-12.3459	-9.84962	-6.72924	-0.48847	5.75229	11.99305	18.23382	24.47458	30.71534	36.95611	43.19687	49.43763
54	-LOGKb	-19.405	140.7652	135.3378	139.7683	135.3378	129.7995	118.7231	107.6467	96.57025	85.49382	74.41739	63.34096	52.26453	41.1881	30.11166
55	DIPOLE M	1.87	14.23755	13.70756	14.1402	13.70756	13.16677	12.08517	11.00358	9.921985	8.840391	7.758796	6.677202	5.595608	4.514014	3.432419
56	SUM INT	66.06841	3792.293	3606.711	3758.206	3606.711	3417.341	3038.602	2659.862	2281.123	1902.384	1523.644	1144.905	766.1655	387.4261	8.68677
57	Tb-Tm	100	1065.428	1019.652	1057.02	1019.652	972.9419	879.5218	786.1017	692.6816	599.2615	505.8414	412.4212	319.0011	225.581	132.1609
58	MOL. WT.	18.016	540.5732	519.6581	536.7316	519.6581	498.3163	455.6326	412.9489	370.2652	327.5814	284.8977	242.214	199.5303	156.8466	114.1629
59	COUNT	455.9075	6.013063	6.653167	6.130633	6.653167	7.306335	8.612669	9.919004	11.22534	12.53167	13.83801	15.14434	16.45068	17.75701	19.06335
60	logben/h2	0	48.85314	46.59992	48.43928	46.59992	44.30071	39.70229	35.10387	30.50546	25.90704	21.30862	16.71021	12.11179	7.513373	2.914956
61	Tm	0	-647.657	-612.888	-641.271	-612.888	-577.41	-506.453	-435.496	-364.539	-293.582	-222.626	-151.669	-80.7119	-9.75505	61.20178
62		488.4694	7578.575	7206.853	7510.299	7206.853	6827.545	6068.929	5310.312	4551.696	3793.08	3034.464	2275.848	1517.232	758.6161	
63	B	1.803739	0.150853	0.149801	0.15066	0.149801	0.148728	0.146582	0.144436	0.14229	0.140144	0.137998	0.135852	0.133706	0.131559	0.129413
64	C	0	-2.73985	-2.6066	-2.71538	-2.6066	-2.47063	-2.19868	-1.92673	-1.65478	-1.38283	-1.11088	-0.83894	-0.56699	-0.29504	-0.02309
65	D	7.593541	-0.26337	-0.25231	-0.26134	-0.25231	-0.24102	-0.21844	-0.19586	-0.17328	-0.1507	-0.12813	-0.10555	-0.08297	-0.06039	-0.03781
66	E	2.425988	-1.62013	-1.55029	-1.6073	-1.55029	-1.47903	-1.3365	-1.19397	-1.05144	-0.90891	-0.76638	-0.62385	-0.48132	-0.33879	-0.19626
67	F	-1.60276	-0.47962	-0.4561	-0.4753	-0.4561	-0.43211	-0.38412	-0.33613	-0.28814	-0.24016	-0.19217	-0.14418	-0.09619	-0.0482	-0.00022
68	Tb	100	417.7709	406.7639	415.7492	406.7639	395.5323	373.069	350.6057	328.1424	305.6791	283.2158	260.7525	238.2893	215.826	193.3627
69	SOL PAR	9.4	5.148558	5.507508	5.214487	5.507508	5.873784	6.606336	7.338888	8.07144	8.803992	9.536543	10.2691	11.00165	11.7342	12.46675
70	Tc	374.1	647.1926	634.5366	644.868	634.5366	621.6222	595.7936	569.9649	544.1363	518.3076	492.479	466.6503	440.8217	414.993	389.1644
71	Pc	218.3	148.187	144.0249	147.4226	144.0249	139.7778	131.2837	122.7895	114.2953	105.8012	97.30703	88.81288	80.31872	71.82456	63.3304
72	log AC CC	0	-50.7456	-48.2517	-50.2875	-48.2517	-45.707	-40.6175	-35.528	-30.4386	-25.3491	-20.2596	-15.1701	-10.0806	-4.99115	0.098333
73	log par co	0	22.71465	21.72462	22.53281	21.72462	20.71437	18.69389	16.6734	14.65292	12.63243	10.61194	8.591458	6.570972	4.550487	2.530001
74	Hv	-80.4637	0.01272	0.636	0.1272	0.636	1.272	2.543999	3.815999	5.087998	6.359998	7.631997	8.903997	10.176	11.448	12.72
75	Surf Ten	71.97	6.184969	6.141854	6.17705	6.141854	6.097859	6.009868	5.921878	5.833888	5.745897	5.657907	5.569916	5.481926	5.393936	5.305945
76	LOG visc	-0.0504	7.629654	7.337871	7.576062	7.337871	7.040134	6.444658	5.849182	5.253707	4.658231	4.062756	3.46728	2.871805	2.276329	1.680853
77	log(et/h2o	0	47.60871	45.49581	47.22062	45.49581	43.33979	39.02776	34.71572	30.40369	26.09165	21.77962	17.46759	13.15555	8.843516	4.531481
78	logoil/h2o	0	13.67705	12.99873	13.55246	12.99873	12.30656	10.92222	9.537886	8.15355	6.769213	5.384877	4.000541	2.616205	1.231869	-0.15247
79	logcyh/h2	0	46.86483	44.65023	46.45806	44.65023	42.39044	37.87085	33.35127	28.83168	24.3121	19.79251	15.27293	10.75334	6.233755	1.71417
80	logchl/h2o	0	41.04755	39.17047	40.70278	39.17047	37.25509	33.42433	29.59356	25.7628	21.93203	18.10126	14.2705	10.43973	6.60897	2.778205
81	logcl4/h2o	0	62.81857	59.91348	62.28498	59.91348	56.9491	51.02035	45.09159	39.16283	33.23408	27.30532	21.37656	15.44781	9.519049	3.590293
82																
83		Table IV. Mixtures. Aqueous Binary Sol utions With One Optimized Species. Also, Bina ry Solutior of Optimiz ed Specie														

TABLE 35 Matrix Calculations (Quattro Pro) for Desired Properties

AS=P A^-1AS=A^-1P S=A-1P

If you know values for six properties you desire = P
and you have the property equations for these six properties = A
then take inverse of A = A^-1, multiply by P and S=A^-1P
Now S=the set B, C, D, E and F, insert these in the optimization program
and solve for the groups, leading to the structure.

Table 1, 1	A						P
Bp	66.39	532.5	223.6	-365.4	-250.8	-794.6	88.9965
Log Visc	-0.23	2.44	1.26	-0.94	-8.1	5.06	-0.12626
sol par	8.97	5.22	12.1	-15	-12.9	-7.7	8.88631
Tm	77.2	295	194	-443	339	-406	85.3355
logpc	1.604	3.65	-7.66	-5.74	-0.31	5.09	2.323085
ht of vap	8.197	14.92	9.615	-8.07	-12.8	14.5	8.491805
	A^-1						
	0.000623	-0.36161	0.049664	-0.0034	0.045857	0.075479	
	0.001235	-0.03106	-0.05296	-7E-05	0.004358	0.046885	
	-0.00072	0.063826	0.002686	0.001368	-0.09035	0.009509	
	0.001114	-0.15296	-0.03123	-0.00221	-0.03663	0.048831	
	-0.00048	-0.05979	-0.01959	0.001013	-0.00913	0.015874	
	-0.00094	0.056144	-0.01004	0.00075	0.001055	0.012937	
				S=A^-1P			
				1.000003			
				0.045506			
				-0.061			
				-0.01799			
				-0.009			
				-0.004			
Table 1,2	A						P
Bp	66.39	532.5	223.6	-365.4	-250.8	-794.6	88.9965
Tm	77.2	295	194	-443	339	-406	85.3355
sol par	8.97	5.22	12.1	-15	-12.9	-7.7	8.88631
logpc	1.604	3.65	-7.66	-5.74	-0.31	5.09	2.323085
air	-0.11241	2.273426	17.73686	15.5473	-54.5266	105.4682	-1.9875
tap	-0.18926	0.439808	17.65265	14.10209	-58.6877	98.7817	-1.9875
	A^-1						
	4.2E-05	-0.00528	0.148661	0.022103	0.379247	-0.41584	
	0.001867	-0.00093	-0.00986	0.015534	0.15081	-0.15141	
	-0.00024	0.001304	0.004763	-0.07869	-0.00012	0.008021	
	0.001153	-0.0033	0.025124	-0.04115	0.209887	-0.2243	
	-0.00052	0.000643	-0.00032	-0.01195	0.072642	-0.07648	
	-0.00044	0.000614	-0.0043	0.012809	0.013271	-0.00604	
	S=A^-1P						
	0.998088						
	0.036088						
	-0.06589						
	-0.02268						
	-0.0101						
	-0.00935						
	A						P
Bp	66.39	532.5	223.6	-365.4	-250.8	-794.6	52.31
logoil/h2o'	8.197	14.92	9.615	-8.07	-12.8	14.5	7.87
sol par	8.97	5.22	12.1	-15	-12.9	-7.7	10.09
logac	1.241	5.09	13.54	3.36	6.93	7.39	3.87
air	-0.11241	2.273426	17.73686	15.5473	-54.5266	105.4682	0.075
tap	-0.18926	0.439808	17.65265	14.10209	-58.6877	98.7817	1E-05
	A^-1						
	-0.00567	0.322337	-0.14649	0.076789	-0.75934	0.700681	
	0.000944	0.054214	-0.05679	0.002028	-0.013	0.00894	
	0.000624	-0.0618	0.044254	0.053872	0.043717	-0.03317	
	-0.00283	0.214711	-0.18431	0.102626	-0.67945	0.649154	
	0.000109	-0.03697	0.03137	-0.00013	0.181156	-0.16466	
	0.000341	-0.0412	0.037014	-0.02422	0.195417	-0.18503	
	S						
	1.002491		AS=P	A^(-1)AS=A^(-1)P		S=A^(-1)P	
	-0.09004						
	0.204539						
	0.026477						
	0.04435						
	-0.01195						
Bp	66.39	532.5	223.6	-365.4	-250.8	-794.6	100
Tm	77.2	295	194	-443	339	-406	0
act coef	1.241	5.09	13.54	3.36	6.83	7.39	2
log p. coef	1.604	3.65	-7.66	-5.74	-0.31	5.09	0.33
mol. wt	87.8	149.6	-78.3	-239	20	-386	150
logLC50	-149.42	-302.6	649.93	505.41	27	-448.71	3
	-0.00144	-0.00447	0.137679	-2.17392	0.01108	-0.02532	
	0.000534	-0.00087	0.061079	1.953725	0.001203	0.021975	
	0.000906	0.000808	0.00034	-1.37259	-0.00275	-0.01553	
	-0.00099	-0.00273	0.080906	1.782768	0.0054	0.021132	
	-0.00168	0.00096	0.028579	1.31799	0.002348	0.015514	
	0.000219	0.000224	0.006305	-0.49441	-0.00227	-0.00637	
	0.999545						
	1.066708						
	-0.82146						
	1.524748						
	0.722478						
	-0.48784						

TABLE 36 Properties of Hazardous Compounds

1	CQ	CR	CS	CT	CU	CV	CW	CX	CY	CZ	DA	DB	DC	DD	DE	DF	DG	DH
38			chi	dipole m.	logben/h2	Tm	Tb	sol par	Tc	Pc	log act co	log par co	Hv	log visc	log et/h2o	log oil/h2o	logcyh/h2	logchl/h2o
39	PCE		61.05188	1.270683	3.932776	106.4656	131.5796	9.242444	350.6449	49.939	0.285402	3.065523	9.70502	0.144877	3.958534	2.765583	3.341984	3.487663
40	TCE		58.30954	1.303888	3.07148	90.00866	93.35621	9.116633	297.704	51.66767	0.416493	2.424667	8.648748	-0.08538	3.157093	2.135624	2.568751	2.742799
41	1,1-DCE		59.29422	1.134623	3.26994	85.88439	60.05115	9.248601	273.9064	65.62438	-0.50379	2.563445	7.492637	-0.3096	3.612465	2.480345	3.001235	2.62059
42	CH3CCL3		57.61282	1.411312	2.996044	82.0304	92.05552	9.043735	289.0289	49.15585	0.417876	2.321808	8.626704	-0.04032	3.060752	1.918236	2.449156	2.730932
43	CLCH2CH2CL		51.87297	1.78983	1.394828	85.83037	78.19017	9.83162	275.7493	55.70962	1.490021	1.268327	8.419272	-0.16858	1.633783	0.750952	0.866379	1.506397
44	LINDANE		49.80204	3.746298	1.889874	317.5555	476.9467	15.46128	853.6351	59.35669	7.461392	2.246387	20.16471	1.690129	1.719661	1.160073	0.287971	3.325937
45	CHLOROBENZENE		55.75958	1.728096	2.396915	140.4012	152.7875	10.22591	396.3111	55.24438	2.129572	1.992478	10.19319	-0.03653	2.219725	1.850745	1.785372	2.544329
46	P-DICHLOROBENZE		53.36256	2.241565	1.95431	183.4092	221.0466	11.50321	497.5765	57.98998	3.575251	1.79291	12.22675	0.251588	1.730669	1.512587	1.151396	2.487929
47	CCL4		44.58283	2.931649	-0.1981	164.6558	188.1455	12.57527	448.7438	67.09404	4.478617	0.343007	11.68741	0.221907	0.050328	-0.42818	-1.01453	0.773416
48	CH2CLCHCLCH3		54.36035	1.677078	2.102609	92.96318	104.8106	9.645369	305.3206	49.83079	1.391446	1.794514	9.220743	0.018698	2.240664	1.25144	1.478833	2.126697
49	ENDRIN		41.71542	7.503096	2.142327	628.2705	919.0931	24.61104	1612.86	127.961	13.23665	2.274585	30.722	2.998495	1.372155	1.319398	-0.07433	5.268193
50	HEPTACHLOR		46.09201	6.56135	2.808942	573.9666	847.9307	22.68891	1484.803	112.9217	11.7681	2.841151	29.14229	2.88221	2.17237	1.895562	0.691676	5.405025
51	HEXACHLOROBEN		43.77457	4.295384	0.163857	355.4416	494.0817	16.61247	902.6356	68.97239	9.357975	0.994669	20.36108	1.404065	-0.22552	0.159972	-1.38454	2.262265
52	C16H15CL3O2		59.44151	3.673395	4.714169	380.671	634.2568	13.5169	1046.111	13.63011	8.954352	3.544904	23.26727	1.913344	2.265935	3.221393	2.598448	6.777276
53	DICHLOROMETHAN		54.7636	1.318832	2.049616	47.13108	15.58387	8.655145	187.6073	55.44601	-0.09739	1.577793	6.414209	-0.45371	2.312843	1.241129	1.728548	1.740124
54	C9H7CL3O3 SILVEX		62.49893	1.765686	4.344795	232.2483	338.0527	11.10759	639.8194	36.90553	3.884403	3.691237	15.76207	0.922794	3.684142	3.465017	3.179387	4.654149
55	STYRENE		61.98163	1.032712	3.825143	121.3463	138.4486	8.657829	361.1563	41.15916	1.020867	2.914146	9.66919	-0.09007	3.338559	3.014772	3.237578	3.584455
56	1,2,4-TRICHLORO-b		50.96557	2.755015	1.511695	226.4173	289.3053	12.78053	598.841	60.73558	5.020932	1.593353	14.26034	0.539709	1.241626	1.174435	0.51741	2.431508
57	CLCH2CHCL2		49.47156	2.304379	0.952282	128.8095	146.4786	11.11033	377.0331	58.46256	2.935024	1.068758	10.45408	0.120304	1.1455	0.411628	0.23216	1.450069
58	CHLORDANE		61.21325	3.885634	5.380224	424.6817	714.5316	14.27068	1161.187	11.90599	9.732796	4.101368	25.53194	2.284946	2.760784	3.752119	3.063543	7.584724
59	C8H6CL2O3		62.40413	1.368061	4.07965	182.0782	243.2023	10.01791	509.0008	40.04578	2.536641	3.364594	12.94824	0.448159	3.567051	3.301506	3.200657	4.09033
60	HEPTACHLOREPOX		40.85239	5.935667	0.328259	522.5336	758.6113	21.00136	1311.933	85.18175	13.27202	1.344044	27.63915	2.456172	-0.39452	0.312318	-1.82798	3.367478
61	BENZO(a)PYRENE		79.78144	0.506346	8.567717	321.3075	536.4301	11.62569	883.1595	15.89054	3.713999	7.817387	23.12369	2.554632	8.487714	7.191235	6.824705	7.796686
62	VINYL CHLORIDE		55.53648	1.073295	2.03028	41.48749	-1.22156	8.176485	162.5	52.02646	-0.16081	1.560419	5.926888	-0.5993	2.192717	1.377401	1.761399	1.680386
63	BENZENE		58.16017	1.213926	2.839685	97.42082	84.51194	8.947581	295.0483	52.49591	0.684319	2.192043	8.158507	-0.3253	2.70822	2.189912	2.419757	2.600909
64	CH2CLCHBRCH2BR		49.99102	2.07112	0.968312	107.3101	112.9311	10.64504	324.9478	57.56437	2.352699	1.104232	9.629435	0.032968	1.308199	0.421936	0.323342	1.28556
65	CH2BRCH2BR		52.42821	2.009751	1.677822	125.1036	139.0253	10.80733	370.0832	60.01338	2.162761	1.61137	10.21256	0.119064	1.955672	1.050793	1.029352	1.865016
66	O-DICHLOROBENZE		53.36267	2.24155	1.954323	183.4111	221.0475	11.50328	497.5792	57.99095	3.575241	1.792944	12.2268	0.251596	1.730719	1.512625	1.151416	2.487919
67	ETHYL BENZENE		63.14034	0.986963	4.255075	111.7179	137.7121	8.573342	354.1552	40.72753	0.48803	3.244475	9.760109	0.048383	3.921089	3.192271	3.64482	3.841228
68	C5CL6		44.99339	3.951644	0.322451	329.6566	451.8333	15.81803	638.0983	66.10936	8.62485	1.059646	19.18265	1.226689	-0.05134	0.303596	-1.139	2.1923
69	PENTACHLORPPHE		32.3249	5.081344	-3.16086	405.7987	558.8704	19.47796	980.559	73.63034	13.11434	-0.52925	23.87726	1.976812	-2.70449	-2.19671	-5.09274	-0.23901
70	*TCDD DIOXIN*		44.91933	3.742123	-0.23506	364.7414	553.7613	15.59968	918.5366	26.66343	11.60917	***1.203774***	***23.43168***	1.94783	-0.94386	-0.00687	-2.32354	2.240224
71	TOLUENE		60.64827	1.100822	3.547259	104.5544	111.1202	8.760941	324.5991	46.61243	0.586002	2.718213	8.959844	-0.13813	3.314865	2.690532	3.032037	3.220971
72	TOXAPHENE		51.70064	5.405527	3.662116	506.2923	811.8304	19.7245	1351.407	66.25226	11.17721	3.669468	29.25723	3.157708	2.853143	2.338323	1.255419	5.960639
73	XYLENES		63.14034	0.986963	4.255075	111.7179	137.7121	8.573342	354.1552	40.72753	0.48803	3.244475	9.760109	0.048383	3.921089	3.192271	3.64482	3.841228
74	BENZOBFLUORANT		76.99223	1.218286	8.256517	357.5167	575.0241	12.71331	967.5371	28.78817	4.477461	7.302325	23.41906	2.426646	7.816524	6.907995	6.535215	7.891076

TABLE 37 Distribution Numbers for a Plane Triangular Lattice

Entry	m	g,ηs	Figure	Distribution Numbers $P_{g,\eta_s}^{(m)}$	Multiplicity $\lambda g,\eta s^{(m)}$	Energy
1	a	0		$p0^{(a)}$	1	0
2	a	1		$p1^{(a)}$	1	ε
3	b	0		$p0^{(b)}$	1	0
4	b	1		$p1^{(a)}$	2	ε
5	b	2		$p2^{(b)}$	1	2 ε + w
6	c	0		$p0^{(c)}$	1	0
7	c	1		$p1^{(c)}$	3	ε
8	c	2		$p2^{(c)}$	3	2 ε + w
9	c	3		$p3^{(c)}$	1	2 ε + 3w
10	d	0		$P0^{(d)}$	1	0
11	d	11		$p11^{(d)}$	2	ε
12	d	12		$p12^{(d)}$	2	ε
13	d	2,11		$p2,11^{(d)}$	1	2 ε + w
14	d	2,12		$p2,12^{(d)}$	4	2 ε + w
15	d	2,2		$p2,2^{(d)}$	1	2 ε + w′
16	d	3,1		$p3,11^{(d)}$	2	3 ε + 3w
17	d	3,2		$p3,2^{(d)}$	2	3 ε + 2w + w′
18	d	4		$p4^{(d)}$	1	4 e + 5w + w′

TABLE 38 SYNPROPS
Linear Model, Groups

E:A74: ' -H
E:A75: ' -CH3
E:A76: ' -CH2-
E:A77: '>CH-
E:A78: '>C<
E:A79: ' -CH=CH-
E:A80: ' -CH=CH2
E:A81: '>C=CH2
E:A82: ' -C=-CH
E:A83: ' -C6H5
E:A84: ' =CH-
E:A85: 'AROMAT
E:A86: ' -NAPHTH
E:A87: ' -CYCLOH
E:A88: ' -F
E:A89: ' -CL
E:A90: ' -BR
E:A91: ' -I
E:A92: ' -CF3
E:A93: ' -CCL3
E:A94: ' -OH
E:A95: ' -O-
E:A96: ' -C=0
E:A97: ' -CH=O-
E:A98: ' -COO-
E:A99: ' -COOH
E:A100: ' -NH2
E:A101: ' -NH-
E:A102: ' -N-
E:A103: ' -CN
E:A104: ' -N=
E:A105: 'PYRIDINE
E:A106: ' -NO2
E:A107: ' -CONH2
E:A108: ' -S-
E:A109: ' -SH

TABLE 39A SYNPROPS Linear Model, Property Equations

```
E:N72:   'B/A^.5
E:O72:   (O73/Q73)
E:N73:   'B
E:O73:   +0.121+0.225*Q87-0.07*Q88+0.15*Q89+0.11*Q90-0.12*Q91
E:P73:   'A^(1/2)
E:Q73:   +4.144+6.93*Q87-0.23*Q88-0.03*Q89+1.2*Q90-10.6*Q91
E:N74:   '-H
E:O74:   +B74
E:P74:   'mol wt
E:Q74:   +12.011*Q108+1.0079*Q109+15.9994*Q110+14.0067*Q111+18.998403*Q112+126.9045*Q113+32.06*O111+35.453*O112+79.904*O113
E:N75:   '-CH3
E:O75:   +B75
E:P75:   'E rho g/cc
E:Q75:   +Q82/Q107
E:N76:   '-CH2-
E:O76:   +B76
E:P76:   'LOG(E)
E:Q76:   +0.728+0.57*Q87+2.6*Q88-1.97*Q89-5.1*Q90-2.3*Q91
E:N77:   '>CH-
E:O77:   +B77
E:P77:   'CHI
E:Q77:   +54.58+11.4*Q87-51.8*Q88-16.1*Q89+42.9*Q90+13*Q91
E:N78:   '>C<
E:O78:   +B78
E:P78:   ' -LOGKb
E:Q78:   +0.6*(1.604+3.65*Q87-7.66*Q88-5.74*Q89-0.31*Q90+5.09*Q91)+9.12*(1.33+1.5*Q87+5*Q88-1.5*Q89-10*Q90-20.4*Q91)-2.71
E:N79:   '-CH=CH-
E:O79:   +B79
E:P79:   'DIPOLE M
E:Q79:   +1.33+1.5*Q87+5*Q88-1.5*Q89-10*Q90-20.4*Q91
E:N80:   ' -CH=CH2
E:O80:   +B80
E:P80:   'SUM INT
E:Q80:   @SUM(B74..B109)
E:N81:   '>C=CH2
E:O81:   +B81
E:P81:   'Tb-Tm
E:Q81:   +Q92-Q85
E:N82:   '-C=-CH
E:O82:   +B82
E:P82:   'MOL. WT.
E:Q82:   +87.8+149.6*Q87-78.3*Q88-239*Q89+20*Q90-386*Q91
E:N83:   '-C6H5
E:O83:   +B83
E:P83:   'COUNT
E:Q83:   +0.5*(B81+B101+B102+B103+@SUM(B85..B90)+@SUM(B92..B98))+0.75*(B82)+1*(B78+B80+B83+B91)+1.5*(B84+B79+B99+B100)-0.5*B94
E:N84:   '=CH-
E:O84:   +B84
E:P84:   'logben/h2o
E:Q84:   +1.81+4.4*Q87-13.2*Q88-6*Q89-18*Q91
E:N85:   'AROM
E:O85:   +B85
E:P85:   'Tm
E:Q85:   +77.2+295*Q87+194*Q88-443*Q89+339*Q90-406*Q91
E:N86:   '-NAPTH
E:O86:   +B86
E:P86:   'B rho g/cc
E:Q86:   +Q82/Q106
E:N87:   '-CYCLOH
E:O87:   +B87
E:P87:   'B
```

TABLE 39B SYNPROPS Linear Model, Property Equations

```
E:Q87:   +B74*0.066+B75*0.142+B76*0.076+B77*0.003+B78*-0.075+B79*0.147+B80*0.212+B81*0.147+B82*0.171+B83*0.4
         67+B84*0.088+B86*0.766+B87*0.489+B88*0.078+B89*0.165+B90*0.213+B91*0.302+B92*0.15+B93*0.41+B94*0.20
         2+B95*0.044+B96*0.135+B97*0.219+B98*0.167+B99*0.323+B100*0.167+B101*0.082+B102*-0.006+B103*0.241+B
         104*0.102+B106*0.238+B107*0.444+B108*0.136+B109*0.231+B110*-0.5065
E:N88:   '-F
E:O88:   +B88
E:P88:   'C
E:Q88:   +B74*0.018+B75*-0.02+B76*-0.038+B77*-0.058+B78*-0.076+B79*-0.043+B80*-0.025+B81*-0.043+B82*0.074+B83*
         -0.007+B84*0.002+B86*0.018+B87*-0.148+B88*0.088+B89*0.087+B90*0.095+B91*0.103+B92*0.017+B93*0.015+B
         94*0.324+B95*0.155+B96*0.246+B97*0.244+B98*0.17+B99*0.342+B100*0.269+B101*0.251+B102*0.189+B103*0.2
         69+B104*0.183+B106*0.241+B107*0.499+B108*0.13+B109*0.155+B110*-0.056
E:N89:   '-CL
E:O89:   +B89
E:P89:   'D
E:Q89:   +B74*-0.027+B75*-0.016+B76*0.011+B77*0.053+B78*0.091+B79*0.028+B80*0+B81*0.028+B82*0.027+B83*0.012+
         B84*-0.007+B86*-0.026+B87*0.004+B88*0.009+B89*-0.024+B90*-0.033+B91*-0.056*+0.035*B92-0.009*B93*-0.012
         *B94+0.061*B95+0.061*B96+0.01*B97+0.062*B98+B99*-0.011+B100*0.037+B101*0.095+B102*0.125+B103*-0.007
         +B104*0.031+B106*-0.012+B107*-0.019+B108*0.028+B109*-0.026+B110*0.007
E:N90:   '-BR
E:O90:   +B90
E:P90:   'E
E:Q90:   +B74*-0.019+B75*-0.023+B76*-0.004+B77*0.018+B78*0.043+B79*0.01+B80*-0.009+B81*-0.01+B82*0.002+B83*0.0
         07+B84*0.001+B86*0.024+B87*-0.029+B88*-0.019+B89*-0.012+B90*-0.008+B91*-0.01+B92*-0.037+B93*-0.017+B9
         4*-0.015+B95*0.019+B96*0.023+B97*-0.014+B98*0.015+B99*-0.017+B100*0.027+B101*0.056+B102*0.069+B103*-
         0.023+B104*-0.011+B106*-0.027+B107*-0.039+B108*0.032+B109*-0.011+B110*0.031
E:N91:   '-I
E:O91:   +B91
E:P91:   'F
E:Q91:   +B74*-0.019+B75*-0.015+B76*0.003+B77*0.015+B78*0.034+B79*0.003+B80*-0.015+B81*0.003+B82*-0.012+B83*-
         0.017+B84*-0.003+B86*-0.028+B87*-0.009+B88*-0.02+B89*-0.021+B90*-0.02+B91*-0.031+B92*-0.013+B93*-0.017
         +B94*0.003+B95*-0.022+B96*-0.021+B97*-0.027+B98*-0.027+B99*0.008+B100*-0.014+B101*–0.01+B102*0.014+B
         103*-0.041+B104*-0.02+B106*-0.037+B107*-0.012+B108*-0.02+B109*-0.013+B110*0.028
E:N92:   'CF3
E:O92:   +B92
E:P92:   'Tb
E:Q92:   +66.39+532.6*Q87+223.6*Q88-365.4*Q89-250.8*Q90-794.6*Q91
E:N93:   '-CCL3
E:O93:   +B93
E:P93:   'SOL PAR
E:Q93:   +8.97+5.22*Q87+12.1*Q88-15*Q89-12.9*Q90-7.7*Q91
E:N94:   '-OH
E:O94:   +B94
E:P94:   'Tc
E:Q94:   +248.6+770.4*Q87+343.1*Q88-927*Q89-68*Q90-1810*Q91
E:N95:   '-O-
E:O95:   +B95
E:P95:   'Pc
E:Q95:   +46.3-7.1*Q87+56.2*Q88-243*Q89-51*Q90-230*Q91
E:N96:   '-C=O
E:O96:   +B96
E:P96:   'log AC CO
E:Q96:   +1.241+Q87*5.09+Q88*13.54+3.36*Q89+6.93*Q90+7.39*Q91
E:N97:   '-CH=O
E:O97:   +B97
E:P97:   'log par co
E:Q97:   +1.604+3.65*Q87-7.66*Q88-5.74*Q89-0.31*Q90+5.09*Q91
E:N98:   '-COO-
E:O98:   +B98
E:P98:   'Hv
E:Q98:   +8.197+14.92*Q87+9.615*Q88-8.07*Q89-12.8*Q90+14.5*Q91
E:N99:   ' -COOH
E:O99:   +B99
E:P99:   'Surf Ten
E:Q99:   +L125+41*Q87+L155*Q88+L156*Q89+L157*Q90+L158*Q91
E:N100:  '-NH2
```

TABLE 39C SYNPROPS Linear Model, Property Equations

```
E:O100:   +B100
E:P100:   'log visc
E:Q100:   (-0.23+2.44*Q87+1.26*Q88-0.94*Q89-8.1*Q90+5.06*Q91)
E:N101:   '-NH-
E:O101:   +B101
E:P101:   'log(et/h2o)
E:Q101:   +1.91+3.3*Q87-11*Q88-9.8*Q89-8*Q90+1*Q91
E:N102:   '-N-
E:O102:   +B102
E:P102:   'logoil/h2o
E:Q102:   +1.23+3.5*Q87-10.2*Q88-7.5*Q89+12*Q90-3*Q91
E:N103:   '-CN
E:O103:   +B103
E:P103:   'logcyh/h2o
E:Q103:   +1.29+3.1*Q87-14.2*Q88-7.4*Q89+3*Q90-19*Q91
E:N104:   '-N=
E:O104:   +B104
E:P104:   'logchl/h2o
E:Q104:   +1.77+5.2*Q87-9.2*Q88-3.1*Q89-26*Q91
E:N105:   'PYRIDINE
E:O105:   +B105
E:P105:   'logcl4/h2o
E:Q105:   +1.55+3.3*Q87-12.5*Q88-3.7*Q89-6*Q90-33*Q91
E:N106:   '-NO2
E:O106:   +B106
E:P106:   'vol.Bondi
E:Q106:   +48.9+94.6*Q87-34.5*Q88+89*Q89+11*Q90+35*Q91
E:N107:   '-CONH2
E:O107:   +B107
E:P107:   'v exp
E:Q107:   +90+140*Q87-90*Q88+246*Q89-108*Q90-3*Q91
E:N108:   '-S-
E:O108:   +B108
E:P108:   'C
E:Q108:   +B75+B76+B77+B78+2*(B79+B80+B81+B82)+6*B83+B84+10*B86+6*B87+B92+B93+B96+B97+B98+B99+B103+
          B107
E:N109:   '-SH
E:O109:   +B109
E:P109:   'H
E:Q109:   +B74+3*B75+2*B76+B77+2*B79+3*B80+2*B81+B82+5*B83+B84+8*B86+11*B87+B94+B97+B99+2*B100+2*B101
          +2*B107+B109
E:N110:   'MOLECUL
E:O110:   +B110
E:P110:   'O
E:Q110:   +B94+B95+B96+B97+B107+2*B98+2*B99+2*B106
E:N111:   'S
E:O111:   +B108+B109
E:P111:   'N
E:Q111:   +B100+B101+B102+B103+B104+B106+B107
E:N112:   'CL
E:O112:   +3*B93+B89
E:P112:   'F
E:Q112:   +B88+3*B92
E:N113:   'BR
E:O113:   +B90
E:P113:   'I
E:Q113:   +B91
E:N114:   'loggamma
E:O114:   (1.241+5.09*Q87+13.54*Q88+3.36*Q89+6.83*Q90+7.39*Q91)
E:P114:   'D1
E:Q114:   +1.39E-07/(Q100*10^(-5))
E:N115:   'm.refract
E:O115:   +22.94+52.89*Q87-21.58*Q88+4.21*Q89+83.6*Q90-12.02*Q91
E:P115:   'LC50mg/L
E:Q115:   +Q74*@EXP(-0.5756-2.0173*R74+4.3662*F111+3.3694*H111+0.182*J111-2.9914*L111)
```

TABLE 39D SYNPROPS Linear Model, Property Equations

```
E:N116:   'thermcond
E:O116:   +3.17+3.19*Q87+4.5*Q88+2.1*Q89-13.2*Q90-14.5*Q91
E:P116:   'isothcom
E:Q116:   +11.6-22.6*Q87-10.8*Q88+57.6*Q89+11*Q90+22*Q91
E:N117:   'COST $/g
E:O117:   (0.01/B77)*(Q108*1+Q109*1+Q110*1+Q111*1+Q112*3+Q113*5+O111*2+O112*2+O113*4)
E:P117:   'w
E:Q117:   +0.42857*(Q92/Q94)*(1/(1-(Q92/Q94)))*(@LOG(Q95))
E:N118:   ' MCL
E:O118:   +1000*(@EXP(-(4.5906817+0.6819838*Q87+14.218037*Q88+16.568472*Q89-71.61533*Q90+58.154409*Q91)))
E:P118:   'Zc
E:Q118:   +0.291-0.08*Q117
E:N119:   '  TAP
E:O119:   @EXP(-(-0.18926+0.439808*Q87+17.65265*Q88+14.10209*Q89-58.6877*Q90+99.7817*Q91))
E:P119:   'AIR
E:Q119:   @EXP(-(-0.11241+2.273426*Q87+17.73686*Q88+15.5473*Q89-54.5266*Q90+105.4682*Q91))
E:N120:   ' COM
E:O120:   @EXP(-(-5.08345+1.893833*Q87+17.59624*Q88+21.39851*Q89-102.557*Q90+97.51875*Q91))
E:P120:   ' RES
E:Q120:   @EXP(-(-4.06262+2.084032*Q87+14.7937*Q88+16.6342*Q89-75.3489*Q90+80.87492*Q91))
E:N121:   'drug1
E:O121:   +4.276992+1.539887*Q87+9.439779*Q88-1.43346*Q89+5.149976*Q90+40.43886*Q91
E:P121:   'drug2
E:Q121:   -8.1294-1110.27*Q87+535.1663*Q88-255.649*Q89+2694.279*Q90-2060.02*Q91
E:N122:   'drug4
E:O122:   +95.87415+233.9213*Q87+21.71042*Q88+6.720305*Q89+2086.843*Q90+294.2722*Q91
E:P122:   'drug3
E:Q122:   +3.607096+0.287852*Q87+10.02711*Q88-3.68816*Q89-6.33959*Q90+26.95464*Q91
E:N123:   'drug6
E:O123:   +4.854351-2.69057*Q87+4.847684*Q88-8.01527*Q89+6.459043*Q90-0.42498*Q91
E:P123:   'drug5
E:Q123:   +10.51994+7.018316*Q87-2.32424*Q88+8.363076*Q89-11.9406*Q90+19.66987*Q91
E:N124:   'C-E
E:O124:   +Q88-Q90
E:P124:   'B-D
E:Q124:   +Q87-Q89
E:N125:   'B-C
E:O125:   +Q87-Q88
E:P125:   'sum grps
E:Q125:   @SUM(B74..B109)
```

TABLE 40 SYNPROPS Linear Model, Parameter Calculations

```
E:D111:   @SUM(D73..D110)
E:E111:   'C =
E:F111:   @SUM(F73..F110)
E:G111:   'D =
E:H111:   @SUM(H73..H110)
E:I111:   'E =
E:J111:   @SUM(J73..J110)
E:K111:   'F =
E:L111:   @SUM(L73..L110)

E:B74:    0.204412349784967
E:C74:    0.066
E:D74:    +B74*C74
E:E74:    0.018
E:F74:    +B74*E74
E:G74:    -0.027
E:H74:    +B74*G74
E:I74:    -0.019
E:J74:    +B74*I74
E:K74:    -0.019
E:L74:    +B74*K74
E:B75:    0.229881572290862
E:C75:    0.142
E:D75:    +B75*C75
E:E75:    -0.02
E:F75:    +B75*E75
E:G75:    -0.016
E:H75:    +B75*G75
E:I75:    -0.023
E:J75:    +B75*I75
E:K75:    -0.015
E:L75:    +B75*K75
E:B76:    0.295905505235176
E:C76:    0.076
E:D76:    +B76*C76
E:E76:    -0.038
E:F76:    +B76*E76
E:G76:    0.011
E:H76:    +B76*G76
E:I76:    -0.004
E:J76:    +B76*I76
E:K76:    0.003
E:L76:    +B76*K76
E:B77:    0.375771301473642
E:C77:    0.003
E:D77:    +B77*C77
E:E77:    -0.058
E:F77:    +B77*E77
E:G77:    0.053
E:H77:    +B77*G77
E:I77:    0.018
E:J77:    +B77*I77
E:K77:    0.015
E:L77:    +B77*K77
E:B78:    1.9122966126907
E:C78:    -0.075
E:D78:    +B78*C78
E:E78:    -0.076
E:F78:    +B78*E78
E:G78:    0.091
E:H78:    +B78*G78
E:I78:    0.043
E:J78:    +B78*I78
E:K78:    0.034
E:L78:    +B78*K78
E:B79:    0.333
E:C79:    0.147
E:D79:    +B79*C79
E:E79:    -0.043
E:F79:    +B79*E79
E:G79:    0.028
E:H79:    +B79*G79
E:I79:    0.01
E:J79:    +B79*I79
E:K79:    0.003
E:L79:    +B79*K79
E:B80:    0.333
E:C80:    0.212
E:D80:    +B80*C80
E:E80:    -0.025
E:F80:    +B80*E80
E:G80:    0
E:H80:    +B80*G80
E:I80:    -0.009
E:J80:    +B80*I80
E:K80:    -0.015
E:L80:    +B80*K80
E:B81:    0.333
E:C81:    0.147
E:D81:    +B81*C81
E:E81:    -0.043
E:F81:    +B81*E81
E:G81:    0.028
E:H81:    +B81*G81
E:I81:    0.01
E:J81:    +B81*I81
E:K81:    0.003
E:L81:    +B81*K81
E:B82:    0.23602034374736
E:C82:    0.171
E:D82:    +B82*C82
E:E82:    0.074
E:F82:    +B82*E82
E:G82:    0.027
E:H82:    +B82*G82
E:I82:    0.002
E:J82:    +B82*I82
E:K82:    -0.012
E:L82:    +B82*K82
E:B83:    0
E:C83:    0.467
E:D83:    +B83*C83
E:E83:    -0.007
E:F83:    +B83*E83
E:G83:    0.012
E:H83:    +B83*G83
E:I83:    0.007
E:J83:    +B83*I83
E:K83:    -0.017
E:L83:    +B83*K83
E:B84:    0.249276530901826
E:C84:    0.088
E:D84:    +B84*C84
E:E84:    0.002
E:F84:    +B84*E84
E:G84:    -0.007
E:H84:    +B84*G84
E:I84:    0.001
E:J84:    +B84*I84
```

TABLE 40 SYNPROPS Linear Model, Parameter Calculations (continued)

E:K84:	-0.003
E:L84:	+B84*K84
E:B85:	0.265245838463637
E:D85:	+B85*C85
E:F85:	+B85*E85
E:H85:	+B85*G85
E:J85:	+B85*I85
E:L85:	+B85*K85
E:B86:	0.00141934574841432
E:C86:	0.766
E:D86:	+B86*C86
E:E86:	0.018
E:F86:	+B86*E86
E:G86:	-0.026
E:H86:	+B86*G86
E:I86:	0.024
E:J86:	+B86*I86
E:K86:	-0.028
E:L86:	+B86*K86
E:B87:	0.0731247803863777
E:C87:	0.489
E:D87:	+B87*C87
E:E87:	-0.148
E:F87:	+B87*E87
E:G87:	0.004
E:H87:	+B87*G87
E:I87:	-0.029
E:J87:	+B87*I87
E:K87:	-0.009
E:L87:	+B87*K87
E:B88:	0.215430240462231
E:C88:	0.078
E:D88:	+B88*C88
E:E88:	0.088
E:F88:	+B88*E88
E:G88:	0.009
E:H88:	+B88*G88
E:I88:	-0.019
E:J88:	+B88*I88
E:K88:	-0.02
E:L88:	+B88*K88
E:B89:	0.164917797677181
E:C89:	0.165
E:D89:	+B89*C89
E:E89:	0.087
E:F89:	+B89*E89
E:G89:	-0.024
E:H89:	+B89*G89
E:I89:	-0.012
E:J89:	+B89*I89
E:K89:	-0.021
E:L89:	+B89*K89
E:B90:	0.152009268378816
E:C90:	0.213
E:D90:	+B90*C90
E:E90:	0.095
E:F90:	+B90*E90
E:G90:	-0.033
E:H90:	+B90*G90
E:I90:	-0.008
E:J90:	+B90*I90
E:K90:	-0.02
E:L90:	+B90*K90
E:B91:	0.161494524208325
E:C91:	0.302
E:D91:	+B91*C91
E:E91:	0.103
E:F91:	+B91*E91
E:G91:	-0.056
E:H91:	+B91*G91
E:I91:	-0.01
E:J91:	+B91*I91
E:K91:	-0.031
E:L91:	+B91*K91
E:B92:	0.234205620528963
E:C92:	0.15
E:D92:	+B92*C92
E:E92:	0.017
E:F92:	+B92*E92
E:G92:	0.035
E:H92:	+B92*G92
E:I92:	-0.037
E:J92:	+B92*I92
E:K92:	-0.013
E:L92:	+B92*K92
E:B93:	0.218441243346678
E:C93:	0.41
E:D93:	+B93*C93
E:E93:	0.015
E:F93:	+B93*E93
E:G93:	-0.009
E:H93:	+B93*G93
E:I93:	-0.017
E:J93:	+B93*I93
E:K93:	-0.017
E:L93:	+B93*K93
E:B94:	0.22527038665414
E:C94:	0.202
E:D94:	+B94*C94
E:E94:	0.324
E:F94:	+B94*E94
E:G94:	-0.012
E:H94:	+B94*G94
E:I94:	-0.015
E:J94:	+B94*I94
E:K94:	0.003
E:L94:	+B94*K94
E:B95:	0.233432463759066
E:C95:	0.044
E:D95:	+B95*C95
E:E95:	0.155
E:F95:	+B95*E95
E:G95:	0.061
E:H95:	+B95*G95
E:I95:	0.019
E:J95:	+B95*I95
E:K95:	-0.022
E:L95:	+B95*K95
E:B96:	0.205795358929518
E:C96:	0.135
E:D96:	+B96*C96
E:E96:	0.246
E:F96:	+B96*E96
E:G96:	0.061
E:H96:	+B96*G96
E:I96:	0.023
E:J96:	+B96*I96
E:K96:	-0.021
E:L96:	+B96*K96
E:B97:	0.156386682043277

TABLE 40 SYNPROPS Linear Model, Parameter Calculations (continued)

```
E:C97:    0.219
E:D97:    +B97*C97
E:E97:    0.244
E:F97:    +B97*E97
E:G97:    0.01
E:H97:    +B97*G97
E:I97:    -0.014
E:J97:    +B97*I97
E:K97:    -0.027
E:L97:    +B97*K97
E:B98:    0.206031794348446
E:C98:    0.167
E:D98:    +B98*C98
E:E98:    0.17
E:F98:    +B98*E98
E:G98:    0.062
E:H98:    +B98*G98
E:I98:    0.015
E:J98:    +B98*I98
E:K98:    -0.027
E:L98:    +B98*K98
E:B99:    0.00155855928871511
E:C99:    0.323
E:D99:    +B99*C99
E:E99:    0.342
E:F99:    +B99*E99
E:G99:    -0.011
E:H99:    +B99*G99
E:I99:    -0.017
E:J99:    +B99*I99
E:K99:    0.008
E:L99:    +B99*K99
E:B100:   0.194991346082966
E:C100:   0.167
E:D100:   +B100*C100
E:E100:   0.269
E:F100:   +B100*E100
E:G100:   0.037
E:H100:   +B100*G100
E:I100:   0.027
E:J100:   +B100*I100
E:K100:   -0.014
E:L100:   +B100*K100
E:B101:   0.290415030691074
E:C101:   0.082
E:D101:   +B101*C101
E:E101:   0.251
E:F101:   +B101*E101
E:G101:   0.095
E:H101:   +B101*G101
E:I101:   0.056
E:J101:   +B101*I101
E:K101:   -0.01
E:L101:   +B101*K101
E:B102:   0.406573325816339
E:C102:   -0.006
E:D102:   +B102*C102
E:E102:   0.189
E:F102:   +B102*E102
E:G102:   0.125
E:H102:   +B102*G102
E:I102:   0.069
E:J102:   +B102*I102
E:K102:   0.014
E:L102:   +B102*K102
E:B103:   0.115833509603271
E:C103:   0.241
E:D103:   +B103*C103
E:E103:   0.269
E:F103:   +B103*E103
E:G103:   -0.007
E:H103:   +B103*G103
E:I103:   -0.023
E:J103:   +B103*I103
E:K103:   -0.041
E:L103:   +B103*K103
E:B104:   0.21908155839764
E:C104:   0.102
E:D104:   +B104*C104
E:E104:   0.183
E:F104:   +B104*E104
E:G104:   0.031
E:H104:   +B104*G104
E:I104:   -0.011
E:J104:   +B104*I104
E:K104:   -0.02
E:L104:   +B104*K104
E:B105:   0.265245838463637
E:D105:   +B105*C105
E:F105:   +B105*E105
E:H105:   +B105*G105
E:J105:   +B105*I105
E:L105:   +B105*K105
E:B106:   0.123575259536194
E:C106:   0.238
E:D106:   +B106*C106
E:E106:   0.241
E:F106:   +B106*E106
E:G106:   -0.012
E:H106:   +B106*G106
E:I106:   -0.027
E:J106:   +B106*I106
E:K106:   -0.037
E:L106:   +B106*K106
E:B107:   0.0111887435551548
E:C107:   0.444
E:D107:   +B107*C107
E:E107:   0.499
E:F107:   +B107*E107
E:G107:   -0.019
E:H107:   +B107*G107
E:I107:   -0.039
E:J107:   +B107*I107
E:K107:   -0.012
E:L107:   +B107*K107
E:B108:   0.199574761560419
E:C108:   0.136
E:D108:   +B108*C108
E:E108:   0.13
E:F108:   +B108*E108
E:G108:   0.028
E:H108:   +B108*G108
E:I108:   0.032
E:J108:   +B108*I108
E:K108:   -0.02
E:L108:   +B108*K108
E:B109:   0.156212505944989
E:C109:   0.231
E:D109:   +B109*C109
E:E109:   0.155
```

TABLE 40 SYNPROPS Linear Model, Parameter Calculations (continued)

E:F109:	+B109*E109
E:G109:	-0.026
E:H109:	+B109*G109
E:I109:	-0.011
E:J109:	+B109*I109
E:K109:	-0.013
E:L109:	+B109*K109
E:B110:	1
E:C110:	-0.5065
E:D110:	+B110*C110
E:E110:	-0.056
E:F110:	+B110*E110
E:G110:	0.007
E:H110:	+B110*G110
E:I110:	0.031
E:J110:	+B110*I110
E:K110:	0.028
E:L110:	+B110*K110

TABLE 41 SYNPROPS Hierarchical Model, Groups

B:D73:	'Table III. Property - Structure Conversion Spreadsheet Group parameters, Property Parameters and Opti izer P rogram Results. Clone used for Mixture Table IV.
B:A75:	'coefficient
B:A76:	'intercept
B:A77:	'mol. wt.
B:A78:	'O atoms
B:A79:	'N atoms
B:A80:	'S atoms
B:A81:	'c-c bonds
B:A82:	'arom. c-c
B:A83:	'c=c bonds
B:A84:	'c=-bonds
B:A85:	'c-h bonds
B:A86:	'c-f bonds
B:A87:	'c-cl bonds
B:A88:	'c-br bonds
B:A89:	'c-i bonds
B:A90:	' -oh grps
B:A91:	' -o- grps
B:A92:	' -(c=o)- grp
B:A93:	' -ch=o grps
B:A94:	' -cooh grps
B:A95:	' -coo- grps
B:A96:	' -sh grps
B:A97:	' -nh2 grps
B:A98:	' -nh- grps
B:A99:	' -n- - grps
B:A100:	' =n- grps
B:A101:	' -cn grps
B:A102:	' -conh2 grps
B:A103:	' -no2 grps
B:A104:	'cycles(rings)
B:A105:	'tertiary C
B:A106:	'quatern c
B:A107:	'x-c-x traces
B:A108:	'X3-C-C-X3
B:A109:	'c=c-x traces
B:A110:	'arom fusions
B:P73:	'Zc
B:Q73:	0.23

TABLE 42A SYNPROPS Hierarchical Model, Property Equations

```
B:N74:   'rho (B-E)
B:O74:   +Q86-Q75
B:P74:   'mol wt
B:Q74:   +12.011*Q108+1.0079*Q109+15.9994*Q110+14.0067*Q111+18.998403*Q112+126.9045*Q113+32.06*O111+35
         .453*O112+79.904*O113
B:N75:   'coefficiemts
B:O75:   +B75
B:P75:   'E rho g/cc
B:Q75:   +Q82/Q107
B:N76:   'intercept
B:O76:   +B76
B:P76:   'LOG(E)
B:Q76:   +0.728+0.57*Q87+2.6*Q88-1.97*Q89-5.1*Q90-2.3*Q91
B:N77:   'mol. wt.
B:O77:   +B77
B:P77:   'CHI
B:Q77:   +54.58+11.4*Q87-51.8*Q88-16.1*Q89+42.9*Q90+13*Q91
B:N78:   '# o
B:O78:   +B78
B:P78:   ' -LOGKb
B:Q78:   +0.6*(1.604+3.65*Q87-7.66*Q88-5.74*Q89-0.31*Q90+5.09*Q91)+9.12*(1.33+1.5*Q87+5*Q88-1.5*Q89-10*Q90-2
         0.4*Q91)-2.71
B:N79:   '# n
B:O79:   +B79
B:P79:   'DIPOLE M
B:Q79:   +1.33+1.5*Q87+5*Q88-1.5*Q89-10*Q90-20.4*Q91
B:N80:   '# s
B:O80:   +B80
B:P80:   'SUM INT
B:Q80:   @SUM(B81..B103)
B:N81:   'c-c
B:O81:   +B81
B:P81:   'Tb-Tm
B:Q81:   +Q92-Q85
B:N82:   'arom. c-c
B:O82:   +B82
B:P82:   'MOL. WT.
B:Q82:   +87.8+149.6*Q87-78.3*Q88-239*Q89+20*Q90-386*Q91
B:N83:   'c=c
B:O83:   +B83
B:P83:   'COUNT
B:Q83:   +0.5*(B81+B101+B102+B103+@SUM(B85..B90)+@SUM(B92..B98))+0.75*(B82)+1*(B78+B80+B83+B91)+1.5*(
         B84+B79+B99+B100)-0.5*B94
B:N84:   'c=-c
B:O84:   +B84
B:P84:   'logben/h2o
B:Q84:   +1.81+4.4*Q87-13.2*Q88-6*Q89-18*Q91
B:N85:   'c-h
B:O85:   +B85
B:P85:   'Tm
B:Q85:   +77.2+295*Q87+194*Q88-443*Q89+339*Q90-406*Q91
B:N86:   'c-f
B:O86:   +B86
B:P86:   'B rho g/cc
B:Q86:   +Q82/Q106
B:N87:   'c-cl
B:O87:   +B87
```

TABLE 42B SYNPROPS Hierarchical Model, Property Equations

```
B:P87:  'B
B:Q87:  (9.26*B84+4.28*B85+1.48*B86+6.44*B87+1.3*B88-0.38*B89+13.52*B90-4.14*B91+10.43*B92+1.85*B93-1.11*B
        94-2.43*B95+11.28*B96+9.84*B97-0.46*B98-11.06*B99+3.82*B100+19.11*B101+14.22*B102+8.55*B103+10.45
        *B104+B83*4.36-B82*0.08-50.647*B76+0.2251*B77+0.82*B78+2.31*B80-4.09*B81-1.99*B106-0.99*B107+1.17*
        B79-0.74*B105-0.77*B108-0.47*B109-0.31*B110)*0.01
B:N88:  'c-br
B:O88:  +B88
B:P88:  'C
B:Q88:  (-0.15*B85+6.24*B86+5.7*B87+5.23*B88+4.64*B89+30.3*B90+11.39*B91+27.99*B92-2.04*B93-22.99*B94-19.8
        6*B95+4.44*B96+25.76*B97+22.02*B98+13.95*B99+16.12*B100+29.38*B101-1.59*B102+22.73*B103+9.96*B10
        4-0.12*B105+8.8*B84-0.85*B83-5.605*B76+0.0294*B77-0.32*B78-3.91*B81-1.71*B82-17.62*B108-1.31*B79-0.1
        6*B106-5.7*B107-4.96*B109-6.57*B110+8.1*B80)*0.01
B:N89:  'c-l
B:O89:  +B89
B:P89:  'D
B:Q89:  (0.663*B76-0.0669*B77-0.31*B78+0.55*B79+8.67*B80+3.69*B81+2.77*B82+2.47*B83+0.34*B84-0.83*B85+2.2*
        B86+1.84*B87+3.89*B88+4.78*B89+2.05*B90+11.2*B91+4.59*B92-2.37*B93-6.02*B94-5.91*B95-7.22*B96+6.05
        *B97+13.6*B98+18.39*B99+7.76*B100-1.38*B101-11.7*B102+3.82*B103-10.34*B104+1.44*O105+2.57*B106-0.
        11*B107-0.63*B108-0.74*B109+1.81*B110)*0.01
B:N90:  'oh
B:O90:  +B90
B:P90:  'E
B:Q90:  (0.34*B84-1.05*B85-1.17*B86-0.59*B87-0.37*B88-0.84*B89-0.35*B90+3.89*B91-0.88*B92-1.8*B93-2.71*B94-2.5
        3*B95-5.15*B96+4.02*B97+7.68*B98+9.9*B99+2.55*B100-2.72*B101-8.91*B102-8.35*B103-2.85*B104+1.32*B8
        3+1.27*B82+0.95*B106+3.12*B76-0.005*B77-0.39*B78+4.73*B80+1.67*B81+0.043*B107-0.53*B79+0.3*B105-0.
        44*B107-1.42*B108-0.58*B109-1.4*B110)*0.01
B:N91:  'o
B:O91:  +B91
B:P91:  'F
B:Q91:  (-1.14*B84-0.78*B85-0.66*B86-0.57*B87+0.15*B88-0.24*B89+1.84*B90+0.44*B91-3.56*B92+1.23*B93+1.61*B9
        4+2.67*B95-0.41*B96+0.17*B97+1.62*B98+5.08*B99+1.72*B100-4.5*B101+2.34*B102-1.18*B103-4.2*B104+0.1
        6*B83+1.08*B82+2.8*B76-0.014*B77-0.28*B78-0.58*B80+2.11*B81+0.19*B107-0.3*B79-0.67*B105-0.61*B106-0
        .51*B108+0.31*B109+2.41*B110)*0.01
B:N92:  'c=o
B:O92:  +B92
B:P92:  'Tb
B:Q92:  +66.39+532.6*Q87+223.6*Q88-365.4*Q89-250.8*Q90-794.6*Q91
B:N93:  'ch=o
B:O93:  +B93
B:P93:  'SOL PAR
B:Q93:  +8.97+5.22*Q87+12.1*Q88-15*Q89-12.9*Q90-7.7*Q91
B:N94:  'cooh
B:O94:  +B94
B:P94:  'Tc
B:Q94:  +248.6+770.4*Q87+343.1*Q88-927*Q89-68*Q90-1810*Q91
B:N95:  'coo
B:O95:  +B95
B:P95:  'Pc
B:Q95:  +46.3-7.1*Q87+56.2*Q88-243*Q89-51*Q90-230*Q91
B:N96:  'sh
B:O96:  +B96
B:P96:  'log AC CO
B:Q96:  +1.241+Q87*5.09+Q88*13.54+3.36*Q89+6.93*Q90+7.39*Q91
B:N97:  'nh2
B:O97:  +B97
B:P97:  'log par co
B:Q97:  +1.604+3.65*Q87-7.66*Q88-5.74*Q89-0.31*Q90+5.09*Q91
```

TABLE 42C SYNPROPS Hierarchical Model, Property Equations

```
B:N98:   'nh
B:O98:   +B98
B:P98:   'Hv
B:Q98:   +8.197+14.92*Q87+9.615*Q88-8.07*Q89-12.8*Q90+14.5*Q91
B:N99:   ' -n--
B:O99:   +B99
B:P99:   'Surf Ten
B:Q99:   +L114+41*Q87+L116*Q88+L117*Q89+L118*Q90+L119*Q91
B:N100:  ' =n-
B:O100:  +B100
B:P100:  'log visc
B:Q100:  (-0.23+2.44*Q87+1.26*Q88-0.94*Q89-8.1*Q90+5.06*Q91)
B:N101:  'cn
B:O101:  +B101
B:P101:  'log(et/h2o)
B:Q101:  +1.91+3.3*Q87-11*Q88-9.8*Q89-8*Q90+1*Q91
B:N102:  'conh2
B:O102:  +B102
B:P102:  'logoil/h2o
B:Q102:  +1.23+3.5*Q87-10.2*Q88-7.5*Q89+12*Q90-3*Q91
B:N103:  'no2
B:O103:  +B103
B:P103:  'logcyh/h2o
B:Q103:  +1.29+3.1*Q87-14.2*Q88-7.4*Q89+3*Q90-19*Q91
B:N104:  'cycles(rings)
B:O104:  +B104
B:P104:  'logchl/h2o
B:Q104:  +1.77+5.2*Q87-9.2*Q88-3.1*Q89-26*Q91
B:N105:  'tert C
B:O105:  +B105
B:P105:  'logcl4/h2o
B:Q105:  +1.55+3.3*Q87-12.5*Q88-3.7*Q89-6*Q90-33*Q91
B:N106:  'quat C
B:O106:  +B106
B:P106:  'vol.Bondi
B:Q106:  +48.9+94.6*Q87-34.5*Q88+89*Q89+11*Q90+35*Q91
B:N107:  'x-c-x
B:O107:  +B107
B:P107:  'molvolexp
B:Q107:  +90+140*Q87-90*Q88+246*Q89-108*Q90-3*Q91
B:N108:  'x3-c-x3
B:O108:  +B108
B:P108:  'C
B:Q108:  +0.5*B81+B82+B84+0.25*(B86+B88)+B92+B93+B94+B95+B101+B102+2*B83
B:N109:  'c=c-x
B:O109:  +B109
B:P109:  'H
B:Q109:  +B85+B90+B93+B94+B96+2*B97+B98+2*B102
B:N110:  'arom. fusions
B:O110:  +B110
B:P110:  'O
B:Q110:  +B90+B91+B92+B93+2*B94+2*B95+B102+2*B103
B:N111:  'S
B:O111:  +B96
B:P111:  'N
B:Q111:  +B97+B98+B99+B100+B101+B102+B103
B:N112:  'CL
```

TABLE 42D SYNPROPS Hierarchical Model, Property Equations

```
B:O112:   +B87
B:P112:   'F
B:Q112:   +B86
B:N113:   'BR
B:O113:   +B88
B:P113:   'I
B:Q113:   +B89
B:N114:   'loggamma
B:O114:   (1.241+5.09*Q87+13.54*Q88+3.36*Q89+6.83*Q90+7.39*Q91)
B:P114:   'D1
B:Q114:   +1.39E-07/(Q100*10^(-5))
B:N115:   'm.refract
B:O115:   +22.94+52.89*Q87-21.58*Q88+4.21*Q89+83.6*Q90-12.02*Q91
B:P115:   'D12
B:Q115:   '
B:N116:   'thermcond
B:O116:   +3.17+3.19*Q87+4.5*Q88+2.1*Q89-13.2*Q90-14.5*Q91
B:P116:   'isothcom
B:Q116:   +11.6-22.6*Q87-10.8*Q88+57.6*Q89+11*Q90+22*Q91
B:N117:   'COST $/g
B:O117:   (0.01/B77)*(Q108*1+Q109*1+Q110*1+Q111*1+Q112*3+Q113*5+O111*2+O112*2+O113*4)
B:P117:   'w
B:Q117:   +0.42857*(Q92/Q94)*(1/(1-(Q92/Q94)))*(@LOG(Q95))
B:N118:   ' MCL
B:O118:   +1000*@EXP(-(4.5906817+0.6819838*Q87+14.218037*Q88+16.568472*Q89-71.61533*Q90+58.154409*Q91))
B:P118:   'Zc
B:Q118:   +0.291-0.08*Q117
B:N119:   '  TAP
B:O119:   @EXP(-(-0.18926+0.439808*Q87+17.65265*Q88+14.10209*Q89-58.6877*Q90+99.7817*Q91))
B:P119:   'AIR
B:Q119:   @EXP(-(-0.11241+2.273426*Q87+17.73686*Q88+15.5473*Q89-54.5266*Q90+105.4682*Q91))
B:N120:   ' COM
B:O120:   @EXP(-(-5.08345+1.893833*Q87+17.59624*Q88+21.39851*Q89-102.557*Q90+97.51875*Q91))
B:P120:   ' RES
B:Q120:   @EXP(-(-4.06262+2.084032*Q87+14.7937*Q88+16.6342*Q89-75.3489*Q90+80.87492*Q91))
B:N121:   'drug1
B:O121:   +4.276992+1.539887*Q87+9.439779*Q88-1.43346*Q89+5.149976*Q90+40.43886*Q91
B:P121:   'drug2
B:Q121:   -8.1294-1110.27*Q87+535.1663*Q88-255.649*Q89+2694.279*Q90-2060.02*Q91
B:N122:   'drug4
B:O122:   +95.87415+233.9213*Q87+21.71042*Q88+6.720305*Q89+2086.843*Q90+294.2722*Q91
B:P122:   'drug3
B:Q122:   +3.607096+0.287852*Q87+10.02711*Q88-3.68816*Q89-6.33959*Q90+26.95464*Q91
B:N123:   'drug6
B:O123:   +4.854351-2.69057*Q87+4.847684*Q88-8.01527*Q89+6.459043*Q90-0.42498*Q91
B:P123:   'drug5
B:Q123:   +10.51994+7.018316*Q87-2.32424*Q88+8.363076*Q89-11.9406*Q90+19.66987*Q91
```

TABLE 43 SYNPROPS Hierarchical Model, Parameter Calculations

B:C111:	'B=
B:D111:	(@SUM(D75..D110))*0.01
B:E111:	'C=
B:F111:	@SUM(F75..F110)*0.01
B:G111:	'D=
B:H111:	@SUM(H75..H110)*0.01
B:I111:	'E=
B:J111:	@SUM(J75..J110)*0.01
B:K111:	'F=
B:L111:	@SUM(L75..L110)*0.01
B:B76:	1
B:C76:	-50.647
B:D76:	+B76*C76
B:E76:	-5.605
B:F76:	+B76*E76
B:G76:	0.663
B:H76:	+B76*G76
B:I76:	3.12
B:J76:	+B76*I76
B:K76:	2.8
B:L76:	+B76*K76
B:B77:	+12.011*Q108+1.0079*Q109+15.9994*Q110+14.0067*Q111+18.9984*Q112+32.06*O111+35.453*O112+79.904 *Q113
B:C77:	0.2251
B:D77:	+B77*C77
B:E77:	0.0294
B:F77:	+B77*E77
B:G77:	-0.0669
B:H77:	+B77*G77
B:I77:	-0.0051
B:J77:	+B77*I77
B:K77:	-0.014
B:L77:	+B77*K77
B:B78:	+B90+B91+B92+B93+2*B94+2*B95+B102+2*B103
B:C78:	0.82
B:D78:	+B78*C78
B:E78:	-0.32
B:F78:	+B78*E78
B:G78:	-0.31
B:H78:	+B78*G78
B:I78:	-0.39
B:J78:	+B78*I78
B:K78:	-0.28
B:L78:	+B78*K78
B:B79:	+B97+B98+B99+B100+B101+B102+B103
B:C79:	1.17
B:D79:	+B79*C79
B:E79:	-1.31
B:F79:	+B79*E79
B:G79:	0.55
B:H79:	+B79*G79
B:I79:	-0.53
B:J79:	+B79*I79
B:K79:	-0.3
B:L79:	+B79*K79
B:B80:	+B96
B:C80:	2.31

TABLE 43 SYNPROPS Hierarchical Model, Parameter Calculations (continued)

B:D80:	+B80*C80
B:E80:	8.1
B:F80:	+B80*E80
B:G80:	8.67
B:H80:	+B80*G80
B:I80:	4.73
B:J80:	+B80*I80
B:K80:	0.58
B:L80:	+B80*K80
B:B81:	0
B:C81:	-4.09
B:D81:	+B81*C81
B:E81:	-3.91
B:F81:	+B81*E81
B:G81:	3.69
B:H81:	+B81*G81
B:I81:	1.67
B:J81:	+B81*I81
B:K81:	2.11
B:L81:	+B81*K81
B:B82:	0.000135101747709757
B:C82:	-0.08
B:D82:	+B82*C82
B:E82:	-1.71
B:F82:	+B82*E82
B:G82:	2.77
B:H82:	+B82*G82
B:I82:	1.27
B:J82:	+B82*I82
B:K82:	1.08
B:L82:	+B82*K82
B:B83:	0.00138590445593159
B:C83:	4.36
B:D83:	+B83*C83
B:E83:	-0.85
B:F83:	+B83*E83
B:G83:	2.47
B:H83:	+B83*G83
B:I83:	1.32
B:J83:	+B83*I83
B:K83:	0.16
B:L83:	+B83*K83
B:B84:	3.29730041538257E-05
B:C84:	9.26
B:D84:	+B84*C84
B:E84:	8.8
B:F84:	+B84*E84
B:G84:	0.34
B:H84:	+B84*G84
B:I84:	0.34
B:J84:	+B84*I84
B:K84:	-1.14
B:L84:	+B84*K84
B:B85:	3
B:C85:	4.28
B:D85:	+B85*C85
B:E85:	-0.15
B:F85:	+B85*E85
B:G85:	-0.83
B:H85:	+B85*G85
B:I85:	-1.05
B:J85:	+B85*I85
B:K85:	-0.78
B:L85:	+B85*K85
B:B86:	0
B:C86:	1.48
B:D86:	+B86*C86
B:E86:	6.24
B:F86:	+B86*E86
B:G86:	2.2
B:H86:	+B86*G86
B:I86:	-1.17
B:J86:	+B86*I86
B:K86:	-0.66
B:L86:	+B86*K86
B:B87:	0.00128513133885904
B:C87:	6.44
B:D87:	+B87*C87
B:E87:	5.7
B:F87:	+B87*E87
B:G87:	1.84
B:H87:	+B87*G87
B:I87:	-0.59
B:J87:	+B87*I87
B:K87:	-0.57
B:L87:	+B87*K87
B:B88:	0.00234987372777171
B:C88:	1.3
B:D88:	+B88*C88
B:E88:	5.23
B:F88:	+B88*E88
B:G88:	3.89
B:H88:	+B88*G88
B:I88:	-0.37
B:J88:	+B88*I88
B:K88:	0.15
B:L88:	+B88*K88
B:B89:	2.75951670591105
B:C89:	-0.38
B:D89:	+B89*C89
B:E89:	4.64
B:F89:	+B89*E89
B:G89:	4.78
B:H89:	+B89*G89
B:I89:	-0.84
B:J89:	+B89*I89
B:K89:	-0.24
B:L89:	+B89*K89
B:B90:	0.122937674367141
B:C90:	13.52
B:D90:	+B90*C90
B:E90:	30.3
B:F90:	+B90*E90
B:G90:	2.05

TABLE 43 SYNPROPS Hierarchical Model, Parameter Calculations (continued)

B:H90:	+B90*G90
B:I90:	-0.35
B:J90:	+B90*I90
B:K90:	1.84
B:L90:	+B90*K90
B:B91:	0
B:C91:	-4.14
B:D91:	+B91*C91
B:E91:	11.39
B:F91:	+B91*E91
B:G91:	11.2
B:H91:	+B91*G91
B:I91:	3.89
B:J91:	+B91*I91
B:K91:	0.44
B:L91:	+B91*K91
B:B92:	1.09632730035834
B:C92:	10.43
B:D92:	+B92*C92
B:E92:	27.99
B:F92:	+B92*E92
B:G92:	4.59
B:H92:	+B92*G92
B:I92:	0.88
B:J92:	+B92*I92
B:K92:	-3.56
B:L92:	+B92*K92
B:B93:	2.50590718910703E-06
B:C93:	1.85
B:D93:	+B93*C93
B:E93:	-2.04
B:F93:	+B93*E93
B:G93:	-2.37
B:H93:	+B93*G93
B:I93:	-1.8
B:J93:	+B93*I93
B:K93:	1.23
B:L93:	+B93*K93
B:B94:	0.00209984681720876
B:C94:	-1.11
B:D94:	+B94*C94
B:E94:	-22.99
B:F94:	+B94*E94
B:G94:	-6.02
B:H94:	+B94*G94
B:I94:	-2.71
B:J94:	+B94*I94
B:K94:	1.61
B:L94:	+B94*K94
B:B95:	0.00113187114949053
B:C95:	-2.43
B:D95:	+B95*C95
B:E95:	-19.86
B:F95:	+B95*E95
B:G95:	-5.91
B:H95:	+B95*G95
B:I95:	-2.53
B:J95:	+B95*I95
B:K95:	2.67
B:L95:	+B95*K95
B:B96:	0.002012092270902
B:C96:	11.28
B:D96:	+B96*C96
B:E96:	4.44
B:F96:	+B96*E96
B:G96:	-7.22
B:H96:	+B96*G96
B:I96:	-5.15
B:J96:	+B96*I96
B:K96:	-0.41
B:L96:	+B96*K96
B:B97:	0.000499508872349091
B:C97:	9.84
B:D97:	+B97*C97
B:E97:	25.76
B:F97:	+B97*E97
B:G97:	6.05
B:H97:	+B97*G97
B:I97:	4.02
B:J97:	+B97*I97
B:K97:	0.17
B:L97:	+B97*K97
B:B98:	0.000672014609973495
B:C98:	-0.46
B:D98:	+B98*C98
B:E98:	22.02
B:F98:	+B98*E98
B:G98:	13.6
B:H98:	+B98*G98
B:I98:	7.68
B:J98:	+B98*I98
B:K98:	1.62
B:L98:	+B98*K98
B:B99:	0.0020034950927622
B:C99:	-11.06
B:D99:	+B99*C99
B:E99:	13.95
B:F99:	+B99*E99
B:G99:	18.39
B:H99:	+B99*G99
B:I99:	9.9
B:J99:	+B99*I99
B:K99:	5.08
B:L99:	+B99*K99
B:B100:	0.00130417345205668
B:C100:	3.82
B:D100:	+B100*C100
B:E100:	16.12
B:F100:	+B100*E100
B:G100:	7.76
B:H100:	+B100*G100
B:I100:	2.55
B:J100:	+B100*I100
B:K100:	1.72

TABLE 43 SYNPROPS Hierarchical Model, Parameter Calculations (continued)

B:L100:	+B100*K100
B:B101:	0
B:C101:	19.11
B:D101:	+B101*C101
B:E101:	29.38
B:F101:	+B101*E101
B:G101:	-1.38
B:H101:	+B101*G101
B:I101:	-2.72
B:J101:	+B101*I101
B:K101:	-4.5
B:L101:	+B101*K101
B:B102:	0.0025214986451494
B:C102:	14.22
B:D102:	+B102*C102
B:E102:	-1.59
B:F102:	+B102*E102
B:G102:	-11.7
B:H102:	+B102*G102
B:I102:	-8.91
B:J102:	+B102*I102
B:K102:	2.34
B:L102:	+B102*K102
B:B103:	0.00378433205528902
B:C103:	8.55
B:D103:	+B103*C103
B:E103:	22.73
B:F103:	+B103*E103
B:G103:	3.82
B:H103:	+B103*G103
B:I103:	-8.35
B:J103:	+B103*I103
B:K103:	-1.18
B:L103:	+B103*K103
B:B104:	0
B:C104:	10.45
B:D104:	+B104*C104
B:E104:	9.96
B:F104:	+B104*E104
B:G104:	-10.34
B:H104:	+B104*G104
B:I104:	-2.85
B:J104:	+B104*I104
B:K104:	-4.2
B:L104:	+B104*K104
B:B105:	0
B:C105:	-0.74
B:D105:	+B105*C105
B:E105:	-0.12
B:F105:	+B105*E105
B:G105:	1.44
B:H105:	+B105*G105
B:I105:	0.3
B:J105:	+B105*I105
B:K105:	-0.67
B:L105:	+B105*K105
B:B106:	0
B:C106:	-1.99
B:D106:	+B106*C106
B:E106:	-0.16
B:F106:	+B106*E106
B:G106:	2.57
B:H106:	+B106*G106
B:I106:	0.95
B:J106:	+B106*I106
B:K106:	-0.61
B:L106:	+B106*K106
B:B107:	0
B:C107:	-0.99
B:D107:	+B107*C107
B:E107:	-5.7
B:F107:	+B107*E107
B:G107:	-0.11
B:H107:	+B107*G107
B:I107:	-0.44
B:J107:	+B107*I107
B:K107:	0.19
B:L107:	+B107*K107
B:B108:	0
B:C108:	-0.77
B:D108:	+B108*C108
B:E108:	-17.62
B:F108:	+B108*E108
B:G108:	-0.63
B:H108:	+B108*G108
B:I108:	-1.42
B:J108:	+B108*I108
B:K108:	-0.51
B:L108:	+B108*K108
B:B109:	0
B:C109:	-0.47
B:D109:	+B109*C109
B:E109:	-4.96
B:F109:	+B109*E109
B:G109:	-0.74
B:H109:	+B109*G109
B:I109:	-0.58
B:J109:	+B109*I109
B:K109:	0.31
B:L109:	+B109*K109
B:B110:	0
B:C110:	-0.31
B:D110:	+B110*C110
B:E110:	-6.57
B:F110:	+B110*E110
B:G110:	1.81
B:H110:	+B110*G110
B:I110:	-1.4
B:J110:	+B110*I110
B:K110:	2.41
B:L110:	+B110*K110

TABLE 44 VOC Sort

compound P* mm	M. W.	C ug/L	X in H2O	H atm-m3/m(	Vap Pres ATI	g/liter(gas)	g/meter3-gas	ppm	Expos Limit I	E.L.speciesi	ug/m3=CON	ug/m3refcon	CON/refcon
vinyl chloride de	6.25E+01	3.50E+01	1.01E-08	8.60E-02	8.68E-10	2.22E-09	2.22E-06	5.42E-02	1.00E+03	5.42E-05	2.22E+00	2.80E-02	7.92E+01
&1,1,1-Trichk loroethane	1.33E+02	1.80E+03	2.43E-07	1.72E-02	4.18E-09	2.28E-08	2.28E-05	5.58E-01	1.00E+03	5.58E-04	2.28E+01	3.30E+02	6.91E-02
ccl4	1.54E+02	8.37E+03	9.79E-07	2.30E-02	2.25E-08	1.42E-07	1.42E-04	3.47E+00	1.00E+03	3.47E-03	1.42E+02	6.00E-02	2.36E+03
vinyl chloride de	6.25E+01	5.00E+02	1.44E-07	8.14E-02	1.17E-08	3.00E-08	3.00E-05	7.33E-01	1.00E+01	7.33E-02	3.00E+01	1.99E-02	1.51E+03
Tetrachloroet thylene	1.66E+02	2.60E+04	2.82E-06	1.53E-02	4.32E-08	2.93E-07	2.93E-04	7.17E+00	1.00E+01	7.17E-01	2.93E+02	1.06E+00	2.77E+02
&1,1-dichloro oethylene	9.70E+01	3.20E+03	5.95E-07	1.50E-02	8.92E-09	3.54E-08	3.54E-05	8.65E-01	1.00E+01	8.65E-02	3.54E+01	4.90E-02	7.22E+02
&1,1,1,2-tetra chloroethane	1.68E+02	9.85E+03	1.06E-06	1.10E-02	1.16E-08	7.98E-08	7.98E-05	1.95E+00	7.50E+01	2.60E-02	7.98E+01	1.35E-01	5.91E+02
styrene	1.04E+02	2.80E+02	4.84E-08	9.70E-03	4.70E-10	2.00E-09	2.00E-06	4.89E-02	7.50E+01	6.52E-04	2.00E+00	1.17E-01	1.71E+01
trichloroeth ylene	1.31E+02	7.40E+01	1.01E-08	9.10E-03	9.23E-11	4.96E-10	4.96E-07	1.21E-02	7.50E+01	1.62E-04	4.96E-01	5.00E-01	9.93E-01
METHYLENE oride	9.30E+01	3.40E+01	6.59E-09	3.19E-03	2.10E-11	7.99E-11	7.99E-08	1.95E-03		ERR	7.99E-02	5.20E+00	1.54E-02
cis-1,2-di chloroethene	9.70E+01	1.00E+03	1.86E-07	3.19E-02	5.93E-09	2.35E-08	2.35E-05	5.75E-01		ERR	2.35E+01	7.30E+01	3.22E-01
furan	6.81E+01	5.00E+03	1.32E-06	5.70E-03	7.54E-09	2.10E-08	2.10E-05	5.14E-01		ERR	2.10E+01	3.50E+00	6.00E+00
benzene	7.81E+01	1.36E+04	3.14E-06	6.60E-03	2.07E-08	6.61E-08	6.61E-05	1.62E+00	1.00E+03	1.62E-03	6.61E+01	2.90E-01	2.28E+02
Xylene o,m,p	1.06E+02	2.50E+02	4.24E-08	6.25E-03	2.65E-10	1.15E-09	1.15E-06	2.82E-02	1.00E+03	2.82E-05	1.15E+00	7.30E+02	1.58E-03
&1,1-dichloro oethane	9.90E+01	4.14E+02	7.54E-08	1.54E-02	1.16E-09	4.70E-09	4.70E-06	1.15E-01	1.00E+03	1.15E-04	4.70E+00	3.50E+02	1.34E-02
&1,2-chlorob enzene	1.47E+02	3.31E+02	4.06E-08	3.60E-03	1.46E-10	8.78E-10	8.78E-07	2.15E-02	1.00E+01	2.15E-03	8.78E-01	3.15E+02	2.79E-03
chlorobenzen e	1.13E+02	5.00E+01	8.01E-09	3.50E-03	2.80E-11	1.29E-10	1.29E-07	3.15E-03	1.00E+01	3.15E-04	1.29E-01	2.00E+01	6.45E-03
&1,4-chlorob enzene	1.47E+02	7.63E+02	9.35E-08	3.10E-03	2.90E-10	1.74E-09	1.74E-06	4.26E-02	1.00E+01	4.26E-03	1.74E+00	3.00E-01	5.81E+00
&1,1,1-Trichk loroethane	1.33E+02	7.20E+00	9.72E-10	3.00E-02	2.92E-11	1.59E-10	1.59E-07	3.89E-03	7.50E+01	5.19E-05	1.59E-01	1.10E+03	1.45E-04
chloroform	1.19E+02	2.80E+02	4.23E-08	3.39E-03	1.43E-10	6.99E-10	6.99E-07	1.71E-02	7.50E+01	2.28E-04	6.99E-01	1.10E-01	6.36E+00
&1,2-dichloro obenzene	1.47E+02	5.00E+03	6.13E-07	2.80E-03	1.72E-09	1.03E-08	1.03E-05	2.52E-01	7.50E+01	3.36E-03	1.03E+01	3.15E+02	3.28E-02
&1,3-dichloro obenzene	1.47E+02	5.00E+03	6.13E-07	2.63E-03	1.61E-09	9.69E-09	9.69E-06	2.37E-01		ERR	9.69E+00	3.12E+02	3.11E-02
&1,4-dichloro obenzene	1.47E+02	5.00E+03	6.13E-07	2.37E-03	1.45E-09	8.73E-09	8.73E-06	2.14E-01		ERR	8.73E+00	1.46E-01	5.98E+01
ccl4	1.54E+02	5.00E+02	5.85E-08	2.30E-02	1.35E-09	8.47E-09	8.47E-06	2.07E-01		ERR	8.47E+00	2.69E-02	3.15E+02
&1,2,4-trichlo lorobenzene	1.81E+02	5.00E+03	4.96E-07	2.30E-03	1.14E-09	8.47E-09	8.47E-06	2.07E-01		ERR	8.47E+00	7.00E+01	1.21E-01
vinyl chloride de	6.25E+01	5.00E+01	1.44E-08	1.99E-01	2.87E-09	7.33E-09	7.33E-06	1.79E-01		ERR	7.33E+00	1.19E-02	6.16E+02
Tetrachloroet thylene	1.66E+02	1.00E+04	1.09E-06	1.53E-02	1.66E-08	1.13E-07	1.13E-04	2.76E+00		ERR	1.13E+02	4.70E+00	2.40E+01
&1,1-dichloro oethylene	9.70E+01	5.00E+02	9.29E-08	1.50E-02	1.39E-09	5.53E-09	5.53E-06	1.35E-01		ERR	5.53E+00	2.92E-03	1.89E+03
&1,1,1,2-tetra chloroethane	1.68E+02	5.00E+02	5.36E-08	1.10E-02	5.90E-10	4.05E-09	4.05E-06	9.91E-02		ERR	4.05E+00	1.35E-01	3.00E+01
styrene	1.04E+02	5.00E+02	8.65E-08	9.70E-03	8.39E-10	3.57E-09	3.57E-06	8.74E-02		ERR	3.57E+00	1.17E-01	3.05E+01
&1,2-dichlor oethane	9.90E+01	3.00E+01	5.46E-09	1.20E-03	6.55E-12	2.65E-11	2.65E-08	6.49E-04		ERR	2.65E-02	9.40E-02	2.82E-01
Trichloroethyl ylene	1.31E+02	5.00E+00	6.86E-10	9.10E-03	6.24E-12	3.35E-11	3.35E-08	8.20E-04		ERR	3.35E-02	2.06E-01	1.63E-01
&1,1,2-trichlo loroethane	1.33E+02	2.50E+01	3.38E-09	7.42E-04	2.51E-12	1.37E-11	1.37E-08	3.34E-04		ERR	1.37E-02	6.10E-02	2.24E-01
Toluene	9.21E+01	5.00E+04	9.78E-06	6.70E-03	6.55E-08	2.47E-07	2.47E-04	6.04E+00		ERR	2.47E+02	2.10E+03	1.18E-01
trans-1,2-dicł chloroethene	9.70E+01	3.79E+02	7.04E-08	6.60E-03	4.65E-10	1.84E-09	1.84E-06	4.51E-02		ERR	1.84E+00	7.30E+01	2.53E-02
furan	6.81E+01	5.00E+02	1.32E-07	5.70E-03	7.54E-10	2.10E-09	2.10E-06	5.14E-02		ERR	2.10E+00	3.50E+00	6.00E-01
benzene	7.81E+01	5.00E+02	1.15E-07	5.50E-03	6.34E-10	2.03E-09	2.03E-06	4.95E-02		ERR	2.03E+00	1.21E-01	1.67E+01
Xylene o,m,p	1.06E+02	2.00E+02	3.39E-08	4.33E-03	1.47E-10	6.38E-10	6.38E-07	1.56E-02		ERR	6.38E-01	1.01E+03	6.32E-04
&1,1-dichloro oethane	9.90E+01	5.00E+02	9.10E-08	4.26E-03	3.88E-10	1.57E-09	1.57E-06	3.84E-02		ERR	1.57E+00	3.50E+02	4.48E-03
&1,1,2,2-tet chloroethane	1.68E+02	5.00E+03	5.36E-07	6.44E-03	3.45E-09	2.37E-08	2.37E-05	5.80E-01		ERR	2.37E+01	3.50E+02	6.78E-02
carbondisulfi ide	1.47E+02	5.00E+02	6.13E-08	1.68E-02	1.03E-09	6.19E-09	6.19E-06	1.51E-01		ERR	6.19E+00	3.70E+02	1.67E-02
chlorobenzen e	1.13E+02	2.27E+03	3.64E-07	3.50E-03	1.27E-09	5.85E-09	5.85E-06	1.43E-01		ERR	5.85E+00	2.00E+01	2.93E-01
&1,4-chlorob enzene	1.47E+02	5.00E+02	6.13E-08	3.10E-03	1.90E-10	1.14E-09	1.14E-06	2.79E-02		ERR	1.14E+00	1.46E-01	7.82E+00
&1,1,1-Trichk loroethane	1.33E+02	5.00E+01	6.75E-09	3.00E-02	2.03E-10	1.11E-09	1.11E-06	2.70E-02		ERR	1.11E+00	3.15E+02	3.51E-03
chloroform	1.19E+02	5.00E+02	7.55E-08	2.88E-03	2.17E-10	1.06E-09	1.06E-06	2.59E-02		ERR	1.06E+00	4.30E-02	2.47E+01
&1,2-dichloro obenzene	1.47E+02	5.00E+02	6.13E-08	2.80E-03	1.72E-10	1.03E-09	1.03E-06	2.52E-02		ERR	1.03E+00	3.15E+02	3.28E-03
&1,3-dichloro obenzene	1.47E+02	5.00E+02	6.13E-08	2.63E-03	1.61E-10	9.69E-10	9.69E-07	2.37E-02		ERR	9.69E-01	3.12E+02	3.11E-03
&1,4-dichloro obenzene	1.47E+02	5.00E+02	6.13E-08	2.37E-03	1.45E-10	8.73E-10	8.73E-07	2.14E-02		ERR	8.73E-01	1.46E-01	5.98E+00
&1,2,4-trichlo lorobenzene	1.81E+02	5.00E+02	4.96E-08	2.30E-03	1.14E-10	8.47E-10	8.47E-07	2.07E-02		ERR	8.47E-01	7.00E+01	1.21E-02
ccl4	1.54E+02	5.00E+01	5.85E-09	2.30E-02	1.35E-10	8.47E-10	8.47E-07	2.07E-02		ERR	8.47E-01	2.69E-02	3.15E+01
Tetrachloroet thylene	1.66E+02	5.00E+01	5.43E-09	1.53E-02	8.31E-11	5.64E-10	5.64E-07	1.38E-02		ERR	5.64E-01	1.06E+00	5.32E-01
&1,1-dichloro oethylene	9.70E+01	5.00E+01	9.29E-09	1.50E-02	1.39E-10	5.53E-10	5.53E-07	1.35E-02		ERR	5.53E-01	2.92E-03	1.89E+02
&1,1,1,2-tetra chloroethane	1.68E+02	5.00E+01	5.36E-09	1.10E-02	5.90E-11	4.05E-10	4.05E-07	9.91E-03		ERR	4.05E-01	1.35E-01	3.00E+00
tetrahydrofur ran	7.21E+01	5.00E+03	1.25E-06	1.08E-04	1.35E-10	3.98E-10	3.98E-07	9.73E-03		ERR	3.98E-01	7.00E+00	5.68E-02
Tetrachlorobe enzenes	2.16E+02	5.00E+03	4.17E-07	1.00E-04	4.17E-11	3.68E-10	3.68E-07	9.01E-03		ERR	3.68E-01	1.05E+00	3.51E-01
styrene	1.04E+02	5.00E+01	8.65E-09	9.70E-03	8.39E-11	3.57E-10	3.57E-07	8.74E-03		ERR	3.57E-01	1.17E-01	3.05E+00
&1,2-dichlor oethane	9.90E+01	5.00E+02	9.10E-08	9.14E-04	8.32E-11	3.37E-10	3.37E-07	8.23E-03		ERR	3.37E-01	3.80E-02	8.86E+00
Trichloroethyl ylene	1.31E+02	5.00E+01	6.86E-09	9.10E-03	6.24E-11	3.35E-10	3.35E-07	8.20E-03		ERR	3.35E-01	2.06E-01	1.63E+00
anthracene	1.78E+02	5.00E+03	5.05E-07	8.60E-05	4.35E-11	3.17E-10	3.17E-07	7.75E-03		ERR	3.17E-01		ERR
&1,1,2-trichlo loroethane	1.33E+02	5.00E+02	6.75E-08	7.42E-04	5.01E-11	2.73E-10	2.73E-07	6.68E-03		ERR	2.73E-01	1.05E+05	2.60E-06
trans-1,2-dicł chloroethene	9.70E+01	6.24E+02	1.16E-07	6.32E-03	7.33E-10	2.91E-09	2.91E-06	7.11E-02		ERR	2.91E+00	7.00E+01	4.15E-02
MEK	7.21E+01	5.00E+03	1.25E-06	5.80E-05	7.25E-11	2.14E-10	2.14E-07	5.23E-03		ERR	2.14E-01	7.70E+02	2.78E-04
furan	6.81E+01	5.00E+01	1.32E-08	5.70E-03	7.54E-11	2.10E-10	2.10E-07	5.14E-03		ERR	2.10E-01	3.50E+00	6.00E-02
benzene	7.81E+01	5.00E+01	1.15E-08	5.50E-03	6.34E-11	2.03E-10	2.03E-07	4.95E-03		ERR	2.03E-01	1.21E-01	1.67E+00
Xylene o,m,p	1.06E+02	5.00E+01	8.49E-09	5.10E-03	4.33E-11	1.88E-10	1.88E-07	4.59E-03		ERR	1.88E-01	1.40E+03	1.34E-04
Diethyl phtha alate	2.06E+02	5.00E+03	4.37E-07	4.75E-05	2.08E-11	1.75E-10	1.75E-07	4.28E-03		ERR	1.75E-01	2.80E+03	6.25E-05
&1,1-dichloro oethane	9.90E+01	5.00E+01	9.10E-09	4.26E-03	3.88E-11	1.57E-10	1.57E-07	3.84E-03		ERR	1.57E-01	3.50E+02	4.48E-04
&1,1,2,2-tet chloroethane	1.68E+02	5.00E+02	5.36E-08	3.80E-04	2.04E-11	1.40E-10	1.40E-07	3.42E-03		ERR	1.40E-01	1.75E-02	8.00E+00
&1,2-chlorob enzene	1.47E+02	5.00E+01	6.13E-09	3.60E-03	2.21E-11	1.33E-10	1.33E-07	3.24E-03		ERR	1.33E-01	3.15E+02	4.21E-04

References

Section 1.1

The emergence of computational chemistry, Chapter 2 in *Mathematical Challenges from Theoretical/Computational Chemistry*, http://www.nap.edu/readingroom/books/mctcc/chap2.html.

Section 1.2

Cross-fertilization between mathematics and chemistry, Chapter 2 of *Mathematical Challenges from Theoretical/Computational Chemistry*, http://www.nap.edu/readingroom/books/mctcc/chap3.html.

Richon, A. B., An introduction to molecular modeling, *Mathematics*, 1, 83, 1994. http://www.awod.com/netsci/Science/Compchem/feature01.html.

Section 1.3

Young, D., *Introduction to Computational Chemistry*, http://www.auburn, edu/~youngd2/topics/compchem.html.

Section 1.4

Computational Chemistry and Materials Science (CCM), designed and maintained by the Programming Environment and Training (PET) initiative within the Department of Defense High Performance Computing Modernization Program (DoDHPCM). http://www.arl.hpc.mil/PET/cta/ccm/...ourses/foundations/foundations.html.

Flurchick, K., A Tour of the Foundations of Computational Chemistry and Material Science, Lecture 2: Classical Modeling, http://www.arl.hpc.mil/PET/cta/ccm/...tions/camd_md/node2.html#SECTION 002.

Flurchick, K., A Tour of the Foundations of Computational Chemistry and Material Science, Lecture 3: Quantum Molecular Dynamics, http://www.arl.hpc.mil/PET/cta/ccm/...foundations/qmd_md/qmd.html#SECTION 001.

Flurchick, K., A Tour of the Foundations of Computational Chemistry and Material Science, http://www.arl.hpc.mil/PET/cta/ccm/...thodsApp/MethodsApp/.html#SECTION 001.

Section 1.5

Computational Chemistry at the ARC, http://www.arc.unm.edu/workshop/quantum_chem/chemistry.html.

Section 1.6

Desktop Chemistry and Molecular Modeling Tools is part of Oxford Molecular Projects, http://dreamon.oxmol.co.uk/prods/.

Section 1.7

The Computer Simulation by Molecular Dynamics, Monte Carlo, and Molecular Mechanics of Liquid and Solid Phases, in CCP5 Computer Simulation of Condensed Phases, http://www.dl.ac.uk/CCP/CCP5/main.html.

Section 1.8

Chemical Modeling — Sparc Computational Approach, Overview of SPARC Principles.

Section 1.9

AHSystems Group, http://www2.ari.net/ahsystems/wwwhome.htm.

Section 1.10

Environmental Fate Data Base (ENVIROFATE) Complete: XREF, CASLST, DATALOG, BIOLOG, CHEMFATE, and BIODEG (on magnetic tape), Environmental Protection Agency, Office of Pollution, Prevention and Toxics, Nov., 1994, http://www.ntis.cov/fcpc/cpn5871.htm.

Section 1.11

Materials Science, Materials Science and Engineering for the 1990s, Summary, Conclusions, and Recommendations, http://www.ul.cs.cmu.edu/books/materials_science/0000025.htm.

Section 1.12

Bertalanffy, L. von, *General Systems Theory*, George Braziller, New York, 1968.

Section 1.13

Rogers, T. N. and Kline, A. A., Physical and Environmental Property Data Banks Management System and Simulation, Chemical Properties Tool (StEPP/PPMS), Michigan Technological University (References: Hand et al., Luehrs and Rogers, 1966; Miller, 1997; Miller et al., 1966).

Section 1.14

Kauffman, S. A., Abstract Intelligent Agent and Physical and Environmental Properties, Sept. 13, 1996, http://www.santafe.edu/sfi/People/kauffman/Investigations.html.

Section 1.15

Fujita, T., Iwasa, J., and Hansch, C., *JACS*, 86, 51–55, 1964.

Hansch, C. A., Maloney, P. P., Fujita, T., and Muir, R. M., *Nature*, 194, 178, 1962.

Hansch, C. A., *Acc. Chem. Res.*, 2, 232, 1969.

Hansch, C. A., Kim, D., Leo, A. J., Novellino, E., Silipo, C., and Vittoria, A., *CRC Crit. Rev. Toxicol.*, 19, 185, 1989.

Lyman, W. J., Reehl, W. F., and Rosenblatt, D. H., *Handbook of Chemical Property Estimation Methods: Environmental Behavior of Organic Compounds*. American Chemical Society, Washington, DC, 1990.

Howard, P. H. and Meylan, W. M., Quantitative Structure-Activity Relationships in Environmental Sciences, VII, Chen, F., Ed., Shuurmann, G., SETA Press, Pensacola, FL, 1997.

Richon, A. B. and Young, S. S., An Introduction to QSAR Methodology, http://www.netsci.org/Science/Compchem/feature19.html

Hansch, C., Leo, A., Hoekman, D., and Li, P., A Comprehensive Approach to Structure–Activity Relationships, Department of Chemistry, Pomona College, Claremont, CA.

Coats, E. A., QSAR Challenges and Opportunities: A Commentary, http://www.netsci.org/Science/Compchem/feature13.html.

Van de Waterbeend, H., *The QSAR and Modelling Society*, No. 6, Nov., 1995, [http://www.pharma.ethz.ch/qsar].

Section 1.16

Clark, M., Cramer III, R. D., Jones, D. M., Patterson, D. E., and Simeroth, P. E., Comparative Molecular Field Analysis (CoMFA). 2. Toward Its Use with 3D-Structural Databases, Manuscript of Tripos paper about CoMFA.

Semus, S. F., CoMFA: A Field of Dreams? http://www.netsci.org/Science/Compchem/feature11.html.

Cramer, R. D., Patterson, D. E., and Bunce, J. D., *J. Am. Chem. Soc.*, 110, 5959, 1988.

Section 1.17

Marshall, G. R. and Cramer, III, R. D., *Trends Pharmacol. Sci.*, 9(8), 285, 1988.

Van Drie, J. H., 3D Database Searching in Drug Discovery, http://www.awod.com/netsci/Science/Cheminform/feature06.html.

Waller, C. L. and Kellogg, G. E., Adding Chemical Information to CoMFA Models with Alternative 3D QSAR Fields, http://www.awod.com/netsci/Science/Compchem/feature10.html.

Ho, C. M. W. and Marshall, G. R., DBMAKER: A set of programs to generate three-dimensional databases based upon user-specified criteria, *J. Comput. Aided Mol. Design*, 9, 65–86, 1995.

Introduction of HASL 3D-QSAR methodology, *J. Med. Chem.*, 31, 1396, 1988.

Charifson, P. S., Leach, A. R., and Rusinko III, A., The Generation and Use of Large 3D Databases in Drug Discovery, http://www/awod.com/netsci/Science/Cheminform/feature03.html.

Basak, S. C., Grunwald, G. D., and Niemi, G. J., *3-D Molecular Structure and Chemical Graph Theory*, Plenum Press, New York, 1995.

Todeschini, R., Milano, Chemometrics, 1997, http://disat.umimi.it/chm/dfault.htm.

Section 1.18

Combinatorial Chemistry: A Strategy for the Future, Molecular Connection, March, 1995, http://www.awod.com/netsci/Science/Combichem/feature02.html.

Krieger, J. H., Combinatorial chemistry spawns new software systems to manage flood of information, *Chem. Eng. News*, Feb. 12, 1996.

Xiao, Dong Xiang, Schultz, P. G., LBNL, Materials Sciences Div., 1996, Highlight, CAM, Biomolecular Program, http://www.161.govmsd/PLS/Schultz/96//96-13_PGS-sym.

Section 1.19

Kier, L. B. and Hall, L. H., *Molecular Connectivity in Structure-Activity Analysis*, John Wiley & Sons, New York, 1986.

Kier, L. B., *J. Pharm. Sci.*, 69(7), 807, 1980.

Hall, L. H. and Kier, L. B., *J. Mol. Struct. (Theochem)*, 134, 309–316, 1986.

Bonchev, D., *Information Theoretic Indices for Characterization of Chemical Structures*, John Wiley & Sons, New York, 1983.

Section 1.20

Ho, C. M. W. and Marshall, G. R., *J. Comput. Aided Mol. Design*, 9, 65–86, 1995.

Computer-Assisted Drug/Molecular Design, Software Lists, (Internet).

Balbes, L. M., Mascarella, S. W., Boyol, D. B., A Perspective of Modern Methods in Computer-Aided Drug Design, Reviews in Computational Chemistry, Vol. 4, Lipkowitz, K. B. and Floyd, D. B., Eds., VCH Publishers, 1994.

Section 1.21

The WWW Chemical Structures Database, http://schiele.organik.uni-erlangen.de/services/webmol.html.

Neubuhler, S. K., Thull, U., Altomare, A., Carta, V., Gaillard, P., Carrupt, P., Carotti, B., Test, B., *JMC*, 38, 3874–3883, 1995.

Golenbar, D. E., Vesterman, V. B., Vorpagel, E., Apex-3D, Expert System for Drug Design, http://awod.com/netxci/science/compchem/feature09.html.

Cho, S. J., Cummins, D., Bentley, J., Andrews, C. W., Tropsha, Back to 2D QSAR: Application of Genetic Algorithm to PLS Selection of Topological Indices, http://mmlin1.pha.unc.edu/.

Section 1.22

Balbes, L. M., Mascarella, S. W., and Boyd, D. B., A perspective of modern methods in computer-aided drug design, *Reviews in Computational Chemistry*, Vol. V, K. B. Lipkowitz and D. B. Boyd, Ed., VCH Publishers, N.Y., 1994.

Bugg, C. E., W. M. Carson, and J. A. Montgomery, Drugs by design, *Sci. Am.*, 269, 92–98, 1993.

Sullivan, M., Reshaping the drug discovery environment, *Today's Chemist Work*, 7(1), 11–12, 15, 1988.

Golender, V., Vesterman,V. B., and Vorpagel, E., APEX-3D Expert System for Drug Design, http://www.awod.com/netsci/Science/Compchem/feature09.html.

Advanced Computational Methods for Rational Drug Design by the Center for Computational Engineering at Sandia National Laboratories, October 30, 1992, http://midway.ca.sandia.gov/doe_qc/fwp_intro.html.

Chemical Applications of Topology and Graph Theory, King, R. B., (Ed.), Studies in Physical and Theoretical Chemistry, 28, Elsevier Science publishers, Amsterdam, 1983.

Section 1.23

Boethling, B., Tools for Assessing Chemicals in the Environment and Designing for Biodegradation, U.S. EPA and Georgetown University, Green Chemistry Seminar Series/georgetown.edu/departments/chemistry/gree.

Howard, P. H. and Meylan, W. M., Quantitative Structure-Activity Relationships in Environmental Sciences, VII, Chen, F., Ed., Shuurmann, G., SETA Press, Pensacola, FL, 1997.

Lyman, W. J., Reehl, W. F., and Rosenblatt, D. H. , *Handbook of Chemical Property Estimation Methods: Environmental Behavior of Organic Compounds*, American Chemical Society, Washington, D.C., 1990.

Section 1.24

Hokanson, D. R., Rogers, T. N., Hand, D. W., Miller, M. D., and Crittenden, J. C., A Physical Resource Tool for Air and Water Treatment Unit Operations, Submitted to Environmental Progress, 1977.

Hand, D. W., Gobin, F., Hokanson, D. R., Crittenden, J. C., and Rogers, T. N., Adsorption Simulation Software, Submitted to Environmental Progress, 1997.

Shonnard/Halada/, Pacific NWP@Resource Center, P2 Research Projects Database, EFRAT (Environmental Fate and Risk Assessment), 1998, http://pprc/.org/pprc/rpd/fedfund/epa/epaeed/envir.

DIPPR Pure Component Compilation (DIPPR), http://www.dippr.com/ind_num2.htm#DIPPR.

http://www.beilstein.com/beist_5.html.

Section 1.25

Evaluating the Potential for Safe Substitutes in Pollution Prevention, EPA/600/R-94/170: The Product Side of Pollution Prevention: Evaluating the Potential for Safe Substitutes from RREL, Office of Research and Development, Cincinnati, OH.

Section 1.26

Lindsay, R. B. and Margenau, H., *Foundations of Physics*, Dover Publications, New York, 1957.

Coveny, P. and Highfield, R., *Frontiers of Complexity*, Fawcett Columbine, New York, 1995.

Section 1.27

Reviews in Computational Chemistry, Subject Index, Vol. 6, 1995.

Section 1.28

White, W. B., Johnson, S. M., and Dantzig, G. D., Chemical equilibria in complex mixtures, *J. Chem. Phys.*, 28, 751–755, 1958.

Dabdub, D., Mathematical Modeling of Air Pollution Dynamics and HPCC, http://nicarol.che.caltech.edu/~dabdub.

Seinfeld, J. H., *Air Pollution*, McGraw-Hill, NY, 1975.

Saunders, S. M., Bauch, O. L., Cooke, K. M., Pilling, M. J., and Smur, H. Waite, *Int. J. Chem. Kinetics*, 26, 113–130, 1994.

Computer Programs, HPC, High Performance Computing and Combustion, Combustion and High Performance Computing, http://www.osc.edu/pcrm/chpc.97.html.

Section 1.29

Glen, R. C. and Payne, A. W. R., A genetic algorithm for the automated generation of molecules within constraints, *J. Comput. Aided Mol. Design*, 9, 181–202, 1995, http://alfredo.wustl.edu/camd/282/paper.html.

Section 1.30

Friedler, F., et al., A combinatorial approach for generating candidate molecules with desired properties based on group contribution, *Comp. Chem. Eng.*, 22(6), 809–817, 1998.

Shuzo, Ohe, Session PC5-4, Physics and Chemistry, The Neural Network System of Predicting and Processing Thermodynamic Properties, http://www.codata.race.u-toko.ac.ip/pc5.html.

Leete, 4th International Meeting on DNA-Based Computing of Logically Reversible Computation with Minimal Energy Dissipation, 1998, http://cls.udel.edu/!dna4/.

Meusinger, R. and Moros, R., in Moll, R. Ed., *Software*, Entwicklung in der Chemie-9 Gesellshaft Deutcher Chemiker, Frankfurt, 1995, 205–211.

EPA Programs, UAM Guides User's Guide, http://www.iceis.mcnc.org/uamguide/usersguide/ug.ch1.html.

Section 1.31

Crabtree, E. W. and El-Hawagi, M. M., Synthesis of Environmentally Acceptable Reactions, AIChE Symposium, 90(50), 117, 1995.

Blurock, The Reaction Library manual, *JACS*, 119, 4033–4012, 1997, http://www.risc.unilinz.ac.at/people/blurock/!REACTION/prog/REACTION.toc.html.

Chen and Gasteiger, Agnew. Chem. Intern., Ed., 35, 763–765, 1996, Agnew. Chem., 108, 824–846, 1996.

Ferguson, A. M., Paterson, D. E., Carr, L. D., and Underliner, T. L., *J. Bimolecular Screening*, 1(2), 65, 1996.

Databases, http://www.che.ufl.edu/WWW-CHE/topics/data.html.

Section 1.32

DIPPR Pure Component Compilation (DIPPR), http://www.dippr.com/ind_num2.htm#DIPPR.

Section 1.33

NIST, Physical and Chemical Properties Division, Physical and Chemical Properties, http://properties,nist.gov/.

NIST, Physical and Chemical Properties Division, Technical Highlights, Technical Transfer in Computational Thermochemistry, http://properties.nist.gov./TAR/file32.html.

NIST, Physical and Chemical Properties Division, Technical Highlights, Thermodynamic Data for Environmental Fate and Risk Analysis, http://properties.nist.gov/TAR/file17.html.

NIST, Physical and Chemical Properties Division, Technical Highlights, Thermophysical Properties of Alternative Refrigerants, http://properties.nist.gov/TAR/file13.html.

NIST, Physical and Chemical Properties Division, Technical Highlights, Thermodynamic Data for the Industrial Application of Chemical Compounds, http://properies.nist.gov/TAR/file08.html.

NIST, Physical and Chemical Properties Division, Technical Highlights, Critical Evaluation of Chemical Kinetics Data — New Tools and Old Problems, http://properties.nist.gov/file05.html.

Section 1.34

Brocchini, S., James, K., Tangpasuthadol, V., and Kohn, J., *J. Am. Chem. Soc.*, 119(19), 4553, 1997. Equations of State of Gases, Liquids and Solids, CHM4411 Fall 1995 — Computer Project: Project 1.2, Computer Project: Project 1.2 http://www.chem.ufl.edu/~chapo/chm4411/chm4411_comp12html.html.

Bras, B. and Rosen, D., Computer Aided Design for Demanufacturing and Remanufacturing, 1966, [http://www.srl.gatech.edu/researc/CAD/CAD.abstract.html].

Hougen, D. A. and Watson, K. M., *Chemical Process Principles, Part 2, Thermodynamics*, John Wiley & Sons, NY, 1947.

Section 1.35

The Enigmatic Substance http://www.psc.edu/science/Jordan/Jordan.html matic Substance in Mysteries of Water: Simulating Small Water Clusters.

Sections 1.36 through 1.38

The following six URLs are from Boris Veytsman and Michael Kotelyanskii (1997), and cover equipartition, the harmonic oscillator, and crystals: http://planck. Plmsc.psu.edu/~www/ma...html#SECTION
00020000000000000000000
00030000000000000000000
00040000000000000000000
00025030000000000000000
http://planck.plmsc.psu.edu/~www/ma..
7c-1997/systems/Lecture4/index.html
7c-1997/systems/Lecture3/node3.html

Marcus, Y., *Introduction to Liquid State Chemistry*, John Wiley & Sons, NY, 1977.

Section 1.39

Floudas, C. A., Computer-Aided Systems Laboratory, http://titan.princeton.edu/.

Section 1.40

Benson, S. W., *Thermochemical Kinetics*, John Wiley & Sons, NY, 1976.

Ritter, E. R., THERM User's Manual, Department of Chemical Engineering and Environmental Science, NJIT, 1980.

Section 1.41

Gordon, S. and McBride, B. J., NASA SP-273, 1971.

Section 1.42

Kee, R. J., Rupley, F. M., and Miller, J. A., Sandia National Laboratories Report, SAND89-8003, 1989.

Coltrin, M. E., Kee, R. J., and Rupley, F. M., Sandia National Laboratories Report, SAND90-8003, 1990.

Coltrin, M. E., Kee, R. J., and Rupley, F. M., *Int. J. Chem. Kinet.*, 23, 1111, 1991.

Kee, R. J., Dixon-Lewis, G., Warnatz, J., Coltrin, M. E., and Miller, J. A., Sandia National Laboratories Report, SAND86-8246, 1986.

Section 1.43

Ianni, J., New, Powerful Kinetics Program: KINTECUS (Dec. 20, 1995), http://www.cpma.u-psud.fr/ccl/243.html.

KINTECUS, http://www.ioc.ac.ru/Chemistry/Soft/KINTECUS .en.html.

KINTECUS-2, http://www.cpma.u-psud.fr/ccl/244.html.

Section 1.44

Chemical Kinetics Simulator 1.01 Features, IBM (Internet).

The CKS Package — A Closer Look, IBM (Internet).

Section 1.45

Friedler, F., Varga, J. B., and Fan, L. T., Algorithmic Approach to the Integration of Total Flowsheet Synthesis and Waste Minimization, *AIChE Symp. Ser.*, No. 303, Vol. 90, 88, 1995.

Varga, J. B., Friedler, F., and Fan, L. T., Risk Reduction through Waste Minimizing Process Synthesis, presented at the 21st Annual RREL Research Symposium, Cincinnati, OH, April 4–6, 1995.

Partin, L. R., Application of Combinatorial Algorithms to Flowsheet Synthesis, April 29, 1997 (Internet).

Friedler, F., Varga, J. B., Hobor, P., and Fan, L. T., Combinatorial approach to process synthesis: sectionally continuous cost functions of operating units, *AIChE Annual Meeting*, Section 198, Paper 198c, 1997, http://www1.ch2.ufl.edu/meeting/1997/annual/session/198/c/index.html.

Section 1.46

Labute, P., *Journal of the Chemical Computing Group*, Winter, 1998, QuaSAR-Binary: A New Method for the Analysis of High Throughput Screening Data, http://www.netsci.org/Science/Compchem/feature21.html, presented at the 1998 Charleston Conference, March 9, 1998, submited for publication May, 1988.

Groundwater Modelling and Analysis Software, http://gwrp.cciw.ca/gwrp/software/software.html.

Section 1.47

Dabdub, D., Mathematical Modeling of Air Pollution Dynamics and HPCC, [http://nicarol.che.caltech.edu/~dabdub].

Section 1.48

Chemical Kinetics and Photochemical Data for Use in Stratospheric Modeling, http://remus.jpl.nasa.gov/jpl97/

Section 1.49

Measuring Air Quality, in Measuring Air Quality, The Pollutant Standards Index, OAQPS, Office of Air Quality Planning and Standards, U.S. Environmental Agency, EPA 451/K-94-001.

Section 1.50

Carslaw, N., Jacobs, P., and Pilling, M., Modelling in Leeds, Modelling OH and HO2 in the Marine Boundary Layer, Atmospheric Field Studies Group, http://chem.leeds.ac.uk/Atmospheric/Field/model.html.

Analysis of Atmospheric Hydrocarbons, http://129.11.12.1/Atmospheric/Field/voc.html.

Tropospheric Chemistry Modelling (Internet).

Mike Pilling (Home Page).

Hand, D. W., Yu, J. Crittenden, J. C. Rogers, T. N., and Kline, A. A., Prediction of liquid phase adsorption equilibrium from thermodynamic properties, American Institute of Chemical Engineers, Spring National Meeting, Session #106: Thermophysical Properties for Environmental Applications (Paper 106a), New Orleans, February, 1996.

Section 1.51

U.S. EPA Envirofacts Warehouse, http://www.epa.gov/enviro/index_java.html.

Section 1.52

Domine, D. and Devillers, *J., Sci. Comput. Automation*, 55, March, 1998.

Section 1.53

Hansch, C., Mabay, P. P., Fujita, T., and Muir, R. M., *Nature*, 194, 178–80, 1962.

Combustion and High Performance Computing, http://www.osc.edu/pcrm/chpc.97.html.

Section 1.54

Combustion, Environment and Heating, The Role of High Performance Simulation, http://www.osc.edu/pcrm/poster.html.

HPC, High Performance Computing and Combustion, Combustion and High Performance Computing, http://www.osc.edu/pcrm/chpc.97.html.

Sections 2.2 through 2.4

Bumble, S., Emerging Computer Simulation and Control of Plant Design and Retro-Design of Waste Minimiz-ation/Pollution Prevention in the Late Twentieth and Early Twenty-First Centuries, in *Proceedings Manual*, EPA Region III, Waste Minimization/Pollution Prevention Technical Conference, Philadelphia, June 3–5, 1996.

Section 2.5

Bumble, S., Chart presented before the National Meeting of the ACS, in Atlantic City, NJ, in 1969.

Hougen, O. A. and Watson, K. M., The law of corresponding states, in *Chemical Process Principles*, Part Two, John Wiley & Sons, New York, 1947.

Section 2.6

Hougen, O. A. and Watson, K. M., Group contributions, Chemical Process Principles, Part Two, John Wiley & Sons, New York, 1947.

Section 2.7

Fowler, R. H., *Statistical Thermodynamics*, Statistical Mechanics, Cambridge, 1966.

Section 2.8

Hierarchical Analysis of Molecular Spectra and Intramolecular Dynamics, http://anchi8.chm.anl.gov/g44/review/MJD/HAIVR.htm.

Section 2.9

Vapor Pressure Used for the Design of Equilibrium Stage Processes, ARSoftware, http://www.ari.net/ars/chemsage.html.

Sections 2.10 through 2.16

Cramer, R. D., *J. Am. Chem. Soc.*, 102(6), 1837 and 1843, 1980.

Cramer, R. D., *Quant. Struct.-Act. Relat.*, 2, 7-12, 1983.

Bumble, S., Emerging Computer Simulation and Control of Plant Design and Retro-Design of Waste Minimization/Pollution Prevention in the Late Twentieth and Early Twenty-First Centuries, in *Proceedings Manual*, EPA Region III, Waste Minimization/Pollution Prevention Technical Conference, Philadelphia, June 3–5, 1996.

Section 2.17

Schewe, P. F. and Stein, B., *Phys. News Update*, The American Institute of Physics Bulletin of Physics News, Number 326 (Story 1), June 28, 2997 and Ben Stein.

Sections 2.18 through 2.20

Bumble, S., Emerging Computer Simulation and Control of Plant Design and Retro-Design of Waste Minimization/Pollution Prevention in the Late Twentieth and Early Twenty-First Centuries, in *Proceedings Manual*, EPA Region III, Waste Minimization/Pollution Prevention Technical Conference, Philadelphia, June 3–5, 1996.

Section 2.22

Gorelick, O., Bucky Balls, Compound Chemistry, http://www.geocities.com/ResearchTriangle/Lab/5466/index.htm.

Sections 2.25 through 2.31

Purcell, W. P., Bass, G. E., and Clayton, J. M., *Strategy of Drug Design, A Molecular Guide to Biological Activity*, John Wiley & Sons, New York, 1973.

Section 2.32

Antineoplastic drugs, in *Physicians Desk Reference*, 48th ed., Medical Economics Data Production Company, Montvale, NJ, 1994.

Section 2.33

Jamming the Helix, Academic Press, Daily Insight, January 29, 1998, http://www.apnet.com/insight/01291998/grapha.htm.

Section 2.34

Brown, H. C., Physical Organic Chemistry/Steric Hindrance, The Nobel Prize Internet Archive, 1979 Nobel Laureate in Chemistry, http://www.almaz.com/nobel/chemistry/chemistry/1979a.html.

Section 2.35

Veith, The Role of Ecological Epidemiology in Regulatory Risk Assessment, Meeting, Society for Risk Analysis.

Liss and Slater, Volatization, http://gsf.de/ufis/ufis/model/133/gleichung/69.html.

Whitman, W. G., *Chem. Metall. Eng.*, 146–148, 1923.

Environmental Properties, 132. DIPPR Pure Component Compilation (DIPPR); http://www.dippr.com/ind_num2.htm#DIPPR http://www.apnet.com/insight/01291998/grapha.htm.

Section 2.36

Gombar, V. K., Enslein, K., and Reid, D. A., Computer Assisted Toxicity Assessment: Criteria for Acceptance, http//www.awpd.com/netsci/Science/Special/feature05.html.

Section 2.37

Web Elements, http://www.bonus.com:8080/contour/w...chem/web-elements/nofr-phys/pr.html.

Section 2.38

Mihelcic, J. R., Baillod, C. R., and Crittenden, S. C., *J. Air Water Mat. Assoc.*, 43(1), 97–105, 1993.

Physical Property Data Needs in Clean Manufacturing Design, National Center for Environmental Research and Quality Assurance, Office of Research and Development, U.S. Environmental Protection Agency, March 21, 1997, http://es.epa.gov/ncerqa_abstracts/...rs/cencitt/year2/clean/mullins.html.

Baker, A., An Evaluation of Sorption Models of Organic Compounds onto Activated Sludge, M.S. thesis, Michigan Technological University, Houghton, 1996.

Crittenden, J. C., Possible Research Interactions with cenCITT, in Particular Their Potential Interactions with the Effluent Treatment Design Option Tools, Nishihara, Ltd., Tokyo, Japan, June, 1996.

Crittenden, J. C., Interactions between the cenCITT Project on Effluent Treatment Design Option Tools and MTU's Present Project Water Treatment for the Space Station, NASA Marshall Space Center, Huntsville, AL, January, 1997.

Crittenden, J. C., Hand, D. W., Hokanson, D. R., and Bulloch, J. L., Predicting the performance of fixed-bed granular activated carbon adsorbers, First IAWQ Specialized Conference on Adsorption in Water Environment and Treatment Processes, Shirahama, Wakayama, Japan, November, 1996.

Figueroa, A., Boiler/Combustion Analysis, presented to visiting faculty members of the University of Sonora and University of Chihuahua, Mexico, Michigan Technological University, Houghton, December, 1996.

Figueroa, A. Boiler/Combustion Analysis, presented to Air Products and Chemicals, Inc., Michigan Technological University, Houghton, February, 1997.

Hand, D. W., Yu, J., Crittenden, J. C., Rogers, T. N., and Kline, A. A., Prediction of liquid phase adsorption equilibrium from thermodynamic properties, AIChE, Spring National Meeting, Session 106: Thermopysical Properties for Environmental Applications (Paper 106a), New Orleans, February, 1996.

Hand, D. W., Gobin, F., Hokanson, D. R., Crittenden, J. C., and Rogers, T. N., Adsorption Simulation Software, submitted to *Environmental Progress*, 1997.

Herlevich, D. R., Conceptual Design Stage Chemical Process Evaluation to Optimize Economic, Safety and Pollution Prevention Attributes — Comprehensive Assessment of Heat and Power Cogeneration Processes, M.S. thesis, Michigan Technological University, Houghton, 1996.

Herlevich, J., Miller, M., Parikh, P., Fifueroa, A., Stark, D., and Larson, E., Feasibility Survey of Cogeneration of Steam and Electricity at Michigan Technological University: An Incremental Return Comparison Based on Cost, Safety and Environmental Impact, Michigan Technological University, Houghton, December, 1996.

Hertz, D. W., and Patty, R. M., Multimedia Tool for Safety and Constructability, Interactive Demonstration and Alpha Test, Kelsafe Conference, The M.W. Kellogg Company, Houston, September, 1996.

Hokanson, D. R., Development of Software Design Tools for Physical Property Estimation, Aeration and Adsorption, M.S. thesis, Michigan Technological University, Houghton, 1996.

Hokanson, D. R., Hand, D. W., and Crittenden, J. C., Aeration Systems Analysis Program, submitted to *Environmental Progress*, 1996.

Hokanson, D. R., Rogers, T. N., Hand, D. W., Miller, M. D., and Crittenden, J. C., A physical property resource tool for air and water treatment unit operations, submitted to *Environmental Progress*, 1997.

Kline, A. A., Rogers, T. N., Pintar, A. J., Mullins, M. E., Selover, T., and Thomson, G., AIChE/DIPPR Projects 911 and 912: Developing a new, reliable source of environmental, safety, and health data, AIChE Spring National Meeting, Session 106: Thermophysical Properties for Environmental Applications (Paper 106e), New Orleans, February, 1996.

Luehrs, D. C. and Rogers, T. N., Correlation of the limiting partition coefficient of organic solutes between water and activated carbon with the octanol/water partition coefficient, Great Lakes Regional Meeting of the American Chemical Society, May, 1996.

Miller, M. D., Development of a Physical Property Management System to Support Clean Process Designs, M. S. thesis, Department of Chemical Engineering, Michigan Technological University, Houghton, October 1995.

Miller, M. D., Rogers, T. N ., Hokanson, D. R., Hand, D. W., and Kline, A. A., Development of a physical property management system to support clean process designs, AIChE Spring National Meeting, Session 106: Thermophysical Properties for Environmental Applications (Paper 106d), New Orleans, LA, February, 1996.

Parikh, P.B., Implementation and Application of the Dow Hazard Evaluation Indices in a Computer-Based Environment, M.S. thesis, Michigan Technological University, Fall, Houghton, 1996.

Rogers, T. N., Kline, A. A., and Unger, S. J., *Standard Handbook for Hazardous Waste Treatment & Disposal*, 2nd ed., Harry M. Freeman, Ed., Chapter 6.2: Distillation, McGraw-Hill, accepted for publication, summer 1997.

Shonnard, D. R. and Barna, B. A., Pollution Prevention Education in Chemical Engineering at Michigan Technological University, DOE Pollution Prevention in the 21st Century Conference XII, Chicago, July, 1996.

Shonnard, D. R., Markey, J. M., and Deshpande, P. A., Evaluation of unit operation-specific emission databases, in *Proceedings of the 1995 AIChE Summer National Meeting*, Boston, July,1995.

Shonnard, D. R., Herlevich, J., Jr., Parikh, P., Stark, D., Barna, B. A., Rogers, T. N. , Crowl, D. A. , Kline, A. A., and Radecki, P. P., Pollution assessment software as chemical industry process simulator enhancements, in *Proceedings of the Fifth World Congress of Chemical Engineering*, San Diego, CA, July, 1996.

Suri, R. P. S., Crittenden, J. C., Hand, D. W., and Perram, D. L., Steam regeneration of adsorbents and photocatalytic destruction of organic compounds, in press, *Water Environ. Res.*, 1995.

Toth, T. J. and Barna, B. A., Process optimization with environmental constraints: The Design Option Ranking Tool, AIChE Spring National Meeting, New Orleans, February, 1996.

Toth, T. J. and Barna, B. A., Application of the Design Option Ranking Tool Software for the environmental and economic optimization of a refinery light ends utility system, AIChE Spring National Meeting, New Orleans, February, 1996.

Yy, J., Crittenden, J. C., and Hand, D. W., The correlation of single solute adsorption from aqueous solution, to be submitted, 1997.

Section 2.39

Kier, L. B. and Hall, L. H., *Molecular Connectivity in Structure, Activity, Analysis*, John Wiley & Sons, NY, 1986.

Famini, G. R., Using Theoretical Descriptors in Quantitative Structure Activity Relationships and Linear Free Energy Relationships, http://www.netsci.org/Science/Compchem/feature08.html.

Cerrini, S., Quantitative Descriptors of Molecular Shape and Molecular Symmetry (Internet).

Section 2.40

Wikel, J. H., Dow, E. R., and Heathman, M., Interpretative Neural Networks for QSAR, http://netsci.org/Compchem/feature02.html.

Section 2.41 through 2.53

First Indo-U.S. Workshop on Mathematical Chemistry, Visva-Bharati University, Santiniketan, West Bengal, India, January 9–13, 1998, http://wyle.nrri.umn.edu/India/Abstract.html.

Section 2.54

Adleman, L. H., Molecular computations of solutions to combinatorial problems, *Science*, 266, 1021, 1994.

Aviram, A. and Ratner, M. A., Molecular rectifiers, *Chem. Phys. Lett.*, 29, 277–283, 1974.

Rambidi, N. G., Biomolecular Computer: Roots and Promises, Wayne State University, Department of Computer Science, Int. Research Institute for Mgt. Science, Moscow, Russia, http://www.cs.wayne.edu/~ngr/biocomputing1. html.

Section 2.55

Guillemin, E. A., Synthesis of networks from given parameter matrices, in *Theory of Linear Physical Systems*, John Wiley & Sons, New York, 1963.

Sections 2.56 through 2.59

Hohnsonbaugh, R., *Discrete Mathematics*, Macmillan, New York, 1984.

Welcome to the Flow Net Environment PCMS Lab, Manufacturing Engineering Department, Boston University (Internet).

Section 2.60

Naumann, T. and Lowis, D., HQSAR, A New Highly Predictive QSAR Technique, First Electronic Electronic Conference on Synthetic Organic Chemistry (ECSOC-1), www.mdpi.org/ecsoc/. September, 1–30, 1997.

Section 2.61

Krieger, J. H., Integration Hits Software Tools, *Chem. Eng. News*, October 6, 1997.

Section 2.62

Hsueh and Ching, WA20.4, Intelligent Huristic Applications, Tbu Search, Simulated Annealing and Their Hybrid for the Quadratic Assignment Problems, Informs.

Physical Properties of Inorganics, Plastics, Metals, Semiconductors, Ceramics, Composites and Other Materials, 1.11 Materials Science, Materials Science and Engineering for the 1990s, Summary, Conclusions, and Recommendations, http://www.ul.cs.cmu.edu/books/materials_science/0000025.htm.

Section 2.63

Membrane Research and Technology at McMaster University, Department of Chemistry, http://www.chemistry.mcmaster.ca/membrane/#modelling.

Section 2.64

NFS Benign Chemical Synthesis and Processing, Environmentally Benign Chemical Synthesis and Processing Research on Pollution Prevention at Its Source, A University–Industry Cooperative Program, Program Guideline, March 1, 1992, http://www.nsf.gov/pubs/stis1992/nsf9213/nsf9213.txt.

Section 2.65

NSF 91-108, Research in Electrochemical Synthesis, Program Guideline, September 1, 1991, http://www.nsf.gov/pubs/stis1991/nsf91108/nsf91108.txt.

Section 2.66 through 2.68

Friedler, F. and Fan, L. T., Design of Molecules with Desired Properties by Combinatorial Analysis, 1967, Preprint.

Section 2.69

Pierce, J. R., *Symbols, Signals and Noise*, Harper and Brothers, New York, 1961.

Coding Theory I (Internet).

Berger, T., Rate Distortion Theory, *A Mathematical Basis for Data Compression*, Prentice-Hall, Englewood Cliffs, NJ, 1971.

Section 270

Information Theory IEEE Information Theory Society Home Page, http://www.ifsoc.org/.

Section 2.71

Computer Synthesis of Molecules, Laboratory for Organic Synthesis Hoornaert, http://www.chemie.uni-regenburg.de/external.html.

Adleman, L. H., Molecular computation of solutions to combinatorial problems, *Science*, 266, 1021, 1994.

Section 2.72

Computer-Assisted Molecular Design (CAMD), http://http1.brunel.ac.uk:8080/depts/chem/ch2415/refer. htm.

Section 2.73

Pierre, D. A., *Optimization Theory with Applications*, Dover Publications, New York, 1986.

Section 2.74

Goicoechea, A., Hansen, D. R., and Duckstein, L., *Multiobjective Decision Analysis with Engineering and Business Applications*, John Wiley & Sons, New York, 1982.

Nemhauser, G. L. and Wolsey, L. A., *Interger and Combinatorial Optimization*, John Wiley & Sons, New York, 1989.

Owens, D. H., *Multivariable and Optimal Systems*, Accademic Press, London, 1981.

Mah, R. S., *Chemical Process Structures and Information Flows*, Butterworths, London, 1989.

Pamadimitriou, C. H. and Steiglitz, K., *Combinatorial Optimization: Algorithms and Complexity*, Prentice-Hall, Inc., Englewood Cliffs, NJ, 1982.

Section 2.75

Blurock, E. S., Methods of Computer Aided Synthesis Design, http://riscuni-linz.ac.at/people/blurock/courses/cao.

Speiser, A., *Die Theorie der Gruppen von Endlicher Ordnung*, Dover Publications, New York, 1945.

Pierce, J. R., *Symbols, Signals and Noise: The Nature of Communication*, Harper & Brothers, New York, 1961.

Cooper, L., Bhat, U. N., and LeBlanc, L. J., *Introduction to Operations Research Models*, W. B. Saunders, Philadelphia, 1977.

Reference Book for Radio Engineers, 4th ed., International Telephone and Telegraph Corp., New York, 1956.

Bloch, N. J., *Abstract Algebra with Applications*, Prentice-Hall, Englewood Cliffs, NJ, 1987.

Kauffmann, A., *Graphs, Dynamic Programming, and Finite Games*, Academic Press, New York, 1967.

Reddick, H. W. and F. H. Miller, *Advanced Mathematics for Engineers*, John Wiley & Sons, New York, 1938.

Rothe, R., Ollendorff, F., and Pohlhausen, K., *Theory of Functions as Applied to Engineering Problems*, MIT Press, Cambridge, 1993.

James, H. M., Nichols, N. B., and Phillips, R. S., *Theory of Servomechanisms*, McGraw-Hill, New York, 1947.

Section 2.76

Van Bramer, S., Mathcad Applications in Chemistry, Widener University, Science Division, 1/5/96, http://science.widener.edu/~svanbram/mathcad.html.

Section 2.77

Miller, F. H., *Advanced Mathematics for Engineers*, 3rd ed., John Wiley & Sons, New York, 1938.

Section 2.78

Peitgen, H., Jurgens, H., and Saupe, D., *Chaos and Fractals*, Springer-Verlag, New York, 1992.

Section 2.79

Complex Systems, PSL Survey of Complex Systems, http://www-psych.nmsu.edu/~linda/complex/index.html.

Complexity, Complex Systems & Chaos Theory Organizations as Self-Adaptive Complex Systems, Feb. 15, 1988, http://www.brint.com/Systems.htm.

Section 2.80

Elusive Heat, Chaotic Patterns, Turbulence and Convection, Spiral Chaos, http://www.psc.edu/science/Gunton/gunton.html.

Section 2.82

Bumble, S., Emerging Computer Simulation and Control of Plant Design and Retro-Design of Waste Minimization/Pollution Prevention in the Late Twentieth and Early Twenty-First Centuries, Proceedings Manual. EPA Region III, Waste Minimization/Pollution Prevention Technical Conference, Philadelphia, June 3–5, 1966.

Sections 2.84 through 2.95

Prof. J. Gasteiger, Research, Computer-Chemie-Centrum Erlangen, http://schiele.organik.uni-erlangen.de/ak/tg.htm#topics.

Section 2.96

Laboratory of Mathematical Chemistry and Computer Design, N. D. Zelinsky Institute of Organic Chemistry, Division of Computational Chemistry and Modern Information Technologies, http://www.loc.ac.ru/Laba/lab44.en.html.

Section 2.97

NASA, GRI-MECH, What's New in GRI-Mech 2.11, http://www.me.berkeley.edu/gri_mech/version21/text21.htm#thefiles.

GRI=MECH™ data printed by permission of Gas Research Institute.

Section 2.98

Ritter, E. R., THERM User's Manual. Department of Chemical Engineering and Environmental Science, New Jersey Institute of Technology, Neward, 1980.

Section 2.99

Kee, R. J., Rupley, F. M., and Miller, J. A., Sandia National Laboratories Report, SAND89-8003, 1989.

Coltrin, M. E., Kee, R. J., and Rupley, F. M., Sandia National Laboratories Report, SAND90-8003, 1990.

Coltrin, M. E., Kee, R. J., and Rupley, F. M., *Int. J. Chem. Kinet.*, 23, 1111, 1991.

Kee, R. J., Dixon-Lewis, G., Warnatz, J., Coltrin, M. E., and Miller, J. A., Sandia National Laboratories Report, SAND86-8246, 1986.

Section 2.100

Kee, R. J., Rupley, F. M., and Miller, J. A., Rate Coefficient Format, http://www.me.berkeley.edu/gri_mech/data/k_form.html; Chemkin-II: A FORTRAN Chemical Kinetics Package for the Analysis of Gas Phase Chemical Kinetics, Sandia Report SAND-8009B UC-706, 22–23, Jan., 1993.

Section 2.101

Widman, M. T. and Worden, R. M., Modeling Microbial Chemotaxis in a Diffusion Gradient Chamber, The Influence of Cell Motility on Microbial Competition, http://www.egr.msu.edu/widman1/cmc96/page3.html.

Section 2.102

Harris, T. E., *The Theory of Branching Processes*, Springer-Verlag, Berlin, 1963.

Section 2.104

Hysteresis and Avalanches, http://www.lassp.cornell.edu/sethna/hysteresis/hysteresis.html.

Section 2.105

Energy & Environmental Research Enhancement Program — Research Abstracts, http://www.nrcce.wvu.edu/programs/e&eabs1.html.

Section 2.106

Green, D., Cellular Automata, preprint, Environmental and Information Sciences, Charles Stuart University, 1993, http://life.csu.edu.au/complex/tutorials/tutorial1 .html.

Resource Optimization and Self-Interest: Variations on the Game of Life, http://www.cs.gsu.edu/~matrag/evil.html.

Wolfram, S., Caltech preprint CALT-GS-938. Cellular Automata as simple-organization systems.

Section 2.107

Cooper, L., Bhat, U. N., and LeBlanc, L. J., *Game Theory and Decision Theory, Introduction to Operations Research Models*, W. B. Saunders, Philadelphia, 1977.

Von Neumann and Morgenstern, *Theory of Games and Economic Behavior*, 3rd ed., Wiley, New York, 1964.

Section 2.108

Crabtree, E. W. and El-Halwagi, M. M., Synthesis of environmentally acceptable reactions, *AIChE Symp.*, 90(303), 117, 1995.

El-Halwagi, M. M. and Spriggs, H. D., Mass integration: now a comprehensive methodology for integrated process design, in press, *Chem. Eng. Prog.*, 1997.

Section 2.109

Sheldrake, R., *Recipes for Environmental Fields, The Presence of the Past*, Random House, New York, 1980.

Reddick, H. W. and Miller, F. H., *Advanced Mathematics for Engineers*, John Wiley & Sons, New York, 1938.

Kauffman, S., *The Origins of Order*, Oxford, 1993.

Section 2.111

American Institute of Chemical Engineers, Sponsored Research, Design Institute for Physical Property Data (DIPPR), Project 801, Evaluated Process Design Data; Project 805, Experimental Data on Mixtures; Project 821, Pure Component Liquid Vapor Pressure; Project 851, Critical Properties of Pure Compounds; Project 871, Determination of Pure Component Ideal Gas Heat of Formation; Project 911, Environmental, Safety and Health (ESH) Data Compilation; Project 931, Data Prediction Methods (Internet).

Section 3.1

Bumble, S., Chapter 32, in *Clean Production, Environmental and Economic Perspectives*, K. B. Misra, Ed., Springer-Verlag, New York, 1996.

Bumble, S., Emerging Computer Simulation and Control of Plant Design and Retrodesign of Waste Minimization/ Pollution Prevention in the Late Twentieth and Early Twenty-First Centuries, Proceedings Manual, EPA Region III, Waste Minimization/Pollution Prevention Technical Conference, Philadelphia, June 3–5, 1996.

Bumble, S. and Honig, J. M., *J. Chem. Phys.*, 33, 424, 1960.

Hijmans, J. and de Boer, J., *Physica*, 21, 771, 485, 499, 1955.

Kauffman, S. A., *The Origins of Order*, Oxford University Press, London, 1993.

Kikuchi, R., *Phys. Rev.*, 81, 988, 1951.

Lehninger, A. L., *Biochemistry*, Worth Publishers, New York, 1972.

Ling, G. N., A Physical *Theory of the Living State: The Association-Induction Hypothesis*, Blaisdell Publishing, New York, 1962.

Section 3.2

UNIQUAC, Magnussen, Ramussen, & Fredenslund, Copyright, American Chemical Society, 1981. Used by permission.

Reid, R. C. and Sherwood, T. K., *The Properties of Gases and Liquids*, McGraw-Hill, New York, 1966.

Sections 3.4 through 3.12

Smith, R. D., Senior Toxicologist, EPA, Region III, Periodic Risk-Based Concentration Tables, Philadelphia, PA.

Hippel, A., *Dielectrics and Waves*, Artech House, 1995.

Bumble, S., Chapter 32, in *Clean Production, Environmental and Economic Perspectives*, K. B. Misra, Ed., Springer-Verlag, New York, 1996.

Gibson, J. E., *Nonlinear Automatic Control*, McGraw-Hill, New York, 1963.

Gould, L. A., *Chemical Process Control and Applications*, Addison-Wesley, London, 1969.

Karman, Y. V. and Biot, M. A., *Mathematical Methods in Engineering*, McGraw-Hill, New York, 1940.

Kauffman, S. A., *The Origins of Order*, Oxford University Press, London, 1993.

Lotka, A. J., *Elements of Mathematical Biology*, Dover Publications, New York, 1956.

Mah, R. S., *Chemical Process Structures and Information Flows*, Butterworths, London, 1989.

Moore, G. T., *Emerging Methods in Environmental Design and Planning*, M. I. T. Press, Cambridge, 1968.

Nemhauser, G. L. and Wolsey, L. A., *Integer and Combinatorial Optimization*, John Wiley & Sons, New York, 1989.

Owens, D. H., *Multivariable and Optimal Systems*, Accademic Press, London, 1981.

Pamadimitriou, C. H. and Steiglitz, K., *Combinatorial Optimization: Algorithms and Complexity*, Prentice-Hall, Englewood Cliffs, NJ, 1982.

Pierre, D. A., *Optimization Theory with Applications*, Dover Publications, New York, 1969.

Reddick, H. W. and Miller, F. H., Advanced Mathematics for Engineers, 3rd ed., John Wiley & Sons, Inc., New York, 1938.

Reza, F. and Seely, S., *Modern Network Analysis*, McGraw-Hill, New York, 1959.

Saaty, Y. L., *Modern Nonlinear Equations*, Dover Publications, New York, 1981.

Saaty, T. L. and Bram, J., *Nonlinear Mathematics*, Dover Publications, New York, 1964.

Science Advisory Board to U.S. EPA, (A-101), 3AB-EC-90-021, Reducing Risk: Setting Priorities and Strategies for Environmental Protection, Sept., 1990.

Sethi, S. P. and Thompson, G. C., *Optimal Control Theory, Applications to Management Science*, Martinus Nijhoff, Boston.

Soroka, W. W., *Analog Methods in Computation & Simulation*, McGraw-Hill, New York, 1940.

Wilson, A. G., *Catastrophe Theory and Bifurcation*, University of California Press, Berkeley, 1981, chemistry, http://properties.nist.gov./TAR/file32.html.

Duffus, J. H. and Worth, H. G. J., Fundamental Toxicity for Chemists, Blackwell Science, 1998.

Sections 3.13 through 3.15

Bockhoff, F. J., *Elements of Quantum Theory*, 2nd ed., Addison-Wesley, London, 1976.

Landau, L. D. and Lifshitz, E. M., *Quantum Mechanics*, Pergamon Press, Oxford, 1977.

Bohm, D., *Quantum Theory*, Prentice-Hall, Englewood Cliffs, NJ, 1951.

Section 3.18

Claerbout, J., Multidimensional Recursive Filters via a Helix, http://sepwww.stanford.edu/sep/jon/helix/paper_htm/index.html.

Section 3.20

3D FFT/Deconv, http://pluto.neurologie.uni-duessel...e/analyze/Analyze_7.5/3DFFT7.5.html.

Section 3.21

Wavelets, http://www.ma.ic.ac.uk/~atw/.

Section 3.22

Time Series, http://www.ma.ic.uk/~atw/.

Section 3.23

Sloone, N. J. A. and Wyner, A. D., Eds., *Claude Elwood Shannon: Collected Papers*, New York, 1993.

Molecular Information Theory and the Theory of Molecular Machines, http://www-lmmb.ncifcrf.gov/~toms.

The Economist, Science and Technology, How (and Why) to Find a Needle in a Haystack, http://www-lecb.ncifcrf.gov/~toms/theeconomist/economist.html.

Section 3.24

Figueras, J., Syntree 2.0, A Program for Exploring Organic Synthesis, Computer Consultant, http://www.trinity software.com/syntree.html.

Section 3.25

Moirai: Modeling Biochemistry, http://www.ibc.wustl.edu/moirai/moirai.html.

Section 3.26

Fan, L. T. and Friedler, F., Reaction pathway analysis by a network synthesis technique, Abstract, 1997 Annual Meeting, Session 264, AIChE, http://www1.che.ufl. edu/meeting/1997/annual/session/264/i/index. html.

Section 3.27

TIMECHS, Pollution Prevention by Optimizing Kinetic Studies of Mechanisms on a Spreadsheet, Overview of Chemical Models, http://biomodel.georgetown.edu/model/chmodels.html.

Rodiguin, N. M. and Rodiguina, E. N., *Consecutive Chemical Reactions, Mathematical Analysis and Development,* van Nostrand Company, Princeton, NJ, 1964.

Library for Math Models of Biological Systems, Georgetown University Medical Center, University of Pennsylvania, Vanderbilt University, http://biomodel.georgetown.edu/model/model.html.

Rodiguin, N. M. and Rodiguina, E. N., *Collective Chemical Reactions, Mathematical Analysis and Development*, D. Van Nostrand, Princeton, NJ,1964.

Purpose, http://biomodel.georgetown.edu/model/purpose.html.

Section 3.28

Master Chemical Mechanism User Info., http://www.chem.leeds.ac.uk/Atmospheric/MCM/mcmpro.html.

Section 3.29

Rieckmann, C. and Keil, J., Multicomponent Diffusion and Reaction Kinetics in Three-Dimensional Networks — General Kinetics, Department of Chemical Engineering, Technical e/veroeff/veroeff_41/index.html.

Section 3.30

Programs for Other Physical Properties, Chetah, NIST Electronic Products Offered through the Standard Reference Data Program, http://www.nist.gov/srd/special.html.

Section 3.31

Naser, S. F. and Fournier, R. L., Experience using the lemds software to select extracting solvents for some hazardous wastes, presented at the ACS I&EC Special Symposium, Emerging Technologies for Hazardous Waste Management, Sept. 1992, Atlanta.

Creativity and Innovation: Chemical Substitution, http://www.oak.doe.gov/wmpp/candi.html.

Section 3.32

EPA/600/SR-95/049, SAGE 2.1, K. R. Monroe, Solvent Alternatives GuidE: User's Guide, Air and Engineering Research Laboratory, Research Triangle Park, NC 27711, May 1995.

SAGE Process Alternatives, sagemaster@clean.rti.org, May 13, 1997, All Sage Material, Copyright 1992, Research Triangle Institute, http://clean.rti.org/tp_tp.htm.

Section 3.33

EnviroSense, Integrated Solvent Substitution Data System (ISSDS), Idaho National Laboratory, P.O. Box 1625, Idaho Falls, ID, 83415-3730. Twi@inel.gov.

Section 3.34

Sources of Physical Property Data, http://library.adelaide.edu.au/guide/eng/Che_ll.html.

http://es.inel.gov/issds/index.html.

Cabezas, H. and Zhao, R., AIChE, 1998 Spring Meeting, Session 68, Paper 68g, Designing Environmentally Benign Solvents: Physical Property Considerations, National Risk Management Research Laboratory, U.S. EPA, Cincinnati, OH, March 11, 1998, http://www1.che.ufl.edu/meeting/1998/spring/session/68/g/index.html.

Section 3.35

Computational Thermodynamics and Physical Property Program, Environmental Science Center, PHYSPROP Database, http://esc.syrres.com/~ESC/physprop. htm.

Section 3.36

Laboratory for Applied Thermodynamics and Phase Equilibria: Research, Computer-Aided Design of Technical Fluids, Faculty of Chemical Technology and Materials Science, Delft University of Technology, Julianalaan 136, 2628 BL Delft.

Section 3.37

Computational Fluid Dynamics, http://www.csc.fi/english/cfd.html.

Section 3.38

Web Collection of Physical Property Sites, Links to Related Web Sites, http://info.cas.org/ONLINE/DBSS/beilsteinss.html; http://www.dartmouth.edu/~krescook/ref/matpropdata.htm/; http://www.dartmouth.edu/~krescook/home.html.

Section 3.39

PLG, Inc., http://www.plg-ec.com/.

Section 3.40

Tackling the Scale Problem, Turbulence and Convection, Big Mixups, http://www.psc.edu/science/Majda/majda.html.

Section 3.41

Barton, Dynamic Optimization Formulations for Operating Procedure Synthesis, AIChE Symposium, 1997, Annual Meeting Technical Session 194, Santa Barbara, http://yoric. mit. edu/sgalan/papers/AIChE97Paper.html.

Process Design and Simulation, The Ecosse Group, http://www.chemeng.edu.ac.uk/ecosse/; http://gttserv.lth.rwth-aachen.de/~spltt/other/index.htm.

Section 3.42

KHACE Consulting, Modern Methods of Control System Design for the Process Industries, Kelvin Hales Associates Ltd., Egham, Surrey TW20 9NB, U.K., khales@cix.compulink.co.uk, http://www.mathworks.com/connections/khace Cheminfo, Indiana University, Chemical and Information Sources, Physical and Chemical Property Data Sources, http://www.indiana.edu/~cheminfo.10-06.html.

Section 3.43

Thermodynamic Database, Thermo-Calc Databases, http://www.met.kth.se/tc/tdb/.

Section 3.44

Diffusion Program, http://www.met.ktm.se/dictra/.

Section 3.45

Physical Properties of Mixtures, httm://library.advanced.org/3659/atommole/mixtures.html.

Section 3.46

Cheminfo, Indiana University, Chemical and Information Sources, Physical and Chemical Property Data Sources, http://www.indiana.edu/~cheminfo.10-06.html.

Section 3.47

Computer Modeling of Polymers, Polymer Simulation, http://www.msi.com/info/materials/polymers/index.html.

Section 3.48

Properties of Semiconductor Materials, Semiconductor Applications at CFDRC, http://www.cfdrc.com/~semconductor/.

Metals and Insulators, Quantum Chemistry, Changing States, Metals and Insulators, http://www.psc.edu/MetaCenter/MetaScience/Articles/Klein/Klein.html.

Section 3.49

Theoretical Condensed Matter Physics, Polymer Physics, and Statistical Mechanics, http://physics.bu.edu/bu/TheoConMatPhy.html.

Section 3.50

The SGTE Casebook — Thermodynamics at Work, Edited by Klaus Hack, The Institute of Materials, London, 1996, available through GTT-Technologies, http://gttserv.lth.rwth-aachen.de/~sp/tt/general/sgtebook.htm.

Section 3.51

Program for Material and Process Development, MSE/Computational Thermodynamics/CAMPADA, Updated 11/19/96, Computer-Assisted Materials and Process Development (CAMPADA), http://www.met.kth.se/campada/.

Section 3.53

Variations on the Game of Life: Self-Interest and Resource Optimization, Lauer, M. R., Mitchem, P. A., Gagliano, R. A., Resource Optimization and Self Interest: Variations on the Game of Life, http://www.cs.gsu.edu/~matrag/evil.html.

Section 3.54

Villar, H. O., Affinity Fingerprints: Applications and Implications, http://www.netsci.org/Science/Chemin form/feature07.html.

Section 3.55

Gasteiger, J., Compound Properties Using Data for Sufficiently Similar Chemicals, GDCh, http://schiele.organik.uni-erlangen.de/tagung/10_cic/kuehne/.

Section 3.56

Multsimplex, http://www.multisimplex.com/.

Section 3.58

Structure-Reactivity Maps as a Tool for Reaction Site Descriptions, http://www.fiz-karlsruhe.de/stn/Data bases/cheminfo.html;http://www.fiz-karlsruhe.de/stn/Databases/chemreac.html.

Gifford, E. M., Johnson, M. A., Smith, D. A., and Tsai, Chun-che, Structure-Activity Maps as a Tool for Visualizing Xenobiotic Structure-Reactivity Relationships, http://www.netsci.org/Science/Special/featureo4.html.

Section 3.59

Renner, R., Calculating the cost of natural resource damage, *Environ. Sci. Technol.*, 32(3), 86A–90A, February 1, 1998, http://pubs.acs.org/hotartcl/est/98/feb/renner.html.

Section 3.60

Value of the World's Ecosystem Services, NSF PR 97-36, http://www.nsf.gov/od/lpa/news/press/pr9736.htm.

The value of the world's ecosystem services, *San Diego Earth Times*, June, 1997, provided by Nature, http://www.sdearthtimes.com/et0697/et0697s2.html; also http://darwin.eeb.uconn.edu/Documents/nsf-970514. html.

What's new in nature, Nature, Feb. 1, 1998, Costanza, R., d'Arge, R., de Groot, R., Farber, S., Grasso, M., Hannon, B., Limburg, K., Naeem, S., O'Neill. R. V., Paruelo, J., Raskin, R. G., Sutton, P., and van den Belt, Marjan, The Value of the World's Ecosystem Services and Natural Capital.

Section 3.61

Irkura, K. K., Frurip, D. J., Summary Report, Symposium on Computational Thermochemistry, Computers in Chemistry Division, American Chemical Society, 212th National Meeting, Orlando, FL, August 25–29, 1996.

Bibliography

Adleman, L. H., Molecular computation of solutions to combinatorial problems (Science, 266, 1021, 1994). He uses graph theory for a directed graph with a Hamiltonian path to encode DNA. Another (retro)synthesis program is CAOS (Computer-Assisted Organic Synthesis). It is called igor2.Zip and can be found and downloaded from Belgium's LOSH (Laboratory for Organic Synthesis Hoornaert) hyperlink at http://www.chemie.uni-regensburg.de/external.html on the World Wide Web.

Aho, A. V., Hopcroft, J. E., and Ullman, J. D., *The Design and Analysis of Computer Algorithms*, Addison-Wesley, Reading, MA, 1974.

Alfrey, T., Jr. and Gurney, E. F., *Dynamics of Viscoelastic Behavior in Rheology*, Vol. 1, E. R. Eirich, Ed., Academic Press, New York, 1956.

American Chemical Society, *Chemical Engineering in Medicine*, Advances in Chemistry 118, Washington, D.C., 1973.

Atmospheric Research Group in the School of Chemistry, Tropospheric Chemistry Modelling, http://www.chem.leeds.ac.uk/Atmospheric/MCM/main.html.

Barnhart, E. R., *Physicians Desk Reference*, 42nd ed., Oradell, N.J., 1994.

Benson, S. W., *Thermochemical Kinetics*, John Wiley & Sons, New York, 1976.

Biological Database, http://clogp.pomona.edu/medchem/chem/papers/comp.approach.to.html.

GRI-MECH™ data printed by permission of Gas Research Institute.

Bowman, C. T., Hanson, R. K., Davidson, D. F., Gardiner, W. C., Jr., Lissianski, V., Smith, G. P., Golden, D. M., Frenklach, M., and Goldenberg, M., http://www.me.berkeley.edu/gri_mech.

Bumble, S., Application of Order-Disorder Theory to Gas Adsorption, Ph.D. thesis, Purdue University, West Lafayette, IN, 1958.

Bumble, S., Application of order–disorder theory in physical adsorption, I. Fundamental Equations, *J. Chem. Phys.*, 33, 424, 1960.

C2.QSAR+, http://www.biosym.com/info/products/modules/QSAR_page.html.

Carotti, A., Altomare, C., Cellamare, S., Monforte, A., Bettoni, G., Loiodice, F., Tangari, N., and Tortorella, V., *J. Comput. Aided Mol. Design*, 9, 131–138, 1995.

CHEMINFO, SIRCh: Physical Property Information, Link to Lecture Notes on this topic, http://www.indiana.edu/~cheminfo/ca_ppi.html.

Cho, S. J., Cummins, D., Bentley, J., Andrews, C. W., and Tropsha, A., "Back to 2D QSAR: Application of Genetic Algorithm to PLS Selection of Topological Indices, http://mmlin1.pha.unc.edu/

Clark, D. E., Frenkel, D., Levy, S. A., J. L., Murray, C. W., Robson, B., Waszkowycz, B., and Westhead, D. R., *J. Comput. Aided Mol. Design*, 9, 13–32, 1995.

Clark, R. D. and Cramer, R. D., *CHEMTECH*, 27(5), 24–31, 1997.

CLOGP3 User Guide / CLOGP3 / CLOGP3 Examples / CLOGP3 Algorithm Manager, http://clogp.pomona.edu/medchem/chem/biobyte/manuals/CLOGP_Formatted.html.

Cramer, R. D., *J. Am. Chem. Soc.*, 102(6), 1837–1843, 1980.

Cramer, R. D., *Quant. Struct. Act. Relat.*, 2, 7–12, 1983.

Cramer, C. J., Famini, G. R., and Lowrey, A. H., *Acc. Chem. Res.*, 26, 599, 1993.

Davis, P. J., *Spirals from Theodorus to Chaos*, A. K. Peters, Wellesley, MA, 1993.

Dirac, P. A. M., *The Principles of Quantum Mchanics*, 4th ed., Clarendon Press, Oxford, 1930.

Drug Evaluations, Annual, 1992, Prepared by the Division of Drugs and Toxicology, Department of Drugs, American Medical Association, 1992.

Ferguson, A. M., Patterson, D. E., Carr, C. D., and Underiner, T. L., *J. of Bimol. Screening*, 1(2), 65, 1996.

Friedler, F. K. and Pinter, Z., Combinatorial synthesis of entire chemical processes, *Chemdata*, 88, 526–535, 1988.

Friedler, F., Blickle, T., Gyenis, J., and Tarjan, K., Computerized generation of technological structures, *Comput. Chem. Eng.*, 3, 241–249, 1979.

Friedler, F., Varga, J. B., Feher, E., Fan, L. T., Floudas, C. A., and Pardalos, P. M., *State of the Art in Global Optimization*, Kluwer Academic Publishers, 1996, 609–626.

Gasteiger, J., *J. Am. Chem. Soc.*, 119(17), 4033–4042, 1997.

Gasteiger, J. and Zupan, J., *Angew. Chem. Int. Engl.*, 32, 503, 1993.

Handbook of RCRA Ground-Water Monitoring Constituents, EPA, Office of Solid Waste, Science Applications International Corp., August, 1992.

Harris, T. E., *The Theory of Branching Processes*, Springer-Verlag, Berlin, 1963.

Hijmans, J. and de Boer, J., *Physica*, 21, 471, 485, 499, 1955.

Hill, T. L., *Statistical Mechanics, Principles and Selected Applications*, McGraw-Hill, New York, 1956.

http://biomodel.georgetown.edu/model/model.html.

http://www.mcs.anl.gov/home/compbio/PUMA/Development/puma_des.

http://comp.chem.umn:80/WWW/POLYRATE/POLYRATE.html.

http://losaltos.saic.com/cafta/default.html.

http://www.epa.gov/enviro/html/ef_overview.html.

http://www.epa.gov/enviro/html/mod/index.html.

http://www,epa.gov/enviro/html/sitemap.html.

http://www.epa.gov/enviro/html/efovw/.html, etc.

http://www.me.berkeley.edu/gri_mech/data/nasa_plnm.html for the NASA polynomials for thermodynamic quantities.

http://www.me.berkeley.edu/gri_mech/version21/files/21/grimech211.dat. The reaction file for O H C N AR containing 50 species about 300 reactions with rate constants approved by DOE, NIST, NASA for use with Chemkin calculations.

http://www.me.berkeley.edu/gri_mech/data/reactions/. A reaction cited here as an example is ch2(s)+co2 co+ch2o.html with the equal sign replaced with _ and .html immediately after the last species in the equation.

http://uxh.cso.uc.edu/~carroll/ga.html.

http://www,lassp.cornal.edu/sethna/hysteresis/noise.html. http://www.lassp.cornell.edu/sethna/hysteresis/ReturnPointMemory.html.

http://www.me.berkely.edu/gri_mech/version21/files21/grimech211.dat.

http://schiele.organik.uni-erlangen.de/ak/tg.html#topics.

http:/remus.jpl.nasa.gov/jpl97.

http://www.risc.uni-linz.ac.at/general/; http://www.risc.uni-linz.ac.at/projects/basic/chem/; http://www.risc.uni-linz.ac.at/research/; http://www.risc.uni-linz.ac.at/projects/basic.

http://www.cs.sandia.gov/opt/survey/sa.html; http//www.cs.sandia.gov/opt/survey/ts.html; http://www.cs.sandia.gov/SandiaOpt.html; http://www.cs.sandia.gov/opt/survey/intro.html; http://www.cs.sandia.gov/opt/survey/mip.html; http://www.cs.sandia.gov/opt/survey/interval.html; http://www.cs.sandia.gov/opt/survey/cluster.html; http://www.cs.sandia.gov/opt/survey/ea.html; http://www.cs.sandia.gov/opt/survey/hybrid.html.

http://anchi8.chm.anl.gov/g44/review/MJD/HAIVR.htm.

http://www.apnet.com/inscight/01291998/grapha.htm.

http://library.adelaide.edu.au/guide/eng/Che_ll.html.

http://www.zonge.com/ferro/prope_10.htm.

http://www.vt4.tu-harburg.de/veroeff/veroeff_41/index.html.

http://www.met.kth.se/tc/.

http://esc.syrres.com/~ESC/physprop.htm.

http://www.met.kth.se/tc/tdb/.

http://www.met.kth.se/dictra.

http://www.bonus.com:8080/contour/w...chem/web-elements/nofr-phys/Pr.html.

Hubbard, J. H. and West, B. H., *Differential Equations: A Dynamical Systems Approach, Higher-Dimensional Systems*, Springer-Verlag, New York, 1995.

Inorganic Chemistry and Thermodynamic's Home Page, maintained by Astrid M. C. Barrow, March 28, 1966.

James, C. A., Weininger, D., and Delany, J., Daylight Theory Manual Daylight 4.51, 1997, http://www.daylight.com/dayhtml/doc/theory/theory.html.

JANAF Thermochemical Tables, 2nd ed., Stull, D. R., Prophet, H. et al., Eds., NSDRS, NBS-37, U.S. Government Printing Office, Washington, D.C., 1971.

Kaplan, W., *Operational Methods for Linear Systems*, Addison-Wesley, Reading, MA, 1962.

Kaufman, E. D., *Advanced Concepts in Physical Chemistry*, McGraw-Hill, New York, 1966.

Kee, R. J., Miller, J. A., and Jefferson, T. H., Chemkin: A General-Purpose, Problem-Independent, Transportable, FORTRAN Chemical Kinetics Code Package, SAND80-8003.

Kikuchi, R., Phys. Rev., 81, 988 (1951)

Kikuchi, R., The path probability method, *Prog. Theor. Phy.*, Supple. 35, 1966.

Krieger, J. H., Integration hits software tools, *Chem. Eng. News*, October 6, 1997, http://pubs.acs.org/hotartcl/cenear/971006/integ.html.

Kober, H., *Dictionary of Conformal Representations*, Dover Publications, New York, 1952.

Lawrie, I. D., *A Unified Grand Tour of Theoretical Physics*, Adam Hilger, New York, 1990.

Lehninger, A. L., *Biochemistry*, Worth Publishers, New York, 1972.

Lin, A., QuaSar: the MOE system for QSAR, *J. Chem. Comput. Group*, 1998, http://www.chemcomp.com/feature/qsar.htm.

Ling, G. N., *A Physical Theory of the Living State: The Association-Induction Hypothesis*, Blaisdell, New York, 1962.

Love, A. E. H., *A Treatise on the Mathematical Theory of Elasticity*, 4th ed., Dover Publications, New York, 1944.

Marshall, G. R. and Cramer III, R. D., *Trends Pharmacol. Sci.*, 9(8), 285, 1988.

Massively Parallel Architecture for Air Quality Modelling, 1996, http:/www. csc.ncsu.edu/epa.html.

Mathematical Challenges from Theoretical/Computational Chemistry, http//www.nap.edu/readingroom/books/mctcc/contents.html.

MDL, Molecular Connection, March, 1995, http://www.awod.com/netsci/Science/Combichem/feature 0.2html.

Molecular Evolution of Organelle Genomics and Geneome Bioinformatics Resources, http://megasun.bch.umontreal.ca/other/evolution.html.

Mullins, M. E. and Rogers, T. N., Physical Property Needs in Clean Manufacturing Design, Department of Chemical Engineering, Michigan Technological University, National Center for Environmental Research and Quality Assurance, ORD, EPA, http://es.epa.gov/ncerqa_abstracts/...rs/cencitt/year2/clean/mullins.html.

Naser, S. F. and Fournier, R. L., A system for the design of an optimum liquid-liquid extractant molecule, *Comput. Chem. Eng.*, 15(6), 397–414, 1991.

Neumaier, A., Global and Local Optimization, http://solon.cma.univie.ac.at/~neum/glopt.html.

Peitgen, H., Jurgens, H., and Saupe, D., *Chaos and Fractals*, Springer-Verlag, New York, 1992.

Program for Computational Reactive Mechanisms (PCRM), Ohio Reactive Supercomputer Center, http://www.osc.edu/pcrm.

QSAR: Application of Genetic Algorithms and Partial Least Squares to Variable Selection of Topological Indices, http:/mm/in/ipha.unc.edu/~jin/QSAR/GA_PLS/gaphs.html.

Saaty, T. L. and Bram, J., *Nonlinear Mathematics*, Dover Publications, New York, 1964.

Sakurai, J. J., *Advanced Quantum Mechanics*, Addison-Wesley, London, 1967.

Shaw, A. C., *The Logical Design of Operating Systems*, Prentice-Hall, Englewood Cliffs, NJ, 1974.

Smith, R. D., Senior Toxicologist, EPA, Region III, *Periodic Risk-Based Concentration Tables*, Philadelphia.

Some Available Phylogeny Programs, http://expasy.hcuge.ch/info/phylogen.sof.

SRI, Huestis, D. L., Environmentally Benign Products and Processes, http://www.mpl.sri.com/topics/env-benign.html.

Green, M. R., DIPPR, STN Database Summary Sheet, http://www.tds-tds.com.

Steckler, R., Hu, W., Liu, Y., Lynch, G. C., Garrett, B. C., Isaacson, A. D., Melissas, V. S., Lu, D., Truong, T. N., Rai, S. N., Hancock, G. C., Lauderdale, J. G., Joseph, T., and Truhlar, D. G., *Comput. Phys. Commun.*, 88, 341–343, 1995.

Thermodynamics Research Laboratory, Thermodynamic Data and Property Calculation Sites on the Web, http://www.uic.edu/~mansoori/Thermodynamic.Data.and.Property_html.

Todeschini, R., Milano Chemometrics, 1997, http://disat.umimi.it/chm/default.htm.

Toxicology Handbook, Principles Related to Hazardous Waste Sites Investigations, a presentation by J. S. Heaton, ICAIR, and PRC for the Office of Waste Programs, Enforcement, EPA, 1986.

Tratnyek, P. G., Research Interests of Dr. Paul G. Tratnyek, 1996, http:/www.ese.ogi.edu/ese_docs/tratnyek/research.html.

VanDrie, J. H., Weininger, D., and Martin, Y. C., *J. Comp-Aided Mol. Design*, 3, 255, 1989.

University of Minnesota Biocatalysis/Biodegradation Database, metabolic emphasis, http://dragon.labmed.umn.edu/~lynda/index.html.

User's Guide, Borland Quattro Pro for Windows, Version 5.0, Houghton Mifflin, Scotts Valley, CA, 1991.

Wagener, M. and Gasteiger, J., *Angew. Chem.*, 106, 1245, 1994, *Angew. Chem. Int. Ed. Engl.*, 33, 1189, 1994.

Weast, R. C. and Astle, M. J., Eds., CRC Handbook of Chemistry and Physics 6th ed., CRC Press, Inc., Boca Raton, FL, 1980.

Weber, P. C., Wendoloski, J. J., Pantoliano, M. W., and Salemme, F. R., *J. Am. Chem. Soc.*, 114, 3197–3200.

Index

C

E

F

N

Q

R

T

U

V

W